Disaster Risk

The text offers a comprehensive and unique perspective on disaster risk associated with natural hazards. It covers a wide range of topics, reflecting the most recent debates but also older and pioneering discussions in the academic field of disaster studies as well as in the policy and practical areas of disaster risk reduction (DRR). This book will be of particular interest to undergraduate students studying geography and environmental studies/science. It will also be of relevance to students/professionals from a wide range of social and physical science disciplines, including public health and public policy, psychology, sociology, anthropology, political science, geomorphology and geology.

Irasema Alcántara-Ayala is former Director and current Professor and Researcher at the Institute of Geography of the National Autonomous University of Mexico (UNAM). Her research seeks to understand the root causes and drivers of disaster risk through forensic investigations of disasters, and to promote integrated research on disaster risk. She is particularly interested in bridging the gap between science and policymaking and practice in the developing world.

Christopher Gomez was born in France and, after holding several academic positions in different countries, he took on the Professorship of Sediment Hazards and Disaster Risk (Sabo) at Kobe University in Japan, where he leads the Sabo laboratory. He also holds a research professor position at the Geography Department of Universitas Gadjah Mada in Indonesia, as he is geographically focusing on the Indonesian and Japanese archipelago. His works focus on sediment transfer processes, related hazards as well as the philosophy and ethics of science and scientific research related to hazards and disaster risk.

Ksenia Chmutina is Reader in Sustainable and Resilience Urbanism at Loughborough University, UK. Her research focuses on the processes of urban disaster risk creation and systemic implications of sustainability and resilience in the context of neoliberalism. Ksenia uses her work to draw attention to the fact that disasters are not natural. Ksenia is Co-Host of a popular podcast *Disasters: Deconstructed*.

Dewald van Niekerk is Professor in Geography and Founder and Head of the African Centre for Disaster Studies at North-West University. He is Editor-in-Chief of the international journal *Jàmbá: Journal of Disaster Risk Studies*. His research interests include resilience thinking, complexity, anticipation and disaster risk governance. Dewald is a South African B3 NRF-rated researcher.

Emmanuel Raju is Director of the Copenhagen Centre for Disaster Research (COPE) and Associate Professor of Disaster Risk Management at the Global Health Section at the University of Copenhagen. He is also Extraordinary Associate Professor at North-West University, South Africa. His research interests include urban disaster risk creation, disaster recovery processes, and the intersections of disaster risk reduction and climate change adaptation. Emmanuel is also Co-Editor of *Disaster Prevention and Management* journal.

Victor Marchezini has studied the sociology of disasters in Latin America and Caribbean since 2004 (https://victormarchezini.weebly.com/). He worked as Disaster Risk Analyst in the monitoring room of the Brazilian Warning Center (Cemaden) (2012–2014). Since 2014, he has worked as Researcher at Cemaden, bridging scientists, practitioners, communities and policymakers. One result of this collective effort was the free e-book entitled *Reduction of Vulnerability to Disasters: From Knowledge to Action*. Victor is also Professor at the Doctorate Program on Earth System Science at the National Institute for Space Research (CCST/INPE) and at the Postgraduate Program on Disaster Science (ICT/UNESP). Twitter account: @VMarchezini.

Jake Rom Cadag is Professorial Lecturer at the Department of Geography of the University of the Philippines Diliman, Philippines. His specialties and research interests include disaster risk reduction and management, climate change adaptation and spatial mapping (i.e. cartography and geographic information system). He is skilled in community work and conduct of participatory methods and tools involving different stakeholders and community members, particularly marginalised sectors (i.e. gender and ethnic minorities, older people, children and people with disabilities, amongst others).

JC Gaillard is Ahorangi/Professor of Geography at Waipapa Taumata Rau/The University of Auckland. His work focuses on power and inclusion in disaster and disaster studies. It includes developing participatory tools for engaging minority groups in disaster risk reduction with an emphasis on ethnic and gender minorities, prisoners, children and homeless people. More details: https://jc gaillard.wordpress.com.

Disaster Risk

Irasema Alcántara-Ayala, Christopher Gomez, Ksenia Chmutina, Dewald van Niekerk, Emmanuel Raju, Victor Marchezini, Jake Rom Cadag and JC Gaillard

LONDON AND NEW YORK

First published 2023
by Routledge
2 Park Square, Milton Park, Abingdon, Oxon OX14 4RN

and by Routledge
605 Third Avenue, New York, NY 10158

Routledge is an imprint of the Taylor & Francis Group, an informa business

British Library Cataloguing-in-Publication Data
A catalogue record for this book is available from the British Library

Library of Congress Cataloging-in-Publication Data
Names: Alcántara, Irasema, author. | Gomez, Christopher, author. | Chmutina, Ksenia, author.
Title: Disaster risk / Irasema Alcantara-Ayala, Christopher Gomez, Ksenia Chmutina, Dewald van Niekerk, Emmanuel Raju, Victor Marchezini, Jake Rom Cadag and JC Galliard.
Description: 1 Edition. | New York, NY : Routledge, 2022. | Includes bibliographical references and index.
Identifiers: LCCN 2022022292 (print) | LCCN 2022022293 (ebook) | ISBN 9781138204331 (hardback) | ISBN 9781138204348 (paperback) | ISBN 9781315469614 (ebook)
Subjects: LCSH: Natural disasters. | Disasters. | Disaster relief. | Emergency management.
Classification: LCC GB5014 .A433 2022 (print) | LCC GB5014 (ebook) | DDC 363.34—dc23/eng/20220518
LC record available at https://lccn.loc.gov/2022022292
LC ebook record available at https://lccn.loc.gov/2022022293

ISBN: 978-1-138-20433-1 (hbk)
ISBN: 978-1-138-20434-8 (pbk)
ISBN: 978-1-315-46961-4 (ebk)

DOI: 10.4324/9781315469614

Typeset in Sabon
by Apex CoVantage, LLC

Contents

Figures

Why a textbook on disaster risk?

Chapter 1

Chapter 2

Chapter 3

Chapter 4

Chapter 5

Chapter 6

Chapter 7

Chapter 8

Chapter 9

Chapter 10

Chapter 11

Chapter 12

Chapter 13

Chapter 14

Chapter 15

Chapter 16

Tables

Chapter 3

Chapter 7

Chapter 8

Chapter 9

Chapter 10

Chapter 11

Chapter 14

Chapter 15

Boxes

Chapter 1

Chapter 2

Chapter 3

Chapter 4

Chapter 5

Chapter 6

Chapter 7

Chapter 8

Chapter 9

Chapter 10

Chapter 11

Chapter 12

Chapter 13

Chapter 14

Chapter 15

Chapter 16

Why a textbook on disaster risk?

Disasters seem to be a growing concern worldwide. They are not new issues, but the increasing number of large and smaller events experienced and recorded over the past six decades has been stirring increasing attention amongst a wide range of stakeholders. These include students and scholars, international and civil society organisations, government agencies and, more recently, the private sector.

At the core of this seemingly growing interest for disasters are the media. Driven by their appetite for the "spectacular" and "extra-ordinary" (Bourdieu, 1996), the traditional media, which are TV, radio and newspapers, seem to have found in disastrous events a perfect subject to cover. This increasing focus on disasters, especially the large ones, is sustained by a dominant scientific paradigm that has long depicted these events as extreme and rare (Hewitt, 1983). Humanitarian organisations that respond to these events have also gained massive traction over the past 40 years.

This collusion of interest and appetite for extraordinary events have forced disasters into the policy space. There are now two United Nations agencies focusing on disasters, which are the United Nations Office for Disaster Risk Reduction (UNDRR) and the United Nations Office for the Coordination of Humanitarian Affairs (OCHA), and a number of others that cover some aspects of DRR in their agenda. Many other international organisations, such as the World Bank and the International Federation of Red Cross and Red Crescent Societies (IFRC), have also made DRR a priority after recognising that disasters are slowing development and hampering the benefits of other investments and initiatives.

These organisations have further encouraged national governments to take actions to reduce risks through international treaties, such as the Yokohama Strategy and Plan of Action for a Safer World in 1994, the Hyogo Framework for Action (HFA) in 2005 and, more recently, the Sendai Framework for Disaster Risk Reduction (SFDRR) signed in 2015. Although non-binding, these international agreements have led to dozens of new institutional and legal arrangements at the national level, which, in turn, have stimulated interest across national and local government agencies (Chapter 14).

DOI: 10.4324/9781315469614-1

The question is whether this growing concern for disasters amongst stakeholders of DRR actually reflects a growing problem on the ground (Chapter 1)? Whether the impact of disasters is actually increasing in demographic, social and economic terms (Chapters 2 and 3)? Whether people and societies are becoming more vulnerable in facing an increasingly hazardous world (Chapter 5)? Whether recovery following disasters allows us to reduce the risk on the long term (Chapter 16)? All these questions, which are the core of this textbook, have been central to the field of so-called disaster studies.

A growing field of scholarship

Disaster studies, as a formal field of scholarship, is a century old. It emerged with S. Prince's research on the Halifax disaster in 1917 in Canada. It grew steadily over the subsequent decades of the past century but has, since the beginning of the 21st century, experienced a very fast expansion (Figure 0.1). The enormous amount of contemporary scientific production on disasters is symptomatic of the traction the subject has gained amongst scholars.

The reasons behind this growing academic interest for disasters are manifold. They include large and high-profile events at both the international and national scales, and a genuine commitment to help relieve suffering. For example, the 2004 Boxing Day tsunami disaster in Southeast and South Asia has spurred a large impetus for disaster research at the international level. Similarly, the impact of Hurricane Katrina in 2005 in the United States of America has renewed the interest for such research in the country. Very much like the devastating effects of back-to-back cyclones Ondoy and Pepeng in 2009 have been a turning point for disaster research in the Philippines, the 2010 earthquake has boosted scholarship in Chile. Growing scholarship is also sustained by increasing funding opportunities from government and other organisations, including NGOs, which have to meet the expectations of

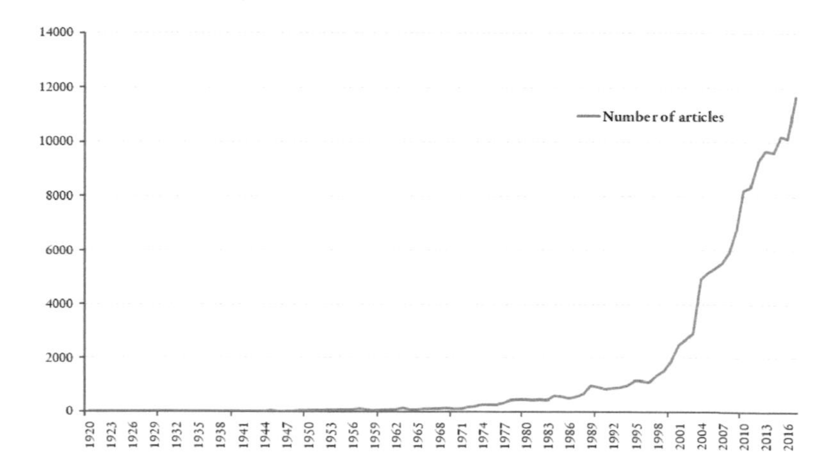

Fig 0.1
Yearly number of journal articles referring to 'disaster' between 1920 and 2018.
Source: Scopus

international treaties, such as the SFDRR, and, therefore, need increasing evidence of both the problems they face and of the outcomes of their initiatives to address them.

These scholars of disaster studies have created multiples networks and research groups within professional organisations. The International Sociological Association's Research Committee 39 and the American Association of Geographers' Specialty Group on Hazards, Risks and Disasters are amongst the most established ones and have inspired many other organisations to follow their path, including, for instance, the Risk and Disasters Topical Interest Group of the Society for Applied Anthropology. These organisations facilitate many discipline-based events that complement a growing number of multidisciplinary disaster studies, conferences and workshops that stir widespread interest amongst researchers.

The increasing number of researchers interested in disaster studies has resulted in a profusion of new outlets to share research outputs. These include more than 80 academic journals, publishing in English only, that focus exclusively on disasters and cognate fields, such as natural hazards (Alexander et al., 2020). There are also a multitude of publishers interested in publishing books that deal with disasters, suggesting that these volumes must sell well amongst the increasing number of scholars. In fact, there are now so many publications on disasters that it is virtually impossible to monitor them all and have an exhaustive perspective of the field.

The state of the field

The field of disaster studies is broad and fragmented, covering both discipline-based research and scholarship transcending traditional academic boundaries in what some call interdisciplinary or transdisciplinary research.

The nature of the object of study, that is disaster, itself transcends the scope of traditional academic disciplines and fields of research. It encompasses the natural and built environments as well as people and society. Therefore, virtually all existing academic disciplines have something to contribute to understanding disasters and reducing risk. The field involves earth and climate scientists interested in understanding natural processes that may be hazardous to societies. It also consists of chemists, biologists and engineers that may contribute to understanding other forms of hazards associated with technology and biosecurity. Disaster studies further involve social scientists from a wide range of disciplines, who investigate the anthropogenic causes of disasters, their impact, recovery processes, and DRR and management initiatives. Finally, disaster scholarship call for the expertise of health and medical scientists whose skills and knowledge are crucial in understanding some disasters associated with pandemics and epidemics as well as the impact of disasters and humanitarian response.

All these disciplines bring in unique expertise and approaches in studying disasters. Some disciplines focus on particular objects/subjects. Volcanologists study volcanic hazards whilst psychologists explore people's behaviour and individual response to disaster. Meanwhile, geographers look at the spatial dimension of physical and anthropogenic processes associated with disasters whilst historians ground these processes in time. These diverse interests and approaches draw on different ontologies or worldviews and call for various epistemologies or ways of knowing that often make collaborations and interactions across disciplines challenging. In fact, the field of disaster studies is clearly characterised by distinct paradigms of scholarship.

On the one hand, there is a long tradition of research grounded in the belief that disasters primarily result from the occurrence of rare and extreme hazards. These hazards are believed to exceed people's and societies' ability to cope and, therefore, call for extraordinary responses from agencies in charge of DRR (Burton et al., 1978). On the other hand, many scientists emphasise that disasters are the consequence of everyday, hazard-independent political, economic and social processes that make people at the margin of society more vulnerable in facing both extreme and everyday hazards (Wisner et al., 2004). This vulnerability paradigm, which emerged in the 1970s, challenges the long-standing predominant position of the so-called hazard paradigm (Chapter 4).

Interestingly, the vulnerability paradigm originally developed as both an ontological and epistemological critique of how we were considering and studying disasters in the earlier decades of the 20th century. It was challenging both our understanding of the causes of disaster, that is to move from disasters as natural to social constructs, and the way we were studying them. In the 1970s, pioneers of the vulnerability paradigms were calling for disaster scholarship grounded in local and indigenous studies that would challenge approaches inherited from the Enlightenment. Nonetheless, two decades into the 21st century, disaster studies still reflect the latter and not so much has changed in the way we study disasters (Gaillard, 2022).

In fact, disaster scholarship is well known amongst cognate fields of studies within the physical, engineering, medical and social sciences for constantly reinventing the wheel (Alexander et al., 2020). Many studies are indeed conducted in ignorance of similar research carried out decades ago. Pioneering works done between the 1930s and 1980s, in particular, are consistently being ignored, although, in the eyes of many, they constitute the golden ages of disaster studies, and, some would argue, our understanding of disasters has not made significant progress since then (Wisner et al., 2015). Recently, diverse initiatives have been developed to bridge this knowledge gap and make key vintage resources known and available to everyone. These include compendia of key publications in the field available as books and book series (Cutter, 1994; Wisner et al., 2015) or online: http://emergency-planning.blogspot.com/2017/02/

what-is-essential-reading-in-disaster.html and http://www.ilankelman.org/disasterarchives.html.

In parallel, disaster studies are also well known for being an applied field of scholarship driven by a genuine commitment to contribute to reducing disaster risk. As such, many research initiatives have been guided by the need to address practical issues observed by scientists or identified by stakeholders of DRR. Far fewer studies have been steered by theoretical reflections and the need to enhance our conceptual understanding of what disasters are and what DRR entails. In fact, the field is recognised by many critical social scientists for being under-theorised, and many would argue that such lack of theoretical grounding has hampered the advancement of the vulnerability paradigm and its application to actual DRR policies and initiatives.

The emergence and growth of academic programmes on disaster risk

The applied nature of disaster studies and its direct contribution to reducing the risk of disaster and making our world better have not only stirred increasing interest amongst scholars and researchers. It has also attracted a growing number of undergraduate and postgraduate students who genuinely wish to make our world better once they graduate. This trend has been observed worldwide, within wealthy and less affluent countries, and has materialised by the proliferation of academic programmes focusing on disasters, DRR, disaster management as well as humanitarian action. As of late 2019, the PreventionWeb databases record 351 academic programmes spread around the world that focus on DRR, disaster management and cognate fields.

These programmes cater to both undergraduate and postgraduate students, although the latter are more common. At the undergraduate level, named trans/multi/interdisciplinary bachelor's degrees are emerging. They cover a range of issues within the spectrum of natural hazards, disaster, DRR, disaster management and humanitarian actions that somehow reflect the persistence of a hazard/event-focused understanding of disasters. Postgraduate programmes are more diverse and specialised at the same time. They include postgraduate diplomas and all sorts of master's degrees, taught and research-based, from one to two years. They usually focus on either natural hazards or even a single hazard, such as earthquakes, DRR, disaster management, emergency management, disaster medicine, humanitarian actions and disaster recovery. The list is long and reflects diverse skills and expertise amongst teaching staff.

Most postgraduate programmes in the broad field of disaster studies assume that students have prior background in either a particular academic discipline or acquired background through previous work experience. The cohorts of students are, therefore, very diverse and include students coming straight from various undergraduate studies as well as practitioners with years or decades of experience in the field.

Oftentimes, these students also come from a large array of countries, many of which experience frequent disasters.

This growing appetite for disaster studies has, therefore, created a profitable market for universities and other academic institutions that are nowadays competing to develop appealing programmes. In doing so, many universities and other academic institutions are working in collaboration with international organisations and/or non-governmental organisations (NGOs), thus reinforcing the practical nature of disaster studies. Donors and aid agencies have also developed a parallel market for scholarships that target the best students from less wealthy countries and encourage them to study disasters in Northern America, Europe, East Asia and Australasia. For example, the Australian and New Zealand aid programmes have clearly identified disasters and DRR as one of their priorities, with particular focus on Southeast Asia and the Pacific. International organisations, aid agencies and NGOs eventually absorb a significant proportion of graduates who otherwise find their way within the expanding market for jobs in the field of DRR.

The need for a textbook

The growing interest for studying disasters involves a large cohort of newcomers to this field of scholarship, including both students and researchers. This textbook aims to provide them with a solid overview of the multiple facets of disaster risk and DRR. Such an overview is important for five main reasons:

1. Per the earlier sections of this introduction, the field of disaster studies is fragmented, and it is important to bring different threads of knowledge together so that one be able to fully understand the multiple dimensions of disaster risk and DRR. This is crucial to locate future studies within the field and build upon existing and relevant research across the silos of disciplines and sub-fields of studies. Sharing of knowledge and mutual learning is sine qua non to move our understanding of disasters forward and avoid re-inventing the wheel again and again.
2. There are so many new materials being published every day in disaster studies that it is, as discussed earlier, impossible to both keep up, and at the same time get to discover pioneer research that appeared in the literature decades ago. A compendium of old and newer knowledge is, thus, necessary to make sure that the latter do not get buried and often forgotten, whilst still emphasising the contribution of contemporary research. This is, here again, essential for students and new researchers to build on a comprehensive overview of the field in historical perspective to move knowledge, policy and practice forward.
3. An overview of the field of disaster studies is similarly important for students and new researchers to fully capture the ontological

and epistemological divide between existing research paradigms and their implications for DRR policy and practice. Such ontological and epistemological grounding is critical for future studies to be theoretically and methodologically sound. Only then, they will have the potential to make a strong contribution to our understanding of disasters as well as DRR.

4 This textbook is designed to support the teaching of disaster studies across the myriad of specific academic programmes discussed in the previous section of this introduction. It particularly intends to level up the knowledge of diverse cohorts of students from different backgrounds and with different professional and life experience. Per the previous three points, the textbook further provides both an overview of the field as well as pointers to where future studies should focus on and how current knowledge should be translated into policy and practice.

5 All foregoing objectives should ultimately contribute to learning from the past to build a safer future. Disaster studies is over 100 years old, and an incredible number of studies have been conducted in all corners of the world on all sorts of topics. Yet disasters continue to occur on a frequent basis. Progress made towards preventing them have been real but somehow limited with regard to the scope of the issues we are dealing with. It is, therefore, essential that newcomers to disaster studies and future policymakers and practitioners in the field of DRR henceforth focus on praxis. We hope this textbook to be a stepping stone in this direction.

Structure of the book

Providing such a broad overview of the multiple facets of disaster risk and DRR can only be a team effort (Wisner et al., 2012). This textbook is a team effort. It combines the expertise of a couple of geomorphologists with knowledge of a wide range of hazards, three human geographers, two political scientists and a sociologist. All authors further work in multidisciplinary institutional environments and collaborate on a daily basis with earth scientists of all sorts, ecologists, archaeologists, architects, engineers, public health experts, anthropologists and lawyers.

The authors' combined expertise allows for this textbook to cover most dimensions of disaster risk and DRR. Nonetheless, it does not claim to be exhaustive. As argued later in the book, understanding disasters is about understanding everyday life – and understanding everyday life in its entire complexity cannot fit in a book. The readers should, therefore, see the subsequent chapters and sections as an introduction to the complexity of disaster risk and DRR policies and practices. The numerous references and extensive bibliography compiled at the end of the book constitute pointers to more specialised knowledge.

This textbook is, therefore, an invitation to further studies and explorations rather than a stand-alone volume.

The book is structured around five main sections that look at distinct dimensions of disaster risk. These sections have not been designed in a silo, and many bridges exist throughout the book. Themes, such as gender or earthquakes, for the sake of examples, run across the different sections and chapters so that readers who are brave enough to engage with the entire volume will be able to appreciate different perspectives on disaster risk. Readers are, therefore, invited to find their own personal ways of reading the book based on their own interests and priorities.

Part I provides an overview of disasters. It includes three distinct chapters. Chapter 1 starts with defining disaster risk and other related concepts. It reviews different definitions and associated underpinning theoretical assumptions. It then provides a critical reflection on different research paradigms. Chapter 2 reviews trends in the occurrence of disasters in time and space, at multiple scales and across hazards over the past century. This chapter also reflects upon the limitations of existing datasets. Chapter 3 explores the environmental, physical, social and economic impacts of disasters associated with a wide range of hazards. It particularly focuses on the unequal impact of disasters across geographical and social spaces in both short and long time frames. It also emphasises the relative importance of both small and large events.

Part II focuses on vulnerabilities and capacities in dealing with hazards. It includes three chapters. Chapter 4 unpacks the root causes of disasters that explain the unequal occurrence and impact of disasters across social and geographical spaces. It looks at multiple forms of vulnerabilities through the lens of quantitative approaches, spatial analysis and various qualitative studies. Chapter 5 focuses more particularly on people's vulnerability. It explores why those affected by disasters live in dangerous places without being able to protect themselves, with particular attention given to minority groups who are disproportionally affected. Chapter 6 underlines that those dealing with hazards and disasters display a wide range of capacities. These skills, resources and knowledge are detailed and articulated in their larger social context.

Part III is dedicated to natural hazards. It is composed of four chapters that each tackle one category of hazards. Chapter 7 discusses the endogenous natural processes associated with seismicity and volcanism. It will explain both physical phenomena and their potential impact for people and societies. Chapter 8 rather explores exogenous natural processes associated with water and gravity, again looking at both physical phenomena and how these can affect people and societies. Chapter 9 focuses on climatic and hydrometeorological hazards. It covers both long-term (e.g. drought) and short-term threats (e.g. cyclones and floods), as well as their potential impact. Finally, Chapter 10 highlights the role of societies in exacerbating natural hazards. It shows how natural hazards are often triggered by people's interaction with their environment.

Part IV explores people's response to hazards and disasters and includes three chapters. Chapter 11 shows that people's behaviours in time of crisis are most often rational and focused on helping fellow survivors, thus pushing back against a wide range of myths commonly found in popular imaginary. Chapter 12 unpacks the concept of resilience and people's ability to overcome the negative impact of disasters. It focuses on different dimensions of people's everyday life following disasters, including their psychological, social, cultural and economic well-being. Chapter 13 addresses the larger issue of societies' long-term recovery in following disaster. It discusses challenges and opportunities for macro-economic recovery, infrastructure reconstruction and ecological restoration in view of reducing future risks of disaster.

Part V finally focuses on DRR and management in three distinct chapters. Chapter 14 provides an overview of DRR policies and practices. It identifies the main principles and components of DRR and underlines good practices at different geographical scales, from international policies and initiatives to national and local actions. Chapter 15 explores the management of actual disasters, including response and relief activities. It will provide an overview of current practices amongst emergency and humanitarian actors and emphasise challenges and opportunities for addressing the needs of affected people and places. Chapter 16 addresses policies and practices designed to foster recovery following both small and large disasters at different scales and across different geographical spaces.

References

Alexander, D., Gaillard, J. C., Kelman, I., Marincioni, F., Penning-Rowsell, E., van Niekerk, D., & Vinnell, L. (2020, in press). Academic publishing in disaster risk reduction: Past, present, and future. Disasters, 43(Suppl. 1).

Bourdieu, P. (1996). *Sur la télévision, suivi de l'emprise du journalisme*. Raisons d'agir.

Burton, I., Kates, R. W., & White, G. F. (1978). *The environment as hazard*. Oxford University Press.

Cutter, S. (1994). *Environmental risks and hazards*. Prentice Hall.

Gaillard, J. C. (2022). *The invention of disaster: Power and knowledge in discourses on hazard and vulnerability*. Routledge.

Hewitt, K. (1983). *Interpretations of calamity, from the viewpoint of human ecology. The risks and hazards series, 1*. Allen Press and Unwin, Inc.

Wisner, B., Blaikie, P., Cannon, T., & Davis, I. (2004). *At risk: Natural hazards, people's vulnerability, and disasters* (2nd ed.). Routledge.

Wisner, B., Gaillard, J. C., & Kelman, I. (Eds.). (2012). *Handbook of hazards and disaster risk reduction*. Routledge.

Wisner, B., Gaillard, J. C., & Kelman, I. (Eds.). (2015). *Disaster risk, Major Work Series, 4 volumes*. Routledge.

PART I

The nature and impact of disasters

CHAPTER 1

Disaster risk

Fig 1.1
The five senses,
Anthropomorphic Landscape,
18th century.
Source: author anonymous

> Disaster risk is a social construct, the outcome of identifiable social processes that transpire over various lengths of time, ranging from centuries to relatively short periods.
>
> Oliver-Smith et al. (2016)

1.1 Introduction

Beliefs associated with the presence of glowing stars and their relative positions have influenced perceptions of luck or randomness since ancient times (Figure 1.1). The word 'catastrophe' derives from the

DOI: 10.4324/9781315469614-3

Greek καταστροφή (*katastrophē* – "disaster", "destruction") with the roots κατὰ (*kata* – "down") and στροφή (*strophē* – "turning"), that is, "a downturn" or a change for the worse.

The term 'disaster', defined as "an event that results in great harm, damage, or death, or serious difficulty", derives from the Greek δυσ- (bad) and ἀστήρ (star); in other words, a calamity blamed on an unfavourable position of a star or planet. For centuries, its meaning has been related to the gods (Drabek, 1991) and forces of nature.

In antiquity, for example, it was widely believed that so-called natural disasters reflected divine anger but also represented responses to, or warnings of, events in the political domain (Kelly, 2004). One of the most famous historical and detailed accounts is the narrative by the Roman historian Ammianus Marcelinus of the earthquake (retrospectively estimated as M > 8) and tsunami that occurred in the Eastern Mediterranean on 21 July AD 365 (Box 1.1), presumed to have

Box 1.1: Historical account of the Eastern Mediterranean earthquake and tsunami (Ammianus Marcelinus, 21 July AD 365)

> Slightly after daybreak, and heralded by a thick succession of fiercely shaken thunderbolts, the solidarity of the whole earth was made to shake and shudder, and the sea was driven away, its waves were rolled back, and it disappeared, so that the abyss of the depths was uncovered and many-shaped varieties of sea-creatures were seen stuck in the slime; the great wastes of those valleys and mountains, which the very creation had dismissed beneath the vast whirlpools, at that moment, as it was given to be believed, looked up at the sun's rays. Many ships, then, were stranded as if on dry land, and people wandered at will about the paltry remains of the waters to collect fish and the like in their hands; then the roaring sea as if insulted by its repulse rises back in turn, and through the teeming shoals dashed itself violently on islands and extensive tracts of the mainland, and flattened innumerable buildings in towns or wherever they were found. Thus in the raging conflict of the elements, the face of the earth was changed to reveal wondrous sights. For the mass of waters returning when least expected killed many thousands by drowning, and with the tides whipped up to a height as they rushed back, some ships, after the anger of the watery element had grown old, were seen to have sunk, and the bodies of people killed in shipwrecks lay there, faces up or down. Other huge ships, thrust out by the mad blasts, perched on the roofs of houses, as happened at Alexandria, and others were hurled nearly two miles from the shore, like the Laconian vessel near the town of Methone which I saw when I passed by, yawning apart from long decay.

Source: Kelly (2004)

Box 1.2: Risk is perceived through different lenses

Fig 1.2
Risk is perceived through different lenses.
Source: author created

been prompted by the sorrow or wrath of the gods on the death of the Emperor Julian who had attempted to reinstate the pagan religion (Stiros, 2001). Cities were destroyed, and thousands of people drowned in coastal regions from the Nile Delta to modern-day Dubrovnik, as Western Crete was lifted above sea level by up to 10 m (Shaw et al., 2008).

With an increase in knowledge and scientific comprehension of natural phenomena and socio-natural processes, and of their associated impacts on societies, the sense of guilt derived from the interpretation of disasters as divine retribution has largely faded out. Nonetheless, disaster is still regarded in the context of the forces of nature, and its unnatural aspects remain to be fully recognised. Although many events are regarded as "natural disasters" (Chmutina & von Meding, 2019), disasters and disaster risk are bound up with the perceptions, judgements, choices, practices, decisions and actions of society, and the way misconceived development creates conditions of vulnerability and exposure (Hewitt, 1983; Quarantelli, 1987, 1989; Blaikie et al., 1994; Cannon, 1994; Twigg, 2004; Wisner et al., 2004; Oliver-Smith, 1998; Oliver-Smith et al., 2016, 2017). Hence, the need to understand disaster risk as a societal process must be considered (Box 1.2 and Figure 1.2).

1.2 The meaning of disaster risk: a departure point

Understanding disaster risk is not an easy endeavour, and perhaps this is partly because, hand in hand with scientific advances, the beliefs and understanding about disasters and risk have changed over time. Definitions vary from cosmological, cultural, technical, theoretical, social and practical to policymaking perspectives. Whereas the concepts of disasters and risk have been used for a long period, the combined term 'disaster risk' has emerged only in recent years.

This book constitutes an opportunity to consider several conceptions of disaster risk; an initial step is offered by the words of Wisner et al. (2004):

> Risk of disaster is a compound function of the natural hazard and the number of people, characterised by their varying degrees of vulnerability to that specific hazard, who occupy the space and time of exposure to the hazard event.
>
> (Wisner et al., 2004, p. 46)

According to those authors, there are three elements included in this concept: disaster risk, vulnerability and hazard. Their interlinkages can be depicted as $R = H \times V$, where R is risk of disaster, H is hazard and V is vulnerability (Figure 1.3).

Cardona et al. (2012, p. 69) suggested, "Disaster risk signifies the possibility of adverse effects in the future. It derives from the interaction of social and environmental processes, from the combination of physical hazards and the vulnerabilities of exposed elements".

Hazards are threatening phenomena, human activities or processes capable of triggering disasters that affect vulnerable persons, communities or environmental systems. They can be of diverse origins. *Natural* hazards are caused by geodynamic or hydrometeorological processes, such as earthquakes, volcanic activity (see Chapter 7),

Fig 1.3
The concept of disaster risk as a function of diverse types of hazards and vulnerability of people.
Source: author created

hurricanes or cyclones (see Chapter 9). *Socio-natural* hazards are associated with some typical phenomena of natural hazards that have an expression or incidence that is socially induced, as they are produced or exacerbated by human intervention in nature; classical examples are certain floods and landslides that are aggravated by urbanisation (see Chapter 10). *Technological* hazards derive from harmful technological, industrial and infrastructure settings, procedures, and failures, such as gas explosions or nuclear radiation. *Biological* hazards are of organic genesis or transmitted by biological vectors – for instance, viruses, such as SARS-CoV-2; bacteria; and mosquito-borne diseases.

Quite often, the terms 'hazard' and 'disaster' have mistakenly been used interchangeably and, for a long time, have been considered synonymous; the role of social factors has been overlooked. However, it is widely recognised that the impact of disasters depends on vulnerability; this refers to the degree of predisposition of people and systems to be affected by different types of hazards due to their particular characteristics, susceptibility and situations, largely derived from social, economic, cultural, political, and institutional processes, conditions and contexts.

Disasters and disaster risk causality is deeply rooted in the history of societies, their structure and organisation. Disaster risk is not wholly explained by nature per se; it requires an understanding of vulnerability, which in turn involves the recognition of root causes and the drivers of disaster risk. Underlying or root causes of disasters are processes or conditions related to historical aspects of development, associated with political, economic and territorial decisions and practices that have occurred throughout history. Quite often, such processes precede by several decades the disaster itself. Drivers of disaster risk are ongoing dynamic and active conditions or processes that create or increase conditions of vulnerability and exposure, often linked to, or rooted in, models of development. Major drivers of disaster risk include urbanisation and overcrowding, deforestation, inequality, poverty, unsafe conditions, illiteracy, insalubrity, lack of planning, corruption, failed risk governance and climate change.

The complex relationship between hazards and vulnerability expresses the long-standing interactions of developing societies and their socio-economic processes with the environment. Hence, disaster risk cannot be defined as a static entity, and it certainly does not appear instantaneously. Disaster risk is a dynamic condition that reflects a state of potential adverse consequences for humanity, built on a series of processes, decisions, and practices constructed and unfolded over time; these have taken place on the planet as a result of asymmetrical linkages associated with social inequalities and disequilibrium within and between human populations and nature.

The concept of disaster risk has been shaped from multidimensional perspectives. Thus, an overview of the trajectory of disaster studies will provide insights into the contributions of diverse disciplines to the construction of a definition. As disaster risk has diverse approaches, both academic developments and changes outside academia have influenced thinking. The major theoretical contributions that have guided the disaster risk community have been conceived within the academic and practitioner sectors. As well as changes in the scope, content and conception of disaster risk and, thereby, of policies to reduce disaster risk, there have also been changes in the ways the main stakeholders have started to make not only interdisciplinary but also transdisciplinary efforts to manage disasters and disaster risk.

The present chapter pulls together the writings of a number of scholars devoted to the understanding of disaster risk through time; hence, it adheres to no particular account but to the desire to further reduce the impacts of disasters. Ultimately, the aim of this book is to be a further step in the never-ending journey to understand, address and manage disaster risk.

1.3 The international agenda: a brief historical account

Almost a century ago, in 1927, the convention and statute establishing an International Relief Union stated under objective 2 of article 2 that in the case of any disaster, in addition to coordination of efforts by relief organisations, the study of measures that might prevent future disasters should be encouraged (League of Nations, 1932).

During the 1960s, disasters, particularly in the developing world, such as earthquakes in Iran and the former Yugoslavia, gave rise to specific disaster-oriented international initiatives; in these, a significant role was taken by the United Nations General Assembly. In 1971, endeavours included the creation of the United Nations Disaster Relief Office (UNDRO) in order

> to establish and maintain the closest co-operation with all organizations concerned and to make all feasible advance arrangements with them for the purpose of ensuring the most effective assistance; to mobilize, direct and co-ordinate the relief activities of the various organizations of the United Nations system in response to a request for disaster assistance from a stricken State; . . . and to *promote the study, prevention, control and prediction of 'natural disasters', including the collection and dissemination of information concerning technological developments*.
>
> (UN Res. 2816, emphasis added)

Since its inception, UNDRO's efforts were primarily focused on disaster response and humanitarian assistance.

Ten years later, in 1981, the need to strengthen the capacity of the United Nations system to respond to natural disasters and other disaster situations was recognised (UN Res. 36/225). In the wake of the increasing impact of disasters, and in consideration of the contribution of the World Commission on Environment and Development (WCED), there was a call for new perspectives at national and international scales in addressing the various factors affecting the environment, including 'natural disasters' (Brundtland Commission, 1987). In 1987, the International Decade for Natural Disaster Reduction was conceived (UN Res. 42/169).

Major concerns regarding the impact of disasters at global scale were emphasised by the WCED in the document *Our Common Future* (Box 1.3):

> During the 1970s, twice as many people suffered each year from "natural disasters" as during the 1960s. The disasters most directly associated with environment/development mismanagement – droughts and floods – affected the most people and increased most sharply in terms of numbers affected. Some 18.5 million people were affected by drought annually in the 1960s, 24.4 million in the 1970s. There were 5.2 million flood victims yearly in the 1960s, 15.4 million in the 1970s. Numbers of victims of cyclones and earthquakes also shot up as growing numbers of poor people built unsafe houses on dangerous ground.
>
> (Brundtland Commission, 1987)

Observance of an International Decade for Natural Hazard Reduction was originally conceived by a group of experts in earth sciences led by Frank Press, who proposed it in a keynote address to the International Association of Earthquake Engineering during the Eighth World Congress on Earthquake Engineering in San Francisco in 1984 (Xie, 1989; Revet, 2009; Schemper, 2019). It came into being with the proclamation of the International Decade for Natural Disaster Reduction (IDNDR) as a resolution of the UN General Assembly, beginning on 1 January 1990 (UN Res. 44/236); this first concerted effort to reduce disasters came with high expectations. The International Framework

Box 1.3: Brundtland Commission (1987)

If people destroy vegetation in order to get land, food, fodder, fuel, or timber, the soil is no longer protected. Rain creates surface runoff, and the soil erodes. When the soil is gone, no water is retained and the land can no longer produce enough food, fodder, fuel, or timber, so people need to turn to new land and start the process all over again.

All major disaster problems in the Third World are essentially unsolved development problems. Disaster prevention is thus primarily an aspect of development, and this must be a development that takes place within the sustainable limits.

Source: adapted from Kelman (2018)

of Action for the IDNDR was discussed by a Scientific and Technical Committee (STC) (24 scientists and technical experts from around the world) and was expressed through the Tokyo Declaration:

> Throughout history, mankind has lived under the threat of natural disasters. Millions of lives have been lost in recent decades, with untold human suffering and property damage as well as setbacks to development efforts. Indeed, the situation is growing worse. Vulnerability to natural disasters is rising due to population growth, urbanisation, and the concentration of industry and infrastructure in disaster-prone areas. But we now have improved capacity to confront the problem. *Fatalism is no longer acceptable*; it is time to bring the full force of scientific and technological advancement to reduce the human tragedy and economic loss of natural disasters. . . . We believe that the Decade is a moral imperative. It is the first coordinated effort *to prevent the unnecessary loss of life from natural hazards*. It also makes practical sense. The Decade is an opportunity for the world community, in a spirit of global cooperation, to use the considerable existing scientific and technical knowledge to alleviate human suffering and enhance economic security. In implementing the Decade, *the vulnerability* of developing countries *must be of special concern*.
>
> (UNCRD, 1989, p. 49, emphasis added)

Specific targets were set by the STC for the year 2000, requiring that all countries would have in place (1) comprehensive national assessments of risks from natural hazards, with these assessments being taken into account in development plans; (2) mitigation plans at national and/or local levels, involving long-term prevention and preparedness and community awareness; and (3) ready access to global, regional, national and local warning systems, and wide dissemination of warnings. Interestingly, though, the programme framework was based on disaster mitigation (Figure 1.4), which was mostly hazard-oriented with a "technofix" status (Dynes, 1990; Merani, 1991; Bunin, 1989; Salter, 1998; Alexander, 2001). "Hence, in the eyes of some practitioners, the IDNDR has assumed the status of a 'technofix', in which the proponents of technology and hard science use it as a justification for generating yet more of the same" (Alexander, 2001, p. 617).

Concerns regarding the potential influence of the IDNDR towards the physicalist or technocratic perspective of disasters, through its emphasis on the study of hazards and technology transfer, led to the foundation of LA RED, the Network of Social Studies on the Prevention of Disasters in Latin America. This contributed to the Yokohama Strategy and Plan of Action for a Safer World, established at the first United Nations World Conference on Natural Disaster Reduction, by incorporating the essence of the Cartagena Declaration on Reducing Vulnerability (LA RED, 1994).

FUNCTIONAL ACTIVITIES

1. Identification of hazard zones and hazard assessment
2. Vulnerability and risk assessment, cost-benefit analysis
3. Awareness at level of decision and policymakers
4. Monitoring, protection and warning
5. Long-term preventive measures
   (a) Non-structural measures
   (b) Structural measures
6. Short-term protective measures and preparedness
7. Early intervention measures

IDNDR

GEOGRAPHICAL LEVELS

1. Local
2. National
3. Regional
4. Global

SUPPORTING ACTIVITIES

1. Education and training of local and national specialists
2. Public education and information
3. Transfer of appropriate technology
4. Application of proven technology
5. Research to develop new technologies and devise new policies

Fig 1.4
IDNDR disaster mitigation matrix.
Source: adapted from Mauro (1995)

The IDNDR had aimed at "reducing the loss of life, property damage, and social and economic disruption caused by natural disasters, through concerted international action, especially in developing countries". The mounting impact of disasters during and after the designated decade stressed the need to assess the progress and strengthen initiatives at national and global scales. Accordingly, it was in the framework of the first World Conference on Disaster Reduction, held at Yokohama, Japan, from 23 to 27 May 1994, when the IDNDR mid-term evaluation took place, and the disaster risk question took on new shapes and slants.

Regardless of the lack of recognition of the need for an integrated approach, "natural disasters" and "environmental catastrophes" were considered by some, from the beginning of the IDNDR, as two sides of the same coin on which connections between humankind and the environment were more than evident; vulnerability of communities increases at the same time as development, with human assets being combined in an increasingly complex and enriched system (Merani, 1991). Nonetheless, the prevailing conception of DRR in the 1990s was still largely centred on hazard mitigation as the key objective.

Ironically, at the dawning of the 2000s, the world saw the result of the impact of the IDNDR, when the mounting impact of disasters worldwide documented by the scientific community led to the recognition that a key issue must be to understand and address the complexity of vulnerability; this had already been noted by Horlick-Jones (1995):

> Vulnerability is a multidimensional entity, and it is not clear how one can measure it without ambiguity or controversy. . . .

> The range of conceptual tools required to address the problems of disaster management therefore transcend the methods of the "exact" sciences associated with traditional "hard" science. High levels of complexity and uncertainty, and the socio-economic and cultural context of these problems must be taken into account, and "softer", more flexible, methods of analysis developed.
>
> (Horlick-Jones, 1995, p. 3)

After the celebration of the IDNDR Programme Forum 1999, when a global multisectoral and interdisciplinary dialogue summarised the efforts undertaken, the UN secretary-general proposed the creation of an inter-agency task force and inter-agency secretariat for disaster reduction, under the direct authority of the under-secretary-general for Humanitarian Affairs; this gave rise to the International Strategy for Disaster Reduction (UNISDR). Amongst its first endeavours were the incorporation of the objectives included within the Johannesburg Plan of Action, the 2004 review of the Yokohama Strategy and the organisation of the Second World Conference on Disaster in Kobe, Japan, in January 2005.

Moved by the consequences of the 2004 Indian Ocean earthquake and tsunami, government representatives and other stakeholders adhered to the Hyogo Declaration and the HFA (2005–2015): to build the resilience of nations and communities to disasters; these were endorsed by the UN secretary-general. The HFA encompassed five priorities: "(1) Ensure that disaster risk reduction is a national and a local priority with a strong institutional basis for implementation; (2) Identify, assess and monitor disaster risks and enhance early warning; (3) Use knowledge, innovation and education to build a culture of safety and resilience at all levels (4) Reduce the underlying risk factors; and (5) Strengthen disaster preparedness for effective response at all levels" (UNISDR, 2005).

The year 2005 was a watershed in disaster studies when the effects of Hurricane Katrina in the USA, a developed nation, involved more than 1,800 deaths, affected 500,000 people and had a total damage worth in the order of US$125 billion (EM-DAT, n.d.). Such consequences arose from multifaceted relationships and a series of pre-existing conditions that pre-ordained the vulnerability in a manner that had not been foreseen by a large part of the society.

In the wake of the repercussions of the 2004 Indian Ocean earthquake and tsunami, moves began in 2006 towards a more social perspective with a broadening research agenda that exposed the need for new approaches for early warning systems (EWSs). It was clear that EWSs needed to have four key ingredients: risk knowledge, monitoring and prediction, dissemination of information, and response (UNISDR, 2006). Nonetheless, overall, the enforcement of EWSs has largely been focused on providing real-time information for people to act in response to the occurrence of a hazard, rather than strengthening the knowledge, awareness and preparedness to understand risk and, therefore, the ability to decrease it (Alcántara-Ayala & Oliver-Smith, 2017). Despite the advances made in the field, there remains the danger that an increasingly specialised technology can be called upon to design EWS, thereby giving undue emphasis to technical viewpoints at the expense of disaster risk understanding and integrated management.

In February 2007, by resolution of the UN secretary-general (UN Res. 62/192), a Global Platform for Disaster Risk Reduction was created; four months later, its first session took place in Geneva and provided a forum for member states and other stakeholders with the following aims: to evaluate the advances made in the implementation of the HFA, to enhance awareness of DRR, to share experiences and learn from good practice, to identify remaining gaps and to identify actions to accelerate national and local implementation.

On 11 March 2011, the Tōhoku earthquake, also known as the Great East Japan earthquake, produced a tsunami that devastated the northeast coast of Honshu on the Japan Trench: there were 19,846 fatalities, some 368,820 people affected, US$210 billion economic damage (EM-DAT, n.d.) and 468,000 persons evacuated. The dimensions of the impact were inconceivable in a country whose disaster prevention strategies included preparedness and awareness, and had been known as the best in the world.

Prompted by concern over the Great Sendai earthquake, another international milestone in the concept of disaster risk occurred in 2015. The growing international interest in the topic highlighted the problem of the increasing impact of disasters in both developed and developing nations; at the United Nations Third World Conference on Disaster Risk Reduction, held in Sendai, Japan, the SFDRR was convened. The SFDRR consists of 7 targets, 4 priorities for action and 13 guiding principles; the priorities for action are as follows: "(1) Understanding disaster risk; (2) Strengthening disaster risk governance to manage disaster risk; (3) Investing in DRR for resilience; and (4) Enhancing disaster preparedness for effective response and to 'Build Back Better' in recovery, rehabilitation and reconstruction" (UNISDR, 2015).

Since the issue of DRR requires inter and transdisciplinary approaches within the sphere of development, linkages amongst SFDRR, the 2030 Agenda for Sustainable Development (United Nations, 2015), the Paris Climate Change Agreement (UNFCC, 2015) and the New Urban Agenda (United Nations Habitat III, 2016) are more than evident (Aitsi-Selmi et al., 2016; Alcántara-Ayala et al., 2017; Munene et al., 2018). It can be argued, however, that this combined set of propositions associated with these ambitious international frameworks, which rely unquestioningly on the use of technology, are likely to have a bigger and more structured impact in theory than in practice. For example, the early stages of the SFDRR paid scant attention to root causes of disasters other than those issues related to barriers to implementation at local level.

Five years after the establishment of the SFDRR, it is not clear how its message has been noted and, most importantly, implemented across sectors and multiscale governments and actors (Figure 1.5). Beyond constructed indicators and progress reports filed by nations, "interpreting words used in a report like that of SFDRR can fuel different, even competing, perceptions and actions about what to do to prepare for foreseeable disasters" (Glantz, 2015). In like manner, it has been recognised that "it will require greater collective human and material resources, and political will than are currently in play, to achieve the targets of the Sendai Framework, reduce risk and build resilience, and to

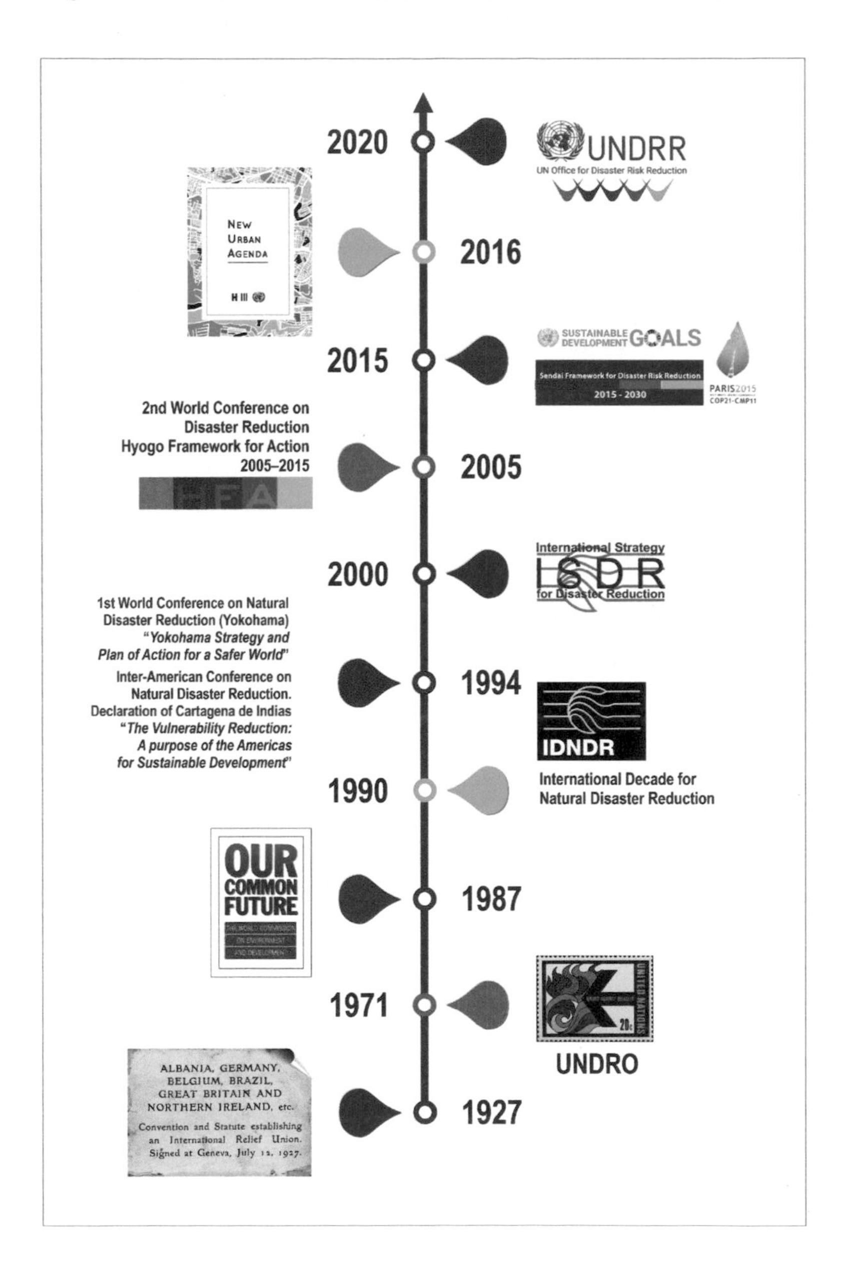

Fig 1.5
Timeline of the international agenda on disaster risk reduction. Source: author created

protect any progress in the achievement of the Sustainable Development Goals (SDGs)" (Mizutori, 2020).

The IDNDR and HFA were largely identical with respect to their common goals, but they differed in terms of transformative outcomes and change. With a remaining decade to accomplish the seven targets of the SFDRR, some could argue that there is still time to witness whether the multidimensions of vulnerability will be afforded increasing attention regarding their significance in the reduction of disaster risk. Yet it seems inescapable that the COVID-19 pandemic will join the series of obstacles to decreasing the vulnerability of people, improving disaster risk governance and achieving a more sustainable future.

1.4 Understanding disaster risk: a journey through the academic perspective

The concepts of disaster and disaster risk have undergone important changes over the past decades. Many academics, practitioners and other stakeholders have defined the concept in ways that reflect their background, experience, orientation, goals, context and time. Distinct academic disciplines have, through time, contributed key interpretations of disasters and disaster risk (Alcántara-Ayala, 2002); only more recently has a broader picture been sought through an integrated approach. In large measure, integrated contributions from the academic community have permeated into the international agenda directed at prevention of disasters and at reduction and management of disaster risk.

Studies of disasters and disaster risk were only formulated a little over a century ago. Before World War I, few initiatives or scientific groups were interested in disasters (Schemper, 2018). According to Quarantelli (1998a), the work of Prince (1920) was the first scientific study of disasters; this was devoted to the understanding of people's behaviour and social change related to the Halifax Explosion in 1917, when a collision between two ships off the coast of Canada detonated a cargo of high explosives. Other single studies by Carr (1932), Kutak (1938) and Sorokin (1942) were published during the 20 years that followed.

Looking at the impacts of disasters in the late 19th and early 20th centuries in the USA, Carr (1932) reflected on the disaster concept as a materialisation of the collapse of cultural protections:

> Not every windstorm, earth-tremor, or rush of water is a catastrophe. A catastrophe is known by its works; that is to say, by the occurrence of disaster. So long as the ship rides out the storm, so long as the city resists the earth-shocks, so long as the levees hold, there is no disaster. It is the collapse of the cultural protections that constitutes the disaster proper.
>
> (Carr, 1932, p. 211)

Although this conception could be more related to the term of 'resilience' commonly used nowadays (UNISDR, 2017), quite clearly, the interpretation of disaster occurrence moves somewhat beyond nature, as he identified three phases in every disaster: (1) the preliminary period, during which the forces that are to cause the ultimate collapse are getting under way; (2) the dislocation and disorganisation phase, involving the collapse of the cultural protections; and (3) the phase of readjustment and reorganisation (Carr, 1932).

Influenced by the disasters associated with the Mississippi flood of 1927 (Barry, 1998; Klein & Zellmer, 2007) and severe drought (Hurt, 1981) during the late 1920s and 1930s in the USA, the seminal work of White (1945) opened the door to an era that sought an understanding of disasters. Fritz (1961) defined a 'disaster' as "an event concentrated in time and space, in which a society or one of its subdivisions undergoes physical harm and social disruption, such that all or some essential functions of the society or subdivision are impaired" (Fritz, 1961, p. 655). In a similar vein, Barton considered disasters as "sudden and violent changes in the physical environment threatening both life and property" (Barton, 1969, p. 53).

To fully understand the notion of disaster, a consideration of the interface between human beings and their environment was required. This recognised that a "natural hazard was defined as an interaction of people and nature governed by the coexistent state of adjustment in the human use system. Extreme events which exceed the normal capacity of the human system to reflect, absorb, or buffer them are inherent in hazard. An extreme event was taken to be any event in a geophysical system displaying relatively high variance from the mean. There is large latitude for classifying human response to extreme events, but the term commonly used to describe a human activity intended to reduce the negative impact of the event was adjustment" (White, 1974, p. 4).

For Hewitt (1998), the 1970s and 1980s were the beginning of reflexive modernisation, as concepts recognised the effect of the conditions of society on the impact of disasters. What Lavell called "classic literature" of this period included a series of studies carried out mainly at the University of Bradford in the UK (Lavell, 2004). There, disasters were regarded as the footprint of the interface between a physical phenomenon and a vulnerable human population (O'Keefe et al., 1976). According to Lavell (2004), these contributions were the core of the structural paradigm, which confronted the behavioural thought derived from the investigations carried out by White in his group (inspired by Barrows's work (Barrows, 1923) and based on behaviourism) and which also differed from those studies of the sociology of disasters developed by Quarantelli and Dynes in North America.

The question of the significance of vulnerability within disaster studies was, thus, intimately related to the manner of its conceptualisation. Friedman (1975) considered that "vulnerability defines the susceptibility of population-at-risk to loss when an event of given intensity occurs". Westgate and O'Keefe (1976) viewed vulnerability as "the degree to which a community is at risk from the occurrence of extreme physical or natural phenomena where risk refers to the pejorative probability of occurrence, and the degree to which socio-economic and socio-political factors affect the community's capacity to absorb and recover from extreme phenomena".

During the 1970s, disaster research led to specific opinions of principle that were contrary to several of those formed by practitioners of disaster relief; "while relief is conceived of as an externally applied operation, the very nature of preparedness and prevention requires internally applied social and economic development programs" (Lewis, 1979, p. 127).

In an attempt to categorise disasters, Allen and co-workers considered that a "disaster involves an extreme phenomenon inflicting damage and death upon a vulnerable human group" and can be classified into four broad types: "sudden natural" (such as floods, hurricanes, earthquakes, volcanic eruptions or fires), "creeping" or "long-term natural" (such as droughts and epidemics), "deliberate man-made" (such as international or civil wars and disturbances), and "accidental" (Allen et al., 1980).

That which had often been relegated to the sidelines came to be seen as central. Pelanda (1981) explained 'disasters' as "the actualization of social vulnerability", and Susman et al. (1983) defined 'vulnerability' as "the degree to which different classes in society are differentially at risk, both in terms of the probability of occurrence of an extreme event and the degree to which the community absorbs the effects of extreme physical events and helps different classes to recover".

There were efforts to capture something more than the so-called naturalness of disasters. For example, the International Ad Hoc Group of Experts (1989), in implementing the International Decade for Natural Disaster Reduction, defined 'disasters' as a "disruption of the human ecology that exceeds the capacity of the community to function normally", and Horlick-Jones et al. (1993) viewed them "as arising from the evolution of socio-technical systems, and the interaction of those systems with their environments".

The conceptual advances in understanding disaster risk cannot be understood without taking account of the seminal contribution of Blaikie et al. (1994). Their work profoundly influenced the ways in which disasters are understood on the premise that disaster risk is driven by vulnerability associated with social, political and economic dimensions that are unfolded over time. Through the Pressure and Release (PAR) model, Blaikie and colleagues depict in a singular way all the ingredients needed to demystify the naturalness versus the social causation of disasters (Figure 1.6).

The Pressure and Release model (PAR model) is based upon the idea of vulnerability to unravel disasters and emphasises how social systems induce disaster occurrence by engendering vulnerability of people; here,

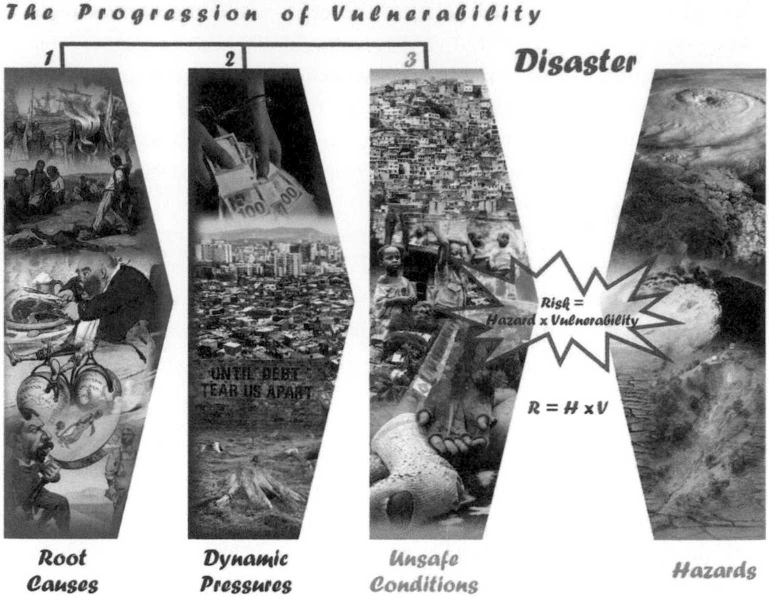

Fig 1.6
Pressure and Release (PAR) model: the progression of vulnerability.
Source: adapted from Blaikie et al. (1994)

'vulnerability' was defined as "characteristics of a person or group and their situation that influence their capacity to anticipate, cope with, resist and recover from the impact of a natural hazard" (Blaikie et al., 1994, p. 11).

Three interrelated processes are central to the PAR model: root causes, dynamic pressures and unsafe conditions. Root causes possess economic, political and structural aspects that have developed throughout the history of a society; on these, dynamic pressures are shaped. Dynamic pressures are interconnected drivers that include elements such as capabilities, demographics, institutional arrangements, ethical frameworks and environmental impacts; these drivers generate unsafe conditions associated with physical, economic, institutional and government contexts, which are of particular significance to the vulnerable sectors of society. Accordingly, the progression of vulnerability to hazards is embedded in the diverse interdependencies and interactions amongst these three processes (Figure 1.6).

In keeping with this line of thought, Lavell (1996) defined 'disasters' as "the product of processes of transformation and growth of society, which do not guarantee an adequate relationship with the natural and built environment that supports it".

From a sociological four-dimensional approach, building on the work of Fritz (1961), Kreps considered disasters as "nonroutine events in societies or their larger subsystems that involve social disruption and physical harm. Among, the key defining properties of such events are (1) length of forewarning, (2) magnitude of impact, (3) scope of impact, and (4) duration of impact" (Kreps, 1998, p. 34). This perspective matches quite well that of the high-magnitude and low-frequency events.

From an anthropological perspective, Oliver-Smith used the term 'disaster' as evolution, unfolding, revelation and innovation, and argued that the complexity of the concept stemmed from the fact that a disaster "is a collectivity of intersecting processes and events, social, environmental, cultural, political, economic, physical, technological, transpiring over varying lengths of time" (Oliver-Smith, 1998). Not only does this perspective consider what a disaster is, but its central component is to consider why, with the further definition of 'disaster' as "a process/event involving the combination of a potentially destructive agent(s) from the natural, modified and/or constructed environment and a population in a socially and economically produced condition of vulnerability, resulting in a perceived disruption of the customary relative satisfactions of individuals and social needs for physical survival, social order and meaning" (Oliver-Smith, 1998). This led to the warning "a disaster is made inevitable by the historically produced pattern of vulnerability, evidenced in the location, infrastructure, socio-political structure, production patterns, and ideology, that characterize a society. The society's pattern of vulnerability is an essential element of a disaster" (Oliver-Smith, 1998).

Vulnerability was also regarded as "the likelihood that some socially defined group in society will suffer disproportionate death, injury, loss or disruption of livelihood in an extreme event, or face greater than normal difficulties in recovering from a disaster" (Handmer & Wisner, 1998).

By viewing disaster as multidimensional, disaster has been regarded (Perry, 1998, p. 211) as "a socially defined occasion, serving as a context for human behaviour, recognised across social time as a radical change in the effectiveness of social structures (norms, practices, beliefs etc.) to meet human needs, and framed in a social change perspective". Disasters do not cause effects; the effects are what we call a disaster (Dombrowsky, 1998).

Parallel to the definitions of disasters, interest in the concept of risk started to emerge more widely. Addition of an exposure component led to the concept of risk as "the probability of a loss, and this depends on three elements, hazard, vulnerability, and exposure. If any of these three elements in risk increases or decreases, then the risk increases or decreases respectively" Crichton (1999).

A more detailed explanation defined 'risk' as "the likelihood, or more formally the probability, that a particular level of loss will be sustained by a given series of elements as a result of a given level of hazard. The elements at risk involve of populations, communities, the built environment, the natural environment, economic activities and services, which are under threat of disaster in a given area" (Alexander, 2000).

Indeed, the stimulus for change in comprehending disaster risk included increasing concern regarding vulnerability, seen as "the degree to which a system, subsystem, or system component is likely to experience harm due to exposure to hazard, either a perturbation or stress/stressor" (Turner et al., 2003). This entailed the understanding of 'hazard' as "threats to a system, comprised of perturbations and stress (and stressors), and the consequences they produce. A perturbation is a major spike in pressure (e.g., a tidal wave or hurricane) beyond the normal range of variability in which the system operates. Perturbations commonly originate beyond the system or location in question. Stress is a continuous or slowly increasing pressure (e.g., soil degradation), commonly within the range of normal variability. Stress often originates and stressors (the source of stress) often reside within the system. Risk is the probability and magnitude of consequences after a hazard (perturbation or stress)" (Turner et al., 2003).

Whereas disaster risk theory had been dominated by theories and models derived by scholars in the Global North, new strands were also appearing in the Global South through the establishment of LA RED in 1992 (LA RED, 1992) (Cardona, 2004; Lavell, 2004; Lavell et al., 2013). Important to stress, though, is that Northern pioneers in disaster risk studies, such as Piers Blaikie, Ian Burton, Terry Cannon, Ian Davis, Kenneth Hewitt, Robert W. Kates and Ben Wisner, have been instrumental in disaster studies in Latin American countries also, and especially by working with or as members of LA RED. Amongst the main contributions of LA RED to disaster risk management (DRM), reference can be made to the following themes that, although conceived from regional experience, have influenced global thought (Lavell et al., 2013; Alcántara-Ayala, 2019):

1. Disasters are not natural but socially constructed (Maskrey, 1993; Mansilla, 1996);
2. Community-based disaster mitigation (Maskrey, 1984);
3. The inherent nexus between disaster risk, development and the environment (Maskrey, 1993; Fernández, 1996);
4. The significance of small- and medium-sized disasters, and extensive and intensive risks (Lavell, 1994; UNISDR, 2009);
5. Socio-natural hazards (Fernández, 1996; Lavell, 1996);
6. Corrective, prospective and compensatory risk management (Lavell & Franco, 1996);
7. DRM at the local level (Wilches-Chaux, 1998; Lavell, 2003; Maskrey, 2011);
8. Integrated disaster risk research and the need for forensic investigations of disasters (Burton, 2010, 2015; Oliver-Smith et al., 2016, 2017; Alcántara-Ayala & Oliver-Smith, 2019);
9. Disaster risk as endogenous to social and economic development versus the imaginary of disasters as exogenous events (Lavell & Maskrey, 2014); and
10. The systemic nature of disaster risk and actionable frameworks for disaster risk governance (Maskrey et al., 2022).

In the past two decades, several authors have expanded the range of analytical dimensions for disasters and disaster risk. For example, analysis of links with development (Cardona et al., 2003) suggested that "disasters should be understood as unsolved development problems since they are not events of nature *per se* but situations that are the product of the relationship between the natural and organizational structure of society". As a component of this, he discussed in more detail vulnerability as "the physical, economic, political or social susceptibility or predisposition of a community to suffer damage in the case that a destabilizing phenomenon of natural or anthropogenic origin occurs" (Cardona, 2004).

Some years later, Lavell and colleagues comprehensively defined 'disasters' for the Intergovernmental Panel on Climate Change (IPCC) as "severe alterations in the normal functioning of a community or a society due to hazardous physical events interacting with vulnerable social conditions, leading to widespread adverse human, material, economic, or environmental effects that require immediate emergency response to satisfy critical human needs and that may require external support for recovery" (Lavell et al., 2012); they further indicated that the definition of 'vulnerability' should be refined to include a wide variety of human needs "as the propensity or predisposition to be adversely affected. Such predisposition constitutes an internal characteristic of the affected element". By incorporating the concept provided by Wisner et al. (2004), they added that vulnerability "includes the characteristics of a person or group and their situation that influences their capacity to anticipate, cope with, resist, and recover from the adverse effects of physical events". Vulnerability, therefore, "is a result of diverse historical, social, economic, political, cultural, institutional, natural resource, and environmental conditions and processes" (Lavell et al., 2012).

Another contribution of this approach was to provide a clear rationale for the existence of vulnerability arising from "the whole range of economic, social, cultural, institutional and political factors that shape people's lives and create the environments that they live and work in . . . vulnerability is highly dynamic, changing in response to many different influences, yet most vulnerabilities remain persistent because they result from deep-rooted social marginalisation, the indifference or incapacity of political and official institutions and the inadequacy of public services" (Twigg, 2015).

Today, the concept of disaster risk encompasses all spheres of interplay between vulnerability, hazards and exposure. "Disaster risk is the likelihood over a specified time period of severe alterations in the normal functioning of a community or a society due to hazardous physical events interacting with vulnerable social conditions, leading to widespread adverse human, material, economic, or environmental effects that require immediate emergency response to satisfy critical human needs and that may require external support for recovery. Disaster risk derives from a combination of physical hazards and the vulnerabilities of exposed elements and will signify the potential for severe interruption of the normal functioning of the affected society once it materializes as disaster" (Lavell et al., 2012).

In the same vein of thought, a good deal of work has recently been devoted to developing concepts that would define relevant factors in further study of disaster risk and integrated disaster risk management (IDRM), particularly those focused in depth on the significance of causality of disaster risk and the necessity to identify and address root causes and disaster risk drivers (Blaikie et al., 1994; Cannon, 1994; Hewitt, 1998; Burton, 2010, 2015; Oliver-Smith et al., 2016, 2017).

The benchmark for DRR is a comprehensive understanding of disaster risk as a complex sphere of socially constructed processes intertwined from the past to the present. Either as objective or as policy goal, it involves "strategic and instrumental measures employed for anticipating future disaster risk, reducing existing exposure, hazard, or vulnerability, and improving resilience. This includes lessening the vulnerability of people, livelihoods, and assets and ensuring the appropriate sustainable management of land, water, and other components of the environment" (Lavell et al., 2012).

Rather than enhancing the well-being of societies, as intended, through skewed development processes, policies are intensifying vulnerability and exposure. Such disjunctions may, in turn, trigger more complex and frequent disasters, as the development policies fail to meet their stated objectives. In these circumstances, it is now widely recognised that DRM should encompass a series of "processes for designing, implementing, and evaluating strategies, policies, and measures to improve the understanding of disaster risk, foster DRR and transfer, and promote continuous improvement in disaster preparedness, response, and recovery practices, with the explicit purpose of increasing human security, well-being, quality of life, and sustainable development" (Lavell et al., 2012).

1.5 Key strands and major paradigms in disaster risk

Theoretical disaster and disaster risk perspectives developed in diverse scientific fields have led to a series of disaster paradigms evolving through time that reflect the spheres of influence concerned in the conceptual and practical worlds (Figure 1.7).

Paradigms of hazard and vulnerability

In an attempt to trace the roots of disaster risk and disasters, Smith (2002) examined published studies, and drawing on much experience in the field, he wrote about the paradigms of hazard: the behavioural and the structural paradigms. More recently, in considering the concept of vulnerability, Wisner (2016) summarised the viewpoints regarding such paradigms, including that of hazard perception.

In the 1950s, White's corpus of work was the first to dominate disaster studies depicting the behavioural or dominant paradigm. Smith

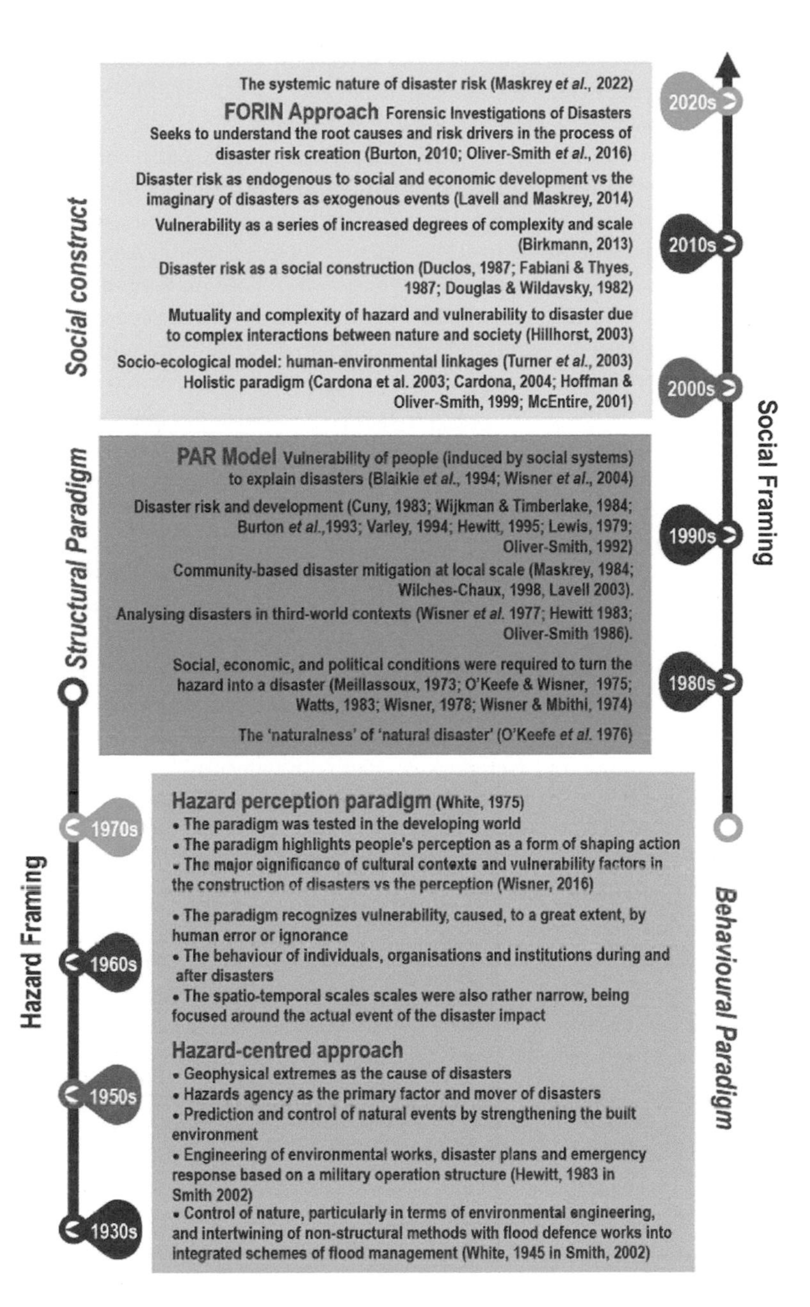

Fig 1.7
Key strands and major paradigms in disaster risk (time limits are relative since paradigms can be parallel over time).

noted the efforts undertaken during the 1930s and 1940s that sought to control nature, particularly in terms of environmental engineering aimed at flood control. Nonetheless, whilst the dominant paradigm was focused on controlling hazards, for White (1945), it was axiomatic that non-structural methods must be combined with flood defence works into an integrated scheme of flood management (Smith, 2002). In a period in which behaviourism was dominant, geophysical extremes were considered the cause of disasters, and prediction and control of natural events by strengthening the built environment seemed to be the most reasonable way to decrease the impact of disasters.

Science and technology were the magic bullet for technical solutions in the developed world, which afterwards could also be exported to the developing nations. Resources were channelled to disaster mitigation approaches in order to control the bad behaviour of nature. Accordingly, in this hazard-centred approach, emphasis was placed on monitoring, forecasting and modelling of geodynamic and hydrometeorological processes, favouring engineering environmental works, in addition to the elaboration of disaster plans and emergency responses based on a military operation (Hewitt, 1983, in Smith, 2002). This, sometimes known as the physicalist and technocratic paradigm (Hilhorst, 2003), "relies on the quantification of risks, the erection of physical structures and the investment of public funds to design-out hazards from the environment" (Smith, 2002, p. 53).

The early emphasis on disaster awareness and perception followed directly from the influence of White (1945). Wisner (2016) chronicled how, in the 1970s, the role of Burton et al. (1978) in understanding disaster also strongly influenced a new paradigm by highlighting people's perception as a form of shaping action. The hazard perception paradigm was tested in the developing world, and it was clear that cultural contexts, in addition to factors associated with vulnerability, such as "bad governance, corruption, skewed access to resources (land, water etc.), and lack of investment in infrastructure and social services", were far more significant in the construction of disasters than the perception itself (Wisner, 2016, p. 6).

The other side of that coin provided an alternative view developed to a great extent by experiences of the so-called Third World (Waddell, 1983 in Smith, 2002) framed in the structural paradigm. Whilst stressing that disasters arise from the global economy, the increase of capitalism and the consequential marginalisation of poor people, rather than from the impact of geophysical events, Smith also pointed out that there are key elements within societies that are not related to the application of science and technology in attempts to control nature. The integral components of these key elements recognised that environmental disasters basically depend on the extent of human exploitation and vulnerable groups, and are not reliant on geodynamic and hydrometeorological processes; disasters occur in areas with rapid environmental and social change; disaster victims are commonly located in unsafe conditions; disaster mitigation does not depend upon structural change, and development is not always beneficial; in seeking sustainable development and DRR, local knowledge should be favoured over technology (Smith, 2002).

This being the case, disaster occurrence is not so much a function of nature, as objectively described by natural scientists, as of the structural factors and related social processes that create vulnerable societies

(Blaikie et al., 1994; Cannon, 1994; Hewitt, 1998; Hilhorst, 2003; Oliver-Smith et al., 2016).

Whilst the conception of disaster prevention in the 1990s was still largely centred on hazard mitigation as the key objective, towards the end of that decade, the mounting impact of disasters in the developing and developed worlds led to the realisation that a consideration of vulnerability was a crucial step towards understanding disaster risk and disasters. In view of the fragmentation of social and natural sciences, and the differences that existed at that time amongst the diverse approaches to disaster risk, the structural paradigm became the baseline for a redirected discourse. Thus, the debate went from a hazard-driven conceptualisation to structural factors and social processes that generate conditions of vulnerability of people to disaster, and this entailed far more than a formal renaming of the concept.

Of the many individuals who contributed to establishing the new paradigm for which vulnerability was the essence, Hewitt (1983) had already provoked intense debate, and Blaikie, Cannon, Davis and Wisner were strongly influential through their book *At Risk: natural hazards, people's vulnerability and disasters* (Blaikie et al., 1994). These paved the way for a new paradigm that was built upon the foundations of the preceding decades and retained most of the experience derived from the seedbed of disaster risk research (Baird et al., 1975; O'Keefe et al., 1976; Davis, 1978; Wisner, 1978, 1979; Lewis, 1979; Pelanda, 1981; Cuny, 1983; Susman et al., 1983; Anderson & Woodrow, 1989; LA RED, 1992; Maskrey, 1993; Cannon, 1993, 1994; Merriman & Browitt, 1993; Wilches-Chaux, 1993; Cutter, 1996; Lavell, 1996, 2004; Quarantelli, 1998b; Mileti, 1999; Cardona, 2004; Oliver-Smith, 1986, 1996, 1998; Hewitt, 1998; Hoffman & Oliver-Smith, 1999; Lavell et al., 2013).

Despite significant efforts dating back to the 1970s and 1980s, at the end of the 1990s and start of the new millennium, 'vulnerability', defined as "the characteristics of a person or group and their situation that influence their capacity to anticipate, cope with, resist and recover from the impact of a natural hazard" (Blaikie et al., 1994), became the most important ingredient within an understanding of disaster risk. The central argument of the structural or alternative paradigm is that vulnerability generates disaster risk and disasters (Maskrey, 1993; Cannon, 1993, 1994; Bolin & Stanford, 1999; Blaikie et al., 1994; Quarantelli, 1998b; Lavell, 2000; Dynes & Rodriguez, 2007; Burton, 2010; Oliver-Smith et al., 2016). This vulnerability paradigm "puts the main emphasis on the various ways in which social systems operate to generate disasters by making people vulnerable" (Blaikie et al., 1994, p. 11).

Fundamental components of the alternative approach articulated a series of ideas, including that disasters express the vulnerability of human society, which is contingent upon social pre-conditions (Britton, 1986) created by social systems and power rather than by natural forces (Blaikie et al., 1994) influenced by values, attitudes and practices (Mileti, 1999). The spirit of this theory is that disaster risk is intrinsic to the multidimensions of vulnerability as it "is deeply rooted, and any fundamental solutions involve political change, radical reform of the international economic system, and the development of the public policy to protect rather than exploit people and nature" (Blaikie et al., 1994, p. 233). Disasters are understood as systemic processes that unfold over time; their causes are profoundly rooted in the history of societies, structures and organisations, and human-environment relations (Oliver-Smith, 1998; Hoffman & Oliver-Smith, 1999).

Renewed strands in disaster risk research frameworks

Partly catalysed by the influx of different disciplines into disaster risk understanding, a number of new strands began to emerge in the 21st century. As approaches regarding disaster and disaster risk have been especially characterised by interdisciplinary work and cooperation, the need to transcend antagonism between these two paradigms, since qualitative and quantitative approaches are complementary to each other, reinforced the direction towards a holistic paradigm (Cardona et al., 2003; Cardona, 2004), previously glimpsed by Hoffman and Oliver-Smith (1999) and McEntire (2001).

Cardona (2004) further argued that within the holistic paradigm of risk and based upon a theoretical framework of complexity that considers both geological and structural variables, as well as those of economic, social, political and cultural character, non-linear relationships of the contextual parameters and the complexity and dynamics of social systems could be adequately assessed; this could guide decision-making in space and time to strengthen the effectiveness of DRM.

Linkages between disasters, disaster risk and development are not new. One major strand of disaster risk research has been directed towards the disaster and development paradigm. It is not hard to see why the concept of disaster risk is so essential to development, as widely documented by a number of authors (Burton et al., 1978; Cuny, 1983; Wijkman & Timberlake, 1984; Varley, 1994; Hewitt, 1995; Lewis, 1979, 1999; Oliver-Smith, 1992; Collins, 2009, 2013; Gaillard, 2010; Thomalla et al., 2018). One strong conclusion to emerge from disaster risk analyses is that disasters are social constructions and manifestations of unresolved development problems, and indicators of skewed unsustainable development processes and practices (Wijkman & Timberlake, 1984; Cardona, 2004; Lavell & Maskrey, 2014).

"Disasters should be understood as unsolved development problems since they are not events of nature *per se* but situations that are the product of the relationship between the natural and organizational structure of society" (Cardona, 2004). Broadening the perspective of the disaster and development paradigm, Collins (2018) stated that achieving DRR, sustainable development and related policy objectives require further investment on three thematic fronts: "(1) building up earlier a human well-being that offsets negative risk; (2) living better with uncertainty; and (3) knowing the nature of barriers to more effective transitions in sustainable development and disaster risk reduction".

Another harmonising perspective of the alternative paradigm built on the foundations of the Pressure and Release model (Blaikie et al., 1994; Wisner et al., 2004, 2011) is the Forensic Investigations of Disasters perspective, the FORIN approach, which further stresses the need to understand the social construction of risk through an understanding of underlying causes and disaster risk drivers. The intention of FORIN is as follows: to identify the causes of disasters that are rooted in society and to encourage the advancement of integrated studies formulated to further our understanding of disaster causation; and to be policy-relevant in order to provide policy options and evidence-based alternatives for improved DRR. Since human agency is the prime factor and agent in the risk and disaster process, expressed at diverse levels of time and space, FORIN aims at promoting integrated and transdisciplinary research that engages all relevant DRR stakeholders to enable a more holistic comprehension of root causes and disaster risk (Figure 1.8) (Burton, 2010, 2015; Oliver-Smith et al., 2016, 2017).

A call for the implementation of a "new paradigm of development" (Lavell & Maskrey, 2014) noted that "a new DRR approach would

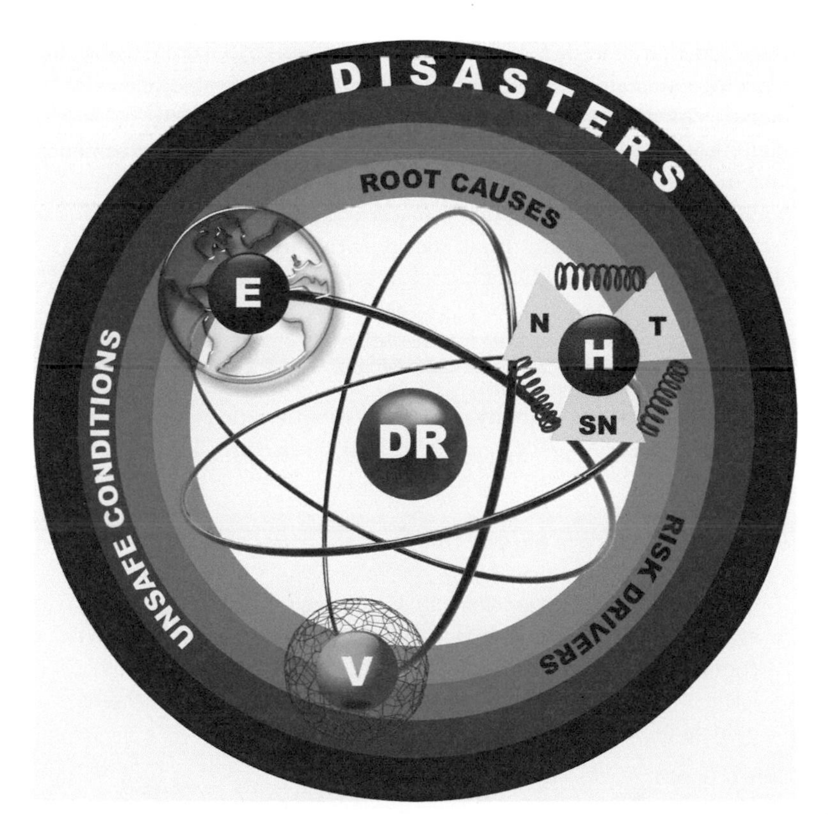

Fig 1.8
Key relationships and processes from the social construction of risk to the social production of disaster. E, exposure; V, vulnerability; H, hazard, with the categories N (natural), T (technological) and SN (socio-natural); and DR, disaster risk.
Source: Oliver-Smith et al. (2016)

necessarily have to take account of its role in the definition of modified or new development parameters" and that DRM could no longer ignore the following:

(a) DRR as a paradigm, however, continues to be driven by the increasingly outdated notion that disasters are exogenous and unforeseen shocks that affect supposedly normally functioning economic systems and societies rather than endogenous indicators of failed or skewed development, of unsustainable and insane economic and social processes and of ill-adapted societies;
(b) The DRR paradigm implies that governments increase their investments in corrective and compensatory risk management;
(c) The emphasis is still on reducing or compensating for disaster losses and damage as opposed to transforming the underlying drivers that generate risk in the first place;
(d) The conceptualization of disasters and risks as objective *things* rather than as inherent characteristics of evolving processes and relationships fosters technocracy and technocratic and bureaucratic approaches to DRR.

(Lavell & Maskrey, 2014)

Last but not least, within the dimensions of paradigms, one of the strongest growths in disaster risk studies has been in the integration and co-production of knowledge (Box 1.4). As disaster risk research aims at reducing vulnerability and exposure, periodic scientific assessments of disaster risk are obligatory (Cutter et al., 2015), and it is essential that there be a paradigm shift in disaster science that favours transdisciplinary system analysis with action-oriented research into DRR co-produced with multiple stakeholders; this will accentuate the inextricable relationship between nature and society within disaster science (Ismail-Zadeh et al., 2017).

1.6 What on earth then is disaster risk?

Disaster risk is the probability or a latent condition that expresses the potential impact of one or more hazards of diverse origin on a group of

Box 1.4: Experts are part of the problem rather than the solution

Experts are part of the problem rather than the solution. Many experts themselves struggle with understanding risks, lack skills in communicating them, and pursue interests not aligned with yours. Little is gained when risk-illiterate authorities are placed in charge of guiding the public.

Source: Gigerenzer (2014, p. 14)

people (or a system) with different degrees of vulnerability, who occupy a territorial space exposed to the effects of such hazards.

If the expected negative effects occur, the condition has become a disaster. Therefore, one of the approaches that can be used to understand disaster risk is to deconstruct a disaster that has already occurred to analyse the underlying characteristics of vulnerability and exposure, along with the dynamics of the hazard(s) that triggered the disaster. As mentioned earlier in the chapter, forensic investigation of a disaster, also known as FORIN, is one of the approaches that can help to understand how disaster risk builds over time; it entails identification and analysis of root causes and disaster risk drivers (Oliver-Smith et al., 2016, 2017).

But why is it important to understand disaster risk? The way in which disaster risk is framed and understood by different stakeholders can influence the nature of the research, risk assessments, interventions and strategies proposed for its proper management. In contrast, a lack of understanding can be used as an excuse not to reduce disaster risk but only to provide post-disaster or emergency actions.

Very often, when the occurrence of one or more hazards is conceived as the primary factor and mover of disasters, governments blame nature for the consequences of disasters and efforts are not focused on prevention but on how to respond after impacts.

The disaster triggered by the earthquake in Haiti on 12 January 2010 is a clear example of how disaster risk is constructed through a series of interdependent and interconnected social, economic, political, institutional and environmental processes that are circumscribed within specific contexts. A forensic study (FORIN) (Oliver-Smith, 2010) formed the basis of a retrospective longitudinal analysis (see Oliver-Smith et al., 2016) to identify the most significant ingredients of disaster risk that preceded the disaster in Haiti.

Haiti shares the Caribbean island of Hispaniola with the Dominican Republic to the east. In the 2009 Human Development Report, Haiti ranked 149th, at the bottom of the medium-development category and fairly close to the low-development category (UNDP, 2009). In 2010, it had a population of 9.9 million people and an annual population growth of 1.5%. With 58.5% of the population living below the national poverty line and 24.5% of the population living on less than US$1.90 a day, it was (and continues to be) the poorest nation in the Western Hemisphere (World Bank, n.d.).

Haiti's gross domestic product (GDP) reached US$11.66 billion in 2010 (whereas the USA GDP for the same year was US$14,992 billion). Agriculture, forestry, hunting and fishing together contributed 20% of the Haitian GDP, whereas industry (manufacturing, mining, construction, electricity, water and gas) contributed 25%. Exports and imports of goods and services accounted for 7% and 37% of the GDP, respectively. Likewise, Haiti's gross national income (GNI) per capita was US$650 in 2010, whereas, for example, that of its neighbour, the Dominican Republic, reached US$4,860 (World Bank, n.d.).

Before the earthquake, more than half of the population did not have access to formal health services or drinking water, 30% of the children already suffered from chronic malnutrition and it is estimated that 40% of the households were food insecure. More than 500,000 children between 6 and 12 years old were not in school, 70% of those who went to school had an educational deficit of more than 2 years and 38% of the population over 15 years was illiterate. Most Haitians lack occupational retirement benefits, social security and savings. Income distribution is predominantly unequal in Haiti: almost half of the national income goes to the richest decile of the population, whereas the last two deciles earn less than 2% of national income (Haiti, Office of the Prime Minister, 2010). Furthermore, also in 2010, only 40 out of every 100 people had cellular mobile phone subscriptions, and only 8.4% of the total population had access to the internet (World Bank, n.d.).

Understanding how conditions of vulnerability and exposure are built

In the article "Haiti and the Historical Construction of Disasters", Oliver-Smith (2010) offers an account of how the consequences of the disaster triggered by the 12 January 2010 earthquake were deeply rooted in the history of colonisation and slavery, and in the complex institutional legacy of past governments that reflects the various dimensions of the social construction of risk. This analysis also provides the perspective necessary to understand the root causes and drivers of disaster risk and disasters.

The case of Haiti is not isolated. Several disasters in Latin America have had their origins in unresolved problems of development (Maskrey, 1996). In fact, these problems have been associated with the postcolonial persistence of deep-seated socio-economic and political inequality rooted in colonial institutions (Frankema & Masé, 2014), in which societies have been debilitated to make the best possible use of their productive resources, particularly human resources (Engerman & Sokoloff, 1997; 2005 in Frankema & Masé, 2014).

On 6 December 1492, Christopher Columbus arrived at the Caribbean island, which he called *La Española*, meaning "the Spanish island", and which was later given the Latin-derived name of Hispaniola. Division of the island in the 1600s eventually led to the recognition of two independent countries, Haiti and the Dominican Republic.

During Columbus's first encounter with the indigenous people of the Caribbean, known collectively as the Taínos, the local people traded parrots, balls of cotton and javelins for European glass beads and bells. Columbus noted in his daily journal that the indigenous people were "very well formed, with handsome bodies and good faces" and had short, coarse hair, "almost like the tail of a horse", and "they should be good and intelligent servants [and] I believe that they would become Christians very easily" (Rogers, 2017).

In the early stages, the most valuable commodities sought in the Indies were gold, pearls and slaves, and the Taínos were enslaved by the Spanish to extract gold. They were decimated by European diseases since they had no inbuilt immunity to viruses and bacteria brought in by the first Spanish settlers in 1493.

In 1697, through the Treaty of Rijswijk, the western third of Hispaniola was ceded from Spain to France. After more than 125 years of Spanish rule and that of other European powers, in the 1660s, the French established Port-de-Paix in the northwest, and the French West India Company took control of the colony, at that time known as Saint-Domingue, which occupied one-third of the island. By the end of the 17th century, African slavery was institutionalised as a mechanism to guarantee the necessary labour to work in the plantations for the export of sugar and smaller amounts of coffee, cacao, indigo and cotton. The multiple forms of dehumanisation that the enslaved Africans faced were the underlying ingredient in the long-term construction of Haiti's vulnerability (Oliver-Smith, 2010).

In 1789, Saint-Domingue (now Haiti) was the wealthiest colony in the world and led the export of sugar and coffee, and the import of slaves (Steckley & Weis, 2017); at that time, it accounted for the production of 40% of all the sugar and 60% of all the coffee consumed in Europe. The beneficiaries were the European planters and their descendants, derived through the system of slave concubines. These children, *mulâtres*, constituted the first national elites of Haiti since they were free and could inherit property and own slaves (Dupuy, 1989, in Oliver-Smith, 2010).

The historical causes of vulnerability are also associated with structural inequities that have prevailed as a result of discrimination in terms of stereotypes of class, colour, gender and race; rights of participation and belonging have been either inequitable or lacking.

These structural disparities can be attributed to the rift between the black Creole-speaking population (the peasantry) and the upper class that ruled the country (the urbanites). The peasants were illiterate and had limited access to power and resources; they were considered *mounandeyo* in Haitian *Kreyol* or "people outside". The urbanites constituted the Francophone elite made up largely of the urban population of Port-au-Prince, who depended on global interests. Despite the necessary symbiotic economic relationship for the transfer of goods and necessities that implied the payment of taxes by the peasantry to the urbanites, the French language was used as a mechanism of exclusion to isolate the peasants both culturally and politically (Trouillot, 1990; Charles, 2021).

Inspired by the French Revolution of 1789 and the proclamation of the Declaration on the Rights of Man and of the Citizen, "men are born and remain free and equal in rights" (République Française, 1789), a series of reforms and resistance movements was begun in 1790 by the free elite people of colour, followed by the Saint-Domingue slave rebellion on the northern plains in 1791 and the brutal battles against Napoleonic forces that ended (Figure 1.9) when Haiti declared its independence on 1 January 1804 and became the first black republic in the world (De Cauna, 2009).

However, to obtain diplomatic recognition from France, Haiti was forced to pay reparations for the loss of "property", mainly in the form of slaves and land, to the sum of 150 million gold francs, although this was later reduced to 90 million. After new threats of invasion by France and an embargo maintained by France, Great Britain and the United States, in 1825, Haiti accepted the terms of the payment. Not until 1947 was the debt finally paid off, by which time, the economy had been destroyed, owing to the repayment of the huge loans that the government signed with American, German and French banks at excessive interest rates. From "the pearl of the Antilles", Haiti had become the poorest country in the Western Hemisphere (Dupuy, 1989, in Oliver-Smith, 2010).

To strengthen export capacity and foreign exchange, further structural inequalities were created between the peasantry and the urbanites when enforcement of the Rural Code compelled the peasantry to supply the workforce for plantations by tying the labourers to plantation land for life (Steckley & Weis, 2017). Inequality, racism and prejudices based on skin colour were also exacerbated when *mulâtres* or light-skinned Haitians held positions of authority, whilst few Haitians of dark skin served in the ministries or army (Trouillot, 1990).

The USA occupation of Haiti from 1915 to 1934, and the formation of the Haitian Army through an act of the United States Congress, intensified the divergences between the state and the nation by aggravating the contrasting socio-economic conditions between social strata and, thereby, increasing the marginalisation of the peasants (McGreevy, 2013; Charles, 2021).

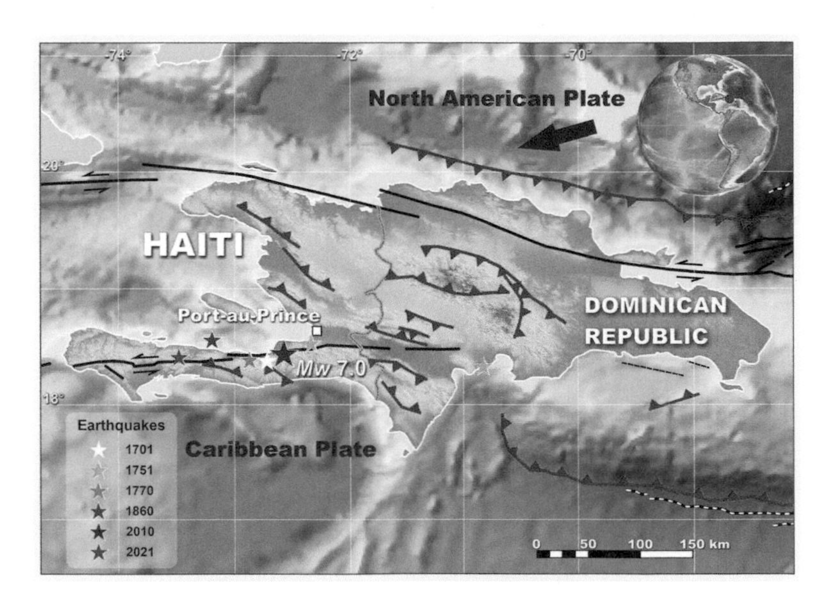

Fig 1.9
Map of Hispaniola (Haiti and the Dominican Republic) and major earthquakes that have affected Haiti in historical time. Fault traces shown as black and red lines (barbed red, thrust; solid black, strike-slip; and black and white, normal). The blue arrow shows the direction of the North American Plate motion relative to the Caribbean Plate.
Source: adapted from Bakun et al. (2012)

Since the beginning of the 19th century, Haiti was marginalised from the global economy and the political arena because of debt incurred when extraction of resources, such as sugar, coffee and indigo, was financed by actors such as the USA and the World Bank (Farmer, 2011; Green & Miles, 2011; Schuller, 2012; Pyles et al., 2018).

The accumulation of power and wealth by the government and elites of Haiti derived from the negotiation of the extraction process with foreign powers, particularly with the USA, whilst national resources have been exhausted. The construction of adequate infrastructure and the establishment of functional institutional frameworks was neglected, whilst the population continued to be submerged in a process of impoverishment linked to brutality, militarism, corruption and mismanagement (Schuller & Maldonado, 2016).

The historical construction of vulnerability in Haiti, and therefore, of disaster risk, has been shaped not only by the financial asphyxia inherited from the cost of its independence, and the continuous processes of impoverishment of peasants and structural inequities, but also through various events that occurred in the country during the last two decades of the 20th century.

François Duvalier ruled in Haiti between 1957 and 1971. After a year of assuming the presidency, he suspended all constitutional guarantees, and in 1964, he proclaimed himself president for life. To stay in power, he ordered the persecution and murder of his political opponents, who were estimated to be in the order of 60,000 people (Aponte, 2010).

From 1968 to 1970, the situation in the country deteriorated alarmingly. "Papa Doc" Duvalier had to repel three invasion attempts by Haitian exiles during that period (Buss, 2008), and the economy collapsed (Eve, 2011). After his death in 1971, he was succeeded as president by his son Jean Claude Duvalier, known as "Baby Doc", who oversaw the accumulation of a greater debt to foreign lenders. His regime was characterised as a kleptocracy, during which, on the order of the US Agency for International Development, the entire pig population of Haiti was slaughtered to avoid the spread of African swine flu virus. This episode exacerbated the conditions of poverty and vulnerability of the rural population since pigs were their accruing asset and source of security and nutrition (Diederich, 1985, in Oliver-Smith, 2010).

To turn Haiti's cities into export production centres for USA companies, the rural economy was dismantled. Baby Doc was persuaded to exploit the urban population as a source of cheap labour by establishing favourable conditions for investment, which led to an almost imperceptible increase in foreign investment in light-assembly manufacturing. The situation that arose through this debacle was exacerbated by the failure to strengthen mass tourism, the multidecadal decline in coffee and sugar exports, and the rise in subsidies for food, such as rice, exported to Haiti by USA agro-industries (Chavla, 2010). Consequently, and despite a rapidly growing population, national agricultural production was largely stagnant (Steckley & Weis, 2017).

The destruction of the rural economy led to massive migration to cities, where the displaced and impoverished rural people inhabited festering slums and hillside shanty towns that lacked services (Lundahl, 2004). Demand by migrants for jobs quickly outstripped supply and exacerbated the impoverishment of high-density populations in vulnerable parts of the cities. Furthermore, as the country has suffered considerable social, economic and political instability, there has been a decrease in the number of companies available to offer jobs (Chavla, 2010, in Oliver-Smith, 2010).

Understanding the hazards

Despite the historical seismicity of the region, where earthquakes had been recorded for the years 1701, 1751, 1770 and 1860 (Scherer, 1912; O'Loughlin & Lander, 2003, in USGS, 2010), Haiti lacked seismograph stations and knowledge of the physical conditions of the soil. Therefore, it has been difficult to fully document the geodynamics of the *Mw* 7.0 earthquake that occurred on 12 January 2010, at an estimated depth of 13 km, 25 km west-southwest of Port-au-Prince and on or near the Enriquillo fault (USGS, 2010) (Figure 1.9).

According to the USGS (n.d.), on 18 October 1751, a major earthquake caused considerable destruction in the Gulf of Azua and also generated a tsunami. On 21 November 1751, a major earthquake destroyed Port-au-Prince but was centred to the east of the city, on the Plaine du Cul-de-Sac. Almost two decades later, on 3 June 1770, a major earthquake destroyed Port-au-Prince again and seemed to be in the west of the city. The consequences of the 1751 and 1770 earthquakes and other minor earthquakes prompted the local authorities to stipulate that

Fig 1.10
Haiti stemmed from the interactions between vulnerability, exposure and hazards.
Source: author created

buildings be constructed from wood, not masonry. On 8 April 1860, a major earthquake occurred and was accompanied by a tsunami.

In the immediate aftermath of the 2010 event, which occurred somewhat to the east of the 1860 event, the earthquake rupture was believed to have been associated with the Enriquillo-Plantain Garden fault; however, subsequent studies have suggested, instead, that it involved the rupture of an unmapped north-dipping sub-parallel fault, now called the Léogâne fault. This also implies that since there was no release of any significant accumulated elastic deformation in the Enriquillo-Plantain Garden fault, this area remains a critical potential source of seismicity for the country and its capital city, Port-au-Prince (Calais et al., 2010), and that there remains the risk of earthquakes that exceed *Mw* 7.2 (Bilham, 2010).

The 2010 earthquake also triggered landslides and rockslides; these were exacerbated by loss of soil and sediment that had occurred through deforestation. Along with these mass movement processes, extensive damage arose because of soil liquefaction at several river deltas that prograde rapidly and are prone to failure. Moreover, the earthquake tremor caused submarine landslides that triggered the most severe tsunami on a local scale. This has drawn attention to the potential for earthquake-generated tsunamis along this and similar fault systems (Hornbach et al., 2010).

Two concatenated tsunamis, triggered by the earthquake, caused wave-induced flooding and damage to local infrastructure. One of them was induced by a coastal submarine landslide along the coastline within the Gulf of Gonâve and caused at least three fatalities at Petit Paradis. The other occurred along the southern coast of Haiti (Fritz et al., 2013).

The configuration of disaster risk in Haiti before the 2010 earthquake: summary

Disaster risk is configured as a result of the interactions and interdependencies amongst vulnerability, exposure and hazards in time and space (Figure 1.10).

The underlying causes of disaster risk are deeply rooted in the history of Haiti. Colonial and postcolonial governance created socio-economic and political inequalities between peasants and urbanites, accompanied by impoverishment, exploitation, marginalisation, class stereotypes and discrimination according to colour, gender and race. Slavery led to the dehumanisation of slaves, and independence brought financial asphyxia.

Additional disaster risk drivers that have arisen in recent decades have included environmental degradation, particularly deforestation, and limited or no access to human basic needs, such as clean water, sewage, electricity, food, healthcare and education. The social construction of risk has been compounded by a lack of urban planning, increasing population growth, informal settlements, widespread malnutrition and hunger, disease, deterioration of the environment, skewed urbanisation, crime, corruption, lack of scientific development and capacity, and a profound crisis in education.

Despite general knowledge about the historical occurrence of earthquakes in Haiti (USGS, 2010), the periods of earthquake recurrence for the region were not fully documented. There were no local scientists involved in the production and dissemination of knowledge about seismicity, and thus, knowledge about hazard was not incorporated into local planning instruments and building codes. The absence of seismic monitoring precluded insight into the potential for future earthquakes. Additionally, Haiti also lacked knowledge and awareness of tsunamis and landslides.

Compounding all these failings, there were no building codes or standards for the design of structures, and seismic forces were not considered in the design of most buildings (Paultre et al., 2013). Most of the population of its capital, Port-au-Prince, lived in poorly constructed houses in densely populated slums (Oliver-Smith, 2010), and the government had made no effort towards IDRM.

This was the context of vulnerability and exposure, together with the tectonic setting and geodynamics of seismicity, in which the Haiti earthquake of 2010 led to disaster expressed as 222,570 deaths, 300,000 persons injured, a population of 1.5 million displaced and 3.7 million people affected (EM-DAT, n.d.). These stark numbers can be comprehended when the social configuration of disaster risk is analysed and understood.

Even 11 years after the disaster triggered by the earthquake of 2010, the high levels of vulnerability and exposure continue to threaten the sustainable future of the people of Haiti. Disaster risk governance is likely to be weakened by political instability, along with the persistence of inequalities, racism, poverty, marginalisation, insecurity, corruption and environmental degradation. As if all these were not enough, on 14 August 2021, a *Mw* 7.2 earthquake occurred along the Enriquillo-Plantain Garden fault zone, ~ 125 km west of the Haitian capital Port-au-Prince and ~ 75 km west of the January 2010 earthquake (USGS n.d.). Official assessment of the consequences is not yet available, but it is to be hoped that they are not as devastating as those of 2010. A local perception of delays in government aid is reported to have led to fighting and demands for assassination; unrest in the face of disaster emphasises the influence of underlying social conditions in shaping the outcome of adverse events (*The New York Times*, 2021).

1.7 Concluding remarks

At the present time, hazard characterisation and mitigation as a dominant all-encompassing objective has been widely, but by no means universally, ousted. It can be argued that the last few years have been marked by a critical re-evaluation and consolidation of previous conceptual frameworks and techniques in seeking to understand the root causes and drivers of disaster risk. Nonetheless, owing to the multitude of factors attached to the concept of 'disaster risk', the term is complex and multifaceted, and it encompasses several dimensions. Consequently, it conveys and involves myriad ideas, interpretations and actions (Box 1.5 and Figure 1.11).

The specific words to depict disaster risk may have been diverse, and the precise arguments reconfigured across disciplines and actors through time, yet an understanding of disaster risk as a societal process provides all relevant stakeholders with a platform to broaden the scope of DRR and IDRM.

However, to understand disaster risk as the advocates of the alternative approach desire, an integrated and transdisciplinary view is necessary. Transdisciplinary research elevates the understanding of root causes and disaster risk drivers to a keystone in the reduction and management of disaster risk.

Indeed, the current wider interpretation of disaster risk now appears to be better understood in academia, even if at the political level it has not yet been fully implemented. DRR can be achieved only if and when it addresses the double burdens of reducing vulnerability and exposure, and of avoiding the creation of new risk on the grounds of favouring

Box 1.5: The term 'disaster risk' conveys and involves myriad ideas, interpretations and actions

Fig 1.11
The term 'disaster risk' conveys and involves myriad ideas, interpretations and actions.
Source: author created

sustainable development. This argument is nowadays crucially important in disaster studies and practice; it is probably fair to say that it has continued to be so ever since the inception of the alternative paradigm has been widely acknowledged.

Moreover, alongside the concern to reduce disaster risk, there has been a growing awareness of new intertwined international frameworks. Conundrums, such as sustainable development goals, the climate change agreement and the urban agenda, must be addressed; efforts towards DRR become entangled with other debates, at the same time as disentangling or examining, and intervening in numerous routes into the core of societal concerns derived from our unsustainable demands.

The continuing failure of a development-oriented DRR strategy to cope successfully with increasing vulnerability and exposure has led to a thorough re-examination of disaster risk governance, as "governance of risk (policies, legislation, and organizational arrangements) still focuses largely on preparing to respond to the hazards and planning for recovery" (Briceño, 2015). It follows that the major problems in disaster risk governance are becoming acute and can no longer be ignored.

'Disaster risk' may be defined, therefore, as the probability or latent state for humankind that involves adverse consequences in the social, economic, political, institutional, cultural and environmental spheres triggered by a hazard or a series of hazards of diverse origin. It is attributed to social constructions and manifestations of unresolved development problems and indicators of skewed unsustainable development processes and practices caused by historically produced intersecting patterns of vulnerability; these are derived from power structures and political and economic systems that have shaped the transformation of societies with an imbalance in the relationship between the natural and the built environment (Alcántara-Ayala, 2020).

The consequences of the global disaster triggered by the COVID-19 pandemic has revealed once again multiscale and multidimensional vulnerabilities. It has also called attention to the complex causality of disaster risk and its interplay with prevailing and emerging hazards across and within territorial and governance scales (Alcántara-Ayala et al., 2021). The map of spatial inequality will inevitably become more complex and multiscalar, and the reassessment of strategies involving development is, now more than ever, of vital importance. In the face of this planetary disaster, our histories are shaped not only by local contexts but also by a global dimension in which all societies will require further strength to cope with what is yet to come. It is hoped that the material found in this book will increase awareness of the dire situation we are in and could serve as an urgent wake-up call across disciplines.

Take-away messages

1. Disaster risk is not easily understood because the concept entails a range of perceptions and perspectives, and also of interests, particularly of an economic nature.
2. Hazards can be natural, socio-natural, technological, biological or anthropogenic in origin. An understanding of specific origins helps to identify the best options for intervention and mitigation, where possible.
3. To understand and reduce disaster risk, the misuse of 'hazard' and 'disaster' as synonymous should be avoided.
4. In large measure, integrated contributions from the academic community have permeated into the international agenda directed at prevention of disasters, and at reduction and management of disaster risk.
5. The social construction of disaster risk is due to diverse choices, practices, decisions and actions of society, and the way misconceived development processes create conditions of vulnerability and exposure, and these are combined with the occurrence of hazards.
6. The impacts of disasters around the world are indicative of the increasing conditions of vulnerability and exposure constructed by societies.
7. Disasters and disaster risk causality are deeply rooted in the history of societies, their structure and organisation. Therefore, disaster risk assessments require not only the understanding and assessment of hazards but also the identification of root causes of disaster and drivers of disaster risk.
8. The complex relationship between hazards and vulnerability expresses the long-standing interactions of developing societies and their socio-economic processes with the environment.
9. Transdisciplinary research elevates the understanding of root causes and disaster risk drivers to a crucial element in the reduction and management of disaster risk.
10. An adequate understanding of disaster risk is a prerequisite for its reduction. Therefore, policymaking and practice should be based on scientific evidence.

To learn more about the topic discussed in this chapter, listen to the *Disasters: Deconstructed* interview with Dr Gonzalo Lizarralde (Figure 1.12).

Fig 1.12
Chapter 1 QR code.

Further suggested reading

Adger, W. N. (2006). Vulnerability. *Global Environmental Change*, *16*(3), 268–281. https://doi.org/10.1016/j.gloenvcha.2006.02.006

Alcántara-Ayala, I., Burton, I., Lavell, A., Mansilla, E., Maskrey, A., Oliver-Smith, A., & Ramírez, F. (2021) [Editorial]. Root causes and policy dilemmas of the COVID-19 pandemic global disaster. *International Journal of Disaster Risk Reduction*, *52*(January), 2021, 101892. https://doi.org/10.1016/j.ijdrr.2020.101892

Alexander, D. E. (2013). Resilience and disaster risk reduction: An etymological journey. *Natural Hazards and Earth System Sciences Discussions*, *1*(2), 1257–1284. https://doi.org/10.5194/nhess-13-2707-2013

Blaikie, P., Cannon, T., Davis, I., & Wisner, B. (1994). *At risk: Natural hazards, People's vulnerability and disasters*. Routledge.

Briceño, S. (2015a). Looking back and beyond Sendai: 25 years of international policy experience on disaster risk reduction. *International Journal of Disaster Risk Science*, *6*(1), 1–7. https://doi.org/10.1007/s13753-015-0040-y

Briceño, S. (2015b). What to expect after Sendai: Looking forward to more effective disaster risk reduction. *International Journal of Disaster Risk Science*, *6*(2), 202–204. https://doi.org/10.1007/s13753-015-0047-4

Burton, I., Kates, R. W., & White, G. F. (1993). *The environment as hazard*. Guilford Press.

Chmutina, K., & Von Meding, J. (2019). A dilemma of language: "natural disasters" in academic literature. *International Journal of Disaster Risk Science*, *10*(3), 283–292. https://doi.org/10.1007/s13753-019-00232-2

Kelman, I. (2018). Lost for words amongst disaster risk science vocabulary? *International Journal of Disaster Risk Science*, *9*(3), 281–291. https://doi.org/10.1007/s13753-018-0188-3

Oliver-Smith, A., Alcántara-Ayala, I., Burton, I., & Lavell, A. (2016). *Forensic investigations of disasters (FORIN). A conceptual framework and guide to research*. Integrated Research on Disaster Risk.

Wisner, B., Blaikie, P., Cannon, T., & Davis, I. (2004). *At risk: Natural hazards, people's vulnerability and disasters*. Routledge.

Wisner, B., Gaillard, J. C., & Kelman, I. (2011). *Routledge handbook of hazards and disaster risk reduction*. Routledge.

References

Aitsi-Selmi, A., Blanchard, K., & Murray, V. (2016). Ensuring science is useful, usable and used in global disaster risk reduction and sustainable development: A view through the Sendai framework lens.

Palgrave Communications, *2*(1), 16016. https://doi.org/10.1057/palcomms.2016.16

Alcántara-Ayala, I. (2002). Geomorphology, natural hazards, vulnerability and prevention of natural disasters in developing countries. *Geomorphology*, *47*(2–4), 107–124. https://doi.org/10.1016/S0169-555X(02)00083-1

Alcántara-Ayala, I. (2019). Time in a bottle: Challenges to disaster studies in Latin America and the Caribbean. *Disasters*, *43*(Suppl. 1), S18–S27. https://doi.org/10.1111/disa.12325

Alcántara-Ayala, I. (2021). Integrated landslide disaster risk management (ILDRiM): The challenge to avoid the construction of new disaster risk. *Environmental Hazards*, *20*(3), 323–344. https://doi.org/10.1080/17477891.2020.1810609

Alcántara-Ayala, I., Burton, I., Lavell, A., Mansilla, E., Maskrey, A., Oliver-Smith, A., & Ramírez, F. (2021, January) [Editorial]. Root causes and policy dilemmas of the COVID-19 pandemic global disaster. *International Journal of Disaster Risk Reduction*, *52*, 2021, 101892. https://doi.org/10.1016/j.ijdrr.2020.101892

Alcántara-Ayala, I., & Oliver-Smith, A. (2017). The necessity of Early Warning Articulated Systems (EWASs): Critical issues beyond response. In K. Sudmeier-Rieux, M. Fernandez, I. Penna, M. Jaboyedoff & J. C. Gaillard (Eds.), *Linking sustainable development, disaster risk reduction, climate change adaptation and migration* (pp. 101–124). Springer. ISBN 978-3-319-33878-1, ISBN eBook: 978-3-319-33880-4

Alcántara-Ayala, I., & Oliver-Smith, A. (2019). Early warning systems: Lost in translation or late by definition? A FORIN approach. *International Journal of Disaster Risk Science*, *10*(3), 317–331. https://doi.org/10.1007/s13753-019-00231-3

Alexander, D. (2000). *Confronting catastrophe—New perspectives on natural disasters* (p. 282). Oxford University Press.

Alexander, D. (2001). *Natural disasters*. Routledge.

Allen, M., Sibahi, Z., & Sohm, E. (1980). Evaluation of the Office of the United Nations disaster relief. *Coordinator*. https://www.unjiu.org/sites/www.unjiu.org/files/jiu_document_files/products/en/reports-notes/JIU%20Products/JIU_REP_1980_11_English.pdf.

Anderson, M., & Woodrow, P. (1989). *Rising from the ashes: Development strategies in times of disaster*. Intermediate Technology Publications.

Aponte, D. (2010). *The Tonton Macoutes: The central nervous system of Haiti's Reign of Terror*. Council on Hemispheric Affairs-COHA. Retrieved March 11, 2010, from http://www.coha.org/tontonmacoutes.

Baird, A., O'Keefe, P., Westgate, K. N., & Wisner, B. (1975). *Towards an explanation and reduction of disaster proneness*. Occasional Papers. University of Bradford, Disaster Research Unit, 11.

Bakun, W. H., Flores, C. H., & ten Brink, U. S. (2012). Significant earthquakes on the Enriquillo Fault system, Hispaniola, 1500–2010:

Implications for seismic hazard. *Bulletin of the Seismological Society of America*, *102*(1), 18–30. https://doi.org/10.1785/0120110077

Barrows, H. H. (1923). Geography as human ecology. *Annals of the Association of American Geographers*, *13*(1), 1–14. https://doi.org/10.1080/00045602309356882

Barry, J. M. (1998). *Rising tide: The Great Mississippi flood of 1927 and how it changed America*. Simon & Schuster.

Barton, A. (1969). *Communities in disaster*. Doubleday Publishing.

Bilham, R. (2010). Invisible faults under shaky ground. *Nature Geoscience*, *3*(11), 743–745. https://doi.org/10.1038/ngeo1000

Blaikie, P., Cannon, T., Davis, I., & Wisner, B. (1994). *At risk: Natural hazards, People's vulnerability and disasters*. Routledge.

Bolin, R., & Stanford, L. (1999). Constructing vulnerability in the first world: The Northridge earthquake in Southern California. In A. Oliver-Smith & S. M. Hoffman (Eds.), *The angry earth: Disaster in anthropological perspective, 1994* (pp. 89–112). Routledge.

Briceño, S. (2015). Looking back and beyond Sendai: 25 years of international policy experience on disaster risk reduction. *International Journal of Disaster Risk Science*, *6*(1), 1–7. https://doi.org/10.1007/s13753-015-0040-y

Britton, N. R. (1986). Developing an understanding of disaster. *Australian and New Zealand Journal of Sociology*, *22*(2), 254–271. https://doi.org/10.1177/144078338602200206

Brundtland Commission. (1987). *Our common future*. World Commission on Environment and Development.

Bunin, J. (1989). Incorporating ecological concerns into the IDNDR. *Natural Hazards Observer*, *14*, 4–5.

Burton, I. (2010). Forensic disaster investigations in depth: A new case study model. *Environment: Science and Policy for Sustainable Development*, *52*(5), 36–41. https://doi.org/10.1080/00139157.2010.507144

Burton, I. (2015). The forensic investigation of root causes and the post-2015 framework for disaster risk reduction. *International Journal of Disaster Risk Reduction*, *12*, 1–2. https://doi.org/10.1016/j.ijdrr.2014.08.006

Burton, I., Kates, R. W., & White, G. F. (1978). *The environment as hazard*. Oxford University Press.

Buss, T. F. (2008). *Haiti in the balance: Why foreign aid has failed and what we can do about it*. Brookings Institution Press.

Calais, E., Freed, A., Mattioli, G., Amelung, F., Jónsson, S., Jansma, P., Hong, S.-H., Dixon, T., Prépetit, C., & Momplaisir, R. (2010). Transpressional rupture of an unmapped fault during the 2010 Haiti earthquake. *Nature Geoscience-NAT GEOSCI, 3*, 794–799.

Cannon, T. (1993). A hazard need not a disaster make: Vulnerability and the causes of "natural" disasters'. In P. A. Merriman & C. W. A. Browitt (Eds.), *Natural disasters: Protecting vulnerable communities* (pp. 92–105). Thomas Telford.

Cannon, T. (1994). Vulnerability analysis and the explanation of "natural disasters". In A. Varley (Ed.), *Disasters development and environment* (pp. 13–30). John Wiley.

Cardona, O. D. (2004). The need for rethinking the concepts of vulnerability and risk from a holistic perspective: A necessary review and criticism for effective risk management. In G. Frerks & D. Hilhorst (Eds.), *Mapping vulnerability: Disasters, development and people*. Earthscan Publications.

Cardona, O. D., Hurtado, J. E., Duque, G., Moreno, A., Chardon, A. C., Velásquez, L. S., & Prieto, S. D. (2003). *The notion of disaster risk: Conceptual framework for integrated management*. Inter-American Development Bank/IDEA Program of Indicators for Disaster Risk Management, National University of Colombia.

Cardona, O. D., van Aalst, M. K., Birkmann, J., Fordham, M., McGregor, G., Perez, R., Pulwarty, R. S., Schipper, E. L. F., & Sinh, B. T. (2012). Determinants of risk: Exposure and vulnerability. In C. B. Field et al. (Eds.), *A special report of working groups I and II of the intergovernmental panel on climate change (IPCC), Managing the risks of extreme events and disasters to advance climate change adaptation* (pp. 65–108). Cambridge University Press.

Carr, L. J. (1932). Disaster and the sequence-pattern concept of social change. *American Journal of Sociology*, *38*(2), 207–218. https://doi.org/10.1086/216030

Charles, J. M. (2021). The cost of regime survival: Political instability, underdevelopment, and (un)natural disasters in Haiti before the 2010 earthquake. *Journal of Black Studies*, *52*(5), 465–481. https://doi.org/10.1177/00219347211012619

Chavla, L. (2010, April 23). *Has the US rice export policy condemned Haiti to poverty?* Hunger Notes.

Chmutina, K., & von Meding, J. A. (2019). A dilemma of language: "Natural disasters" in academic literature. *International Journal of Disaster Risk Science*, *10*(3), 283–292. https://doi.org/10.1007/s13753-019-00232-2

Collins, A. E. (2009). *Disaster and development*. Routledge.

Collins, A. E. (2013). Linking disaster and development: Further challenges and opportunities. *Environmental Hazards*, *12*(1), 1–4. https://doi.org/10.1080/17477891.2013.779137

Collins, A. E. (2018). Advancing the disaster and development paradigm. *International Journal of Disaster Risk Science*, *9*(4), 486–495. https://doi.org/10.1007/s13753-018-0206-5

Crichton, D. (1999). The risk triangle. *Natural Disaster Management*, *102*(3), 102–103.

Cuny, F. (1983). *Disaster and development*. Oxford University Press.

Cutter, S. L. (1996). Vulnerability to environmental hazards. *Progress in Human Geography*, *20*(4), 529–539. https://doi.org/10.1177/030913259602000407

Cutter, S. L., Ismail-Zadeh, A., Alcántara-Ayala, I., Altan, O., Baker, D. N., Briceño, S., Gupta, H., Holloway, A., Johnston, D., McBean, G. A., Ogawa, Y., Paton, D., Porio, E., Silbereisen, R. K., Takeuchi, K., Valsecchi, G. B., Vogel, C., & Wu, G. (2015). Global risks: Pool knowledge to stem losses from disasters. *Nature*, *522*(7556), 277–279. https://doi.org/10.1038/522277a

Davis, I. (1978). *Shelter after disaster*. Oxford Polytechnic Press.

De Cauna, J. (2009). Haïti: l'éternelle révolution: Histoire d'une décolonisation (1789–1804). *Editorial PyréMonde*, 177.

Diederich, B. (1985). Swine fever ironies: The slaughter of the Haitian Black pig. *Caribbean Review*, *14*(1), 16–7, 41.

Dombrowsky, W. R. (1998). Again and again: Is a disaster what we call a disaster? In E. L. Quarantelli (Ed.), *What is a disaster: Perspectives on the question* (pp. 19–30). Routledge.

Drabek, T. E. (1991). The evolution of emergency management. In T. E. Drabek & G. J. Hoetmer (Eds.), *Emergency management: Principles and practice for Local Government* (pp. 3–29). ICMA.

Dupuy, A. (1989). *Haiti in the world economy: Race, class and underdevelopment since 1700*. Westview Press.

Dynes, R. R. (1990). Social concerns and the IDNDR. *Natural Hazards Observer*, *14*, 7.

Dynes, R. R., & Rodriguez, H. (2007). Finding and framing Katrina: The social construction of disaster. In D. L. Brunsma, D. Overfelt, & J. S. Picou (Eds.), *The sociology of Katrina: Perspectives on a modern catastrophe* (pp. 23–33). Rowman & Littlefield Publishing Group.

EM-DAT. (n.d.). http://emdat.be/human_cost_natdis

Engerman, S. L., & Sokoloff, K. L. (1997). Factor endowments, institutions, and differential paths of growth among New World economies: A view from economic historians of the United States. In S. Haber (Ed.), *How Latin America fell behind. Essays on the economic histories of Brazil and Mexico, 1800–1914* (pp. 260–304). Stanford University Press.

Engerman, S. L., & Sokoloff, K. L. (2005). *Colonialism, inequality and long-run paths of development*. NBER Working Paper 11057.

Eve, R. C. (2011). *Histoire d'Haïti. La première république noire du Nouveau Monde*. EditionsPerrin.

Farmer, P. (2011). *Haiti after the earthquake*. Public Affairs.

Fernández, M. (Ed.). (1996). *Ciudades en riesgo*. LA RED and USAID.

Frankema, E., & Masé, A. (2014). An island drifting apart. Why Haiti is mired in poverty while the Dominican republic forges ahead. *Journal of International Development*, *26*(1), 128–148. https://doi.org/10.1002/jid.2924

Friedman, D. G. (1975). *Computer simulation in natural hazard assessment*. Institute of Behavioral Science, University of Colorado Boulder.

Fritz, C. (1961). Disasters. In R. K. Merton & R. A. Nisbet (Eds.), *Contemporary social problems* (pp. 651–694). Harcourt, Brace & World.

Fritz, H. M., Hillaire, J. V., Molière, E., Wei, Y., & Mohammed, F. (2013). Twin tsunamis triggered by the 12 January 2010 Haiti earthquake. *Pure and Applied Geophysics*, *170*(9–10), 1463–1474. https://doi.org/10.1007/s00024-012-0479-3

Gaillard, J. C. (2010). Vulnerability, capacity and resilience: Perspectives for climate and development policy. *Journal of International Development*, 22(2), 218–232. https://doi.org/10.1002/jid.1675

Gigerenzer, G. (2014). *Risk savvy. How to make good decisions*. Penguin Books, 322 pp.

Glantz, M. H. (2015). The letter and the spirit of the Sendai framework for disaster risk reduction (a.k.a. HFA2). *International Journal of Disaster Risk Science*, 6(2), 205–206. https://doi.org/10.1007/s13753-015-0049-2

Green, R., & Miles, S. (2011). Social impacts of the 12 January 2010 Haiti earthquake. *Earthquake Spectra*, 27(1_suppl1)(Suppl. 1), 447–462. https://doi.org/10.1193/1.3637746

Handmer, J., & Wisner, B. (1998). Conference report. Hazards, globalization and sustainability. *Development in Practice*, 9(3), 342–346.

Hewitt, K. (1983). *Interpretations of calamity from the viewpoint of human ecology*. Allen & Unwin.

Hewitt, K. (1995). Sustainable disaster? Perspectives and powers in the discourse of calamity. In J. Crush (Ed.), *Power of development* (pp. 115–128). Routledge.

Hewitt, K. (1998). Excluded perspectives in the social construction of disaster. In E. L. Quarantelli (Ed.), *What is a disaster?* (pp. 75–91). Routledge.

Hilhorst, D. (2003). Responding to disasters: Diversity of bureaucrats, technocrats and local people. *Journal of Mass Emergencies and Disasters*, *21*(3), 37–55.

Hoffman, S. M., & Oliver-Smith, A. (1999). Anthropology and the angry earth: An overview. In A. Oliver-Smith & S. M. Hoffman (Eds.), *The angry earth: Disaster in anthropological perspective* (pp. 1–16). Routledge.

Horlick-Jones, T. (1995). Prospects for a coherent approach to civil protection in Europe. In T. Horlick-Jones, A. Amendola, & R. Casale (Eds.), *Natural risk and civil protection*. E and F N Spon.

Horlick-Jones, T., Fortune, J., & Peters, G. (1993). Vulnerable systems, failure and disaster. In F. A. Stowell, D. West, & J. G. Howell (Eds.), *Systems science*. Springer.

Hornbach, M. J., Braudy, N., Briggs, R. W., Cormier, M., Davis, M. B., Diebold, J. B., Dieudonne, N., Douilly, R., Frohlich, C., Gulick, S. P. S., Johnson, H. E., Mann, P., McHugh, C., Ryan-Mishkin, K., Prentice, C. S., Seeber, L., Sorlien, C. C., Steckler, M. S., . . . Templeton, J. (2010). High tsunami frequency as a result of combined strike-slip faulting and coastal landslides. *Nature Geoscience*, *3*(11), 783–788. https://doi.org/10.1038/ngeo975

Hurt, D. R. (1981). *An agricultural and social history of the dust bowl.* Nelson Publishing, 214 pp.

International ad Hoc Group of Experts. (1989). *Implementing the international decade for natural disaster reduction. A report to the Secretary-General of the United Nations, Geneva, Switzerland.* United Nations.

Ismail-Zadeh, A. T., Cutter, S. L., Takeuchi, K., & Paton, D. (2017). Forging a paradigm shift in disaster science. *Natural Hazards, 86*(2), 969–988. https://doi.org/10.1007/s11069-016-2726-x

Kelly, G. (2004). Ammianus and the great tsunami. *Journal of Roman Studies, 94*, 141–167. https://doi.org/10.2307/4135013

Kelman, I. (2018). Lost for words amongst disaster risk science vocabulary? *International Journal of Disaster Risk Science, 9*(3), 281–291. https://doi.org/10.1007/s13753-018-0188-3

Klein, C. A., & Zellmer, S. B. (2007). Mississippi River stories: Lessons from a century of unnatural disasters. *SSRN Electronic Journal, 60*(3), 1471–1538. https://doi.org/10.2139/ssrn.1010611

Kreps, G. A. (1998). Disaster as systemic event and social catalyst: A clarification of subject matter. In E. L. Quarantelli (Ed.), *What is a disaster: Perspectives on the question* (pp. 31–55). Routledge.

Kutak, R. I. (1938). The sociology of crises: The Louisville flood of 1937. *Social Forces, 17*(1), 66–72. https://doi.org/10.2307/2571151

LA RED (1992). *Research agenda and constitution.* ITDG Publishing.

LA RED (1994, January–July). Declaración de Cartagena. *Desastres y Sociedad, 2*(2). LA RED, Bogota.

Lavell, A. (1994). Prevention and mitigation of disasters in Central America: Social and political vulnerability to disasters at the local level. In A. Varley (Ed.), *Disasters, development and environment.* Belhaven Press.

Lavell, A. (1996). *Degradación ambiental, riesgo y desastre urbano. Problemas y Conceptos: Hacia la Definición de una Agenda de Investigación.* En Fernández María Augusta. Ciudades en Riesgo, LA RED, USAID.

Lavell, A. (2000). Desastres y desarrollo: Hacia un entendimiento de las formas de construcción social de un desastre. El caso del huracán Mitch en Centroamérica. In N. Garita & J. Nowalski (Eds.), *Del desastre al desarrollo humano sostenible en Centroamérica. BID-Centro Internacional para el Desarrollo Humano Sostenible* (pp. 7–45). San José.

Lavell, A. (2003). *La gestión local del riesgo: nociones y precisiones en torno al concepto y la práctica.* Centro de Coordinación para la Prevención de los Desastres Naturales en América Central (CEPREDENAC). PNUD.

Lavell, A. (2004). *LA RED: Background, training and contribution to the development of concepts, studies and practice in disaster risk and disasters in Latin America: 1980–2004.* (LA RED: Antecedentes, formación y contribución al desarrollo de los conceptos, estudios y

la práctica en el tema de los riesgos y desastres en América Latina: 1980–2004). LA RED de Estudios Sociales en Prevención de Desastres en América Latina (in Spanish).
Lavell, A., Brenes, A., & Girot, P. (2013). The role of LA RED in disaster risk management in Latin America. In *World social science report. Changing global environments* (pp. 429–433). UNESCO Publishing.
Lavell, A., & Franco, E. (Eds.). (1996). *Estado, Sociedad y Gestión de los Desastres en América Latina*. LA RED, Facultad Latinoamericana de Ciencias Sociales (FLACSO) and Intermediate Technology Development Group, Soledad Hamann.
Lavell, A., & Maskrey, A. (2014). The future of disaster risk management. *Environmental Hazards*, *13*(4), 267–280. https://doi.org/10.1080/17477891.2014.935282
Lavell, A., Oppenheimer, M., Diop, C., Hess, J., Lempert, R., Li, J., Muir-Wood, R., & Myeong, S. (2012). Climate change: New dimensions in disaster risk, exposure, vulnerability, and resilience. In C. B. Field et al. (Eds.), *A special report of working groups I and II of the intergovernmental panel on climate change (IPCC), Managing the risks of extreme events and disasters to advance climate change adaptation* (pp. 25–64). Cambridge University Press.
League of Nations. (1932). *Convention and statute establishing an International Relief Union*. League of Nations Treaty Series, Number 3115, Volume CXXXV, 1932–1933.
Lewis, J. (1979). The vulnerable state: An alternative view. In L. Stephens & S. J. Green (Eds.), *Disaster assistance: Appraisal, reform and new approaches* (pp. 104–129). New York University Press.
Lewis, J. (1999). *Development in disaster-prone places: Studies of vulnerability*. Intermediate Technology Publications.
Lundahl, M. (2004). *Sources of growth in the Haitian economy*. Inter-American Development Bank, Regional Operations Department II.
Mansilla, E. (Ed.). (1996). *Desastres: Modelo para armar*. LA RED and Tercer Mundo Ed.
Maskrey, A. (1984, November 12–16). *Community based disaster mitigation*. Proceedings of the International Conference on Disaster Mitigation Program Implementation, Ocho Rios, Virginia Polytechnic Institute.
Maskrey, A. (Comp.). (1993). *Los desastres no son naturales*. Bogotá: LA RED, tercer mundo editores.
Maskrey, A. (1996). *Terremotos en el trópico húmedo*. LA RED/ITDG.
Maskrey, A. (2011). Revisiting community-based disaster risk management. *Environmental Hazards*, *10*(1), 42–52. https://doi.org/10.3763/ehaz.2011.0005
Maskrey, A., Lavell, A., & Jain, G. (2022). *The social construction of systemic risk: towards an actionable framework for risk governance*. GAR2022, Contributing Paper.
Mauro, A. (1995). Stop disasters: The newsletter of the UN international Decade for Natural Disaster Reduction. In T. Horlick-Jones, A.

Amendola & R. Casale (Eds.), *Natural risk and civil protection* (pp. 511–515). European Commission, E and F N Spon, Chapman & Hall.

McEntire, D. A. (2001). Triggering agents, vulnerabilities and disaster reduction: Towards a holistic paradigm. *Disaster Prevention and Management*, *10*(3), 189–196. https://doi.org/10.1108/09653560110395359

McGreevy, J. (2013). *Haitian disaster vulnerability as a coupled social-ecological system, Furthering Perspectives*, *6*, 57–67.

Merani, N. S. (1991). The international decade for natural disaster reduction. In A. Kreimer & M. Munasinghe (Eds.), *Managing natural disasters and the environment*. World Bank.

Merriman, P. A., & Browitt, C. W. A. (1993). *Natural disasters: Protecting vulnerable communities*. Thomas Telford.

Mileti, D. (1999). *Disasters by design: A reassessment of natural hazards in the United States*. Joseph Henry Press.

Mizutori, M. (2020). Reflections on the Sendai framework for disaster risk reduction: Five years since its adoption. *International Journal of Disaster Risk Science*, *11*(2), 147–151. https://doi.org/10.1007/s13753-020-00261-2

Munene, M. B., Swartling, Å. G., & Thomalla, F. (2018). Adaptive governance as a catalyst for transforming the relationship between development and disaster risk through the Sendai framework? *International Journal of Disaster Risk Reduction*, *28*, 653–663. https://doi.org/10.1016/j.ijdrr.2018.01.021

Office of the Prime Minister. (2010). *Haiti earthquake PDNA: Assessment of damages losses, general and sectoral needs*. Annex to the action plan for national recovery and development of Haiti.

O'Keefe, P., Westgate, K., & Wisner, B. (1976). Taking the naturalness out of natural disasters. *Nature*, *260*(5552), 566–567. https://doi.org/10.1038/260566a0

Oliver-Smith, A. (1992). Disasters and development. *Environmental and Urban Issues*, *20*, 1–3.

Oliver-Smith, A. (1996). Anthropological research on hazards and disasters. *Annual Review of Anthropology*, *25*(1), 303–328. https://doi.org/10.1146/annurev.anthro.25.1.303

Oliver-Smith, A. (1998). Global challenges and the definition of disaster. In E. L. Quarantelli (Ed.), *What is a disaster: Perspectives on the question* (pp. 177–194). Routledge.

Oliver-Smith, A. (2010). Haiti and the historical construction of disasters. *NACLA Report on the Americas*, *43*(4), 32–36. https://doi.org/10.1080/10714839.2010.11725505

Oliver-Smith, A., Alcántara-Ayala, I., Burton, I., & Lavell, A. M. (2016). *Forensic Investigations of Disasters (FORIN): A conceptual framework and guide to research (IRDR FORIN Publication No. 2)*. Integrated Research on Disaster Risk, 56 pp.

Oliver-Smith, A., Alcántara-Ayala, I., Burton, I., & Lavell, A. M. (2017). The social construction of disaster risk: Seeking root causes. *International Journal of Disaster Risk Reduction*, *22*, 469–474. https://doi.org/10.1016/j.ijdrr.2016.10.006
Oliver-Smith, T. (1986). *The martyred city: Death and rebirth in the Andes*. University of New Mexico Press.
O'Loughlin, K. F., & Lander, J. F. (2003). *Caribbean tsunamis, a 500-year history from 1498–1998*. Kluwer Academic Publishers.
Paultre, P., Calais, É, Proulx, J., Prépetit, C., & Ambroise, S. (2013). Damage to engineered structures during the 12 January 2010, Haiti (Léogâne) earthquake. *Canadian Journal of Civil Engineering*, *40*(8), 777–790. https://doi.org/10.1139/cjce-2012-0247
Pelanda, C. (1981). *Disaster and sociosystemic vulnerability*, Preliminary Paper no. 68. Disaster Research Centre, the Ohio State University.
Perry, R. W. (1998). Definitions of disaster and a theoretical superstructure for disaster research. In E. L. Quarantelli (Ed.), *Defining disasters* (pp. 197–217). Routledge.
Prince, S. H. (1920). *Catastrophe and social change: Based upon a sociological study of the Halifax disaster* [Doctoral dissertation, Columbia University, Department of Political Science].
Pyles, L., Svistova, J., Ahn, S., & Birkland, T. (2018). Citizen participation in disaster recovery projects and programmes in rural communities: A comparison of the Haiti earthquake and Hurricane Katrina. *Disasters*, *42*(3), 498–518. https://doi.org/10.1111/disa.12260
Quarantelli, E. L. (1987). What should we study? Questions and suggestions for researchers about the concept of disasters. *International Journal of Mass Emergencies and Disasters*, *5*, 7–32.
Quarantelli, E. L. (1989). Conceptualizing disasters from a sociological perspective. *International Journal of Mass Emergencies and Disasters*, *7*(3), 243–251.
Quarantelli, E. L. (1998a). *What is a disaster? Perspectives on the question*. Routledge.
Quarantelli, E. L. (1998b). *Lessons learned from research on disasters*, Preliminary Paper 133. Disaster Research Center, University of Delaware.
République française. (1789). *Déclaration des droits de l'homme et du citoyen de 1789*. http://www.legifrance.gouv.fr/Droit-francais/Constitution/Declaration-des-Droits-de-l-Homme-et-du-Citoyen-de-1789
Revet, S. (2009). Les organisations internationales et la gestion des risques et des catastrophes naturelles, *Les études du ceri*, *157*, 8.
Rogers, C. (2017). Christopher who? *History Today*, *67*(8), 38–49.
Salter, J. (1998). Risk profiling in disaster management methodology. In H. M. Selby (Ed.), *The inquest handbook* (pp. 170–185). The Federation Press, 248pp.
Schemper, L. (2018). Transnational expertise on natural disasters and

international organizations: Historical perspectives from the interwar period. In A. Schneiker, C. Henrich-Franke, R. Kaiser, & C. Lahusen (Eds.), *Transnational expertise* (pp. 29–54). Nomos.

Schemper, L. (2019). Science diplomacy and the making of the United Nations International Decade for Natural Disaster Reduction. *Diplomatica*, *1*, 243–267.

Scherer, J. (1912). Great earthquakes in the Island of Haiti. *Bulletin of the Seismological Society of America*, *2*(3), 161–180. https://doi.org/10.1785/BSSA0020030161

Schuller, M. (2012). *Killing with kindness: Haiti, international aid, and NGOs*. Rutgers University Press.

Schuller, M., & Maldonado, J. K. (2016). Disaster capitalism. *Annals of Anthropological Practice*, *40*(1), 61–72. https://doi.org/10.1111/napa.12088

Shaw, B., Ambraseys, N. N., England, P. C., Floyd, M. A., Gorman, G. J., Higham, T. F. G., Jackson, J. A., Nocquet, J. -M., Pain, C. C., & Piggott, M. D. (2008). Eastern Mediterranean tectonics and tsunami hazard inferred from the AD 365 earthquake. *Nature Geoscience*, *1*(4), 268–276. https://doi.org/10.1038/ngeo151

Smith, K. (2002). *Environmental hazards: Assessing risk and reducing disaster* (3rd ed.). Routledge.

Sorokin, P. A. (1942). *Man and society in calamity: The effects of war, revolution, famine, pestilence upon human mind, behavior, social organization and cultural life*. Dutton.

Steckley, M., & Weis, T. (2017). Agriculture in and beyond the Haitian catastrophe. *Third World Quarterly*, *38*(2), 397–413. http://doi.org/10.1080/01436597.2016.1256762

Stiros, S. C. (2001). The AD 365 Crete earthquake and possible seismic clustering during the fourth to sixth centuries AD in the eastern Mediterranean: A review of historical and archaeological data. *Journal of Structural Geology*, *23*(2–3), 545–562. https://doi.org/10.1016/S0191-8141(00)00118-8

Susman, P., O'Keefe, P., & Wisner, B. (1983). Global disasters, A radical interpretation. In Hewitt, K. (Ed.), *Interpretations of calamity* (pp. 263–283). Allen and Unwin Inc.

The New York Times (2021). *Patience runs thin in Haiti quake zone as fights erupt for cash and food*. Retrieved August 20, 2021, from https://www.nytimes.com/2021/08/20/world/americas/haiti-quake-aid-fights.html.

Thomalla, F., Boyland, M., Johnson, K., Ensor, J., Tuhkanen, H., Gerger Swartling, Å., Han, G., Forrester, J., & Wahl, D. (2018). Transforming development and disaster risk. *Sustainability*, *10*(5), 1458. https://doi.org/10.3390/su10051458

Trouillot, M. R. (1990). *Haiti State against nation: Origins and legacy Duvalierism*. Monthly Review Press.

Turner, B. L., Kasperson, R. E., Matson, P. A., McCarthy, J. J., Corell, R.

W., Christensen, L., Eckley, N., Kasperson, J. X., Luers, A., Martello, M. L., Polsky, C., Pulsipher, A., & Schiller, A. (2003). A framework for vulnerability analysis in sustainability science. *Proceedings of the National Academy of Sciences of the United States of America, 100*(14), 8074–8079. https://doi.org/10.1073/pnas.1231335100

Twigg, J. (2004). *Disaster risk reduction: Mitigation and preparedness in development and emergency programming*. Overseas Development Institute.

Twigg, J. (2015). *Disaster risk reduction*. Overseas Development Institute.

UNCRD (1989, April 13). *Challenges of the international decade for natural disaster reduction (IDNDR)*. Report and summary of proceedings of the International Symposium on "Challenges of the IDNDR". United Nations Centre for Regional Development.

UNFCC (United Nations Framework Convention on Climate Change). (2015). *The Paris Agreement*. http://unfccc.int/paris_agreement/items/9485.php

UNISDR (United Nations International Strategy for Disaster Reduction). (2005). *Hyogo framework for action 2005–2015: Building the resilience of nations and communities to disasters Geneva.*

UNISDR (United Nations International Strategy for Disaster Reduction). (2006). *Development International early warning systems: A checklist*. Report from the Third International Conference on Early Warning, Geneva.

UNISDR (United Nations International Strategy for Disaster Reduction). (2009). *Terminology*. UNISDR.

UNISDR (United Nations International Strategy for Disaster Reduction). (2015). *Sendai framework for disaster risk reduction 2015–2030*. UNISDR.

UNISDR (United Nations International Strategy for Disaster Reduction). (2017). *Report of the open-ended intergovernmental expert working group on indicators and terminology relating to disaster risk reduction*. UNISDR.

United Nations. (2015). Sustainable development goals. http://www.un.org/sustainabledevelopment/sustainable-development-goals/

United Nations Development Programme. (2009). *Human development report 2009: Overcoming barriers: Human mobility and development*. New York. http://hdr.undp.org/en/content/human-development-report-2009.

United Nations Habitat. (2016). *The new urban agenda*. http://habitat3.org/wp-content/uploads/New-Urban-Agenda-GA-Adopted-68th-Plenary-N1646655-E.pdf, *III*.

United States Geological Survey. (2010). The MW 7.0 Haiti earthquake of January 12 2010: USGS/EERI advance reconnaissance team report. https://pubs.usgs.gov/of/2010/1048/of2010-1048.pdf

United States Geological Survey. (n.d.). M 7.2–13 km SSE of Petit Trou

de Nippes, Haiti. https://earthquake.usgs.gov/earthquakes/eventpage/us6000f65h/executive Retrieved August 14, 2021.
UN Res. 2816 (XXVI). *Assistance in cases of natural disaster and other disaster situations.*
UN Res. 36/225. *Strengthening the capacity of the United Nations system to respond to natural disasters and other disaster situations.* https://undocs.org/en/A/RES/36/225
UN Res. 42/169. *International decade for natural disaster reduction.* https://undocs.org/en/A/RES/42/169.
UN Res. 44/236. *International strategy for disaster reduction.* https://undocs.org/A/RES/44/236
UN Res. 62/192. *International strategy for disaster reduction.* https://undocs.org/en/A/RES/62/192
Varley, A. (Ed.). (1994). *Disasters, development and environment.* Wiley.
Waddell, E. (1983). Coping with frosts, governments and disaster experts: Some reflections based on a New Guinea experience and a perusal of the relevant literature. In K. Hewitt (Ed.), *Interpretations of calamity from the viewpoint of human ecology* (pp. 33–43). Allen & Unwin.
Westgate, K., & O'Keefe, P. (1976). *Natural disasters: An intermediate text.* Bradford Disaster Research Unit, University of Bradford.
White, G. F. (1945). *Human adjustment to floods.* Research Papers 29. Department of Geography, University of Chicago, 225 pp.
White, G. F. (1974). Natural hazards research: Concepts, methods, and policy implications. In G. F. White (Ed.), *Natural hazards: Local, national and global* (pp. 3–16). Oxford University Press.
Wijkman, A., & Timberlake, L. (1984). *Natural disasters; acts of god or acts of man?* Earthscan Publications.
Wilches-Chaux, G. (1993). La vulnerabilidad global. In A. Maskrey (Ed.), *Los desastres no son naturales. LA RED—Tercer mundo editores* (pp. 9–50). Bogotá.
Wilches-Chaux, G. (1998). Auge, Caída y Levantada de Felipe Pinillo, Mecánico y Soldador o Yo Voy a Correr el Riesgo. In Delta (Ed.), *LA RED and intermediate* Technology Development Group, Ed Delta.
Wisner, B. G. (1978). An appeal for a significantly comparative method in disaster research. *Disasters*, 2(1), 80–82. https://doi.org/10.1111/j.1467-7717.1978.tb00070.x
Wisner, B. G. (1979). Flood prevention and mitigation in the People's Republic of Mozambique. *Disasters*, *3*(3), 293–306. https://doi.org/10.1111/j.1467-7717.1979.tb00155.x
Wisner, B. G. (2016). *Vulnerability as concept, model, metric, and tool. Oxford research encyclopedia of natural hazard science.* https://doi.org/10.1093/acrefore/9780199389407.013.25
Wisner, B. G., Blaikie, P., Cannon, T., & Davis, I. (2004). *At risk: Natural hazards, People's vulnerability and disasters* (2nd ed.). Routledge.

Wisner, B. G., Gaillard, J. C., & Kelman, I. (2011). *Routledge handbook of hazards and disaster risk reduction*. Routledge.

World Bank (n.d.). *Haiti country profile*. Retrieved August 12, 2021, from https://data.worldbank.org/country/HT

Xie, L. (1989, April 13). *How do we evaluate IDNDR*. Challenges of the IDNDR, Report and Summary of Proceedings of the International Symposium on "Challenges of the IDNDR", Yokohama, Japan, UNCRD Meeting Report Series, 32. United Nations Centre for Regional Development, Nagoya, Japan.

CHAPTER 2

Where and when disasters occur

Fig 2.1
Village in Gifu Prefecture. Source: photo courtesy of Christopher Gomez (2019)

Figure 2.1 shows a village in Gifu Prefecture (Japan), where the temple (on the slope on the right side) was built over a landslide deposit, where the rest of the houses are located in the lower part. In Japan, very often, temples were built over hazardous zones to keep the monsters of the mountain at peace. Disasters that have been encountered by the local communities are often remembered as a dragon or some monsters awakening. The reader will note that in comparison with Western tradition, where demons were slayed, the Japanese concept is more about "living together" (*Kyozon*), and the monster may awake one day (photo courtesy of Christopher Gomez, 2019).

2.1 Introduction

When asked the question of where and when disaster happens, looking at a world map and a calendar (typhoon season etc.) is a temptation that must be resisted.

DOI: 10.4324/9781315469614-4

Geopolitical boundaries are inherited lines drawn by a group of empires and mostly by the Europeans as a result of the colonial era, with no real connections to the people living on the ground, and very often, they only contribute to hiding the reality of individuals and communities. It also stops us from asking the pertinent questions about the where and when of disaster. The calendar has the same place, as it contributes to focus on the hazard disenfranchising what is certainly at the heart of disasters: us.

Consequently, and although those databases are increasingly reliable, we must resist the temptation to draw conclusions by comparing the statistics by country, like the ones found in the – yet excellent – EM-DAT (Figures 2.2a and 2.2b).

Another game we need to be wary of is the one that rules the interactions between one nation and another, the one helping and the one being helped (Box 2.1). The real endgame is often a very different one than what is being given through the media and the commercials run by NGOs. The visible help that is offered by one country to another is often just a tool in a political game. Moreover, like Batman puts on his superhero suit at night to solve the problems that his own company has imposed on Gotham City by day, numerous countries and corporations like to play heroes, pretending to solve the problems that they created in the first place (although those heroic actions are only to be seen in regions where resources or strategic interests exist – altruism is not a well-spread idea yet). This is typical of what has happened across the African continent with European postcolonial powers notably. Therefore, when you analyse where and when a disaster occurs, it is important to remember what are the historical roles that have brought one region or one community into a disaster, and what is the real input of one country, one community or another. It is, thus, easy to state that one country is prone to disasters whilst another is not and so forth, making judgment values on one community, one country. But if you scrape the surface, you will see that in a lot of cases, external influences and forcing have made some communities less resilient, more disaster-prone, as we will see in the following paragraphs.

Avoiding those pitfalls of postcolonialism, in this chapter, the questions of where and when disasters occur are presented in two different sections. A first section will present (1) the differences between rural and urban disasters; (2) the concept of centre-periphery, of polarisation; and (3) the influence of the types of hazards on the spatiality of disasters. The second section investigates the question of when disasters occur, combining hazards and community realms around the concept of adequacy. At its core, two concepts will be presented, namely, (1) the role of demographic transitions and changes and (2) the role of ill-development.

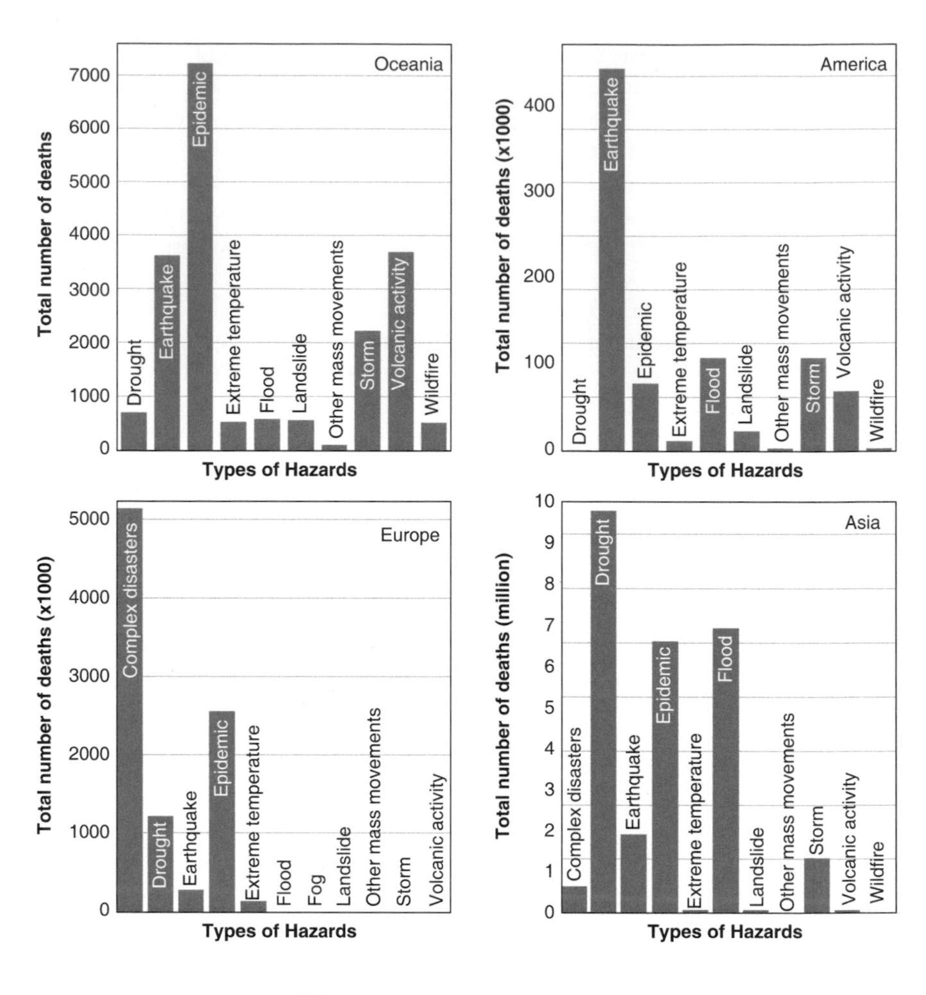

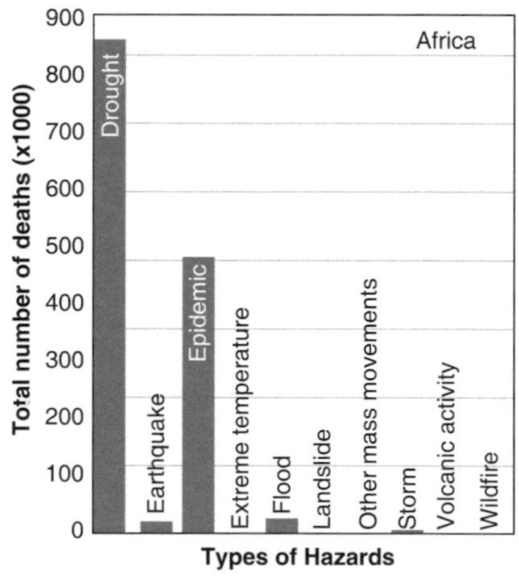

Fig 2.2
Distribution of death for the period 1900–2018, over the five continents (Antarctica is of little interest here), and divided by types of hazards. Source: adapted from the International Disaster Database, www.emdat.be/

Box 2.1: Hazards, disaster and data

In the 1990s and early 2000s, researchers would start carrying laptop with a few mega hard drives and exchange data on floppy disks of 800 kb to a standard 1.44 Mb, soon to be extended to an out-of-that-world 703 Mb with CDs. Today, for half the size of a CD, a solid-state hard drive gives you 1 TB (1 TB is 1,000,000 Mb) so that you carry in your pocket the equivalent of 1,420 CDs or 694,444 floppy disks. This exponential explosion of data storage has come with an ever-increasing amount of data and databases on hazards and disasters. However, data is not knowledge, nor does it make resilience.

Increasingly large datasets are used to measure loss, and those measurements are made through different types of metrics, such as the deaths per million people in a geographical area, or the number of deaths per a given number of individuals, or the average number of deaths per period. The number of casualties can also be measured against a type of hazard or within a cultural community. But more than the direct losses and impacts, large datasets spanning over a long period also allows communities, governments and researchers to measure the longer-term role of a disaster on the trajectory of a community.

Data being created by a plethora of agencies, the question of the quality of the data and whether one dataset can be compared with another also arises. The first builders of datasets are the national governmental bodies and the insurance companies, then come all the international bodies, such as the United Nations, the World Bank and, finally, the different NGOs, such as the Red Cross or the Red Crescent. It appears clearly from this short list that those organisations were put in place by a restricted group of Western countries and that they may have difficulties reflecting the realities of countries, such as North Korea, China or Russia, for instance, countries that are politically and economically different than the US or some part of Western Europe, from which those organisations are mostly piloted. These databases, such as the EM-DAT, is used by governments and those groups, whilst media and the insurance industry rely more on the NatCat database.

There is, therefore, major datasets available, feeding the political system and the media (mostly Western media, and political and industrial units), and it is easy to see how those can be biased and serve the political goal of a group of countries rather than reflect on the reality of the whole world. One entity that has received such critics, for instance, is the World Bank or the International Monetary Fund. And even without malignant intent, data are often collected based on an existing view and specific objectives, which in turn brings bias in the created dataset.

2.2 Urban and countryside disasters, and how one relates to the other

Urban disasters

Undeniably, the world population is increasingly urban. In 1990, 45% was urban; by 2005, half of the population (50%) was urban; and by 2020, projections predict that more than 60% of the world population will be living in cities (data from the World Bank). Furthermore, the sheer size of urban agglomerations has been rising very rapidly. In 1950, only New York exceeded 11 million inhabitants; today, in 2017, 20 boast the same number of dwellers and more; and this number is supposed to be rising to 24 by 2025. Out of these 24 megacities, 17 will be in the so-called developing countries – although those predictions can be derailed when humanity is facing pandemics, like COVID-19, for instance. The world will, thus, have an increasing number of people in an increasing number of megacities, concentrating potential vulnerabilities in very limited space. To the question of where disasters occur, and are most likely to occur in the future, the answer is obviously cities, regardless of whether the disaster originated from the city itself or outside, in the surrounding countryside connected to the city (not only geographically).

Cities are, thus, prone to disasters, as they concentrate economic and political assets as well as populations. But they are also prone to hazards because of their very fabric. On 17 January 1995, the *Hanshin-Awaji Daishinsai*, also known as the Kobe earthquake, occurred at 5:46 a.m. local time, when inhabitants started to light the fire for the morning cooking in their timber-framed and wood-cladded homes. Consequently, the high concentration of houses over the unconsolidated sediment turned the 20-second Mw 6.9 shaking into a huge brazier, which would have not happened in a less densely populated area. Similarly, the 2011 earthquake sequence that hit Christchurch (New Zealand) has sent brick walls and facades falling on the streets, on individuals; and cars and buses were also crushed by falling masonry. In both cases, the modalities of construction, the presence of multistorey buildings around relatively narrow streets, typical of a city landscape, play a central role in casualties and socio-economic impacts. It is, therefore, a city disaster. The same earthquake would not have had the same effect on a crop field and a single timber-framed farm. As civil engineers often say, it is not earthquakes that kill people, but buildings do.

As cities are networked systems, they need and rely on its surrounding countryside and smaller cities that it polarises. Like men, no city is an island.

Disasters in the city can, thus, be the result of disasters within the city (like Christchurch, New Zealand, in 2011) or from disasters in the areas that they polarise. The vulnerability of cities to events occurring in the countryside is, thus, due to their networked nature, which draws upon rural and other sub-level forms of urban areas to function. From these polarised areas, cities draw resources, such as power, food and water. It is the city's "watershed" (to use the terminology of Dr R. Shaw). Indeed, the city tends to concentrate all the resources in the manner a watershed does with water. Often, it also uses this space to keep highly polluting industries and dumpsites away from sight. Thus, when this polarised space is in a disaster position, the cities it serves can then face some of the aftershocks as well and even face disaster themselves. For instance, if you consider the relation between Tokyo (Japan) in March 2011 and the far-field region of Fukushima, where the nuclear power plants used to generate electricity for Tokyo, it is easy to understand the direct consequences of the 2011 Fukushima explosion on Tokyo, which suddenly lacked electricity. The power plant accident occurred as a consequence of the Tōhoku earthquake and tsunami, combined with an ill-designed structure and the absence of consideration for the warning that Japanese scientists, like Professor Imamura, had already brought forward, as his research had shown that approximately 800 years ago, the Jogan tsunami had had the same order of magnitude. The explosion led to the exposure of the Japanese population to radioactive fallout both in Fukushima and even in Tokyo Metropolitan area, about 200 km south of it. From this event, Tokyo suffered a deep energy dilemma, as the power plants in Fukushima were not producing any electricity, and all the train system and electric power system were threatened. Any further failure would have had catastrophic consequences for Tokyo and Japan as a whole. Because of the polarising role of Tokyo and its reliance on a watershed for, notably, its energy, even events that do not hit the city can directly impact and weaken the city, and can contribute to disasters.

Disasters in the polarised countryside

Continuing with the example of the 2011 earthquake and tsunami in the Tōhoku area, we can state that even if the space and economy are closely linked to Tokyo City, the event was not a direct disaster for Tokyo but one for the Tōhoku area. Tokyo dwellers were back in the office the very next day after 11 March, attempting to return to a "normal". The city and the countryside polarised by the city are, thus, linked through a two-way relationship, but these relationships are one where the power of the city is over the countryside, very much like how the feudal castle dominated the land around it. The partnership between a city and its watershed is, therefore, skewed. The polarisation of a city watershed has numerous other detrimental impacts on the rural area. To continue with the Japanese example, the country is experiencing one of the steepest demographic transitions in the world, since the end of World War II, with a rapidly ageing population. This issue is particularly acute in the countryside where young working-age individuals choose, more often than not, to move to large cities and preferably to Tokyo, movement for which there is a terminology signifying that one goes "up" to Tokyo, showing the social ascension associated with this spatial movement (*Tokyo ni agaru*) – interestingly, the same wording for the same reasons has been chosen by populations all over the world, like in France, where the French people say, “monter a Paris”. This concept expresses very well the idea of the power relation between those two spaces. In the polarised watershed, this has two important impacts: (1) the drainage of fresh workers to the profit of larger cities limits the amount of governmental funding and taxpayer money available to improve hard engineering and maintain those structures; (2) the lack of young people can have a cruel effect during the emergency period that characterises a disaster.

When a disaster occurs outside the city, there are numerous external factors that influence how this space will fare during a disaster. Arguably, if the same 2011 disaster had occurred in the 1950s or the 1960s, when the Japanese countryside was still vibrant with a young population, faring through the disaster time and the reconstruction would have been very different. The presence of working-age population would have meant access to bank loans and mortgages to reconstruct, workers to rebuild faster and stronger local communities with more means of expressions of how they would have liked to see their living space rebuilt (Figure 2.3).

Some specific issues appeared due to the skewed age pyramid of the population: in 2011, during the Tōhoku earthquake and tsunami, the evacuation of elderly was notably difficult because of their reduced mobility and their need to rely on the help of others to reach a safe place. In hospitals, notably, workers mentioned that due to the earthquake, the lifts of the hospitals had stopped and that carrying upstairs patients and elderly had not been possible for everyone. A high proportion of elderly passed away in their home due to their inability to manoeuvre in the stairs.

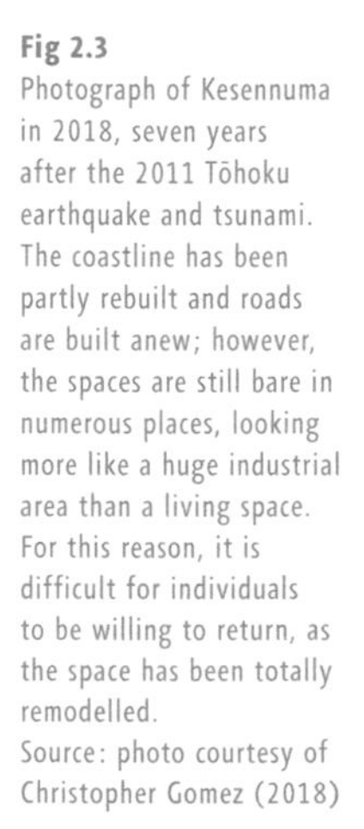

Fig 2.3
Photograph of Kesennuma in 2018, seven years after the 2011 Tōhoku earthquake and tsunami. The coastline has been partly rebuilt and roads are built anew; however, the spaces are still bare in numerous places, looking more like a huge industrial area than a living space. For this reason, it is difficult for individuals to be willing to return, as the space has been totally remodelled.
Source: photo courtesy of Christopher Gomez (2018)

Fig 2.4
The Sendai Plain photographed at the end of 2012. This suburban area on the seaside of Sendai City was stripped bare by the tsunami. The last remaining trees on the horizon are the remains of the coastal pine tree forest.
Source: photo courtesy of Christopher Gomez (2012)

Several years on, the regeneration process of the region is also difficult (Figure 2.4) because there aren't enough young people to go and work on rebuilding worksites, and there aren't enough individuals ready to start a new life in an environment that was already in demographic difficulty before the 2011 event. The polarisation of manpower away from this countryside has, thus, been a strong explaining factor in the disaster. Numerous issues of the ageing population became very acute, as elderly could not restart anew, elderly could not receive a bank loan to rebuild, and communities had to be then scattered over several urban centres in different regions of Japan. In such a case, the disaster occurs as a highlighter of already existing issues, whilst those issues are contributing to the disaster.

Rural disasters

In the previous paragraphs, we defined the rural space as countryside, as opposed to the city, but there are different levels of polarisation, and the rural space can be defined on its own, yet it is a space that is often synonymous with low economic income and marginalisation, from the cultural, intellectual, and economic and political power instances. (Although COVID-19 has regenerated an interest in countryside living. In 2020, for the first time since World War II, Tokyo has seen more people leaving the capitals than those entering.) This vision is notably fuelled by a crisis of the rural values in most developed nations.

In rural areas, because population is sparse, the overall impact of rural disasters is often less catastrophic and receives less mediatisation; however, spatially extensive disasters, such as droughts, have rippling effects on the local, regional and, eventually, global economy. And compared to disasters involving populated areas that are close to the economic and political power, the effects are often very long-lasting (notably because they attract less attention). For instance, the Dust Bowl droughts and dust storms – locally named black blizzards or black rollers – that occurred at the junctions of Oklahoma, Texas, Kansas, New Mexico and Colorado states in the USA between 1935 and 1938 had devastating effects on the American economy. It resulted from the combination of poor agricultural practices, which resulted in Aeolian erosion – which today is being addressed by proper dryland farming practices. Dust Bowl farmers were forced to abandon their farms, unable to make the bank repayments on their mortgages, with the total amounting to more than present-day US$400 million. The ecological disaster was to hamper any economically viable production for several years, and the Roosevelt administration started to address the issue as soon as 1933 and developed several plans to stabilise the economy and curb the malnutrition issues. During the 1930s decade alone, an approximated 3.5 million inhabitants of the Central Plains were forced to move out of the region, and those who stayed faced malnutrition and pulmonary diseases. In 1940, the major black rollers were over, and most of the migration had occurred, but the region had not recovered from the disaster yet. The eroded land was still unrecovered, and for the following decade, there was no significant modification of agricultural practices to recover the eroded land. The local agricultural heritage – a world that is traditionally conservative in Western countries – and the failure of numerous local and regional banks due to the issues in the 1930s resulted in a disaster that had its grip on the region for arguably more than 20 years. The results of the Dust Bowl disaster were amplified by the Great Depression the US was going through since the financial crash of 1929. And the severity of rural disasters is often linked to factors that are external to the rural world.

For instance, both the Asakura heavy rainfall disaster of 2017 and the 2011 tsunami in the rural bays of Miyagi and Iwate Prefecture have been strongly affected by the socio-economic and demographic of Japan as a whole. In both cases, rural exodus has been an issue for several years, and in an already shrinking population, the lack of young individuals has been, and remains, an issue. Therefore, when debris flow and the river floods, and when the tsunami struck, respectively, the Asakura and the Tōhoku regions, rural areas have been struggling to face up to the challenge of recovery. In this case, it is not the mere lack of funding but the reality of the lack of younger generations ready to restart a new life in the countryside in the aftermath of those events. There is, therefore, not enough hands to rebuilt infrastructures and not enough taxpayers to maintain a complex infrastructure system, and the lack of a new generation hampers the passing-on of vernacular knowledge. Communities and their cultures are endangered, thus, by the disaster because of the structural trajectory of the country as a whole, rather than just the effects of the tsunami.

Finally, and related to the previously mentioned idea of the dichotomy between urban and rural areas, and the fact that numerous workers go up to the capital city, those who left the rural areas for the city still hold an affective relation the rural areas where they come from. In Japanese, it is termed as *furusato* (expression for the place where one is born and that is tainted with good old time feelings, in the Japanese language, as the place where the heart often is). This affective relation gives an important role to existing and inherited places, buildings and culture. Consequently, in the aftermath of the 2011 earthquake and tsunami in Japan, *enquetes* have revealed that people who left with the disaster did not find much taste to go back to a place that was not to this image anymore. The main problem to people's return was not the replacement of infrastructure and housing but more about the disappearance of the very soul of a place. And centralised reconstruction did not do so well in preserving this cultural heritage. This affective attachment was trumped by economic relations, however, for people living from the land. They showed a strong attachment to the land itself, where they lived, where they worked and where their ancestors came from. Consequently, even in Fukushima Prefecture, older generations made the choice to come back, even with the radiations. In a rural disaster, then, the place and the positioning of individuals and communities in relation to the space exert a strong control on the recovery process during and after a disaster. The concept of centre-periphery, dichotomy rural-urban, is very much like an onion with many layers to peel.

Consequently, rural disasters are the product of socio-economic, cultural and demographic dynamics of the place; the country and hazards will produce very different disaster in rural areas and in cities. Rural disasters may tend to have a slower recovery, and there is also the obligation to sometimes live with the scars of an event for long periods (Figure 2.5).

Fig 2.5
Photograph in the Asakura area (North Kyushu) one year after the heavy rainfall event that triggered the large landslide seen on the right and the debris flow from which the fan can be seen underneath. The riverbed has been damaged, and consolidation work is also encroaching the landscape and creating the particle haze that can be seen on the photograph. One year on in the disaster, communities were still living with the scars of the event.
Source: photo courtesy of Christopher Gomez (2019)

2.3 Centre-periphery and disasters

The earlier paragraphs addressed the urban and the rural areas, and then one type of "centre-periphery", but there is more to this concept and how it interplays with disaster; and it will be investigated in the coming lines. Whether it is at the local level, between trendy parts, or the economic heart of a city and its suburban areas; or at the regional level, between the polarisation of a city and the surrounding – for instance, French geographers have coined the term 'Paris and its green desert' to mark how the capital is attracting people, funds and attention away from the surrounding countryside – or at the international level, where countries attract or detract funds, workers and so forth, the multiscale spaces we live in are all polarised. The nature of the poles and peripheries can be tangible, or they can be just imagined or perceived. In term, this spatial dynamic does influence hazards and disasters' spatiality (Box 2.2). This concept is rooted in Marxist geography and sociology, where peripheries are defined by centres, assuming colonialist visions of the rest of the space. The same snobbism exists between wealthy city dwellers towards lower socio-economic classes living in the far suburbs, and between wealthy, so-called developed countries towards developing countries.

Box 2.2: Mobile or not mobile in the face of disaster, this is the question

Whether you consider a hydrometeorology hazard (Chapter 9), or an endogenous hazard (Chapter 7), or an exogenous hazard (Chapter 8), one of the potential options is to stay out of harm's way. With Hurricane Sandy or Katrina, more than one million people were displaced, but a week after the event, more than 600,000 had no place to immediately go back to, and a month later, more than 270,000 were still living in temporary accommodation.

Forced displacement contrasts with individuals who are mobile and can resettle to restart or continue a normal life after disaster strikes. Research on the Katrina disaster has shown that the ability to return home or to have a stable home was a factor well correlated with health and economic stability of low-income families, whilst higher-income families could afford to restart anew in a different setting and to rebuild. This dichotomy also appears on the school benches, with one-third of the forced-displaced children falling at least one year behind in the school programme. This results from the time in temporary accommodation but also the inability to return to a normal home and school environment several months to years after Katrina stuck the Gulf region.

A similar issue arose in the aftermath of the 2011 earthquake and tsunami in Japan, where, like for Katrina, income and wealth inequalities determined the mobility and the ability to return to a normal life. This issue is particularly acute with the population displaced due to the Fukushima nuclear reactor explosion. People were asked to jump on buses rapidly and evacuate the city they were living in, leaving everything behind with no possibility to ever go back, if not for sporadic visits in full radiation-protection gears. For elderly without any income, the very meagre retirement fund provided by the Japanese government does not allow them to reconstruct their lives nor buy a new home, and numerous have been parked into small tower apartments. As the Japanese law also sets the price of buildings to decrease over time – one way to keep the economy running and to have buildings that meet the newest anti-seismic norms – elderly who lost their homes did not receive then enough compensation to think about rebuilding. This issue also extends to younger families and individuals who had a home mortgage. Even after the tsunami took their homes away, they remained with the mortgage, but insurance compensation is often less than half the original price, even a couple of years after buying the dwelling, in such a way that returnee often have had to take on a double mortgage to rebuild a home.

This resulted into a displaced population, with a strong community sense, being dispatched all over Japan into impersonal apartments for those who could not afford to rebuild back, and for the younger generations who could take on another mortgage, it created an immense financial strain on them, which will have long-lasting impacts on their life. Any other event, from an aftershock, in the Tōhoku region during the next coming 20 years would have unprecedented effects on the local communities. Once again, those who could afford to rebuild back away from the Tōhoku area and keep a regular income are the ones who were better off, as it was for the Hurricane Katrina disaster in the US.

Basic centre-periphery issue at the local scale

One of the best examples of a centre-periphery dichotomy, which occur within urbanised areas, is most certainly the doughnut effect that divide the suburban from the urban space. It divides not only space but also income, education, race, cultures and so forth. In this model, a rich centre, concentrating wealthy demographics, is surrounded by groups often working for the centre but who can't afford a central living. An outer doughnut then can ring this ring of poverty with higher-income earners looking for a greener lifestyle. Often the different outer rings are then radially divided, with often a street or a highway bridge separating economically wealthy from lower-income areas, like Berkeley rubbing shoulders with Oakland outside San Francisco, California, with simply a highway for a border. In Paris, the surrounding outer rings are traditionally less wealthy on the eastern outer ring and the north, and it is the same in London. This division is now historical due to the dominating winds that were bringing most of the pollution on the less-wealthy suburbs and cleaning the air of the wealthy suburbs. The original doughnut idea is, therefore, more complex in details, with areas with more or less sugar-coating, but it has kept its original specificity of separating socio-cultural and economic groups within one city. In Christchurch City, New Zealand, the city plan is atypical, as the lower-income areas of Aranui, Linwood and New Brighton are located near the coast, whereas the wealthy areas of Fendalton and Ilam are to the west of the city, where the airport is also located. In Christchurch, the location of the good agricultural land against the swampy land of the east have generated this division that persists today, and in the aftermath of the February 2011 earthquake, which struck the city and took the lives of 185 and injured about 2,000, most of the damages to private dwellings occurred in the eastern lower-income suburbs, and compared to the wealthier suburbs, remediation took more time, media concentrated further on the central city and, as more residents had no insurance, the economic blow was more acute. This does not mean that the lower-income neighbourhoods were just more vulnerable. There have been numerous local neighbours helping, and grassroots activity trumped some of the economic distribution effects, movements that were less important in the western, more affluent suburbs. Furthermore, a large proportion of the local Māori population tends to concentrate in the lower-income neighbourhood, mostly due to a different set of cultural values that have left them aside of a mainstream economy dominated by Western values; however, the local *iwi* (the word designating the Māori tribes in New Zealand – in Christchurch and the South islands Ngāi Tahu or Kāi Tahu – is the *main iwi*) provided financial and other forms of help, outside of the official governmental framework. This manner of helping one another and looking after one another even sparked a campaign named "random act of kindness" and "Are you ok", giving more weight to the horizontal relations than the verticals.

Although it is accurate to recognise the increased vulnerability of the periphery, it is as important to recognised that informal and replacement strategies exist, even outside standard and official framework. For instance, the aftermath of the 2010 volcanic eruption of Merapi Volcano that shook the Central region and the Yogyakarta autonomous region of Java Island impacted one of the peripheral LGBT groups, the *warias* (Balgos et al., 2012), who, not content of being in the role of the marginalised group and victim, also offered psychological relief by providing some form of normality and relaxing experiences to the rest of the community in the shelters, providing salon services (haircuts, *krembat* head massages etc.). This LGBT experience, however, is not equal in every country. In the aftermath of the 2011 earthquake and tsunami in Japan, they were segregated, mocked and almost used as scapegoat to express anger. Furthermore, in a country where conforming to the norm and obeying the rules have long been a replacement to most forms of individual ethics, the sudden and often forced coming out of LGBT during and in the aftermath of the disaster did not lead to positive experiences nor to an active positive role in the disaster recovery process (Yamashita et al., 2017).

Regional-scale and international-scale centre-periphery from an ethical perspective

The previous sets of paragraphs presented the centre-periphery concept at the local scale and how it interplays with disaster distribution from a geographical perspective. This concept of centre-periphery can take also non-tangible forms to control the spatiality of disasters.

Hazards that occur on planet Earth, be it earthquakes, storms and hurricanes, are not bounded by human political and cultural borders; however, the human side of disasters is bounded by those limits (Box 2.3). Every year, the eastern end of the Pacific Ocean is battered by typhoon, which often starts to the north of the Philippine Isles, and through different trajectories, the typhoon either goes through the Philippines, Taiwan, China, Hong Kong, South Korea and Japan, or most of them. However, the countries and places where the typhoon becomes a disaster or only a weather event are determined by the governance at the country and regional level. This division does not only exist in terms of human structures, it also influences deeply who we are and how we think about one another.

Box 2.3: Ethical problems in disaster-impacted areas

A couple of years ago, I was in a boardroom at the Gadjah Mada University in Indonesia with the director of international relations and some of the university leaders to have a short meeting with foreign European researchers going to Indonesia for a major disaster research project. I remember very vividly when one of the European colleagues was telling the Indonesian team that it was very good that they had dug holes in the ground to put instruments in but that the Indonesians should get their instruments out and let the professionals (aka the Europeans) do the work and collect the data because "they didn't know what they were doing". How, past the year 2000, such blatantly postcolonial flavoured feelings of superiority could still be on the lips or in the heart of individuals (individuals claiming themselves to be left-wing politically) was one of the worst shocks of my (still short) career. Ethically, there were so many things that were wrong. In the aftermath of the 2011 earthquake in Christchurch, New Zealand, researchers and practitioners also sometimes took the time to shield locals and individuals going through a hardship from the researchers who were coming on a short and intensive data collection campaign. Is it right to have external individuals using time and resources for their own research, when those same resources should be mobilised for the community experiencing disaster (Gaillard & Gomez, 2015)? The next problem is then the generation of irrelevant research from individuals and groups, who are not in a relevant framework. For instance, it has been written from European shores that the 2011 New Zealand disaster in Christchurch and the 2011 disaster in Japan were comparable and in a similar setting. The question of relevance, postcolonialism and ethics then comes to the forefront (Gomez & Hart, 2013).

In Japan, in the aftermath of the 2017 and 2018 heavy rainfall events that have battered North Kyushu and West Japan, respectively, with rainfall events exceeding 800 mm in 24 hours, triggered numerous landslides, debris flows and various forms of mud flows and floods in the surrounding valleys and development downstream of the mountains where those events occurred. Like for numerous disasters in Japan, the volunteers organised themselves and drove from across Japan to offer a hand in cleaning up the streets and houses from the mud, and bringing relief goods – also gathered in a grassroots manner. Despite distances and differences between the regions of Japan, Japanese people have a strong sense of belonging – a feeling partly fuelled by the reality of being an island nation – and the nation has long been organised through different ward, quarter and neighbourhood organisations, which also played an important role during the war. Generous donations are also typical of the aftermath of such events. This reality, however, is strongly bounded by the political borders. One could wonder why people go and help strangers several hundred to thousands of kilometres away from their own home, but even if the distance is shorter, they are most likely to go and help the population in a neighbouring country. Such behaviour is rooted in ethics and our sense of belonging, and an onion layer sense of ethics at the personal level. For instance, any reader owning a pet knows that he or she will be more saddened by the disappearance of his or her pet rather than the passing of a (human) stranger two or three blocks away from home. Even comparing a human to an animal, it seems that we can show more empathy for an animal. In the eyes of the law, if we could only save a pet animal or a human, it would not look good in court to take the stance to save a pet animal instead of a human being. Nevertheless, outside of this framework, in a hypothetical situation, we may choose to save our house pet rather than some strangers. Without getting into ethical theories, this is the result of a personal ethic that works as onion layers, where we cherish and protect our closest family first, then friends and then acquaintances. This onion layer is controlled by an affective distance as well as a physical distance. This is the main reason why NGOs are having so many difficulties, even using shocking television commercials to get our attention about unfolding disasters on the other side of the planet. The world hunger in some remote places of Africa just stops when we hit the off on the remote controller. Consequently, on top of the administrative barriers, the natural tendency of human beings to care more for individuals located close to them help dividing places and countries where a disaster may occur and where a disaster may not. It also explains why one will help people close by but will not feel compelled to do so when an event occurs outside one's own region and country.

2.4 Hazard control over a disaster's spatio-temporal realm

The time and space control that hazards have over disasters depend on their types, whether they are endogenic (earthquake and volcanoes) that tend to be fixed over prescribed spatial areas (at the margin of tectonic plates, for instance) and with a period of returns relatively long compared to atmospheric and waterborne hazards. Atmospheric hazards, on the contrary, are almost ubiquitous around the globe, and they can occur over a short timescale and often with seasonal patterns. Compared to one or two generations earlier, science has made sufficient progress so that we know where and eventually when (with some degrees of certitude) the next hazard will occur. For instance, Japan tracks typhoons over the Pacific Ocean using a set of weather satellites, and combined with near-real-time simulations, they provide an estimate of the rainfall intensity and amount as well as wind velocities. They have a similar system for earthquakes, with an automatic countdown – when there is sufficient time – telling anyone with a cell phone, a TV or a radio receiver that a given number of seconds is left before a jolt arrives. The timing is usually within a few tens of seconds, rarely up to a minute. This time span is, however, sufficient to stop the gas, for instance, or anything in a building that would be best not to be shaken. However, those technologies are only available to a few countries with the necessary financial backbone to develop the expensive communication system and sensor networks that are necessary. Hazards first control over the spatio-temporal realm of a disaster is, therefore, very closely linked to the economy and the politics of the country where the event occurs, and this control is not through the natural environmental process but through the reality of vulnerability and preparedness.

The well-known equation multiplying vulnerability by the hazards to get the risk or the disaster risk could be completed further by stating that a hazard is very much constrained and modified by other hazards that can occur concomitantly or previously. For instance, in the Canterbury Region of New Zealand, the Waimakariri River has been the origin of most of the floods that have swept Christchurch City, which entrenched itself behind levees and floodwalls along the floodplain. The worst flood that originated from the Waimakariri River was in 1957, when more than 320 mm of precipitation fell over the Arthur Pass area in 24 hours. The floodwaters grew and made their way from the upper catchment to the lower. The peak discharge was measured to be 3990 $m^3 \cdot s^{-1}$ (Reinsfield, 1995).

The control of one hazard over another one, and subsequent influence on disasters

Both scientific and community experiences of hazards, and especially those of high-return period, are based on previous events and the realm that the environment does not change drastically between events so that predictions can be made from previous events. This is not the case with extreme events, like large earthquakes, tsunamis or Plinian volcanic eruptions, which can modify the landscapes, change river flowing directions, clog valleys and modify the coastlines, acting as a reset button. In such case, one hazard has profound influence on the forthcoming hazards and associated disasters. For instance, despite historical experience of river floods in Christchurch, a new type of flood occurred in the aftermath of the 2011 Canterbury earthquake sequence. Portions of the city's ground level went down by up to 1 m locally, with notable lateral spreading, riverbank sliding and so forth; thus, gravity-driven storm water pipes were damaged and bent both horizontally and vertically. Consequently, floodwater started to rise in the depressions around the city and at the locations of the storm water pipes; instead of driving the floodwater away, it made a water pond due to the change in the pipes' slope angles. This process was worsened due to the Avon River (the main river in Central Christchurch) being clogged by the lowering of the land, bringing the base level (the sea level) virtually higher so that river water was more difficult to evacuate. The lowering of the land also brought the groundwater level closer to the land surface, and in consequence, the soil did lose some of its ability to absorb rainwater, increasing surface runoff and reducing the time it takes the surface water to contribute to the peak flow. The Christchurch earthquake, therefore, changed the dynamic of the land surface and groundwater water cycle, and in consequence, the dynamic and the spatial distribution of floodwaters. It is because new mechanics are put in place that unforeseen disaster risk increases (Box 2.4).

Box 2.4: Ecological model of disasters – the eco-ethical perspective

Etkin and Stefanovic (2005) have presented an eco-ethical model where human beings are enclosed in the natural environment, and their relation is a love/hate relationship with this environment being an essential source of natural resources, whilst at the same time the origin of hazards (social vulnerability to natural hazards versus exploitation of natural resources, to reuse the authors' terms). They have linked the vulnerability to disaster (the authors used the term 'natural disaster', but we will stay clear of this association of terms), triggering human response in the form of DRR, which then modifies the relation between the human and the natural environment. Their model is also interesting because they relate the concepts of sustenance, economic growth and environmental degradation to potential resource depletion, leading to hazards, which in turn can lead to disaster.

The model is, therefore, not deterministic but shows a complex interaction between the environment and the choices that humans make in this environment as drivers of further hazards and disasters (or not). Etkin further wrote,

> Environmental values and the nature of the relationship between humans and nature play a crucial role in the nature of the

> feedbacks. . . . When nature is not valued, or when the links between human and natural environments are discounted, then hazards are ultimately made worse or vulnerability is increased, although short-term benefits may accrue to social systems.
>
> (Etkin, 2016, p. 216)

This is why it is an eco-ethical perspective; the model leading the human community considered to be resilient has to also do the right choice for nature itself. A typical example that can be found in several countries over the world is the one of deforestation on mountain slopes, leaving slopes bare and, in turn, leading to mud flows and debris flows. Source: Etkin, D., Stefanovic (2005)

Widespread slow-onset hazards: the example of droughts in Asia

One of the most difficult hazards to work against, in order to prevent disasters, are droughts because their unfolding often occur through periods of months to years. "A drought disaster is caused by the combination of both a climate hazard – the occurrence of deficits in rainfall and snowfall – and a societal vulnerability – the economic, social and political characteristics that render livelihoods susceptible in the region influenced by the deficits" (Barlow et al., 2006, p. 1). Droughts, like its antithesis, heavy rainfall, operates through two main controls: the severity or intensity, and the recurrence.

Because droughts have very different effects depending on the target concerned, the occurrence probability of a disaster for similar hazard conditions are very different. A meteorological hazard can become an agricultural, hydrologic or socio-economic disaster in some parts of the world and be barely noticeable in other parts of the world. In Asia, for instance, China has the largest number of droughts since 1977, followed by India. Since 1999, the number of droughts has been on the rise in Asia, with more than 30 droughts in 2000, and more than 35 in 2001, both years being more than twice as much as the highest number of droughts per year during the period 1975–2000. Barlow et al. (2006) identified common features of drought disasters in Asia to be the persistence and severity of the drought, and the median precipitations falling less than 75% for a period of at least three consecutive months. They also show that anomalies in the climate since 1999 are most probably responsible for persistent droughts and that the trends present in Asia have been found in similar climatic zones around the globe.

The difficulty in dealing with creeping hazards like droughts is the fact that they often don't have a clear triggering time, and it will develop slowly until the resilience capacity has been fully spent. Through its history, human beings have been well prepared to act upon fast and rapid hazards and threats before they become a disaster. At the origin, one can imagine that our ancestors had to act quickly to defend their caves against wild animal intrusions and competition for resources and shelters (be it a bear or any other animal), but slowly creeping disasters are nothing our DNA prepares us against, and dealing with those events is difficult. This is especially true for droughts because they usually affect widespread spaces with severity gradients so that members in a network cannot get help from other members from the close network because they are being affected by the drought as well and may just have enough to get by (the gradient issue). If one takes the example of a flood, the surplus of water is confined to a zone, with a limit, and the rest is dry land. The drought is different in this aspect. The second reason why widespread droughts are an issue is the fact that humans tend to settle close to drinkable water (except a few communities adapted to life in deserts), and as communities, we are not used to transporting water over several tens of kilometres to sustain ourselves. Water transport comes from a necessity, and we are so reliant on it that it is often not a solution that can be considered. For this set of reasons (creeping, having no clear trigger, being widespread and having no long-distance water transport habits for most human beings), widespread droughts often lead to food crisis and human disasters.

Comments on the spatiality of climate change-related disasters

The first lesson one should remember from this first part is that the spatiality of disasters is not bound to continents or geographic location per se, but it is a social construct that divides countryside versus city disasters, and the relation and the role of one to the other will define very different realms leading to a disaster; the countryside of Java Island and the countryside of rural Massif Central in France or the rich agricultural land – close to the political power and close to the financial help of the European Union with lobbying groups – are all very different and will play a central role in defining disasters. Furthermore, the space is the expression of relations of powers between countries and between regions of the world, and also regionally and locally. Those relations of power also have a strong imprint on the spatiality of disasters.

In the case of the spatial distribution of disasters that can be at least partly imputed to the effects of climate change, we have observed a major difference compared to other disasters. The most striking segment is the lack or absence of direct relation between the original cause (although we are not looking at a single source that can be pinpointed but at several, spanning over several decades and centuries) and the area impacted. The teleconnections between an event and its impact, thanks to the atmospheric vectors, means that the economic excess of a group of countries can have impacts and trigger disasters in areas of the world where the population have absolutely no role in the anthropogenic climate change. For instance, the tropical islands of Kiribati, which is mostly at sea level, has seen its shoreline shrink rapidly in recent years and seawater bubbling on the centre part of the islands, announcing the disaster to come, but it is not this small island nation that is causing its problems.

2.5 When disasters occur

The temporality of disasters is complex. The first element is that it depends on the temporality of hazards that are characterised by a recurrence interval for different intensity, duration of events and the possibility to have several hazards occurring concomitantly or triggering one another as cascading hazards. As Ben Wisner wrote, "A hazard has a time-space geography, involving the probabilities of events of significant magnitude to cause potential damage of differing magnitudes over geographical space" (Wisner, 2004, 93). Furthermore, those temporalities can be also seen as a rhythm, which, paired with other hazard rhythms, can produce enhanced hazards that are difficult to predict from the examination of previous time series. The second element that comes into play is the human time, in the sense of (a) when in the history of men the hazard occurred and (b) what was the logic, understanding and mitigation strategies against natural hazards. Did the hazard occur when society believed that hazards were divine punishment and were fatalist about it, or when they believed that civil engineering and concrete could solve any natural hazard issue, like the French government used to in the 1970s, or are we in an adaptation strategy where constructions are meant to be destroyed at each hazard, like communities of the Pacific who traditionally would move around the island during the year to shelter themselves from typhoon and weather hazards, knowing that any construction battered by the elements would be vowed to be flattened down. Then, comes the question of when in a calendar year, a hazard occurs (e.g. is a typhoon occurring in the middle of winter when there are no crops, or is it in summer or autumn, just before the harvest season?). The socio-economic impact, for a very same event, would then be very different. Finally, one can zoom in even further to the week and hour scale. The 1994 *Awaji-Daishinsai* or Kobe earthquake had such a destructive impact due to the fires that raged through the timber-framed wooden house quarters, turning the city into flames for days. The timing of the earthquake has been essential to understanding these consequences: Kobe inhabitants, just before six o'clock had their stoves turned on for cooking the morning breakfast. It has been very clearly demonstrated that the morning cooking was the cause of the numerous fires. In 2011, with the Tōhoku earthquake, and in 2018, with the Osaka earthquakes, both events occurred during office hours, during the week, and mass media were consequently filled with images of people attempting to reach their home, as the whole railway transport network came to a stop.

Those divisions of time are natural divisions that are universal – although one could argue that history is not linear, and it does not progress the same way nor in the same direction for all communities on Earth. In contrast of natural time, we can then consider socio-cultural time, which can be counted from events that occur in a community. Is a hazard occurring during wartime or in the immediate aftermath? Is a hazard occurring in a corrupted community, which, unfortunately, might have used emergency funds for other purposes, or are we in a time of division within a community when internal conflicts can hamper disaster recovery and resilience? It was the case with the 2004 tsunami that hit the surrounding of the Indian Ocean in 2004. As a civil war was raging between the national authorities and the local army, tensions were high and migrants that fled the conflicts certainly made a large portion of the victims in Banda Aceh. Furthermore, when international NGOs attempted to provide help, they hit a brick wall. As their agencies and contacts were all concentrated in Jakarta City, the capital of Indonesia, on Java Island, the coming of the Javanese with the international funding to organise a territory they had been fighting for and against until the very day the tsunami occurred made the recovery effort very difficult, if not haphazard, because the local Batak people (ethnic majority in North Sumatra) were not ready to take orders from Javanese people.

The question of temporality, timing, boiled down to the formulation of when disaster occurs, is an issue of adequacy. Is the governance adequate for a resilient community? Are the soft and hard engineering adequate against the hazards? Does the conflict-displaced population have adequate knowledge and understanding of disaster risks in the new living location? And so forth. For instance, scientists have argued that the flourishing of human civilisation, like the Mesopotamian around 7,000 BP, is partly due to climatic stability, and so was the fall of the Mesopotamian civilisation that ensued from repetitive severe droughts. More than thinking about the timing and the length of the drought as the answer to when the disaster happened, it might be more appropriate to think that it occurred when the Mesopotamian community was not adequate or adequately prepared to the drought. The same series of droughts occurring today at the beginning of the 21st century is very unlikely to bring down a civilisation. This emphasises the fact that there are no natural disasters and that disasters occur when a community is in a position of inadequacy (Box 2.5).

Those inadequacies can arise from different positions and evolution of a community in its environment and also with regard to other communities (e.g. the issue of colonialism). In the second part of this chapter, some of those positions will be presented, but it is not an exhaustive approach.

When a community is in demographic transition

The change, and especially the rapid change, in demographic distribution and relation to space is one of the timings for disasters to occur. Given a

Box 2.5: The concepts of emergence and self-organisation in disasters

When a disaster occurs, emergence is said to occur when entities (governmental, private, non-governmental organisations and private individuals) interact and work together towards resilience, doing a part it usually does not in the organisation outside disaster time. This emergence is the sign that a disaster has occurred in a system. "The involvement of all stages of governmental authorities, from the local and the state up to the federal institutions, is one sign for the occurrence of a disaster. Each authority has its own area of responsibility according to its position in the disaster relief system" (Lichtenegger, 2009, p. 14). Emergence is then characterised by group, structural and task emergence, as well as the combination of organisation types. An easy image of the latter appears clearly in the emergency operation centres, where different groups and agencies work together, when they would traditionally not. Emergence is then only possible in a system when the blocks of that system have the ability to self-organise. The ability to self-reorganise to overcome an issue has been recognised as an essential element to resilience. In the contrary, countries and entities with little flexibility, which do not show the ability to self-organise when necessary, are, like in evolution theory, bound for extinction.

resilient community that transitions into a different system (e.g. a rural community seeing its workforce leave for the city), the in-between is the time when the community becomes vulnerable. In O'Keefe et al. (2015), the authors recollect personal experience that summarise the role of transforming environment and living on hazards and disaster triggering. When Phil O'Keefe was making erosion observations in Kenya, he discovered that this apparently natural hazard was actually the result of men leaving the village to work in the city and not putting the necessary work in maintaining the terrace walls to limit erosion and associated hazards phenomena. He concluded that the causes of sediment hazards and disasters were demographic. In Japan, a once important rural population has been drastically depleted due to population ageing and younger generations moving to the city. This results in terraced land once covered by clay for rice culture now replaced by tree plantations with collapsed terraces and tree roots destructuring the once semi-impermeable topsoil. In consequence, the rainfall water penetrates much faster in the soil, and the trees increase the mass on the slopes, all contributing to higher mass movements and debris flow disaster risk. Scientists and media alike are then quite prompt to blame climate change and soil conditions, but rural population depletion and ageing are arguably some of the triggers. This shows that the models of resilience that contemporary societies have built are anchored in the stability in the system and is not well fitted for rapidly changing communities, especially if those changes are not planned in advance. One can, therefore, wonder what are going to be the effects of climate change on communities that will be forced into unplanned transitions.

When ill-development occurs or is forced upon

Ill-development can be conceived as development that is not adapted or in the best interest of the local communities and the environment they live in. This situation can occur in the framework of colonialism, for instance, when the development models are imported and not centred nor best adapted. This issue often arises when power relationship within groups forces one community that was well adapted to its environment into modifications that increase vulnerability (Figure 2.6). A very visible example is the construction of structures and buildings. For instance, the French type of buildings in Haiti or the red-brick British type of buildings in Christchurch (New Zealand) were all not developed for countries prone to earthquakes, and those structures collapsed in catastrophic ways. During the 2011 Canterbury earthquake sequence, several studies have emphasised the importance to integrate development and sustainable development and disaster prevention (Collins, 2009) together to mitigate disaster.

It has been demonstrated that the process of development – not in the sense of inspiring to a Western model – needs to occur through community empowerment and community-based approaches for solving environmental, sustainability and disaster issues. Along those lines, Rajib Shaw (2009) provides different examples on the importance of those points and the necessity to integrate local governance in the process. In other words, disasters are more prone to occur when development and sustainable development does not integrate DRR, and when it happens from a strictly top-down management system. Efforts to mitigate disasters from external sources (such as NGOs) are also prone to failure when occurring as disembedded and sting operations. For instance, in the aftermath of the 2006 earthquake in the Bantul area,

Fig 2.6
Forced development from outside impetus having little knowledge of the reality of natural hazards. Source: cartoon courtesy of Christopher Gomez

South of Yogyakarta, in Indonesia, discussion with international NGO operators clearly showed that a number of them were not professionals, only came for a short period with an execution mandate (i.e. a task to accomplish stated by the NGO, which, in the present case, was to build water wells) and had little, if not no, language and cultural skills. It was very clear that the building of wells was not a priority for the local community because wells were not damaged and that the disconnect with local governance was more a burden on the local communities than a welcomed contribution.

When disaster occurs during armed conflicts

Armed conflicts vampirise both resources and humans, putting pressure on systems, breaking social networks and so forth, until one or the two antagonists collapse. Consequently, in armed conflict settings, natural hazards can have further effects on a community by increasing vulnerability or by creating new hazards (e.g. voluntary droughts to impact the opponent resources or engineered drowning of areas), eventually leading to disasters or worsening them (Box 2.6). Marktanner et al. (2015) found that disasters occurring in regions where armed conflicts were unfolding or occurred anytime from ten years before a disaster, the casualties were on average 40% superior than the death in any other location with no armed conflict. Furthermore, about 14% of the disaster-related casualties, for the second half of the 20th century, are linked to an armed conflict.

Armed conflicts and other political instabilities can also occur as the results of disasters and pressures on a community. To take a well-known historical example in France, the crops of 1788 and 1789 had a particularly low yield, and in a political system where the economy of the land was ready to move into a market economy, it has been argued that the drought disasters were most certainly one of the foundations for the French Revolution and the bloodbath that then ensued both in France and perpetuated through Europe by the armies of Napoleon. It is most likely that the political power was about to change hands, as the holders of the economic power usually harness the first for their benefits (like it is the case today), but it was the spark that started the armed conflict.

The opposite results can also be observed when armed conflicts can be stopped – even temporarily – in the case of disaster occurring. It has been the case of the armed conflict between the GAM (the Sumatra Liberation Army) and the Indonesian National Army. Both parties were fighting in the Sumatran mountain chain, the Barisan, and it was only with the 2004 Boxing Day earthquake and tsunami, which took the lives of more than 100,000 in the area of Banda Aceh alone, that the independentists offered a ceasefire.

When disasters occur in locations where international help is necessary, or where refugees from foreign places are located (i.e. UN camps following wars in the like of the ISIS exactions in the Middle East), or where an international armed conflict is unfolding, individuals call upon the International Law Commission to oversee international human rights. From a law perspective, the first difficulty to action is the lack of a stable and agreed upon definition of what a disaster is (Bartolini, 2018). The second challenge is the existence of alternative frameworks and law systems. Very often, the International Disaster Response Law and the International Human Rights Law and the International Humanitarian Law act as a disconnected alternative, although better efficiency would be achieved by combining them (Venturini, 2012).

Box 2.6: When armed conflicts and disasters collide

For the period of 2005–2009, an estimated 50% of the communities that experienced disaster were also riddled by the impacts of armed conflicts. According to the UNISDR, this proportion can even rise to 80% for selected single years. Although it would be tempting to conclude that armed conflicts can lead to disasters and that disasters can be the fire starter of armed conflicts, experience across different communities show that this is absolutely not the case. For instance, the 2004 Boxing Day tsunami drove the independence army of Sumatra (GAM) to temporarily stop the conflict with the National Army of Indonesia, where fights had been raging in the Barisan mountain (the mountain chain along the island). From another vantage point, however, the long-lasting conflict had driven numerous migrants out of the Barisan mountain into the city of Banda Aceh, where they had settled in informal dwellings. And during the tsunami, after the sea receded, a number of them went to gather seashells, not linking the sea retreat to the arrival of the tsunami. In the Mentawai Islands, in the island of Simeulue, local communities had a memory and a knowledge of tsunamis, and when the sea retreated, they ran away to high ground. In other words, the conflict between the GAM and the national army had driven individuals and communities out of the mountain to the seaside, and this forced transfer was arguably one of the reasons that led to a death toll of over 200,000 for the North Sumatra alone.

According to the Disaster-Conflict Interface report by the United Nations Development Programme, the concomitance of disasters and armed conflicts also impact the ability of communities to cope with forthcoming disasters, but that slow-onset and long-lasting disasters – such as droughts – create the worst results compared to the coupling of conflicts with rapid small-onset hazards, like debris flows or localised floods. In turn, conflicts are a problem for NGOs to positively interact with communities that are experiencing disasters, as it hinders recovery activities. Furthermore "the international community has a funding blind spot when it comes to DRR in fragile and conflict affected states, despite these countries having the greatest need" (www.preventionweb.net/publications/view/48864).

Consequently, Peters and Kelman formulated this complexity as follows: "Disaster research, conflict research, and peace research have rich and deep histories, yet they do not always fully intersect" (Peters & Kelman, 2020).

When there is a disconnect between soft and hard engineering
The incomprehension or the misconception of the public towards existing hard structures is one of the weaknesses that can lead to disasters. Indeed, DRM sees a combination of hard engineering, with structures like dikes, walls and culverts, and soft engineering, which is the education, the participatory mapping methods, the design charrettes and so forth; but the lack of the second can lead to poorer performances of the hard-engineered structures.

For instance, during the 2011 tsunami that hit Tōhoku's coast and flew up the rivers, videos have shown individuals staying on top of the shoreline protections or on top of the river levees, looking at the flow rising and eventually sweeping over the protection. In such cases, the hard-engineered structure performed to a given level as it was designed, but population going to see the tsunami and staying on the structures thinking that the structures would protect them was a misjudgement. Better information, preparation and education on what the structure could do, and the trap they could become if one stayed on one of them, had obviously not been sufficient. Helicopter footage of a truck racing away on one of the river's stopbank and being surrounded and then swallowed by the tsunami water was a good example of this issue.

In South Japan, informal interviews conducted in 2004–2005, in Shimabara City, where the Unzen Volcano is, did show that people living in the shadow of the huge check dam system that belts the volcano had a false sense of security thinking that the dam would even stop further eruption's pyroclastic flows (Figure 2.7).

Fig 2.7
Disconnect between the reality of the engineered structures and the perception of the local population. This drawing was made after a discussion with a local farmer in Shimabara City (Japan) just beneath Unzen Volcano.
Source: cartoon courtesy of Christopher Gomez

It is, therefore, very important to have a good connection between hard and soft engineering, and in the case of Mount Unzen, walls and check dams do not stop pyroclastic flows, rather the opposite. In 2006 and in 2010, at Mount Merapi in Indonesia, the village of Kali Adem was twice recovered by pyroclastic flow deposits, partly because the check dam built against lahar played the role of a ramp propelling the valley-confined pyroclastic flow sole layer into the interfluve where the village was located.

Disasters' systemic disruption timing

The occurrence of disaster disturbs the networked system, such as the banking system, the medical system and the distribution of various goods, such as gasoline. Disturbance in the banking system results in subsequent impacts on households buying basic items, including food and water, but it also disturbs businesses that can't provide payment solutions to customers and their suppliers. In turn, this situation disrupts the recovery process and increases the unfolding of the disaster. This issue is interestingly more acute in so-called developed countries than in their developing counterparts. For instance, South Korea relies mostly on virtual money payment systems, using one's smartphone, credit card or debit card. Japan, a country that relies heavily on cash payments, even in 2019, has vowed to expand the digital payment system to meet the needs of foreigners coming to the country for the Olympic Games of 2020. However, in a country prone to earthquake, tsunami and other large-scale hazards, the paralysis of the banking system could have disastrous effects without the use of physical currency.

In the case of the medical system, a service that is used more rarely, the occurrence of a disaster has effects that change through the unfolding of the disaster. During the emergency period, the medical system becomes clogged and triage is put in place, depending on the severity of a disaster and the connection or not to a network of hospitals and other medical providers. If a network is present and is not disrupted, patients will be dispatched through a group of medical service providers, but otherwise, the emergency period lasts longer, and the clogged hospitals then have difficulties to accept normal time patients who try to receive medical attention. Similar impacts concern all the emergency services (ambulances, firefighters and the police).

Disasters after the disaster

Having experienced various disasters, either as a professional or as a private individual, for instance, I was living in Christchurch, New Zealand, when the Canterbury earthquake sequence shook the city, destroyed my apartment building, led me to fear for my wife's life and destroyed most of my wife's and my memories packed in boxes that had just arrived from the previous country we were living in, I understood that a disaster is not an event on TV – it is everything during and, more especially, afterwards. In the six years my wife and I lived in the city, we both felt that the disaster was more than the event on the day; it was the "trajectory through abnormal life" (terminology after Wisner et al., 2004). Furthermore, as my wife (from Japan) and I (from France) were uprooted newcomers, lacking a local network of support and an understanding of the community systems, we both lived the disaster as a six-year unfolding event rather than a one-shot event. Cloke and Conradson (2018) have captured the transition that occurs during a long-lasting disaster with the necessity to create transition spaces and affective atmospheres. The authors have labelled this city as a "post-disaster city", as it is going through the recovery process, but if the community as a whole is going through recovery, at the individual level, experiences are still one of a disaster. Indeed, the efforts towards resilience can only last for a given time, until the individual course to resilience can be derailed towards psychological difficulties.

Conclusive comments of the second section

The question of when disasters occur is multifaceted and was certainly not addressed exhaustively, as it would deserve a volume of its own, but some of the transdisciplinary concepts at the core of the present handbook were presented, from the questions of stress on a community, would it be war or other forms of external pressures (e.g. ill-development), to the questions of timing and duration. Finally, I briefly introduce the question of transition and how the road to recovery of a community can still be the road to a non-recovery and ongoing disaster for some individuals within this community.

2.6 Concluding remarks

Disasters' spatio-temporal distribution is, of course, conditional to hazards occurring, but the major control is us. The spatial variation of disasters depends a lot on the level of concentration of the population (is an earthquake occurring underneath a ten-people group or under a several-million-inhabited urban areas). It also depends on the community's trajectory: is the community living in a natural habitat with almost no constructed structures (in which case, an earthquake might be just a rumble of the ground without any affect), or is the community living in lavish brick homes (which may resemble a lavish graveyard after an earthquake), or is the community living in structures like the traditional Indonesian *Jogglo* (timber and bamboo structure very resistant to earthquakes) or in a Japanese urban centre with numerous anti-seismic buildings? Then there are the effects to a community, such as war or previous disasters, and their timing. For instance, numerous Sumatrans who fled the civil war in the Barisan Mountains went to live in the coastal city of Banda Aceh, where, without any knowledge of the sea, they were the first preyed upon by the tsunami waves of 2004. Modifications, transitions and forced transitions in a community group are also reasons for disasters to occur, like the people from the

Mentawai Islands forced to the shore by Christian missionaries and, later on, by the governmental bodies to control the population. This forced movement put people at risk of tsunamis, when community experience and knowledge had driven them to the mountain, notably, to escape tsunamis.

Finally, this chapter separated *when* from *where* for presentation reasons and tried to explain the different aspects of the spatio-temporal realms of disasters, but the two are, of course, interconnected, and in many cases, the question of when and where could have almost been presented interchangeably.

Take-away messages

1 Disasters are not natural, and they occur as the result of inadequacy amongst individuals, communities, policies and potential environmental phenomena that can be defined as hazards;
2 Rural disasters, in the shadow of major cities, are often longer-lasting and receive less interest than the urban counterparts;
3 City disasters are to be related to its own fabric, and they have far-reaching impacts due to their networked realms;
4 Stresses that impact a community, such as war or disempowered groups (women in some countries), can be seen as the direct cause of disasters;
5 The over-reliance on hard engineering can create worse disasters if the goal of the structures is not well-understood by the local communities; and
6 Foreign influence and power exercised on a local community can lead to profound changes, which are often not in line with avoiding disasters (e.g. Mentawai Islands in Indonesia).

To learn more about the topic discussed in this chapter, listen to the *Disasters: Deconstructed* interview with Dr Lee Bosher (Figure 2.8).

Fig 2.8
Chapter 2 QR code.

Further suggested reading

Bullard, R. D., & Wright, B. (2009). *Race, place, and environmental justice after Hurricane Katrina: Struggles to reclaim, rebuild and revitalize New Orleans and the Gulf Coast.* Westview Press.

Etkin, D. (2016). *Disaster theory—An Interdisciplinary approach to concepts and causes*. Elsevier 359 p.

Etkin, D., & Stafnovic, I. L. (2005). Mitigating natural disasters: The role of eco-ethics. In *Mitigation of Natural Hazards and Disasters. International Perspectives*, 135–138

Freudenburg, W. R., Gramling, R. B., & Laska, S. B. (2009). *Catastrophe in the making: The engineering of Katrina and the disaster of tomorrow*. Island Press.

Gaillard, J. C., & Gomez, C. (2015). Post-disaster research: Is there gold worth the rush? *Jàmbá: Journal of Disaster Risk Studies*, 7(1). https://doi.org/10.4102/jamba.v7i1.120

Gomez, C., & Hart, D. E. (2013). Disaster gold rushes, sophisms and academic neocolonialism: Comments on 'Earthquake disasters and resilience in the global North'. *Geographical Journal*, *179*(3), 272–277. https://doi.org/10.1111/geoj.12028

Lichtenegger, G. (2009). *The role of Self-Organization in Disaster Relief Operations, Theory and Simulation* [Doctoral Thesis] *at the Department of Engineering and Business Informatics of Graz University of Technology*.

Peters, L. E. R., & Kelman, I. (2020). Critiquing and joining intersections of disaster, conflict, and peace research. *International Journal of Disaster Risk Science*, *11*(5), 555–567. https://doi.org/10.1007/s13753-020-00289-4

References

Balgos, B., Gaillard, J. C., & Sanz, K. (2012). The warias of Indonesia in disaster risk reduction: The case of the 2010 Mt Merapi eruption in Indonesia. *Gender and Development*, *20*(2), 337–348. https://doi.org/10.1080/13552074.2012.687218

Barlow et al. (2006). Drought disaster in Asia. In Arnold et al. (Eds.), *Natural disaster hotspots—Case studies*. Disaster Risk Management Series of The International Bank for reconstruction and Development and The World Bank 1:21.

Bartolini, G. (2018). A taxonomy of disasters in international law. In F. Z. Giustiniani, E. Sommario, F. Casolari, & Batrolini (Eds.), *Routledge handbook of human rights in disasters* (pp. 10–26). Routledge.

Cloke, P., & Conradson, D. (2018). Transitional organisations, affective atmospheres and new forms of being-in-common: Post-disaster recovery in Christchurch, New Zealand. Transactions of the Institute of British Geographers. *Transaction Publishing of the British Geographical Society*, *43*(3), 360–376. https://doi.org/10.1111/tran.12240

Collins, A. (2009). *Disaster and development*. Routledge Publishing.

Marktanner, M., Mienie, E., & Noiset, L. (2015). From armed conflict to disaster vulnerability. *Disaster Prevention and Management*, *24*(1), 53–69. https://doi.org/10.1108/DPM-04-2013-0077

O'Keefe, P., O'Brien, G., & Jayawickrama, J. (2015). Disastrous disasters: A polemic on capitalism, climate change, and humanitarianism. In *Hazards, risks and disasters in society* (pp. 33–44). Academic Press.

Reinsfield, I. (1995). Evidence for high magnitude floods along the Waimakariri River, South Island, New Zealand. *Journal of Hydrology (New Zealand), 34*, 95–110.

Shaw, R. (2009). Community based approaches of sustainable development and disaster risk reduction. In Uitto and Shaw (Eds.), *Disaster risk reduction* (pp. 207–214). https://doi.org/10.1007/978-4-431-55078-5_13

Shaw, R., Srinivas, H., & Sharma, A. (2009). *Urban risk reduction: An Asian perspective—Community, environment and disaster risk management volume 1*. Emerald Books.

Venturini, G. (2012). Disasters and armed conflict. In A. Guttry, M. Gestri, & G. Venturinin (Eds.), *International disaster response law* (pp. 251–266). TMC Asser Press.

Wisner, B., Blaikie, P., Cannon, T., & Davis, I. (2004). *At risk—natural hazards, people's vulnerability and disasters*. Routledge.

Yamashita, A., Gomez, C., & Dombroski, K. (2017). Segregation, exclusion and LGBT people in disaster impacted areas: Experiences from the Higashinihon Dai-Shinsai (Great East-Japan Disaster). *Gender, Place and Culture, 24*(1), 64–71. https://doi.org/10.1080/0966369X.2016.1276887

CHAPTER 3

The impact of disasters

Fig 3.1
Church still damaged three years after the 2015 Fundão dam collapse in Bento Rodrigues district, city of Mariana, Brazil. Source: photo courtesy of Victor Marchezini, 1 October 2018

3.1 Introduction

There are different ways to define what a disaster is and what its impacts are (Figure 3.1). Definitions are always context- and history-specific, and are inseparable from power relations, changing over time. Societies have had a long history in coping with calamities, and they have created their own interpretations and forms of framing the impact of disasters. For instance, due to the cascading impacts in the aftermath of the 1755 earthquake in Lisbon, Portugal, the philosophers Voltaire and Jean Jacques Rousseau discussed their views about the disaster and the human nature. This disaster also generated other forms of expression, such as the engravings of Jacques-Philippe Le Bas. These different expressions allow societies to understand some aspects of their history in relation to disasters' impacts. In Latin America, for instance, this history of disasters was analysed by the three volumes of *Historia y desastres en America Latina* edited by LA RED (Acosta, 1996, 1997, 2008).

DOI: 10.4324/9781315469614-5

Disasters can impact societies in different ways. One of the first scientific studies to analyse and detail these impacts was the dissertation of Samuel Henry Prince, defended in 1920. This sociologist studied the slow process of reorganisation in the aftermath of the Halifax City explosion in Canada (Prince, 1920). Another important work was carried out by the Russian sociologist Pitirim Sorokin in the 1940s. Sorokin studied the ways that famine, pestilence, war and revolution modify social organisation and cultural life. This author questioned, "What, if any, typical changes do calamities introduce into the normal processes of migration and mobility and how do they transform the social, political, and economic structure of a society?" (Sorokin, 1942, p. 106). For this sociologist, calamities are one of the potential agents of sociocultural change, and also reveal that social activities are tougher and more indestructible than we usually think: "If man cannot live without bread, it is equally true that he lives not by bread alone" (Sorokin, 1942, p. 82).

The field of anthropology is also involved in the study of the impacts of disasters. The 1951 eruption of Mount Lamington in Papua New Guinea motivated Cyril Belshaw, in 1951, and Felix Keesing, in 1952 to study the social organisation of the Orokaiva people and their adaptations to disaster impacts. In 1956, Anthony Wallace, rooted in the particular context of North America, proposed a "disaster syndrome" psycho-cultural model in which disaster victims proceed through states of isolation, euphoria, altruism, criticism and normality (Faas & Barrios, 2015). Wallace (1956, p. 1) characterised disasters as "extreme situations" that involve not just impact but also the threat of "an interruption of normally effective procedures for reducing certain tensions, together with a dramatic increase in tensions". In this interpretation, the disaster is the disruption of stability that had an external agent as cause, which later required social readjustments (Perry, 2018).

As discussed in Chapter 1, the question about what a disaster is has motivated multiple debates within the same discipline, between different disciplines as well as in various regions and sociocultural contexts (Maskrey, 1993; Lavell, 1994; Quarantelli, 1998; Perry, 2018).

Major earthquakes happened across the globe in the 1970s, revealing the vulnerability of some groups of people more than others in the initial emergency and the later long-term recovery. In May 1970, a 7.9 magnitude earthquake shook the Pacific Coast of Peru, destroying and damaging infrastructure, including towns and villages, and causing nearly 70,000 deaths. Near the Andean town of Yungay, the shaking earth loosened a glacial formation, creating a massive 50 million cubic metre avalanche that killed approximately 6,000 people, leaving only 300 survivors of a previous population of roughly 4,500. These impacts were also explained by the 500-year history of human-environment relations (Oliver-Smith, 2012). "The earthquake, which devastated the north central coastal and Andean regions of Peru, can be seen as an event which in certain respects began almost five hundred years ago with the conquest and colonization of Peru and its consequent insertion as a colony into the developing world economic system, which has resulted in the severe underdevelopment of the entire region" (Oliver-Smith, 2012, p. 103). The Spanish conquerors' approach to settlement location was to place towns at the confluence of rivers where they were prone to floods and landslides, increasing the likelihood of the people being affected by unsafe buildings and infrastructure during earthquakes.

Another example of a natural event revealing an inequitable social impact was the major earthquake of 1976 in Guatemala, known as the class-quake. The physical event – the earthquake – didn't destroy the well-built houses of the middle and upper classes in Guatemala City, but it devastated the surrounding slums and nearby impoverished towns, which resulted in high mortality rates of the Mayan Indians (Wisner et al., 2004). The impacts of this catastrophe contributed to reflection about the way they were explained, their causes and their spatial and temporal scales. In the same year as the earthquake, the classic article "Taking the 'Naturalness' Out of 'Natural' Disasters" (O'Keefe et al., 1976, p. 566) was published, stating that "the increased vulnerability of people to extreme physical events can be seen as intimately connected with the continuing process of underdevelopment recorded throughout the world". In this approach, disasters are seen not as a natural event, nor as an agent with rationality, nor as an extraordinary or exceptional event (Hewitt, 1983), which causes the impacts to the development. Disasters are caused by maldevelopment projects that explain why the impacts happen and are unequally distributed. This processual approach to social production of disaster risk is important in order to shed light on the ways people understand the types of impacts verified in disasters and their expression in the short and long term.

The conditions that explained the disaster risk are important in determining the human, material, economic and environmental losses and impacts, which can be immediate and localised, or can last for an extended period and across spatial scales. The SFDRR 2015–2030 (UNISDR, 2015) used some types of terms to classify disasters. The next section will explain them.

3.2 Thinking about the types of disasters and their impacts: everyday, cascading and protracted typologies

Usually, the classification of the types of disasters is made referring to their scale and the frequency of hazards to which they are related. The 2017 UNISDR terminology considers the following types (Table 3.1).

The 2017 UNISDR terminology offers definitions for disaster damage and impacts. Disaster damage "occurs during and immediately after the disaster. This is usually measured in physical units (e.g., square meters of housing, kilometres of roads, etc.), and describes the total or partial destruction of physical assets, the disruption of basic services and damages to sources of livelihood in the affected area". Disaster impact (Figure 3.2 and Figure 3.3) is the "total effect, including negative effects

Table 3.1
Types of disasters recognised by UNISDR
Source: **UNISDR (2017)**

Type	*Definition*
Small-scale disaster	This is a type of disaster only affecting local communities, which require assistance beyond the affected community.
Large-scale disaster	This is a type of disaster affecting a society, which requires national or international assistance.
Frequent and infrequent disasters	This depends on the probability of occurrence and the return period of a given hazard and its impacts. The impact of frequent disasters could be cumulative or become chronic for a community or a society.
Slow-onset disaster	This is defined as one that emerges gradually over time. Slow-onset disasters could be associated with, for example, drought, desertification, sea level rise or epidemic disease.
Sudden-onset disaster	This is one triggered by a hazardous event that emerges quickly or unexpectedly. Sudden-onset disasters could be associated with, for example, earthquake, volcanic eruption, flash flood, chemical explosion, critical infrastructure failure or transport accident.

(e.g., economic losses) and positive effects (e.g., economic gains), of a hazardous event or a disaster. The term includes economic, human and environmental impacts, and may include death, injuries, disease and other negative effects on human physical, mental and social well-being" (UNISDR, 2017).

These damages and impacts vary across geographical spaces. The UNISDR report "Economic Losses, Poverty and Disasters", based on the data from the EM-DAT, showed unequal impacts of climate-related and geophysical hazards across geographical spaces between 1998 and 2017. In this period, more than 1 million people died and almost 4.5 billion people were injured, displaced or in need of emergency assistance: on average, 130 people per million died in disaster-affected areas in the poorest nations, compared to just 18 people per million in high-income countries. According to the report, whilst economic losses might be concentrated in high-income countries, the human cost

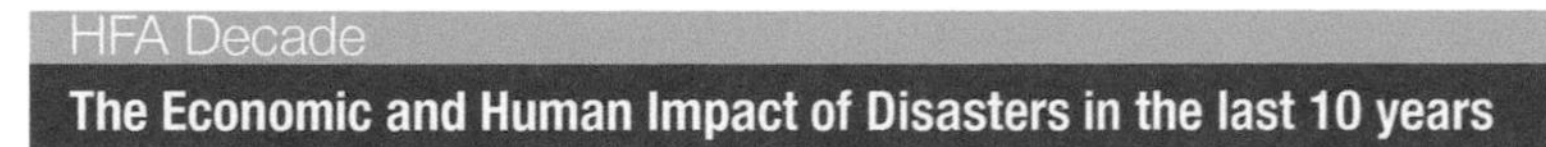

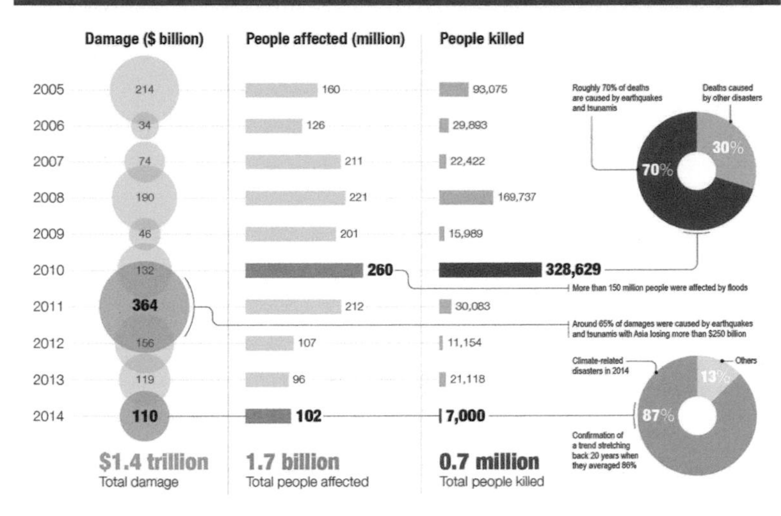

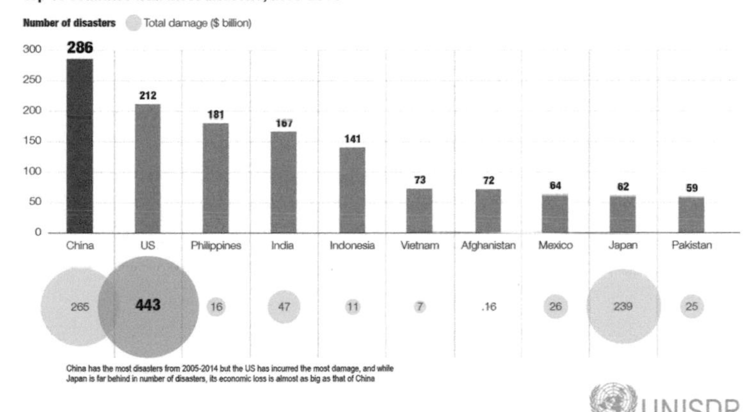

Fig 3.2
The economic and human impact of disasters in the last ten years.
Source: UNISDR, https://www.unisdr.org/files/42862_economichuman impact20052014unisdr.pdf

of disasters falls overwhelmingly on low- and lower-middle-income countries (Wallemacq & House, 2018). Moreover, inequality is even greater than available data suggest because under-reporting is one of the problems of data governance in low-income countries: whilst high-income countries reported losses from 53% of disasters between 1998 and 2017, low-income countries only reported losses in 13% of disasters (Wallemacq & House, 2018). The reported disasters and their consequent impacts can be classified using different classifications. UNISDR (2015) suggests two ways to classify these disasters: intensive and extensive disasters (Box 3.1).

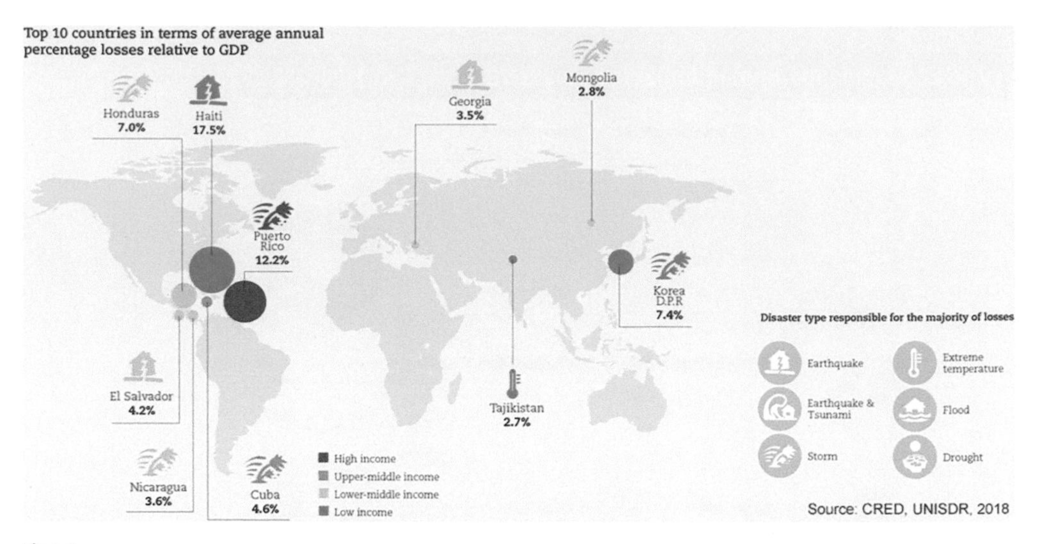

Fig 3.3
Top ten countries in terms of average annual percentage losses relative to GDP.
Source: CRED, UNISDR (2018), https://www.flickr.com/photos/isdr/16111599814

Box 3.1: Intensive and extensive disasters: how do we differentiate them?

Two concepts are important to understand these numbers about disasters: their intensive and extensive patterns. Intensive is used to describe the high-severity, mid- to low-frequency disasters, mainly associated with major hazards. Since 1990, more than 45% of disaster mortality has been concentrated in intensive disasters, such as the 1991 Cyclone Gorky in Bangladesh, the 2004 Indian Ocean tsunamis, the 2008 Cyclone Nargis in Myanmar and the 2010 earthquake in Haiti (UNISDR, 2015b). The economic losses from intensive disasters are usually evaluated by international organisations, governments, insurance industry and NGOs. However, the economic losses and damages of extensive disasters are largely uncounted (UNISDR, 2015b).

Extensive disasters are used to characterise the high-frequency but low-severity losses, mainly recurrent, but not exclusively associated with highly localised hazards. In general, extensive disasters are not captured by global risk modelling, nor are the losses reported internationally. Extensive disasters account for up to 40% of economic losses, mainly in low- and middle-income countries (UNISDR, 2015b). The analysis of 85 databases accounted for a total of US$94 billion. According to the UNISDR (2015b), the extensive disasters are responsible for the majority of disaster morbidity and displacement, and represent an ongoing erosion of development assets, such as houses, schools, health facilities, roads and local infrastructure. The costs of these extensive disasters are not visible and tend to be underestimated, as they are usually absorbed by low-income households and communities and small businesses (UNISDR, 2015b).

Mortality and housing destruction are the two variables used to define the threshold between intensive and extensive disaster losses. Statistically, the threshold is fixed as follows:

- Mortality: less than 30 people killed (extensive); 30 or more killed (intensive); and
- Housing destruction: less than 600 houses destroyed (extensive); 600 or more houses destroyed (intensive).

The EM-DAT database, which has been in use for years, covers the intensive disasters around the world. More recently, the UNISDR has recognised another methodology that also captures the losses and damages in extensive disasters and small-scale disasters. The DesInventar methodology was modified to a new version called DesInventar-Sendai, which is a tool that helps analyse disaster trends and impacts in a systematic manner (Box 3.2).

Box 3.2: What is DesInventar? How does it work?

Until the mid-1990s, systematic information about the occurrence of small- and medium-impact disasters, and their disaggregated data, was not available in most countries in the world.

From 1994 on, the creation of a common conceptual and methodological framework was begun in Latin America and the Caribbean (LAC) by a group of researchers, academics and institutional actors linked to the Network of Social Studies in the Prevention of Disasters in Latin America (*Red de Estudios Sociales en Prevención de Desastres en América Latina* – LA RED).

LA RED conceptualised a system of acquisition, collection, retrieval, query and analysis of information about disasters of small, medium and greater impact, based on pre-existing official data, academic records, newspaper sources and institutional reports in nine countries in LAC. The developed conceptualisation, methodology and software tool is called Disaster Inventory System – DesInventar (*Sistema de Inventario de Desastres*).

DesInventar is a conceptual and methodological tool for the generation of National Disaster Inventories and the construction of databases of damage, losses and, in general, effects of disasters. It includes a software product with two main components:

i the administration and data entry module, which is a relational and structural database through which the database is fed by filling in predefined fields (space and temporal data, types of events and causes, sources) and by both direct and indirect effects (deaths, houses, infrastructure, economic sectors); and,

ii the analysis module, which allows access to the database by queries that may include relations amongst the diverse variables of effects, types of events, causes, sites, dates and so forth. This module allows representation of those queries with tables, graphics and thematic maps.

Intensive and extensive disasters are two important classifications, and the DesInventar methodology is also useful for data collection, analysis and visualisation. However, scientific literature also provides recent and insightful ideas to classify disasters that might be useful in understanding the types of impacts and their magnitude, frequency and particularities. The reflection about the everyday disasters, cascading disasters, neglected disasters and protracted crises are important for the discussion about the impacts and the ways to formulate policies to reduce them.

Everyday disasters

Everyday disasters, such as violent conflict, illness, hunger, displacement, epidemics and water and air pollution, are ordinary occurrences; and often, they do not receive the same visibility as intensive disasters. Permanent human mobility is one example of everyday disasters. It refers to the population movements – voluntary or forced, assisted or spontaneous, long or short distance, long or short term (Guadagno, 2016). Migrants represent 14% of the global population, and almost 200 million people were forced to move as a consequence of disaster from 2009 to 2015 (IDMC, 2015). The situation in 2019 was still worrying. Between January and June 2019, there were about 10.8 million new displacements – 7 million triggered during disasters and 3.8 million by conflict and violence (IDMC, 2019). Even more worrying is the fact that "many of the countries with the highest number of displacements associated with conflict and violence were the same as in the first half of 2018" (IDMC, 2019, p. 3).

Other types of everyday disasters are not being discussed by the SFDRR. The capitalist economic system and its processes of extraction, production, exchange and consumption are creating unprecedented levels of pollution (Matthewman, 2015). Outdoor air pollution, for instance, is responsible for more than 4 million deaths each year. According to Li et al. (2019), the main culprits are fine particles with diameters of 2.5 micrometres or less (PM2.5) that can penetrate deep into the lungs, heart and bloodstream, where they can cause disease and cancer. Besides these invisible disasters that sometimes are not recognised by people who face them, there are other everyday disasters that make sense for people but are not at the centre of international and national recommendations for DRR. The 2019 edition of the *Views from the Frontline* shared important preliminary findings collected from 28,864 community responders around the world (up until 23 October 2019) (Box 3.3). The five main consequences/impacts were as follows: disease/health effects (10.63%), crop damage (9.9%), economic and livelihood losses (9.47%), building destruction (8.88%) and loss of life (8.58%) (Global Network of Civil Society Organisations for Disaster Reduction, 2019). These consequences/ impacts can be combined, thereby producing cascading disasters.

Cascading disasters

Cascading disasters have emerged as a new paradigm in disaster studies (Alexander & Pescaroli, 2019), and they represent an approach that also contradicts a simple linear model to comprehending disaster impacts.

Box 3.3: Views from the Frontline

The Views from the Frontline (VFL) programme is the largest independent global review of DRR at the local level and, to date, has gathered the views of more than 85,000 people across 129 countries. The VFL programme was launched in 2009 by the Global Network of Civil Society Organisations for Disaster Reduction (GNDR) and involved over 500 organisations across 69 countries to carry out a local and participatory monitoring process to measure the progress of UN and government achievements towards DRR. The VFL programme had also editions in 2011 and 2013, and in 2019, a new survey was carried out.

Between 2014 and 2018, GNDR implemented the Frontline programme, which shifted from asking local actors about progress of the UN framework targets (VFL's aims), to more open-ended questions about their priority threats, the consequences of those threats, the actions they thought were needed and the barriers they were facing in reducing risk. This allowed communities to raise issues not necessarily falling within the definition of 'natural hazards' but take a more holistic perspective of risk to include threats related to climate change, poverty and instability (GNDR, 2019). In particular, respondents from the most marginalised groups discussed the need to prioritise small-scale recurrent threats that are unreported and unsupported. Following awareness raising, these everyday disasters became a focus for intergovernmental bodies, such as UNISDR, featuring in the Global Assessment Reports for DRR and the new SFDRR (GNDR, 2019).

Where vulnerabilities interact and overlap, and/or one hazard can act upon another, escalation points are produced and can create secondary effects with greater impacts than the original trigger (Alexander & Pescaroli, 2019). Three examples can help to understand the escalation of impacts developed during complex forms of crisis: the eruption of the Icelandic volcano Eyjafjallajökull in Europe, the Tōhoku triple disaster in Japan and the tailings dam collapse in Mariana, Brazil.

The first example is in April 2010, when the ash from the Icelandic volcano Eyjafjallajökull led to the grounding of civil aviation at 70% of Europe's airports for almost a week. However, many other impacts of this complex crisis were unnoticed (Alexander, 2013). Eight and a half million passengers were stranded, and the demand for ground and sea transportation and hotel accommodation increased. "Critical supplies (such as bone marrow for transplants) could not be air freighted. Industries that are dependent on air transportation suffered major losses, and these included horticultural and agricultural enterprises, not merely airlines and airport service companies" (Alexander & Pescaroli, 2019, p. 4).

The second example is the Tōhoku triple disaster. On 11 March 2011, the M9 earthquake struck the northeast of Japan, killing about 100 people. Additionally, the earthquake triggered a tsunami resulted in the deaths of 19,000 people and left almost 2,500 people missing. Moreover, a critical infrastructure was in the path of the tsunami. The Fukushima Daichi nuclear power plant was damaged and released radiation that provoked radioactive contamination and serious impacts not only in the whole country but also in the automotive production in European plants (Alexander & Pescaroli, 2019), to give some examples of cascading impacts.

The third cascading disaster example refers to the breaking of the Fundão dam in Brazil (see the opening figure, Figure 3.1, in this chapter). On 5 November 2015, the Fundão tailings dam collapsed in Mariana City, Brazil, unleashing 43 million m^3 of tailings (80% of the total contained volume), generating 10 m high waves of mud, killing 19 people and causing irreversible environmental damage to hundreds of watercourses in the basin of the Doce River and associated ecosystems (Carmo et al., 2017). Forty downstream municipalities were affected, and hundreds of thousands of people (including indigenous tribes) were left without access to clean water.

Neglected disasters and protracted crisis

Some of these cascading disasters and their impacts can be visible or unnoticed, be neglected or not, be finished or protracted. There are at least two important concepts to capture the extent and permanence of disaster impacts: neglected disasters and protracted crisis. Neglected disasters are not limited to the perception of the salience of specific events but "extends to the understanding of and attention to the issue of why people suffer as well as who suffers from disasters and who does not" (Wisner & Gaillard, 2009, p. 155). Neglect can be manifested across spatial and temporal scales, by different actors, for many reasons. An example of a neglected disaster occurred in 1973–1974 in Ethiopia where 200,000 people died during droughts and famines (Kelman & Glantz, 2014). Due to political pressure from this disaster, the government of Ethiopia created a commission that gathered information to try to establish a warning system. Ten years later, a second disaster occurred during droughts and famines in the same country. In response, in 1985, the US government created the Famine Early Warning System (FEWS), which is responsible for monitoring emerging food security issues in countries of Africa, Central America and the Caribbean, the Middle East, and Central Asia (Grasso, 2014).

Neglected disasters can be accentuated in the long term due to root causes and dynamic pressures, leading to protracted crises. According to the Food and Agriculture Organization (FAO, 2010), protracted crisis situations are characterised by recurrent disasters and/or conflict, longevity of food crises, breakdown of livelihoods, and insufficient institutional capacity to respond to, and mitigate, the threats to the population or provide adequate levels of protection. In protracted crisis, a significant proportion of the population is acutely vulnerable to death, disease and disruption of livelihoods over a prolonged period. In 2010, three criteria were used to classify a country or an area in protracted crisis: (i) duration or longevity of crisis, the threshold being eight years or more; (ii) aid flows: the proportion of humanitarian assistance received by the country as a share of total assistance – countries are defined as being in protracted crisis if they have received 10% or more of their official development assistance as humanitarian aid since 2000 – and (iii) economic and food security status, if the country appears on the list of Low-Income Food-Deficit Countries (LIFDCs). In 2010, 22 countries were identified as being in protracted crisis (or containing areas in protracted crisis), which represented more than 166 million undernourished people, nearly 40% of the population of these countries and nearly 20% of all undernourished people in the world (FAO, 2010).

Disasters can attract or be neglected by media coverage during the emergency phase. Disaster narratives tend to reinforce hegemonic forces of society, so the construction of disaster through images and numbers is highly political (Kondo et al., 2012; Button, 2012). The way public and private institutions, media, and other social actors define and frame the disaster is important because definitions orient practices and policies to deal with these impacts. Those actors define what are classified as 'social problems' and shape the solutions fabricated by institutions to solve them (Dombrowsky, 1998). Numbers, words and images matter in the field of cognitive battles, and the social agents don't have the same quantity and forms of capital for the creation of social reality (Bourdieu, 1991) to represent publicly what a disaster is. The next section will focus on the qualitative analysis of types of damage and impacts at the local scale.

3.3 Looking at types of damage and impacts in the short and long term

The SFDRR aims for the substantial reduction of disaster risk, loss of life and impacts on livelihoods and health, as well as on the economic, physical, social, cultural and environmental assets of persons, businesses, communities and countries (UNISDR, 2015a). This section will point out some impacts in intensive and extensive disasters. These include environmental, physical, cultural, political, ecological, social and economic impacts.

Disaster impacts can be analysed, and sometimes measured and estimated, according to different types of methods, tools and databases. The availability of these databases, methods and tools varies according to the type of hazards and countries. Floods, for instance, have many models and tools to estimate impacts, especially in developed countries (Dolman et al., 2018). However, loss assessments conducted on the ground are still necessary, and the time frame to measure the losses and damages is an important question to be addressed. For instance, in the aftermath of the September 2017 Hurricane Maria in Puerto Rico, USA, Kishore et al. (2018) pointed out the indirect causes of death resulting from delayed healthcare or from the worsening of chronic conditions. These authors surveyed 3,299 randomly chosen households across Puerto Rico to produce an independent estimate of causes of mortality after the hurricane for which the official death toll was 64 people. Survivors were asked about displacement, infrastructure loss and causes of death. The authors calculated excess deaths by comparing the estimated post-hurricane mortality rate with official rates for the same period in 2016 (from 20 September through 31 December). The authors' household-based survey suggests that the number of excess deaths related to Hurricane Maria in Puerto Rico was more than 70 times the official estimate.

Oftentimes, the number of human deaths is not sufficient to capture the magnitude, extension and diversity of environmental, physical, cultural, political, ecological, social and economic impacts. The Fundão dam collapse in Mariana City, Brazil, released 43 million m^3 of tailings, generating mud waves that instantaneously killed 19 people. The tailings caused irreversible environmental damage to hundreds of watercourses in the basin of the Doce River and associated ecosystems. The tailings directly impacted 135 identified semi deciduous seasonal forest fragments, resulting in 298 ha of vegetation suppression (Carmo et al., 2017), and damaged 863.7 ha of Permanent Preservation Areas associated with watercourses in protected areas. In addition, 294 creeks were affected by the tailings. Out of the 806 buildings directly affected by the tailings, at least 218 were completely destroyed. These were residences, public buildings, commercial real estate, historic churches and ancient farms distributed amongst ten districts of five different municipalities. Bento Rodrigues, a district of Mariana City, just 6 km from the Fundão dam, was the most damaged with 84% of the affected buildings totally destroyed. Areas of cultural heritage also suffered greatly (see the opening figure, Figure 3.1, in this chapter). The damage included at least 2 archaeological sites, 6 places of historical and cultural interest and more than 2,000 pieces of sacred material heritage. One of the main cultural heritage assets irreversibly affected was the São Bento chapel, an 18th-century building surrounded by stone walls. Considering that the disaster occurred in one of the most important regions for biodiversity conservation, it is estimated that the loss was significant (Fernandes et al., 2016). Tons of fish from 21 different species died. Isolated reports identified the deaths of large mammals, such as the South American tapir (*Tapirus terrestris L.*), as well as turtles, birds, amphibians and invertebrates (Carmo et al., 2017). Oftentimes, these long-term impacts are intensified by political decisions and continue several years after the hazard manifestation, as we can see in the case of the L'Aquila earthquake (Box 3.4 and Figure 3.4).

Box 3.4: The tenth anniversary of the L'Aquila earthquake

This episode of the *Disasters: Deconstructed* podcast is a conversation between five Italian disaster researchers that brought a critical social science perspective to discuss the political aspects of disaster recovery policies and interests in the long term of the L'Aquila catastrophe.

Fig 3.4
QR code for Box 3.4. Available at https://disastersdecon.podbean.com/e/friday-special-laquila/

Box 3.5: Voices of Youth on COVID-19 – Part 1 – Stories of Impact

This episode of *Disasters: Deconstructed* podcast is a conversation with Asian-Pacific youth living through COVID-19. Young people shared their local stories of impacts of the long-term catastrophe.

Fig 3.5
QR code for Box 3.5. Available at https://disastersdecon.podbean.com/e/voices-of-asia-pacific-youth-on-covid-19-part-1-stories-of-impact/

Disaster impacts can affect different audiences and sectors. Young people, for instance, represent 50–60% of those affected by disasters (UNICEF, 2012). Many were affected during earthquakes, including when they were at school. In the 1988 Armenian Spitak earthquake, more than 17,000 students died in schools after the school buildings collapsed. In the 2001 Bhuj earthquake in India, 971 students and 31 teachers perished. During the 2005 earthquake in Pakistan, about 19,000 children died, most part in collapses of school buildings. In 2008, more than 10,000 children died during the earthquake in Sichuan, China (Earthquake Engineering Research Institute, 2006; Tuladhar et al., 2014). Not only earthquakes trigger impacts on the educational sector. Floods and landslides affected more than 1,800 schools in 311 Brazilian municipalities from January to March 2004 (Marchezini et al., 2018). In 2006, a landslide reached an elementary school in the Philippines, and 245 children and teachers died. In the same year, floods damaged 30 schools in Nyando district, Kenya (Ochola et al., 2010). The impacts of disasters are not restricted to the hazard manifestation, but they can also be accumulated in the long term, as reported in the long-term catastrophe of the COVID-19 (Box 3.5 and Figure 3.5).

Rural people are also exposed to disaster impacts. Drought is one example of hazards that can trigger losses. For instance, in the Brazilian semi-arid region (SAB), about 28 million people are exposed to droughts. The SAB contains more than half of the country's family-based rural farms (Santos et al., 2015), and 80% of these small farmers practiced subsistence agriculture in a rainfed system (Brazilian Institute of

Geography and Statistics, 2009). Between 2012 and 2017, the Brazilian semi-arid region was strongly affected by a prolonged drought, classified as the most intense event of the last 30 years (Brito et al., 2017; Cunha et al., 2018). Considering the accumulated impacts between 2012 and 2016, about 1,100 municipalities were affected (about 20 million people affected per year), mainly in relation to the water supply and losses of the agro-productive systems. The payment of crop guarantee insurance reached almost US$1.5 billion (Cunha et al., 2019).

People living in tropical grasslands, savannahs and forests are also prone to suffer losses from wildfires. Most wildland fires occur in tropical grasslands and savannahs (86%) and a smaller amount in forests (11%) (Mouillot & Field, 2005). Every year, wildfires burn 330–431 M ha of global vegetation (Giglio et al., 2010), causing great loss of fauna and flora. Human lives are also lost during wildfires. For instance, 173 people perished during the 2011 wildfires in Victoria, Australia. Frequently, people also lose shelters, crops, farm animals, grazing area and others. (De Groot & Flannigan, 2014).

Critical infrastructure in several sectors are also susceptible to different types of hazards (Koks et al., 2019), and/or they can become a hazard, a source of potential losses and damages, as previously discussed in the examples of the nuclear plant during the Tōhoku disaster. The Tōhoku disaster revealed several impacts on lifeline facilities, such as water supply, gas, electricity, communications, broadcasting facilities, for which the losses were estimated at ¥0.6 trillion. Social infrastructure facilities – rivers, roads, ports, sewage works and airports – accounted for ¥2.2 trillion in losses (Kazama & Noda, 2012). Some tools, such as the Post-Disaster Needs Assessment (Box 3.6) help to evaluate these losses and damages according to different sectors.

Box 3.6: Post-Disaster Needs Assessment

The Post-Disaster Needs Assessment (PDNA) tool was developed by the United Nations Development Group, the World Bank and the European Union, and aims to assess the extent of a disaster's impact, define the needs for recovery, and as a consequence, serve as the basis for designing a recovery strategy and guide donors' funding. According to UNDP, PDNA looks ahead to restoring damaged infrastructure, houses, livelihoods, services, governance and social systems, and includes an emphasis on reducing future disaster risks and building resilience (UNDP, 2019). The PDNA has specific guidelines for 18 specific sectors:

1 agriculture, livestock, fisheries and forestry;
2 commerce;
3 community infrastructure;
4 culture;
5 disaster risk reduction;

6 education;
7 employment, livelihood and social protection;
8 environment;
9 gender;
10 governance;
11 health;
12 housing;
13 macroeconomic impact of disasters;
14 manufacturing;
15 telecommunications;
16 tourism;
17 transport; and
18 water and sanitation.

Disaster impacts can be analysed through the lens of lack of access to resources (Table 3.2). The lack of access can exist before the hazards happen and can be intensified after them. In the aftermath of a hazard, survivors have a series of needs. For Chambers (1989), the meaning of losses can take many forms, such as being or becoming physically weaker, economically impoverished, socially dependent, humiliated or psychologically harmed. Social contexts and the recovery policies can shape the meaning of these losses and impacts, which are mediated by the lack of access to different types of resources. The PAR framework offers important insights to reflect on the lack of access to natural, physical, human, social, economic and political resources. Lack of access to natural resources can be exemplified by the lack of arable land, potable water, biodiversity resources and so forth. Lack of social resources can be measured by the existence of marginalised groups and individuals, violence and paramilitary groups. Lack of economic resources is related to poor access to the market, low income levels, limited access to formal credit, unemployment and so forth.

If there are some groups that face lacking of access to resources, there are others who are in advantageous positions to use resources to maximise their capitalist interests in the wake of disasters (Klein, 2007; Schuller, 2008). After the 2004 tsunami in the Indian Ocean, Klein (2007) reported that most tsunami-struck countries used their Armed Forces to control and impose buffer zones, preventing survivors from rebuilding their homes on the coasts. In order to receive food rations and small relief allowances, survivors were moved to temporary shelters patrolled by soldiers. When the "fishing families returned to the spots where their homes once stood, they were greeted by police who forbade them to rebuild" (Klein, 2007, p. 387). Officially, governments said the buffer zone was a safety measure against tsunamis. But resorts were completely exempted from the buffer-zone rule: "Hotels were being encouraged to expand onto the valuable oceanfront where fishing

Table 3.2
Lack of access in the aftermath of hazards
Source: adapted from Wisner et al. (2012)

Type of resources	*Examples*
Natural	Lack of arable land Lack of potable water Lack of biodiversity resources
Physical	Dangerous locations Unprotected buildings and infrastructure Lack of health units, schools and public services Lack of adequate temporary shelters
Human	Fragile health Limited skills and formal education
Social	Marginalised groups and individuals Limited social networks Paramilitary groups Violence
Economic	Poor access to the market Low income levels Limited access to formal credit Unemployment
Political	Poor social protection Lack of political voice Domestic conflicts

people had lived and worked" (Klein, 2007, p. 388). Also in Chile, after the 2010 earthquake-tsunami, the coastal city of Talcahuano was redeveloped: the fishing villages were displaced by restaurants and touristic infrastructure (Vasquez & Marchezini, 2020).

The access to resources can be differentiated by the social context and characteristics of hazards (extension, frequency etc.) as well as by social issues, such as gender, sexual minorities, age, race, ethnic, rural/urban/island/forest, people with disabilities and mobility status (refugees, homelessness), or by the combination/intersection of these and other socially defined drivers. For instance, in India, the Aravanis faced serious gender discrimination before and after the 2004 tsunami (Pincha & Krishna, 2008). This sexual and gender minority cannot be classified and explained using a two-gender category, and the systemic rejection they have suffered has pushed them into extreme poverty. Aravanis' tsunami victims were not recorded in official statistics of the 2004 catastrophe, nor did their family members receive any government compensation, and they still lack access to safe housing, citizenship documents, secure livelihoods, disaster preparedness, capacity building and so forth. (Pincha & Krishna, 2008). Also in the aftermath of the 2005 Hurricane Katrina, there were examples of gentrification: "New Orleans became a city that could be (re)constructed on neoliberal principles of capitalist utility", limiting the possibilities of return "for the city's African American working class" (Barrios, 2010, p. 595). Hoffman (2012, p. 192) stated that opportunism is also an impact in disasters:

> the number of opportunists who flocked into the place. . . . The same phenomenon was reported in Yungay after the avalanche [1970], in Florida after Hurricane Andrew [1992]; and, in Oakland after the fire [1991]. Architects and builders from far-flung communities arrived to take advantage of the destruction.

It is important to analyse the different types of impacts; how they are distributed and differentiated according to the social, cultural and political contexts; and also how these impacts are changing in the short and long term. For instance, six years after the 1991 Oakland firestorm, some houses stood in a state of perpetual abandonment, not all survivors returned to their jobs or received their insurance settlements (Hoffman, 2012). In Nicaragua, after the eruption of the Casita Volcano in 1998, many Managua residents migrated to Costa Rica because there was no work in Nicaragua. Eight years after the 2005 flood in Feliz Deserto, Brazil, families were still waiting for the reconstruction of their homes. A similar situation was reported in Peru, six years after the 2007 earthquake: more than 14,000 affected families in Cañete, Pisco, Chinca and Ica were still sleeping in tents. In Haiti, three years after the 2010 earthquake, 360,000 Haitians were still living in tent camps and another 78,000 remain sheltered in schools and churches, many of them threatened by forced evictions and gender violence (Marchezini, 2014).

Disasters also cause impacts on cultures, music, poetry, audiovisual ways of expression (Oliver-Smith & Hoffman, 2002; Marchezini, 2015). These issues can be used as resources to reflect on the impacts of disasters. Societies have used audiovisual media to express events and phenomena. Dr Gemma Sou and Dr Felix Aponte-Gonzalez, for instance, conducted ethnographic research following 16 low-income Puerto Rican families in the aftermath of the 2017 Hurricane Maria. Gemma later decided to turn this research about the impacts into a 20-page graphic novel in English and Spanish (Box 3.7 and Figure 3.6).

Major disasters and catastrophes can also generate impacts on academic life. For instance, when Hurricane Katrina happened in the USA in 2005, the Social Science Research Council (SSRC) decided to create a task force to investigate the social dimensions of the response with a range of projects dealing with displaced persons and the effects on public-housing residents. The SSRC organised a forum on its website exploring a number of topics related to the following: structures of vulnerability, including the race, class, gender and age of those suffering most; political projects that have distorted the pursuit of homeland security; bias that has sent federal resources disproportionately to rural areas and suburbs rather than cities; media coverage of the disaster; response from the North American public; philanthropic and

Box 3.7: *After Maria* comic: what happened after the storm passed? How did the people recover and what were the long-term and hidden impacts?

On 20 September 2017, the biggest storm in Caribbean history, Hurricane Maria, struck the Caribbean island of Puerto Rico, causing over US$30 billion in damage. For one year, Dr Gemma Sou and Dr Felix Aponte-Gonzalez conducted ethnographic research to find out how 16 low-income Puerto Rican families recovered from Hurricane Maria. Dr Gemma turned this research into a 20-page graphic novella. Although the graphic novella, *After Maria*, tells the story of a fictional family, it is based on the experiences that tie together all the Puerto Rican families Gemma spoke to. Through this story, it is possible to find out how disaster-affected families use their limited resources to recover from the social, cultural, economic and psychological impacts of hazards. The story highlights how and why different family members experience disasters differently from one another – based on gender, age and race.

Source: Sou and Cei Douglas (2019)

Fig 3.6
QR code for Box 3.7. Available at https://hummedia.manchester.ac.uk/institutes/hcri/after-maria/after-maria-eng-web.pdf

charitable responses; physical infrastructure on which cities depend (and its vulnerabilities); implications of the Iraq War; problems of oil dependency and related infrastructures; environmental policy and global warming, wetlands management and so forth; costs of privatization and cuts in government capacity; leadership at every level; law enforcement and public order; economic implications of catastrophic events; and comparisons to the 2004 Asian tsunami, to 9/11 in New York, to earlier hurricane disasters in the US and so forth (SSRC, 2019). Interestingly, a code of conduct for disaster-zone research (Gaillard & Peek, 2019) was not one of the main topics of the SSRC's forum (Box 3.8).

3.4 Concluding remarks

This chapter discussed and provided examples of environmental, physical, social and economic impacts of disasters and their interactions. It showed the unequal impact of disasters across geographical and social spaces on both short and long time frames, including both small and

Box 3.8: When research adds impacts to the affected people

"Researchers working in disaster zones, with people whose culture might be different from their own, need to know how to interact with survivors as well as local officials and scholars, without adding to those people's problems", warned JC Gaillard and Lori Peek, two disaster researchers who have been studying this topic in different disasters around the world. Unfortunately researchers can also offer additional impacts in disasters, but they are not learning the lessons that they recommend to be learnt. In 2004, after the deadly Indian Ocean earthquake and tsunami, an influx of foreign scientists rushed to the region to collect data and besieged fatigued locals with requests for interviews (Missbach, 2011). Some years later, during the 2013 Typhoon Haiyan in the Philippines and the 2017 Hurricane Harvey in the USA, the immediate concerns were to secure housing, food, clothing and education, but the affected people were also deluged with questionnaires by researchers (Gaillard & Peek, 2019).

To avoid these impacts, Gaillard and Peek (2019) argue that disaster research needs a culture shift to think about the need of a code of conduct for disaster-zone research. The authors cited the example of the International Association of Volcanology and Chemistry of the Earth's Interior, which has formulated guidelines on the roles and responsibilities of local and outside scientists, local authorities and the media.

large events. Moreover, it also shed light on alternative ways to think about disasters, including their everyday forms, cascading impacts and neglected forms that can become protracted crises. This chapter dialogues with other sections of this book, which offer insightful concepts, theories, methods and approaches that complement the ways disaster impacts can be analysed.

Take-away messages

1 Concepts are important to guide your research about the disaster impacts;
2 There are unequal impacts in disasters across geographical and temporal scales;
3 You should analyse environmental, physical, social, cultural and economic impacts; and
4 Use quantitative and qualitative methods to analyse disaster impacts.

To learn more about the topic discussed in this chapter, listen to the *Disasters: Deconstructed* interview with the UN Major Group for Children and Youth, and all the partners and supporters of the Asia Pacific Researchers, Practitioners, Policy-Makers in Dialogue with Children and Youth Project (Figure 3.7).

Fig 3.7
QR code for Chapter 3.

Further suggested reading

Gibson, T., & Wisner, B. (2016). "Lets talk about you . . .": Opening space for local experience, action and learning in disaster risk reduction [An international journal]. *Disaster Prevention and Management*, *25*(5), 664–684. https://doi.org/10.1108/DPM-06-2016-0119

Oliver-Smith, A. (1994). Reconstrucción después del desastre: Una visión general de secuelas y problemas. In A. Lavell (Org.), *Al. Norte del Rio Grande. Cidade do Panamá: Red de Estudios Sociales en Prevención de Desastres en América Latina* (pp. 25–40). http://www.desenredando.org/public/libros/1994/anrg/anrg_cap02-RDDD_oct-8-2002.pdf.

Peek, L. (2008). Children and disasters: Understanding vulnerability, developing Capacities, and promoting resilience—An introduction. *Children, Youth and Environments*, *18*(1), 1–29.

SSRC. (2019). *Understanding Katrina*. https://items.ssrc.org/category/understanding-katrina/

Valencio, N. F. L. S., Siena, M., & Marchezini, V. (2011). *Abandonados nos desastres*. https://site.cfp.org.br/wp-content/uploads/2011/12/abandonadosedesastreISBN.pdf. Conselho Federal de Psicologia.

References

Acosta, V. G. (Ed.). (1996). *Historia y desastres en America Latina*. Centro de Investigaciones y Estudios Superiores en Antropología Social.

Acosta, V. G. (Ed.). (1997). *Historia y desastres en America Latina*. LA RED.

Acosta, V. G. (Ed.). (2008). *Historia y desastres en America Latina*. Centro de Investigaciones y Estudios Superiores en Antropología Social/Red de Estudios Sociales en Prevención de Desastres en América Latina. LA RED, Mexico.

Alexander, D. E. (2013). Volcanic ash in the atmosphere and risks for civil aviation: A study in European crisis management. *International Journal of Disaster Risk Science*, *4*(1), 9–19. https://doi.org/10.1007/s13753-013-0003-0

Alexander, D. E., & Pescaroli, G. (2019). What are cascading disasters? *UCL Open Environment*, *1*, 03. https://doi.org/10.14324/111.444/ucloe.000003

Barrios, R. E. (2010). You found us doing this, this is our way: Criminalizing second lines, super Sunday, and habitus in post-Katrina New Orleans. *Identities*, *17*(6), 586–612. https://doi.org/10.1080/1070289X.2010.533522

Bourdieu, P. (1991). *Language and symbolic power*. Polity Press.

Brito, S. S. B. et al. (2017). Frequency, duration and severity of drought in the Brazilian Semiarid. *International Journal of Climatology*, *38*(2), 517–529.

Button, G. V. (2012). The negation of disaster: The media response to oil spills in Great Britain. In A. Oliver-Smith & S. M. Hoffman (Eds.), *The angry earth: Disaster in anthropological perspective* (pp. 150–174). Routledge.

Carmo, F. F., Kamino, L. H. Y., Júnior, R. T., Campos, I. C, Carmo, F. F., Silvino, G., Castro, K. J. S. X., Mauro, M. L., Rodrigues, N. U. A., Miranda, M. P. S., & Pinto, C. E. F. (2017). Fundão tailings dam failures: The environment tragedy of the largest technological disaster of Brazilian mining in global context. *Perspectives in Ecology and Conservation*, *15*(3), 145–151. https://doi.org/10.1016/j.pecon.2017.06.002

Chambers, R. (1989). Editorial introduction: Vulnerability, coping and policy. *IDS Bulletin*, *20*(2), 1–7. https://doi.org/10.1111/j.1759-5436.1989.mp20002001.x

Cunha, A. P. M. do A., Marchezini, V., Lindoso, D. P., Saito, S. M., & Alvalá, R. C. dos S. (2019). The challenges of consolidation of a drought-related disaster risk warning system to Brazil. *Sustentabilidade em Debate*, *10*(1), 43–76. https://doi.org/10.18472/SustDeb.v10n1.2019.19380

Cunha, A. P. M. do A., Tomasella, J., Ribeiro-Neto, G. G., Brown, M., Garcia, S. R., Brito, S. B., & Carvalho, M. A. (2018). Changes in the spatial-temporal patterns of droughts in the Brazilian Northeast. *Atmospheric Science Letters*, *19*(10), 1–8. https://doi.org/10.1002/asl.855

De Groot, W. J., & Flannigan, M. D. (2014). Climate change and early warning systems for wild land fire. In Z. Zommers & A. Singh (Eds.), *Reducing disaster: Early warning systems for climate change* (pp. 127–151). Springer.

Dolman, D. I., Brown, I. F., Anderson, L. O., Warner, J. F., Marchezini, V., & Santos, G. L. P. (2018). Re-thinking socio-economic impact assessments of disasters: The 2015 flood in Rio Branco, Brazilian Amazon. *International Journal of Disaster Risk Reduction*, *31*, 212–219. https://doi.org/10.1016/j.ijdrr.2018.04.024

Dombrowsky, W. (1998). Again and again: Is a disaster we call a "disaster"? In E. Quarantelli (Ed.), *What is a disaster? Perspectives on the question* (pp. 19–30). Routledge.

Earthquake Engineering Research Institute. (2006). *Learning from earthquakes—The Kashmir earthquake of October 8 2005: Impacts in Pakistan*. https://www.eeri.org/lfe/pdf/kashmir_eeri_2nd_report.pdf, Retrieved June 3, 2015

Faas, A. J., & Barrios, R. E. (2015). Applied anthropology of risk, hazards, and disasters. *Human Organization*, *74*(4), 287–295. https://doi.org/10.17730/0018-7259-74.4.287

Fernandes, G. W., Goulart, F. F., Ranieri, B. D., Coelho, M. S., Dales, K., Boesche, N., Bustamante, M., Carvalho, F. A., Carvalho, D. C., Dirzo, R., Fernandes, S., Galetti, Jr., P. M., Millan, V. E. G., Mielke, C., Ramirez, J. L., Neves, A., Rogass, C., Ribeiro, S. P., Scariot, A., & Soares-Filho, B. (2016). Deep into the mud: Ecological and socio-economic impacts of the dam breach in Mariana, Brazil. *Natureza and Conservação*, *14*(2), 35–45. https://doi.org/10.1016/j.ncon.2016.10.003

Food and Agriculture Organization. (2010). *Countries in protracted crisis: What are they and why do they deserve special attention?* Retrieved November 15, 2019, from http://www.fao.org/3/i1683e/i1683e03.pdf

Gaillard, J. C., & Peek, L. (2019). Disaster-zone research needs a code of conduct. *Nature*, *575*(7783), 440–442. https://doi.org/10.1038/d41586-019-03534-z

Giglio, L., Randerson, J. T., Van der Werf, G. R., Kasibhatla, P. S., Collatz, G. J., Morton, D. C., & DeFries, R. S. (2010). Assessing variability and long-term trends in burned area by merging multiple satellite fire products. *Biogeosciences*, *7*(3), 1171–1186. https://doi.org/10.5194/bg-7-1171-2010

Global Network of Civil Society Organizations for Disaster Reduction. (2019). *Views from the frontline*. https://vfl.world/explore-vfl-data/. GNDR.

Grasso, V. F. (2014). The state of early warning systems'. In Z. Zommers & A. Singh (Eds.), *Reducing disaster: Early warning systems for climate change* (pp. 109–126). Springer.

Guadagno, L. (2016). Human mobility in the Sendai framework for disaster risk reduction. *International Journal of Disaster Risk Science*, *7*(1), 30–40. https://doi.org/10.1007/s13753-016-0077-6

Hewitt, K. (1983). The idea of calamity in a technocratic age. In K. Hewitt (Ed.), *Interpretations of calamity* (pp. 3–32). Allen & Unwin.

Hoffman, S. M. (2012). The worst of times, the best of times: Toward a model of cultural response to disaster. In A. Oliver-Smith & S. M. Hoffman (Eds.), *The angry earth: Disaster in anthropological perspective* (pp. 176–203). Routledge.

IBGE, & Brazilian Institute of Geography and Statistics. (2009). *O censo agropecuário 2006 e a agricultura familiar no Brasil*. Muscular Dystrophy Association.

Internal Displacement Monitoring Centre. (2015). *Global estimates 2015: People displaced by disasters*. http://www.acnur.org/fileadmin/Documentos/Publicaciones/2015/10092.pdf. Retrieved August 1 2018. Internal Displacement Monitoring Centre.

Internal Displacement Monitoring Centre. (2019). *Internal displacement from January to June 2019*. http://www.internal-displacement.org/

sites/default/files/inline-files/2019-mid-year-figures_for%20website%20 upload.pdf Retrieved November 1 2019

Kazama, M., & Noda, T. (2012). Damage statistics (Summary of the 2011 off the Pacific Coast of Tohoku Earthquake damage). *Soils and Foundations*, *52*(5), 780–792. https://doi.org/10.1016/j.sandf.2012.11.003

Kelman, I., & Glantz, M. H. (2014). Early warning systems defined. In Z. Zommers & A. Singh (Eds.), *Reducing disaster: Early warning systems for climate change* (pp. 89–108). Springer.

Kishore, N., Marqués, D., Mahmud, A., Kiang, M. V., Rodriguez, I., Fuller, A., Ebner, P., Sorensen, C., Racy, F., Lemery, J., Maas, L., Leaning, J., Irizarry, R. A., Balsari, S., & Buckee, C. O. (2018). Mortality in Puerto Rico after hurricane Maria. *New England Journal of Medicine*, *379*(2), 162–170. http://doi.org/10.1056/NEJMsa1803972

Klein, N. (2007). *The shock doctrine: The rise of disaster capitalism*. Picador.

Koks, E. E., Rozenberg, J., Zorn, C., Tariverdi, M., Vousdoukas, M., Fraser, S. A., Hall, J. W., & Hallegatte, S. (2019). A global multi-hazard risk analysis of road and railway infrastructure assets. *Nature Communications*, *10*(1), 2677. https://doi.org/10.1038/s41467-019-10442-3

Kondo, S., Yamori, K., Atsumi, T., & Suzuki, I. (2012). How do "numbers" construct social reality in disaster-stricken areas? A case of the 2008 Wenchuan earthquake in Sichuan, China. *Natural Hazards*, *62*(1), 71–81. https://doi.org/10.1007/s11069-011-0038-8

Lavell, A. (Ed.). (1994). *Al Norte del Rio Grande*. Red de Estudios Sociales en Prevención de Desastres en América Latina.

Li, X., Jin, L., & Kan, H. (2019). Air pollution: A global problem needs local fixes. *Nature*, *570*(7762), 437–439. https://doi.org/10.1038/d41586-019-01960-7

Marchezini, V. (2014). La producción silenciada de los "desastres naturales' en catástrofes sociales". *Revista Mexicana de Sociología*, *76*(2), 253–285.

Marchezini, V. (2015). Social recovery in disasters: The cultural resistance of luizenses'. In M. Companion (Ed.), *Disaster's impact on livelihood and cultural survival: Losses, opportunities and mitigation* (pp. 293–303). CRC Press Press/Taylor & Francis Group.

Marchezini, V., Aguilar Muñoz, V. A., & Trajber, R. (2018). Vulnerabilidade escolar frente a desastres no Brasil. *Territorium*, *25*(II), 161–178. https://doi.org/10.14195/1647-7723_25-2_13

Maskrey, A. (Ed.). (1993). *Los desastres no son naturales*. LA RED.

Matthewman, S. (2015). *Disasters, risks and revelation*. Palgrave Macmillan.

Missbach, A. (2011). Ransacking the field? Collaboration and competition between local and foreign researchers in Aceh. *Critical Asian Studies*, *43*(3), 373–398. https://doi.org/10.1080/14672715.2011.597334

Mouillot, F., & Field, C. B. (2005). Fire history and the global carbon budget: A 1ox 1o fire history reconstruction for the 20th century. *Global Change Biology*, *11*(3), 398–420. https://doi.org/10.1111/j.1365-2486.2005.00920.x

Ochola, S. O., Eitel, B., & Olago, D. O. (2010). Vulnerability of schools to floods in Nyando River catchment, Kenya. *Disasters*, *34*(3), 732–754. https://doi.org/10.1111/j.1467-7717.2010.01167.x

O'Keefe, P., Westgate, K., & Wisner, B. (1976). Taking the naturalness out of natural disasters. *Nature*, *260*(5552), 566–567. https://doi.org/10.1038/260566a0

Oliver-Smith, A. (2012). The brotherhood of pain: Theoretical and applied perspectives on post-disaster solidarity. In A. Oliver-Smith & S. M. Hoffman (Eds.), *The angry earth: Disaster in anthropological perspective* (pp. 204–225). Routledge.

Oliver-Smith, A., & Hoffman, S. M. (2002). Introduction: Why anthropologists should study disasters. In S. M. Hoffman & A. Oliver-Smith (Eds.), *Catastrophe and culture: The anthropology of disaster* (pp. 03–22). School of American Research Press.

Perry, R. W. (2018). Defining disaster: An evolving concept. In H. Rodriguez, W. Dooner & J. Trainor (Eds.), *Handbook of disaster research, handbook of sociology and social research* (pp. 1–22). Springer.

Pincha, C., & Krishna, H. (2008). Aravanis: Voiceless victims of the tsunami. *Humanitarian Exchange*, *41*, 41–43.

Prince, S. H. (1920). *Catastrophe and social change based upon a sociological study of the Halifax disaster* [Dissertation, Columbia University]. http://www.gutenberg.org/files/37580/37580-h/37580-h.htm

Quarantelli, E. L. (Ed.). (1998). *What is a disaster: Perspectives on the question*. Routledge.

Santos, D. P. A., & Vidal, D. L. (2015). Realidade territorial de unidades familiares no semiárido brasileiro. *Tempo Social*, *28*, 55–83.

Schuller, M. (2008). Deconstructing the disaster after the disaster: Conceptualizing disaster capitalism. In N. Gunewardena & M. Schuller (Eds.), *Capitalizing on catastrophe: Neoliberal strategies in disaster reconstruction* (pp. 17–27). Alta Mira Press.

Sorokin, P. (1942). *Man and society in calamity: The effects of war, revolution, famine, pestilence upon human mind, behavior, social organization and cultural life*. E. P. Dutton, and Company.

Sou, G., & Cei Douglas, J. (2019). *After Maria: Everyday recovery from disaster*. University of Manchester (graphic novella). https://www.hcri.manchester.ac.uk/research/projects/after-maria/.

SSRC. (2019). *Understanding Katrina*. https://items.ssrc.org/category/understanding-katrina

Tuladhar, G., Yatabe, R., Dahal, R. K., & Bhandary, N. P. (2014). Knowledge of disaster risk reduction among school students in Nepal.

Geomatics, Natural Hazards and Risk, 5(3), 190–207. https://doi.org/10.1080/19475705.2013.809556
United Nations Children's Fund. (2012). *UNICEF and disaster risk reduction*. Retrieved June 6, 2014, from http://www.unicef.org/malaysia/UNICEF_and_Disaster_Risk_Reduction.pdf
UNISDR (United Nations International Strategy for Disaster Reduction). (2015a). *Sendai framework for disaster risk reduction 2015–2030*. UNISDR.
UNISDR (United Nations International Strategy for Disaster Reduction). (2015b). *Making development sustainable: The future of disaster risk management. Global assessment report on disaster risk reduction*. United Nations Office for Disaster Risk Reduction.
UNISDR (United Nations International Strategy for Disaster Reduction). (2017). *Terminology: Basic terms of disaster risk reduction*. Retrieved August 1, 2019, from https://www.unisdr.org/we/inform/terminology#letter-e
Vasquez, J. R. S., & Marchezini, V. (2020). Procesos de recuperación posdesastre en contextos biopolíticos neoliberales: Los casos de Chile 2010 y Brasil 2011. *ÍCONOS Revista de Ciencias Sociales*, *66*, 131–148.
Wallace, A. F. C. (1956). *Human behavior in extreme situations*. National Research Council, National Academy of Sciences.
Wallemacq, P., & House, R. (2018). *Economic losses, poverty and disasters*. Retrieved August 1, 2019, from https://www.preventionweb.net/files/61119_credeconomiclosses.pdf
Wisner, B., Blaikie, P., Cannon, T., & Davis, I. (Eds.). (2004). *At risk: Natural hazards, people's vulnerability and disasters*. Routledge.
Wisner, B., & Gaillard, J. C. (2009). An introduction to neglected disasters. *Jàmbá: Journal of Disaster Risk Studies*, 2(3), 151–158. https://doi.org/10.4102/jamba.v2i3.23
Wisner, B., Gaillard, J. C., & Kelman, I. (2012). Framing disaster: Theories and stories seeking to understand hazards, vulnerability and risk. In B. Wisner, J. C. Gaillard, & I. Kelman (Eds.), *The Routledge handbook of hazards and disaster risk reduction* (pp. 18–34). Routledge.

PART II

Vulnerabilities and capacities

CHAPTER 4

Why do disasters occur?

4.1 Why do disasters occur?

For centuries, people have been living with disasters; disasters made or broke civilisations and affected our cultures and behaviours. Disasters destroy livelihoods and cost billions of dollars when it comes to recovery and reconstruction. Hardly a week goes by without media reporting a story about yet another (not so) freak weather events. Floods in Japan or fires in the USA have been extensively covered in news broadcasts (whilst other disasters, especially those in low-income countries, such as Bangladesh, Nepal or Malawi, have been largely ignored; Figure 4.1). Disasters are thus becoming more and more prominent on political and media agendas as the damages are increasing.

Fig 4.1
The gap in equality.
Source: photo courtesy of Ksenia Chmutina

DOI: 10.4324/9781315469614-7

Box 4.1: Are disasters ever natural?

Fig 4.2
QR code for Box 4.1. In this episode of *Disasters: Deconstructed* podcast, the listeners share their views on whether disasters are actually natural. Not everyone agreed – and this is why this conversation is so important.
Available at https://disastersdecon.podbean.com/e/season-1-episode-2-are-disasters-ever-natural/

But are disasters unavoidable; are they a natural phenomenon (Box 4.1 and Figure 4.2)? The answer is no. Too often, disasters are described as natural, as they are perceived to be caused by hazards of natural origin (such as extreme weather, geophysical phenomena or epidemics). However, it is important to recognise that these so-called natural disasters are rarely natural because there tends to be many important human-induced factors that have converted the hazard into a disaster. This argument is not new: in a letter to Voltaire in 1756, Rousseau reflected on the 1755 Lisbon earthquake describing it as something more than an earthquake, noting “that nature did not construct twenty thousand houses of six to seven stories . . ., and that if inhabitants of this great city were more equally spread out and more lightly lodged, the damage would have been much less and perhaps to no account” (in Master & Kelly, 1992, p. 110). This argument has been emphasised further over the years (see, for example, Ball, 1975; O’Keefe et al., 1976; Hewitt, 1983; Oliver-Smith, 1986; Cannon, 1994; Smith, 2005; Kelman, 2010; Chmutina et al., 2019), highlighting that disasters result from the combination of natural hazards and social and human vulnerability, including development activities that are ignorant of local hazardous conditions. By calling disaster natural, we are putting the responsibility for failures of development on freak natural phenomena or acts of God.

So why do disasters occur? There is a growing understanding that disaster is not just an occurrence of a hazard: an earthquake that happens in an uninhabited area is not typically considered a disaster. A disaster is commonly defined as “A serious disruption of the functioning of a community or a society at any scale due to *hazardous events interacting with conditions of exposure, vulnerability and capacity*, leading to one or more of the following: human, material, economic and environmental losses and impacts” (UNISDR, 2018, authors’ emphasis). UNISDR (2018) defines a ‘natural hazard’ as

"A natural process or phenomenon that may cause loss of life, injury or other health impacts, property damage, social and economic disruption or environmental degradation", with an annotation that "Severe hazardous events can lead to a disaster as a result of the combination of hazard occurrence and other *risk factors*" (authors' emphasis). The roles of severity and risk factors are significant here: severity signifies a substantial departure from a mean or trend, and the fundamental determinants of hazards comprise location, time, magnitude and frequency; risk factors are related to the threshold determined by the combination of the lowest limit at which physical forces can cause damage (Alexander, 2000). It is also important to recognise that risk is a complex notion. Whilst 'risk' is traditionally defined as a combination of an impact and a likelihood, it is also important to consider political will to deal with risks, capacity to map and assess frequency of hazard events, susceptibility to loss across a range of population groups and sectors, and capacity to take actions to prepare and to mitigate, to monitor results of actions, to learn from successes and failures and to maintain vigilance and foresight through more quiescent periods, when public interest will decrease (Wisner, 2018). These elements are summed up in an equation-like mnemonic (Wisner et al., 2012, pp. 19–21):

$$DR \approx \left[H \times (V/C) - M\right],$$

where DR stands for disaster risk, H for the specific hazard probability, V for vulnerability, C for localised and individual capacity for self-protection and recovery and M for social protection provided by the state, ideally complementing and supplementing C and never blocking or diluting it (which, in reality, happens quite often). These variables are dynamic and are affected by global economic and geopolitical change and climate instability, national political and economic cycles and other abrupt changes as well as local context.

Vulnerability is "the conditions determined by physical, social, economic and environmental factors or processes which increase the susceptibility of an individual, a community, assets or systems to the impacts of hazards" (UNISDR, 2018); it is a complex characteristic derived from socio-economic factors (Cannon, 1994) that expose people to harm and limit their ability to anticipate, cope with and recover from harm. Understanding vulnerability is the key to understanding why disasters occur – and why disasters do not affect all communities and societies equally. Vulnerability not a condition; it is a social construct. Understanding vulnerability unravels that disaster impacts are disproportionately felt by the poor and/or other social groups that have been pushed to the margins of society and lack access to resources and the means of protection available to those with higher levels of socio-economic or political power. Vulnerability, thus, reflects how power and resources are shared within society. This will be discussed further in Chapter 5.

Disasters are a product of economic, social, cultural and political processes, thus highlighting that whilst a hazard cannot be prevented, disasters can be. Earthquakes, droughts, floods, storms, landslides and volcanic eruptions are natural hazards; they lead to deaths and damages (i.e. disasters) because of human acts of omission and commission rather than the act of nature. This chapter will be building on the vulnerability approach that helps our understanding of the underlying causes of disaster – and how the daily life of people and the constraints and threats they faced, including their economic and political marginality, play an important role in it.

4.2 Underlying causes of disasters

Many people live in hazard-prone areas despite the existence of measurable and known levels of risk. Cities have traditionally been founded close to the coast or on the river, thus, offering access to superior economic opportunities (albeit, in some cases, short term); cities have also grown around fertile volcanic or floodplain soils. Cities play an important role in driving development as they concentrate much of the national economic activity, government, commerce and transportation, and provide crucial links with rural areas, between cities and across international borders. Urban living is, thus, often associated with higher levels of literacy and education, improved health, greater access to social services, and enhanced opportunities for cultural and political participation (UN, 2014). However, whilst some enjoy the economic opportunities, others cannot even meet the basic needs. Many of urban dwellers lack resources that would allow them to adapt to or cope with a hazard or move to safer locations (Alexander, 2001).

A hazard becomes a disaster because its impact threatens the lives and livelihoods of people. A disaster does not happen unless people and cities are vulnerable due to marginalisation, discrimination, and inequitable access to resources, knowledge and support. These vulnerabilities are further – intentionally or unintentionally – enhanced by deforestation, rapid urbanisation, environmental degradation and climate change. Moreover, vulnerabilities are too often enhanced not because the information about dealing with hazards does not exist but because decision makers (and those responsible for the development of the built environment) do not use this information appropriately (or at all).

4.2.1 Social underlying causes

Society itself creates unequal exposure to risk by making some (groups of) people more prone to hazards than the others. Inequality, poverty, political ideology, class and power relations are the root causes of vulnerabilities that turn natural hazards into disasters, making some more vulnerable than others. In other words, vulnerability represents political, economic, physical or social susceptibility of a person/group of people to damage from natural hazards. A series of extreme – but often permanent – conditions exist that make livelihood activities extremely fragile for certain social groups (Cardona, 2003). This means that within the same society, a hazard affects different people differently. For instance, normal existence of an impoverished black community or a poor Hispanic migrant community along the coast of New Orleans differs dramatically from the normal of a white, upper-class community in New York (Donner & Rodriguez, 2008). The normal for the society's poor and disadvantaged often does not cover even the basic needs.

Wisner (2013) notes that vulnerable are those who are "also politically marginal (no voice in decisions that affect them), spatially marginal (resident in urban squatter settlements or in remote rural locations), ecologically marginal (livelihoods based on access to meager natural resources or living in degraded environments), and economically marginal (poor access to markets)" (p. 258). The socio-economic, political and cultural and historic causes discussed subsequently are, thus, intertwined and interdependent.

Socio-economic causes

Inequality increases the exposure to risk; this is reflected in power distribution in terms of class, gender, ethnicity, sexuality, disability and so on (discussed in more detail in Chapter 5). It is a process of feedbacks, which serve to further entrench and ingrain certain people/groups in a cycle of disadvantage, when pre-existing conditions are exacerbated rather than created by a natural hazard.

The economic system allocates income and access to resources, which affects people's intellectual and physical abilities, which consequentially affects their ability to build resilience to natural hazards. Economic marginality is rooted in allocation of land and resources as well as market dynamics that exclude or burden some, whilst benefiting others. In many places, marginality is driven through exploitation of people for their cheap labour or commodities in order to achieve economic growth, or through economic globalisation that has been driving many people (e.g. small farmers) off their lands because they cannot compete with the imported goods.

Poverty is the single most important factor that determines vulnerability (Cannon, 2008). The 2009 Typhoon Ondoy affected the poor in the Philippines disproportionately. Those hit hardest were those who had been self-employed before the typhoon, including fisherfolk, farmers, small-business owners and workers employed in the informal sector. Their households suffered long term, and their livelihoods were disrupted because they were forced to shift to less capital-intensive (and less profitable) occupations.

Low-income urban settlements often have limited social resources, that is, a lack of extended family structure, established networks of contacts or strong relationships of trust (Sanderson, 2000). Thus, inequality is manifested through an access to basic – but not only economic – needs (i.e. goods, services, cultural satisfactions that are needed for adequate survival). If a livelihood merely (or does not) provide sufficient basic needs, then the provision of self- or social protection is highly unlikely, leading to an increase in vulnerability.

However, what turns hazards into disasters is not simply a question of money and resources; vulnerability is not the same as poverty. Although the livelihood and self-protection is strongly linked to wealth and income, a person's gender or ethic roots, for instance, may alter their self-protection capacity despite having a reasonable livelihood (Cannon, 1994).

Political and institutional causes

The roots of social vulnerability are located in power-laden social relations and processes (Sun & Faas, 2018). The political system allocates resources for structural preparedness for and mitigation of hazards – yet the process of decision-making and implementation is often technocratic and reactive and does not take into account the needs of many people. In many cases, a government makes decisions and takes actions that should be aimed at mitigating hazard risk and decrease vulnerability (but in reality, often, only the former is addressed). The options given to people in terms of DRM are then imposed on those at risk rather than informed by them. Many DRM activities are embedded into international agendas, but peripheral status in the global political economy of some countries, combined with the consequences of long-term conflict and battles for power as well as the desire to develop, lead to increasing ecological and social neglect. State power is also closely linked to the impacts of policies and legislations, as the abuse of the state power often leads to neglect of governmental functions, such as the enforcement of building and safety codes, provision of basic infrastructure as well as corruption. By depriving the most marginalised of what little money and resources they have, corruption exacerbates poverty, and at higher social levels, it may keep underserving elites in power (Hoogvelt, 1976). The devastating outcome of the 2008 Sichuan earthquake in China (when 75,000 people, including 900 children in just one school, were killed) was strongly suspected on corruption construction (Lewis & Kelman, 2012).

Political causes are important to consider because people who are spatially isolated (e.g. living in informal settlements), or are poor in terms of depleted ecological legacy or livelihood resources, frequently have limited voice and access to formal institutions, which makes them politically marginalised. Political marginality is interconnected with the social; it often reflects favouritism practiced by ruling parties and historically developed divisions of national territory (Wisner, 2010).

Violent conflicts (often rooted in a political will to achieve a certain goal) contribute to the underlying causes of disasters. In conflict situations, an increasing proportion of casualties are civilians, which increases overall social vulnerability as conflicts result not only in death and injuries but also in disruptions to livelihoods. This leads to displacement of many people, whose vulnerability increases further as they face unknown risks in unknown environments.

For instance, violence towards the Rohingya in Myanmar from late August 2017 caused the mass displacement of over 650,000 people to the south-eastern hilly region of Bangladesh. The majority of the displaced have since been residing in overcrowded temporary shelters on a site prone to rainfall-triggered landslides, flash flooding and cyclones. The flooding increases the risk of waterborne diseases, and the landslides cut road communications when transportation into the camps is already difficult. The overcrowded makeshift huts with poor sanitation and water supply, and the lack of healthcare facilities, are also likely to contribute to an increase in disease (Ahmed et al., 2018).

Past conflicts also increase institutional weaknesses, as they have a direct impact on social welfare and access to infrastructure, as they often destroy irrigation systems, electricity supply, water and sanitation facilities, health and education facilities and so on (Wisner, 2012). These settings create everyday politics, with the norms of which people adjust to, comply or evade (Heijmans, 2017). Some construction practices occur because of those who seek to benefit themselves. Self-seeking expenditure, when those with access to public money manipulate funds that could otherwise be applied to development, is unfortunately a common cause of disasters. Instead of being invested in DRR activities (for which the money would have been originally intended), these funds are put towards projects that are used as a way for personal gain. The abuse of entrusted power for personal gain inevitably leads to increased vulnerability (Lewis & Kelman, 2012). The only high-rise building – a 17-storey Wei Guan Golden Dragon Building – that collapsed in a 2017 earthquake in the Taiwanese city of Tainan, was reinforced with empty oil cans that were packed inside some of the concrete beams. While an investigation has been launched into whether the building's developer had cut corners – and the developer was eventually arrested – this does not matter for the 31 people who died as a result of the collapse.

Cultural and historical causes

Vulnerability is also rooted in historic governance – and the way it impacts contemporary governance. The contemporary vulnerability often originates from the processes of invasion, conquest and colonisation that subsequently provided the structure for later development models and social hierarchies in postcolonial societies. Rooted in histories of colonisation and domination of lands and people, some current postcolonial governments as well as non-governmental organisations reproduce colonial patterns of displacement, dislocation and disadvantage (Box 4.2, Figure 4.4 and Box 4.3). They reproduce vulnerability by perceiving indigenous communities as inherently vulnerable and ignoring important environmental knowledge and a nuanced understanding of hazards to which the vulnerable populations are exposed (Sun & Faas, 2018). A case in point is the 2010 Haiti

Fig 4.3
Extensive damage to housing caused by the Haitian earthquake in 2010.
Source: WEDC

Box 4.2: The role of imperialism and colonialism in disasters

Fig 4.4
QR code for Box 4.2. Available at https://disastersdecon.podbean.com/e/s4e9-coloniality-disasters/.

This episode of *Disasters: Deconstructed* podcast focuses on the causes of disasters rooted in imperialism and colonialism. Dr Danielle Rivera from University of California, Berkeley, focuses on the problems with event-centric narratives of disaster that do not reflect the fact that disaster impacts are shaped by structural violence.

earthquake (Figure 4.3): it exposed that vulnerability is rooted in the heritage of a colonial system (slavery and economic exploitation), the continuing influence of foreign powers in Haiti's domestic affairs starting immediately after its independence, and a state that is rather geared towards serving the political, social and business elite instead of the best interests of the public (Schuller, 2016). Missing leadership also became evident during the response and recovery after the earthquake. The coordination mechanism of humanitarian aid was not sufficient,

Box 4.3: Risk rooted in colonial era weighs on Bahamas's efforts to rebuild after Hurricane Dorian

When Hurricane Dorian made landfall on Great Abaco Island in the Bahamas on 1 September 2019, it packed winds of up to 185 miles per hour and caused a 20-foot storm surge, bringing generational devastation. The cost to the Bahamas has been estimated to be up to US$7 billion – more than half of the country's annual economic output.

But not all structures and communities in Dorian's path were equally affected: whilst structural failure was widespread, houses intentionally built to resist high wind and storm surge fared much better. The problem is that not everyone has access to a house that can weather a storm like Dorian. The different ways in which Abaco and Grand Bahama – and their residents – were affected by the same event is yet another example of how disaster impacts are rooted in the historical development of society.

In the Bahamas, we see this kind of accumulated risk most clearly amongst the Haitian diaspora and Haitian Bahamians, who are stigmatised and face many barriers to full participation in society. The most catastrophic damage from Dorian occurred in communities, like the Mudd – a shanty town housing the nation's largest Haitian immigrant community – where land is not owned by residents and daily survival is paramount. People there trade the risk presented by massive hurricanes for the necessity of a place to live.

When Europeans arrived in 1492, they committed atrocities against the indigenous peoples that lived there. The Caribbean was rapidly turned into a site to sustain and protect colonial circulations of goods, money and slaves. Between the 16th and 19th centuries, an estimated 5 million Africans were enslaved and transported to the Caribbean. Half ended up in British territorial possessions, such as the Bahamas. Colonisation created the conditions for the chronic levels of risk that we see today amongst the descendants of enslaved people. Whilst slavery was abolished in these territories in the 1830s, most descendants of slaves remained indebted and were forced to undertake low-wage agricultural labour for mostly white absentee landowners. Inequalities, injustices and discrimination were, thus, institutionalised in the colonies and remain largely in place within now-independent societies.

Alongside invasion, conquest and colonisation, contemporary vulnerabilities in the Bahamas reflect laissez-faire historical attitudes towards addressing long-term risk. This is the foundation of contemporary structures of governance, society and the economy – and a big part of why, today, poor Bahamians, Haitians and Haitian Bahamians struggle for survival.

Source: This box is based on the article by von Meding, J., Prevatt, D. and Chmutina, K., 2019. "Risk rooted in colonial era weighs on Bahamas' efforts to rebuild after Hurricane Dorian". *The Conversation*, 9 December 2019. Available at https://theconversation.com/risk-rooted-in-colonial-era-weighs-on-bahamas-efforts-to-rebuild-after-hurricane-dorian-125548 leading to chaos and primarily centralised location of aid (DKKV, 2012), and, later on, of reconstruction efforts. The vulnerability has been further exacerbated by corruption and the lack of governmental leadership accompanied by a missing legal framework and weak law enforcement (Oxfam, 2010). The national government has not been considering the requirements of people who are most in need and does not take responsibility for securing food supply or improving of the weak sanitation and health system, or providing civil protection. In October 2013, the Haitian government launched the country's first national housing policy in a bid to address the shortage of 500,000 new homes by 2020. The government-led new housing projects have, however, been negatively affected by corruption and mismanagement, as well as by poor coordination and lack of funds. Disputes over land ownership have proved a major obstacle, too, as the earthquake destroyed the majority of title deeds and land registry records (Lowenstein, 2015).

Western approach to science has also been driving the assumption that local knowledge is inferior, and thus, Western truth needs to be translated into – or sometimes replaced by – vernacular practices. This is done not only through donor organisations but also through local authorities that accept Western knowledge as legitimate. This again is rooted in historical power relationship (Mascarenhas & Wisner, 2012).

Dominating culture can also play into the risk creation. Whilst culture is critical in terms of risk perception and communication (and is often the main driver of building resilience), underlying cultural issues may depend upon the degree of integration of one group within the other as well as legacies of rejection and exclusion. Many traditional cultural perceptions of disaster risk are seen as fantastical and are often seen as source of differentness and otherness. Interventions into everyday lives (that are often driven by culture), thus, modify perceptions of dangerous conditions and increase exposure and vulnerability (Hewitt, 2012). For instance, despite the government instruction to evacuate during the 2006 Merapi Volcano eruption, some of the Javanese community members that lived on the slopes of the volcano refused to do so. The community has been carrying out annual offerings to the volcano following their traditions: this demonstrates the community's belief regarding the relationship between god and human (Lavigne et al., 2008). This is difficult to challenge as in this case disasters are seen as unavoidable.

4.2.2 Spatial underlying causes: disaster risk reduction or disaster risk production?

Space is a common feature of all disasters: although a pattern of disasters dynamically evolves, hazards have distinct geographical distributions patterns. In areas prone to multiple hazards, the impact of one can be higher than the other, depending on particular characteristics of population (Cannon, 1994). For instance, the same group may deal better with earthquakes that have historically affected the areas than with floods that may occur because of the recent inappropriate urban development. Spatially expressed vulnerability can combine elements of biophysical and social vulnerability but within specific geographical areas (where social groups and all the characteristics of a place are located). The emphasis with social space, therefore, may include those places that are most vulnerable (Cutter, 1996).

Physically unsafe places, however, do not always intersect with vulnerable populations (Cutter et al., 2003). For example, in an area of high physical risk to flooding, economic losses might be large, but equally, the population may have significant safety nets, such as insurance to absorb the flood hazards. Recent years have seen a growing trend in insured losses: the average annual cost of insured claims from disasters has increased eightfold since 1970 (up from some US$5 billion in the 1970s and 1980s to over US$40 billion in 2010) (CRED, 2019).

A scale of a disaster does not have to be extraordinary. A moderate flood event can also have a significant impact with a long time span for recovery on socially vulnerable populations. Technical and spatial planning interventions are often seen as necessary in order to prepare a certain locality to an impact of a hazard; these include early warning systems, flood control measures, remote sensing, or building and planning regulations focusing on protecting people and assets. However, spatial and technological solutions are often costly and do not protect everyone equally; in some cases, they actually enhance vulnerability of some groups living in certain locations in order to increase resilience of other – often more affluent – groups. The ability of people to protect themselves depends on their livelihood strength (Cannon, 1994). For example, quality and location of a home often depends on income and saving. Spatial causes should, thus, be explored as interrelating social processes operating on local and wider scales, as this is the combination that constitutes vulnerability relative to specific groups or communities in space and time.

Urbanisation

The rapid urbanisation experienced in some low- and middle-income countries creates a number of challenges different to those faced in higher-income countries where rates of urbanisation may have stabilised (Figure 4.5 and Box 4.4). A number of established risk-reducing approaches that can manage increases in exposure (Bene, 2013) are established around the world, and many citizens barely engage in risk management as it is often assumed that the government will provide support (Satterthwaite, 2008). Nevertheless, there are still many places where governments' capacities are restricted, and the majority of the most vulnerable population does not formally participate in the city's governance mechanisms. In many of these expanding cities, the state is unable to regulate urban development or to provide the necessary infrastructure to adequately support the population increase (Chmutina & Bosher, 2017).

Fig 4.5
Urbanised landscape, Manila, Philippines.
Source: photo courtesy of Ksenia Chmutina

Box 4.4: The origins of urbanisation

In the mid-1850s, Barcelona was on the brink of collapse. A busy industrial port, the city's density had been increasing throughout the industrial revolution. Barcelona was playing the role of the capital of Europe – yet its population of 187,000 still lived in a tiny area, confined by its medieval walls.

With a density of 856 inhabitants per hectare, the mortality rates were rising: life expectancy had dropped to 36 years for the rich and just 23 years for the working class. The walls were becoming a health risk, almost literally suffocating the city dwellers. So it was decided to demolish the walls. But now, the city faced the next problem: the government had to design and manage the sudden redistribution of an overflowing population.

The then unknown Catalan engineer Ildefons Cerdà introduced a radical expansion plan for a large, grid-like district outside the old walls. His plan consisted of a grid of streets that would unite the old city with seven peripheral villages (which later became integral Barcelona neighbourhoods, such as Gràcia and Sarrià). The united area was almost four times the size of the old city (which was around 2 km^2). Cerdà's meticulous scientific study considered a city not just as an efficient cohabiting space but as a source of well-being. He calculated the volume of atmospheric air one person needed to breathe correctly. He detailed professions the population might do and mapped the services they might need, such as marketplaces, schools and hospitals. He concluded that, amongst other things, the narrower the city's streets, the more deaths occurred.

In the process, Cerdà also invented the word 'urbanisation' – a word which did not exist in Spanish, French or English vocabularies before, but which is so widely used today.

Source: This box is based on an article by Baucells, M., 2016, "Barcelona's unloved planner invents science of 'urbanisation'"; available at https://www.theguardian.com/cities/2016/apr/01/story-cities-13-eixample-barcelona-ildefons-cerda-planner-urbanisation

Urbanisation creates and magnifies vulnerabilities that are unique to large cities due to their high concentration of people, their ecological footprint and the development and planning processes (Johnson et al., 2013). Unplanned or inadequately managed urban expansion leads to rapid sprawl, pollution and environmental degradation, together with unsustainable production and consumption patterns (UN, 2014).

Arguably, the governance driver has been playing the largest role in driving urbanisation – and therefore, the risks associated with it. Neoliberal reforms have been a great motivator for the intense urban growth (Johnson et al., 2013). Whilst there are a large number of advantages for the inhabitants of large cities (e.g. improved economic development, easier access to basic services, a comparatively rich cultural life), with increasing social polarisation, segmentation and fragmentation, the number of people that are excluded from these benefits is growing.

With rural livelihoods undermined, many rural people have swelled the density of cities but have not enjoyed the benefits of the urban life – and have often lived in hazardous locations exposed to hazards. Their livelihoods are at risk due to their informal status, impeding their labour, tenure and political rights, as well as poor living environment and an over-dependence on the cash economy (Chmutina & Bosher, 2017). The increased polarisation between rich and poor creates parallel societies, with increased crime rates, urban violence and social unrests (UN, 2014). Widening inequalities also tend to be more starkly visible in urban than rural areas, sometimes with the wealthiest areas of cities often neighbouring slums.

Informal livelihoods and practices

For urban poor, cities are dangerous places: many poor urban dwellers live on the worst quality land on the edges of ravines, on flood-prone embankments, on slopes liable to mudslide or collapse, and in densely packed areas where fires easily start (Sanderson, 2000). Many of the urban poor have to engage in informal activities and establish informal livelihoods (Figure 4.6), as informal economy allows for a diversity of ways to earn income to acquire resources. However, those resources can come at a high price; the poorest often pay more than their better-off neighbours for basic services. Land tenure and ownership are other issues: squatters and slum dwellers endure dangerous conditions to be close to sources of income, whilst in the rental sector, many families may share crowded, poor-quality, illegally divided tenements.

Following the 2004 Indian Ocean tsunami that killed more than 220,000 in the Aceh province, the Indonesian government announced that all families affected by the earthquake and tsunami would be entitled to reconstruction assistance – but this was targeted at the house owners, leaving housing needs of about 70,000 displaced squatters and renters, many of them living in informal settlements, overlooked (McCallin & Schrerer, 2015).

Those living in informal areas are increasingly left to provide their own water, energy and food supply, which affects the overall performance of the infrastructure (e.g. illegal connections to the electricity grid) and has negative environmental and health effects. Coupled with poor adaptability to and coping capacity with a fast-growing number of urban dwellers, constant underinvestment exposes an increasingly deteriorating infrastructure to environmental stresses. Degradation and deterioration are also exacerbated by inappropriate waste disposal leading to blocked drainage and therefore flooding, illegal electrical connections leading to fires, or inadequate water disposal causing structural instabilities (Wamsler, 2004).

Lack of appreciation of the local context and lack of appropriate training may also lead to worsening the existing infrastructure. For instance, building roads and other paved areas may prevent rain from infiltrating into the soil, thus, producing accelerated runoff rates, which

Fig 4.6
Self-built settlement in Canaan, Haiti.
Source: photo courtesy of Gonzalo Lizarralde

can overwhelm the existing drainage systems and lead to flooding. This happens because, too often, DRR projects are top-down and technology-driven: they may utilise foreign engineers and other professionals and use technologies, which supplant local knowledge and local labour. In doing so, applying inappropriate technologies and processes is likely to disengage local stakeholders with the development process and increase the vulnerability of an asset, as some of the context-specific knowledge may not be readily available. Local coping capabilities, available skills and appropriate technologies should, thus, always be considered when new technological approaches are being used.

Many economic activities often take place informally, frequently encouraging corruption. Informal labour is often employed in construction sector, thus making is difficult to ensure that the built environment is constructed safety. The earliest known written (albeit rather draconian) building code is the Code of Hammurabi, in ancient Mesopotamia, dating back to 1772 BC; it states that "If a builder builds a house for someone, and does not construct it properly, and the house which he built falls in and kills its owner, then that builder shall be put to death". A key objective of any building code and building regulation is life safety; this is why it is important to make sure that each building code is context specific. Building regulations seek to ensure that the policies set out in the relevant legislation are carried out. They contain the rules for building work in new and altered buildings to make them safe and accessible, and limit environmental damage. Building codes, therefore, provide a good basis for incorporation of the DRR measures. Non-compliance with building regulations is a challenge. Whilst building regulations are introduced in most countries globally, their enforcement remains an issue. This creates vulnerabilities of the assets as well as affects the economic rights of those involved in the construction process (i.e. the labourers) and those using a building (i.e. the residents and workers). For instance, in Nepal, the informal sector plays an important role in Nepal's construction sector: only 5% of individually constructed buildings undergo professional engineering design and supervision (in addition to limited planning control and land management) (Chmutina & Rose, 2018). The majority of construction workers employed by registered contractors are engaged through informal contracts (where the responsibility is agreed verbally, thus, not imposing any liability). The labourers are paid daily wage rates, which do not include any social benefits. Whilst this informal construction system offers great flexibility in terms of speed of building and lower costs, the quality of buildings often suffers, and there is a lack of professional accountability, with the client bearing all the risks and responsibilities (Jha, 2002).

Land-use planning

The idea that the impact of a disaster could be reduced not only through structural solutions but also by understanding and affecting people's land-use decisions as well as their awareness and perception of risk was brought forward in the 1940s (White, 1945). Land-use planning is considered as complementary tool to engineering hazard management approaches (Box 4.5). Whilst the argument for land-use regulation is intellectually convincing and accepted by policymakers in principle, the actual planning and permitting are prone to the influence of locally and, sometimes, globally powerful interests; the best intentions of land-use planning are often overruled by variances to permitted land uses and lack of enforcement (Wisner, 2018).

Box 4.5: How England's broken planning system has created (not reduced) the risk of floods

The floods that affected England in the autumn of 2019 have been described as unprecedented or even "biblical" events, often with the misguided assumption that they were unavoidable or unpredictable. That is not the case. Over the past few decades, development practice in England has led to more than 300,000 homes being built in high flood risk areas. In this sense, the planning system has actually created (not reduced) flood risk.

Urban planning in England is highly regulated and has often been accused of constraining development or, in some cases, stymieing private sector investments. A 2006 government policy statement attempted to direct development away from areas at highest risk – in simple terms, the intention was to promote building appropriate things in appropriate locations. However, in 2012, the then coalition government published a new National Planning Policy Framework for England, which replaced existing policy and meant there was no longer clear guidance to prevent building in floodplains. To complicate matters further, the Growth and Infrastructure Act in 2013 released large areas of greenfield land for development. The act effectively gave developers a right to submit major planning applications directly to central government, and thus, proposals and decisions could evade not only communities but also local planning authorities. As a result of these legislative changes, there is now a better chance of vulnerable homes being built in flood-prone areas.

The average proportion of new dwellings built in areas of high flood risk has fluctuated annually between 7% and 11%, with some regions, such as London, Yorkshire and Humber and the East Midlands, regularly surpassing these averages. This adds up to more than 300,000 new homes being built since 1989 that are at risk of flooding. Despite a plethora of guidance for planners and apparent restrictions on developers, this building persists.

Over this period, the continued free-market development of floodplains in England has had an unexpected effect. Developers have increasingly been using floodplains to build social housing for low-income families, homes for the elderly/disabled, as well as schools and hospitals. One 2009 study identified 2,374 schools and 89 hospitals in flood-prone areas of England. Planning policy has, thus, caused some of the most vulnerable members of society to occupy highly flood-prone areas.

Source: This box is based on the article by Bosher, L., 2019, "How England's broken planning system has created (not reduced) the risk of floods". *The Conversation*, 21 November 2019. Available at https://theconversation.com/how-englands-broken-planning-system-has-created-not-reduced-the-risk-of-floods-127287

Neo-liberal reforms have been a great motivator for the intense growth in urban populations and have produced an ideological trilogy of competition, deregulation and privatisation. Such ideology is hostile to all forms of spatial regulation, including urban and regional planning, environmental policy, and economic development policies. Powerful interests have suggested that what is needed is complete reliance on market mechanisms for planning and regulation of urban processes. Regulatory controls have simultaneously been reduced (or ineffectively applied) to enable the free market to work, meaning that disaster risks (and other environmental concerns) have been often poorly considered in urban development decisions (Chmutina et al., 2017). Houston, where zoning regulations were not implemented and the hydrology of the city was ignored as it developed, is a case in point. In 2017, Hurricane Harvey devastated the city, where thousands of acres of wetlands that had previously helped absorb rainfall were paved, and in combination with poor flood infrastructure, these actions rendered the city flood prone and unable to sustain the heavy rainfalls that came with the storm.

Land and housing may be reassigned or misappropriated during disaster (Box 4.6 and Figure 4.7). Proof of home or land ownership

Box 4.6: Infrastructure justice

Fig 4.7
QR code for Box 4.6. Available at https://disastersdecon.podbean.com/e/s4e6-infrastructure-justice/

In this episode of *Disasters: Deconstructed* podcast, Dr Marccus Hendricks, Environmental Planning Professor at the University of Maryland and Director of the Stormwater Infrastructure Resilience and Justice (SIRJ) Lab, explains the role of infrastructure in societal justice.

may also be lost, taken or destroyed. In some cases, where customary rights are frequently more dominant than statutory rights, formal proof of ownership or occupation might be rare to begin with. Thus, providing the evidence of the loss of shelter is often difficult for those affected by a disasters; this creates serious difficulties for people when seeking to rebuild and restart their livelihoods. Many shelter recovery programmes require people to demonstrate security of tenure through legal proof of ownership; such restrictive approach to eligibility for assistance excludes large numbers of people, particularly the people who are most vulnerable and, arguably, the most in need, including renters and people living in informal settlements (IFRC, 2018).

Environmental degradation

The examples of the role that environmental degradation plays in creating disaster risks are multiple. The risks may arise from environmental degradation or environmental modification that creates hazards or enhances their impacts.

Deforestation contributes to flooding, landslides and soil erosion. For instance, illegal logging in Pakistan increased the impacts of floods and landslides: in 2010, entire villages had been washed away, 1,600 people had died, 2 million had been made homeless and dissent was leading to political instability. In addition, the felled trees also became a hazard when they were swept away by floodwaters: they were destroying bridges and blocking access to the affected areas (Lewis & Kelman, 2012).

Similarly, infrastructure development in hazard-prone areas that lead to complete change in the environment may have detrimental impact. In Vietnam, thousands have been displaced over the 30 years of hydropower development that degraded the environment and forced many ethnic minority communities into an ever more tenuous situation. Hydropower projects in Lai Chau (completed 2016) and Son La (completed 2012) were designed to maximise profit, and the development approach does not change – even though many marginalised people are routinely killed during disasters that has been enhanced by these projects (von Meding & Thai, 2017).

This section demonstrated the role of space in creating and reducing vulnerability. This role is significant and, thus, has to be carefully considered in the context of DRR and creation.

4.3 Approaching vulnerability

The vulnerability approach to disasters became prominent in the 1980s, challenging hazard-centred and perception-centred approaches. It has been not just focusing on vulnerability seen as weakness, dependency and victimhood but including consideration of self-awareness of capacities, with people living with risk being “powerful claimants with rights, rather than poor victims or passive recipients” (Heijmans, 2004, p. 127) who have valuable knowledge, experience and skill to contribute to social protection and risk reduction (Wisner, 2016). Whilst some (groups of) people are more vulnerable than others, we must not assert that, for instance, women or the vision impaired are vulnerable groups or should be considered at risk in an absolute sense.

The HFA (2005–2015) emphasised that the impacts of disasters on social, economic and environmental conditions should be examined through a system of indicators that will allow assessing vulnerability. A wide range of vulnerability assessments exists: these range from sectoral to multidimensional, demonstrating the distribution of the vulnerability indicators used and disaggregating by sex, age, family size, location and so forth. Measuring vulnerability allows for defining where the greatest need is and setting priorities, determining actions, monitoring progress and analysing trends, measuring effectiveness of mitigation approaches, anticipating undesirable states, informing policymakers and practitioners, alerting the public and raising awareness, stimulating discussion and gaining funding (Birkmann, 2006). However, whilst several methodologies exist, they are often ex ante, limited to specific sectors and largely focused on hazards and risks (whilst neglecting capacities) (UNDRR, 2019). Largely, the approaches are either deterministic (based on modelling) or parametric (using available data to build an impression of the vulnerability of a system) (Balica et al., 2013). The former estimates the vulnerability of a particular place by assessing risk to life or damage based on physical vulnerability, or by assuming a homogeneous vulnerability of the entire population; it relies on a significant amount of detailed topographic, hydrographic and economic data, but often neglects the social dimensions of risk and spatial variation. The latter generally consists of vulnerability metrics based on a large number of assumptions and requires a sensitivity analysis, reliable sources and the subjective manner of interpreting the results; it does, however, allow better understanding vulnerability of a certain demographic group (Mavhura et al., 2017).

Some approaches take a life cycle as a foundation; they allow the identification of risk factors for each group that inform the long-term consequences. For instance, a setback in early childhood has compounding effects throughout the rest of a person's life, in terms of growth, job and social status, and the uncertainties involved with growing older and the transmission of vulnerability to the next generation. Such an approach requires timely and continuous investment to effectively protect those groups whose vulnerability profiles make them more exposed to risks (UNDRR, 2019).

There are many challenges in assessing vulnerability: often, the assessments are driven by a policy decision that only looks at a certain dimension of vulnerability (e.g. women living in disaster-prone areas), largely overlooking other dimensions and – more importantly – root causes (discussed in detail in Chapter 5). Most of the assessments are carried out as projects by international organisations, NGOs and the private sector; once the project is ended, the findings are not always systematically integrated into the overall DRM process. Moreover, external experts do not often engage local communities and vulnerable populations/individuals into the assessments, thus making decisions based on stereotypes of vulnerability rather than measured vulnerability (UNDRR, 2019). Measuring vulnerability is difficult (if not impossible), as vulnerability is not static (for instance, one's vulnerability may increase as they get older). A wide range of quantitative and qualitative approaches to measuring vulnerabilities has been developed, nevertheless; some of which are described in the following sections.

4.3.1 Quantification of vulnerability

Quantitative approaches to vulnerability are based explicitly or implicitly on a model of disaster risk founded on a simple assertion: risk is a function of exposure to a hazard, susceptibility or sensitivity to harm or loss, degree of personal or social protection enjoyed, and capacity to cope or adapt to the impact of the hazard. Many efforts have been taken to develop an index of vulnerability that takes these core components into account, although some see susceptibility, protection and capacity as merely the components of vulnerability. The vulnerability indices are informed by vulnerability indicators that are a proxy for vulnerability characteristics grouped into vulnerability categories (e.g. education could be seen as a category, with education level and access to education being the characteristics). Vulnerability index is used to quantify social vulnerability by measuring individual categories (e.g. education, age and gender) that are aggregated to produce community-level results, and community categories (e.g. population growth and infrastructure quality) are not disaggregated (UNDRR, 2019).

Many of these approaches employ large databases at national, sometimes at sub-national and municipal scales, with variables screened deductively to see which come closest to measuring processes that the case study literature suggests are crucial to vulnerability and/or capacity to cope and to recover (Wisner, 2016). Examples of quantitative approaches to vulnerability include the following:

Social Vulnerability Index has been used to understand the patterns of spatial distribution of mortality. The variables are based on the census information and include socio-economic status (income, poverty, employment and education variables); household composition (age, single parenting and disability variables); minority status (race, ethnicity and English language proficiency); and housing/transportation (housing structure, crowding and vehicle access variables) (Flanagan et al., 2011). Its aim is not to capture profiles of single individuals or buildings; instead, it shows the potential vulnerability of a certain locality, as a profile of general typified demographic profiles, settlement patterns and infrastructure information (Fekete, 2009).

Prevalent Vulnerability Index measures socio-economic fragility and lack of resilience. It employs 24 variables, including population living on below US$1 per day; dependents as a proportion of working-age population; UNDP's Human Poverty Index; social disparity (concentration of income measured by GINI coefficient); Human Development; Gender-Related Development Index; insurance of infrastructure and housing; and hospital beds per 1,000 people (Cardona & Carreño, 2013). These variables allow us to characterise prevalent vulnerability conditions reflected in exposure in prone areas, socio-economic weaknesses and lack of social resilience in general (Cardona, 2005). In addition to these variables, the index also considers national financial impact of a disaster, the impact of localised disaster, and socio-economic fragility and lack of resilience. Cardona (2005) points out that vulnerability is the result of inadequate economic growth, on the one hand, and deficiencies that may be corrected by means of adequate development processes – all of which can be assessed using the index.

WorldRiskIndex uses surrogates available in 171 countries in order to combine exposure (i.e. percentage of population exposed to hazards), susceptibility (population living without adequate sanitation and water supply, population living in slums, population living on less than US$1, undernourished population, dependency ration, GDP per capita and GINI index), coping capacity (corruption index, failed state index, number of doctors and hospital beds per 10,000 people, and percentage of insurance penetration), and adaptive capacity (literacy rate, school enrolment, gender parity in education, life expectancy at birth, public spending on healthcare, state management of natural resources, and state of biodiversity), together to yield a measure of risk. Vulnerability is considered a function of susceptibility, coping capacity and capacity to adapt (UN, 2014). The issue with this approach is that it relies on availability of data, and much of the data that is needed to understand the underlying causes of disasters is often not collected.

Predictive indicators of vulnerability are selected as outcome-based data (e.g. mortality) from EM-DAT. The indicators include economic well-being, health and nutrition, education, physical infrastructure, institutions, governance, conflict and social capital, geographical and demographic factors, dependence on agriculture, natural resources and ecosystems, and technical capacity; with proxies correlated with outcomes of risk. This data may be used to help the understanding of risk and vulnerability, but the authors note that improvements in such data are often needed. The indicators can be used to identify areas for intervention in order to reduce the likelihood and severity of negative outcomes from future climate hazards associated with climate variability and change (Adger et al., 2004).

ENSURE vulnerability and resilience assessment framework provides an interpretation of the relationship between vulnerability and related concepts with a prevention orientation. It establishes a basis for an integrated assessment of vulnerability before a disaster, aiming to aid decision makers and citizens to take appropriate measures. Vulnerability is assessed through four themes related to the natural environment, the built environment, the critical infrastructures and the social/economic systems (Menoni et al., 2012).

Spatial analysis, GIS and modelling aids our understanding of physical configuration as a cause of a disaster by studying the links between the locations where disasters have been reported and the physical attributes of those locations. The vulnerability is estimated based on data that provides information about social (based on census information), physical (i.e. proximity to unsafe location) and economic (usually based on indicators on unemployment and annual income) elements, and critical infrastructure (i.e. number of critical infrastructure elements for or each dissemination area's polygonal area), according to dissemination areas falling within each hazard zone. GIS system allows carrying out the visualisation of the results, but the accuracy of the method depends upon the actual distribution of demographics and infrastructure within each dissemination area. These approaches, however, are useful as analytical tools to identify hotspots of vulnerability, prioritise spatial risks and support spatial decision-making regarding risk-based land-use planning and implementation of DRR strategies (e.g. Armenakis & Nirupama, 2013). A GIS-based approach can show the spatial and temporal evolution of dynamic processes through static maps and matrix of spatial information, as well as the factors that control their behaviour in order to analyse the potential scenarios and to evaluate the impact on buildings and manage them properly (Moe et al., 2000). Through the implementation of hazard classification, some of GIS and modelling approaches allow for identification of critical facilities and the loss estimation for specific hazards when different types of buildings are selected. They provide both a spatial data infrastructure design for collecting, storing and managing critical facilities information, and vulnerability assessment procedures for structural and operational components (e.g. Li et al., 2000).

There is a wide range of quantitative approaches to vulnerability, and the examples provided in this chapter are not exhaustive. Whilst they do not give an answer to addressing the underlying causes of disasters, they have their benefits: global indexing is useful for international organisations and donors, as it aids the willingness to anticipate likely expenditures and to plan interventions. At a national scale, these approaches may be used by governments for efforts in prevention and risk reduction, or least in planning for the worst-case scenario. Quantitative approaches can inform decisions about financial management, insurance and reinsurance, land-use planning, and public and private infrastructure. However, their main limitation is in its iterative thinking that is largely based on the information that is (easily) available (Wisner, 2016). Quantitative indices are descriptive: whilst causation can be seen through data correlation, it strips away all social relations, treating population as a unit rather than as individual human beings. Moreover, most of the indicators are unidimensional (e.g. focusing on gender *or* age), thus not allowing to see the relationship between indicators – and identifying the most vulnerable.

4.3.2 Qualifying social vulnerability

In order to understand the underlying causes of disasters, a more fine-grained knowledge that incorporates people's perceptions and their local knowledge, interaction between social, economic, political, administrative, legal, technological and environmental elements, as well as the complexity, fragility and informality, is needed. A number of qualitative approaches to vulnerability have attempted to provide frameworks for considering vulnerability and asking questions that help to understand susceptibility to loss and harm as well as obstacles to recovery.

Pressure and Release (PAR) model: One of the most common approaches to vulnerability is the Progression of Vulnerability framework, of Pressure and Release model (Blaikie et al., 1994, with many of the ideas dating back to Ian Davis's work in the 1970s.); an example of its use is demonstrated in Figure 4.8.

Root causes may be remote geographically from the local site of vulnerability, such as an investment decision by a distant corporation. They may be remote in time, such as the history of colonialism. Dynamic pressures are normally decadal-scale trends involving business cycles, population dynamics, land use and governance. They translate

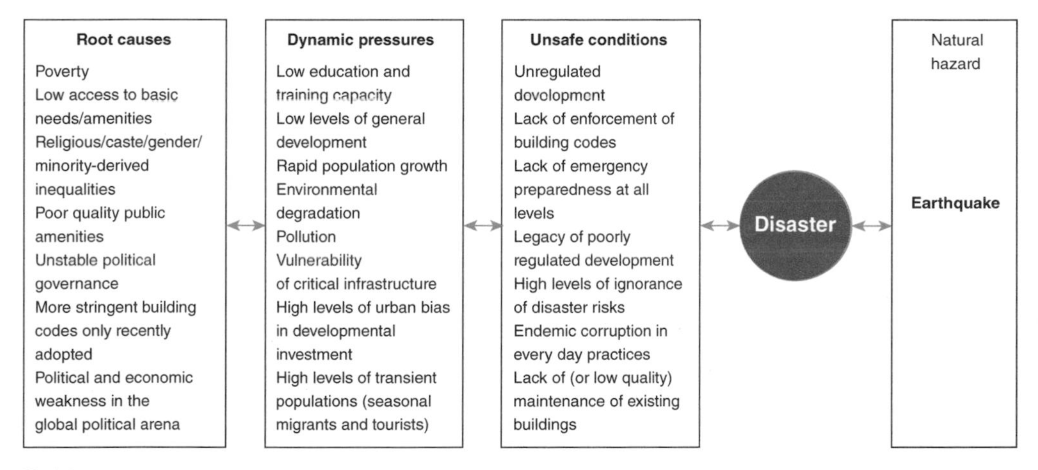

Fig 4.8
Pressure and Release model of the 2015 Gorkha earthquake.
Source: authors

or transmit root causes to local scale and present moment, where they produce unsafe conditions and fragile livelihoods. PAR model shows the relationships amongst these processes and the intersection of scale over time.

Five Components of Vulnerability and Their Main Determinants model (Figure 4.9) has been introduced by Terry Cannon (2008), building up on the PAR model and on the ideas of the Social Determinants of Disaster model.

Cannon breaks vulnerability into five categories: livelihood strength and resilience, well-being and baseline status, self-protection, social protection, and governance – revealing the linkages more dynamically than in the PAR model and pointing out the disconnects that explain the persistence of vulnerability.

Turner's *vulnerability framework* (Turner, Kasperson et al., 2003) aims to facilitate comprehensive vulnerability analysis for sustainability science (still with a focus on natural hazards). The coupled human-environment system is central to the framework, and it is "place-based", recognising the variation in hazards spatially, as well as the complexity and non-linearity of the system (Figure 4.10). The key elements included are as follows:

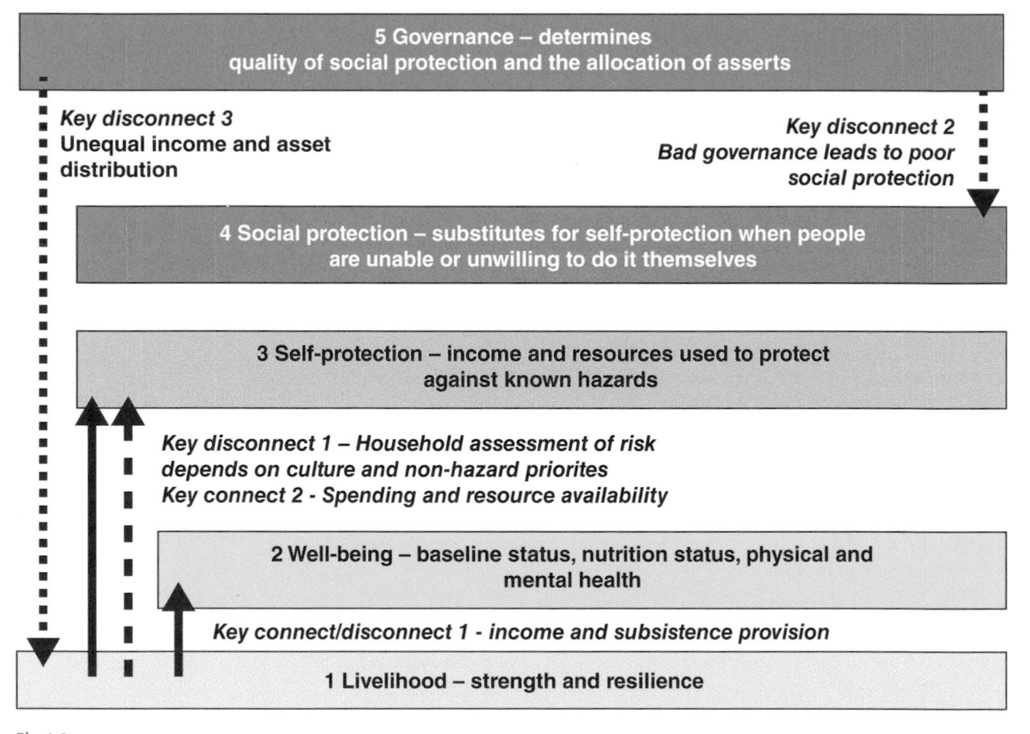

Fig 4.9
Five components of vulnerability and their main determinants.
Source: adapted from Cannon (2008)

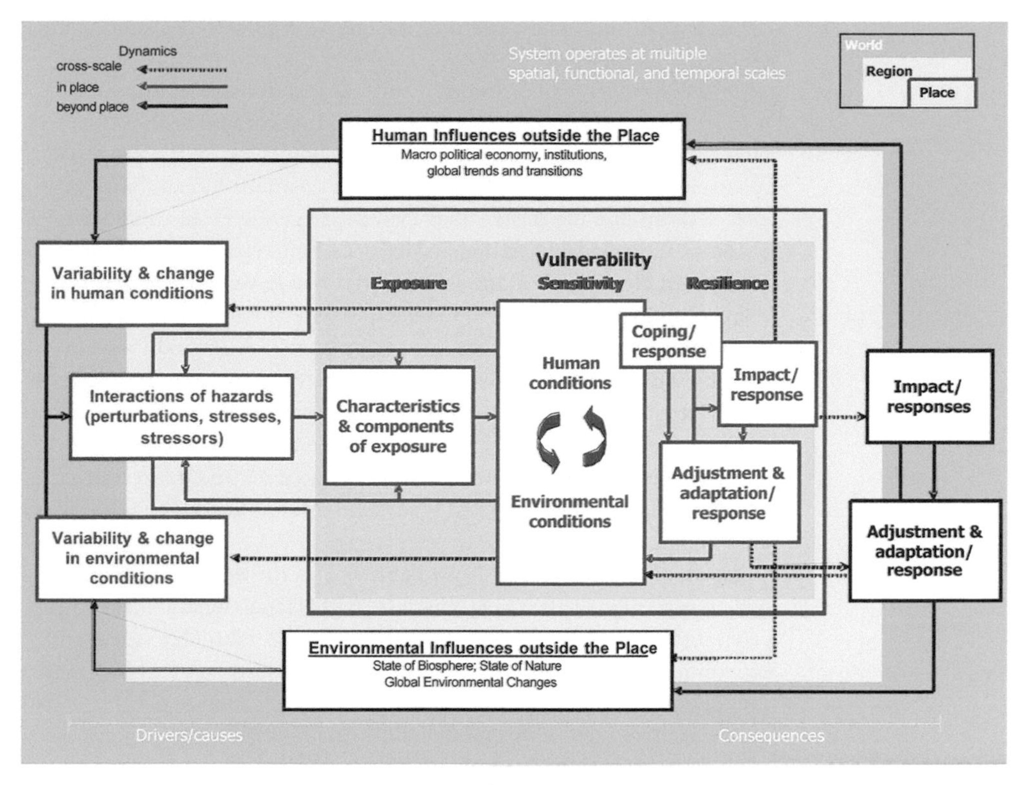

Fig 4.10
Vulnerability framework. Components of vulnerability identified and linked to factors beyond the system of study and operating at various scales.
Source: adapted from Turner II et al. (2003); Copyright (2003) National Academy of Sciences, USA

- Perturbations and stressors (i.e. hazards) can be multiple and interacting, allowing for multi-hazards, and also considers the sequencing;
- Exposure, including characteristics (for example, frequency and duration) and components (for example, households, states and ecosystems);
- Sensitivity of the coupled system to the exposure;
- Capacity of the system to cope or respond (i.e. resilience);
- Adjustments/adaptations (system restructuring after responses); and
- "Nested scales and scalar dynamics of hazards, coupled systems, and their responses" (p. 8075).

Vulnerability analysis using this model may be most useful for decision-making when considering differential vulnerability between components and subsystems, stochastic and nonlinear elements that lead to unexpected or surprise outcomes, the role of institutions in acting as stressors or affecting system sensitivity and resilience, identifying and testing causal structures, developing metrics and measures and developing institutional structures for linking results to decision-making (Turner et al., 2003).

FORIN model takes a forensic approach to disaster risk, exploring it from the social construction perspective. It takes into account physical and biological processes, but largely focuses on the process of development, seeing it as a locus of risk creation. Vulnerability is analysed through the themes that include triggering events, exposure of social and environmental elements, social and economic structures of exposed communities, and institutional and governance elements (Oliver-Smith et al., 2016). FORIN explores vulnerability at two levels: descriptive and analytical. The latter asks questions around loss and damage under determined hazard and exposure conditions, focusing on the different distribution of impacts. The former explores delineation of derived risk drivers (i.e. dynamic processes): population growth and distribution, urban and rural land-use patterns and processes, environmental degradation and ecosystem service depletion, and poverty and income distribution.

Hazards-of-Place model is probably one of the most recognised and widely used models. It is very abstract, with social vulnerability indices generated on the basis of literature around social fabric (i.e. socio-economic, political and institutional causes of social vulnerability) and roughly corresponding census data surrogates to generate (Cutter, 1996). Large datasets used in this model allow fine-tuning to the circumstances in a variety of countries; however, the social fabric elements are not well defined (Figure 4.11).

This model is informed by Social Vulnerability Index (SoVI) (Cutter & Morath, 2013) that are based on various socio-economic and physical elements. This index is, however, descriptive: causes and effects show some evidence of statistical correlation, and quantification remains problematic, as some measurements are difficult (if not impossible) to quantify.

Vulnerability and Capacity Assessment identifies vulnerabilities and capacities at the grassroots level (i.e. communities and single households)

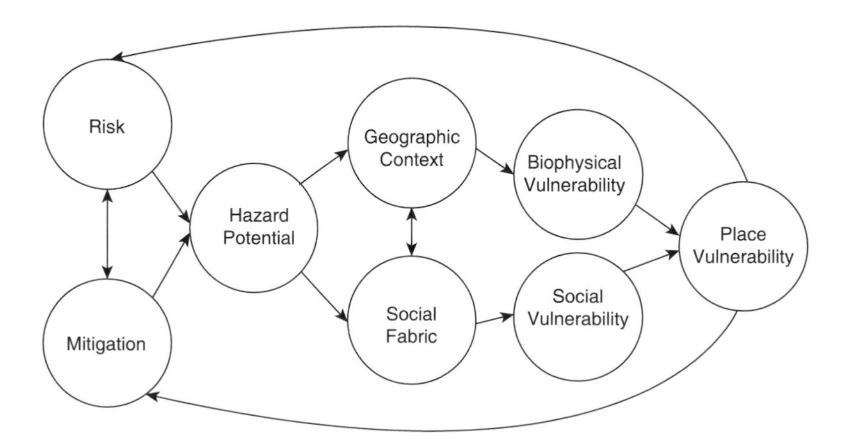

Fig 4.11
Hazards-of-place model. Source: adapted by authors from Cutter (1996)

through qualitative participatory methodologies. This local- and community-based focus allows people to identify and understand their own level of risk as well as local priorities and intervention options (IFRC, 2006). *Community or citizen-based risk assessments* have in recent years become a popular way of approaching social vulnerability assessment. Participatory models in which communities are actively engaged in the research process bridge the gap between the community knowledge, knowledge produced through research and translation of this research into interventions and policies. Such an approach recognises that community is not a setting or location but a social entity with a sense of identity and shared fate, and should, thus, be involved in assessing their own vulnerability and addressing their own priorities for increasing their capacity to prepare for, cope with and mitigate the effects of disasters (Wisner, 2016). However, often, such approaches put more emphasis on revealed vulnerabilities and local risk perception but do not differentiate between drivers and root causes. This model will be further discussed in Chapter 6.

Detecting Disaster Root Causes Framework (DKKV, 2012) examines pre-disaster conditions as well as the DRM activities, looking at disasters as a complex crisis that goes beyond a single event. The framework is informed by the underlying causes rooted in development (criteria related to the national level of development, such as resources and capacities, equipment, knowledge, and education), governance (criteria related to international and national policies, and governance issues with regard to DRM, such as corruption or lack of leadership), awareness and perception (i.e. culture, tradition, or religious norms and beliefs), political environment (i.e. civil war impacts or political instabilities), and physical and environmental conditions (i.e. natural climate variability, climate change and geophysical conditions) inform the framework.

The models described in this section cannot be used for measuring vulnerability – instead, they all aim at assessing it. Such assessments are delivered through asking questions that may challenge some of the decisions that lead to disaster risk creation.

4.4 Concluding remarks

We now have enough scientific information to understand why hazards occur. But why do disasters occur? In order to answer this question, it is important to answer many other whys: Why do many poor people live in hazard-prone areas? Why is the coping capacity of some groups of people more limited than the others? Why is power and access to resources not distributed equally? These and other whys help us explore the underlying causes that turn a hazard into a disaster.

A hazard may have a smaller or a bigger impact on people depending on the level of their vulnerability. Many people have become very resilient, being able to bear losses or adapt to risk in whatever manner is necessary. It is, thus, important to understand the social and economic factors that lead to a disaster rather that focus on a hazard. Similar hazards (in terms of intensity) may have very different outcomes: one may turn into a disaster (in terms of loss of life and assets, as well as disruption to livelihoods and future well-being), whereas the other would cause minor disruption and easy recovery.

Vulnerability has been given many contested definitions, and the various causes of vulnerability have been sought out. As this chapter demonstrated, there is a wide range of models that attempt to understand how vulnerability interacts with other processes that, when combined together, turn a hazard into a disaster. Vulnerability has been and will be quantified, indexed and mapped as an aid to decision-making at various scales; and it has been and will be qualitatively assessed as an aid for cooperation and mobilisation. As perfectly summarised by Birkmann and Wisner (2006, p. 7), "The term measuring vulnerability does not solely encompass quantitative approaches, which is what first comes to mind. It also seeks to discuss and develop all types of methods able to translate the abstract concept of vulnerability into practical tools to be applied in the field. If one takes the bare bones, simple definition as 'subject to harm', then the questions rapidly proliferate and become concrete and situational: What kind of harm? Harm from what? How often? Recoverable or treatable harm? Avoidable harm? and above all, Under what economic, social, and political conditions?".

It is important to remember that understanding and reducing vulnerability does not demand accurate predictions of the incidence of extreme events (Sarewitz et al., 2003). Until we focus on transforming a system based on greed, conflict and destruction of nature, we will not be able to address the root causes of vulnerability. It is also important to remember that disasters are never natural – and that by blaming the nature for creating disasters, we allow ignorance, carelessness and greed to be masked by focusing on natural processes (Chmutina et al., 2019). It is, thus, important to remind ourselves that disasters occur not because of the nature but because of us, human beings – and to paraphrase Heijmans (2001), they will keep occurring until our governments stop ignoring the social and political origins of disasters.

Take-away messages

1 Disasters are a result of a combination of hazards and social and human vulnerability, including development activities that are ignorant of local hazardous conditions. Thus, disasters are not natural.
2 Inequality, poverty, political ideology, class and power relations are the root causes of vulnerabilities that turn natural hazards into disasters, making some more vulnerable than others.
3 In areas prone to multiple hazards, the impact of one hazard can be higher than the other, depending on particular characteristics of population.

4 Technical and spatial planning interventions are often seen as necessary for DRM. However, spatial and technological solutions are often costly and do not protect everyone equally; in some cases, they actually enhance vulnerability of some groups living in certain locations in order to increase resilience of other – often more affluent – groups.
5 There is a wide range of qualitative and quantitative assessments of vulnerability. But reducing vulnerability does not require accurate predictions of the incidence of extreme events. Instead, it requires focusing on transforming a system of oppressions and marginalisation (i.e. the root causes of vulnerability).

To learn more about the topic discussed in this chapter, listen to the *Disasters: Deconstructed* interview with Dr David Prevatt (Figure 4.12).

Fig 4.12
QR code for Chapter 4.

Further suggested reading

Blaikie, P., Cannon, T., Davies, I., & Wisner, B. (1994). *At Risk: Natural Hazards, People's vulnerability, and Disaster*. Routledge.
Bosher, L., & Chmutina, K. (2017). *Disaster risk reduction for the built environment: An introduction*. Wiley.
Cutter, S. L. (1996). Vulnerability to environmental hazards. *Progress in Human Geography*, *20*(4), 529–539. https://doi.org/10.1177/030913259602000407
Kelman, I. (2020). *Disaster by Choice: How our actions turn natural hazards into catastrophes*. Oxford University Press.
Lowenstein, A. (2015). *Disaster capitalism*. Verso.
Lizarralde, G. (2021). *Unnatural disasters*. Columbia University Press.
O'Keefe, P., Westgate, K., & Wisner, B. (1976). Taking the naturalness out of natural disasters. *Nature*, *260*(5552), 566–567. https://doi.org/10.1038/260566a0
Oliver-Smith, T. (1986). *The martyred city: Death and rebirth in the Andes*. University of New Mexico Press.

References

Adger, W. N., Brooks, N., Bentham, G., Agnew, M., & Eriksen, S. (2004). *New indicators of vulnerability and adaptive capacity*. Technical Report 7. Tyndall Centre for Climate Change Research.

Ahmed, B., Orcutt, M., Sammonds, P., Burns, R., Issa, R., Abubakar, I., & Devakumar, D. (2018). Humanitarian disaster for Rohingya refugees: Impending natural hazards and worsening public health crises. *Lancet. Global Health*, *6*(5), e487–e488. https://doi.org/10.1016/S2214-109X(18)30125-6

Alexander, D. (2000). *Confronting catastrophe*. Terra Publishing.

Alexander, D. (2001). *Natural disasters*. Routledge.

Armenakis, C., & Nirupama, N. (2013). Estimating spatial disaster risk in urban environments. *Geomatics, Natural Hazards and Risk*, *4*(4), 289–298. https://doi.org/10.1080/19475705.2013.818066

Balica, S. F., Popescu, I., Beevers, L., & Wright, N. G. (2013). Parametric and physically based modelling techniques for flood risk and vulnerability assessment: A comparison. *Environmental Modelling and Software*, *41*, 84–92. https://doi.org/10.1016/j.envsoft.2012.11.002

Ball, N. (1975). The myth of the natural disaster. *Ecology*, *5*(10), 368–369.

Bene, C. (2013). The climate change—Migration—Urbanisation Nexus: Workshop report. *Evidence Report. IDS*, *15*.

Birkmann, J. (Ed.). (2006). *Measuring vulnerability to natural hazards: Towards disaster resilient societies*. United Nations University Press.

Birkmann, J., & Wisner, B. (2006). Measuring the un-measurable: The challenge of vulnerability. Source 5. *Security*. United Nations University Press, Institute for Environment and Human.

Blaikie, P., Cannon, T., Davis, I., & Wisner, B. (1994). *At risk: Natural hazards, people's vulnerability and disasters*. Routledge.

Cannon, T. (1994). Vulnerability analysis and the explanation of "natural disasters". In A. Varley (Ed.), *Disasters, development and environment*. Wiley & Sons.

Cannon, T. (2008). *Reducing people's vulnerability to natural hazards: Communities and resilience*. Research Papers. World Institute for Development Economics Research, 2008/34.

Cardona, O. D. (2003). The need for rethinking the concepts of vulnerability and risk from a holistic perspective: A necessary review and criticism for effective risk management. In G. Bankoff, G. Frerks, & D. Hillhorst (Eds.), *Mapping vulnerability: Disasters, development and people*. Earthscan Publications.

Cardona, O. D. (2005). *A system of indicators for disaster risk management in the Americas*. Prepared for: 250th Anniversary of the 1755 Lisbon Earthquake. https://www.unisdr.org/2005/HFdialogue/download/tp3-paper-system-indicators.pdf

Cardona, O., & Carreño, M. (2013). System of indicators of disaster risk and risk management for the Americas: Recent updating and application of the IDB-IDEA approach. In J. Birkmann (Ed.), *Measuring vulnerability to natural hazards* (2nd ed., pp. 251–276). United Nations University Press.

Chmutina, K., & Bosher, L. (2017). Rapid urbanisation and security: Holistic approach to enhancing security of urban spaces. In R. Dover,

H. Dylan, & M. Goodman (Eds.), *The Palgrave handbook of security, risk and intelligence* (pp. 27–45). Palgrave Macmillan.

Chmutina, K., & Rose, J. (2018). Building resilience: Knowledge, experience and perceptions among informal construction stakeholders. *International Journal of Disaster Risk Reduction*, *28*, 158–164. https://doi.org/10.1016/j.ijdrr.2018.02.039

Chmutina, K., von Meding, J., & Bosher, L. (2019). *Language matters: Dangers of the "natural disaster" misnomer*. Contributing paper to the Global Assessment Report (GAR). UNDRR. https://www.preventionweb.net/publications/view/65974, *2019*.

Chmutina, K., von Meding, J., Gaillard, J. C., & Bosher, L. (2017). *Why natural disasters aren't all that natural*. OpenDemocracy. Retrieved September 14, 2017, from https://www.opendemocracy.net/ksenia-chmutina-jason-von-meding-jc-gaillard-lee-bosher/why-natural-disasters-arent-all-that-natural.

CRED. (2019). *The international disaster database*. https://www.emdat.be/

Cutter, S. L. (1996). Vulnerability to environmental hazards. *Progress in Human Geography*, *20*(4), 529–539. https://doi.org/10.1177/030913259602000407

Cutter, S. L., Boruff, B. J., & Shirley, W. L. (2003). Social vulnerability to environmental hazards. *Social Science Quarterly*, *84*(2), 242–261. https://doi.org/10.1111/1540-6237.8402002

Cutter, S. L., & Morath, D. (2013). The evolution of social vulnerability index. In J. Birkmann (Ed.), *Measuring vulnerability to natural hazards* (2nd ed., pp. 304–321). United Nations University Press.

Donner, W., & Rodriguez, H. (2008). Population composition, migration and inequality: The influence of demographic changes on disaster risk and vulnerability. *Social Forces*, *87*(2), 1089–1114. https://doi.org/10.1353/sof.0.0141

Fekete, A. (2009). Validation of a social vulnerability index in context to river-floods in Germany. *Natural Hazards and Earth System Sciences*, *9*(2), 393–403. https://doi.org/10.5194/nhess-9-393-2009

Flanagan, B. E., Gregory, E. W., Hallisey, E. J., Heitgerd, J. L., & Lewis, B. (2011). A social vulnerability index for disaster management. *Journal of Homeland Security and Emergency Management*, *8*(1), article 3. https://doi.org/10.2202/1547-7355.1792

German Committee for Disaster Reduction. (DKKV). (2012). *Detecting disaster root causes—A framework and an analytic tool for practitioners* (p. 148). DKKV publication.

Heijmans, A. (2001). *Vulnerability: A matter of perception*. Working Paper 4. Benfield Greig Hazard Research Centre, University College London.

Heijmans, A. (2004). From vulnerability to empowerment. In G. Bankoff, G. Frerks, & D. Hilhorst (Eds.), *Mapping vulnerability: Disasters, development and people* (pp. 115–127). Earthscan Publications.

Heijmans, A. (2017). 'Vulnerability': A matter of perception? In I. Kelman, J. Mercer, & J. C. Gaillard (Eds.), *The Routledge handbook of disaster risk reduction including climate change adaptation* (pp. 48–49). Routledge.

Hewitt, K. (Ed.). (1983). *Interpretations of calamity from the viewpoint of human ecology*. Allen & Unwin.

Hewitt, K. (2012). Ch. 8. Culture, hazard and disaster. In B. Wisner, J. C. Gaillard, & I. Kelman (Eds.), *The Routledge handbook of hazards and disaster risk reduction*. Routledge.

Hoogvelt, A. M. M. (1976). *The sociology of developing societies*. Macmillan.

International Federation of Red Cross and Red Crescent Societies. (IFRC). (2006). *What is a VCA? An introduction to vulnerability and capacity assessment*. IFRC.

International Federation of Red Cross and Red Crescent Societies. (IFRC). (2018). *Leaving no one behind*. World Disasters Report.

Jha, K. K. (2002). *Informal labour in the construction industry in Nepal*. International Labour Organization, Sectoral Activities Programme, Working Paper 187.

Johnson, C., Bosher, L., Adekalan, I., Jabeen, H., Kataria, S., Wijitbusaba, A., & Zerjav, B. (2013). *Private sector investment decisions in building and construction: Increasing, managing and transferring risks*. Working Paper for the Global Assessment Report 2013 on Disaster Risk Reduction. UNISDR.

Kelman, I. (2010). *Natural disasters do not exist (natural hazards do not exist either)* version 3, 9 July 2010 (version 1 was 26 July. 2007). http://www.ilankelman.org/miscellany/NaturalDisasters.rtf

Lavigne, F., de Coster, B., Juvin, N., Flohic, F., Gaillard, J. C., Texier, P., Morin, J., & Sartohadi, J. (2008). People's behaviour in the face of volcanic hazards: Perspectives from Javanese communities, Indonesia. *Journal of Volcanology and Geothermal Research*, *172*(3–4), 273–287. https://doi.org/10.1016/j.jvolgeores.2007.12.013

Lewis, J., & Kelman, I. (2012, June 21). *The good, the bad and the ugly: Disaster risk reduction (DRR) versus disaster risk creation (DRC)* (1st ed.). PLOS Currents: Disasters. https://doi.org/10.1371/4f8d4eaec6af8

Li, Y., Brimicombe, A. J., & Ralphs, M. P. (2000). Spatial data quality and sensitivity analysis in GIS and environmental modelling: The case of coastal oil spills. *Computers, Environment and Urban Systems*, *24*(2), 95–108. https://doi.org/10.1016/S0198-9715(99)00048-4

Lowenstein, A. (2015). *Disaster capitalism*. Verso.

Mascarenhas, A., & Wisner, B. (2012). Ch. 5. Politics: Power and disasters. In B. Wisner, J. C. Gaillard & I. Kelman (Eds.), *The Routledge handbook of hazards and disaster risk reduction*. Routledge.

Master, R. D., & Kelly, C. (Eds.). (1992). *The collected writings of Rousseau*, 3. University Press of New England.

Mavhura, E., Manyena, B., & Collins, A. E. (2017). An approach for measuring social vulnerability in context: The case of flood hazards in Muzarabani district, Zimbabwe. *Geoforum*, *86*, 103–117. https://doi.org/10.1016/j.geoforum.2017.09.008

McCallin, B., & Schrerer, I. (2015). *Urban informal settlers displaced by disasters: Challenges to housing responses* [International Displacement Monitoring Centre report]. http://www.internal-displacement.org/sites/default/files/publications/documents/201506-global-urban-informal-settlers.pdf

Menoni, S., Molinari, D., Parker, D., Ballio, F., & Tapsell, S. (2012). Assessing multifaceted vulnerability and resilience in order to design risk-mitigation strategies. *Natural Hazards*, *64*(3), 2057–2082. https://doi.org/10.1007/s11069-012-0134-4

Moe, K. A., Skeie, G. M., Brude, O. W., Løvås, S. M., Nedrebø, M., & Weslawski, J. M. (2000). The Svalbard intertidal zone: A concept for the use of GIS in applied oil sensitivity, vulnerability and impact analyses. *Spill Science and Technology Bulletin*, 6(2), 187–206. https://doi.org/10.1016/S1353-2561(00)00038-4

O'Keefe, P., Westgate, K., & Wisner, B. (1976). Taking the naturalness out of natural disasters. *Nature*, *260*(5552), 566–567. https://doi.org/10.1038/260566a0

Oliver-Smith, A., Alcántara-Ayala, I., Burton, I., & Lavell, A. (2016). *Forensic investigations of disasters: A conceptual framework and guide to research*. IRDR.

Oliver-Smith, T. (1986). *The martyred city: Death and rebirth in the Andes*. University of New Mexico Press.

Oxfam. (2010). *Haiti: A once-in-a-century chance for change. Beyond reconstruction: Re-envisioning Haiti with equity, fairness, and opportunity*. Oxfam Briefing Paper 136.

Sanderson, D. (2000). Cities, disasters and livelihoods. *Environment and Urbanization*, *12*(2), 93–102. https://doi.org/10.1177/095624780001200208

Sarewitz, D., Pielke, Jr., R., & Keykhah, M. (2003). Vulnerability and risk: Some thoughts from a political and policy perspective. *Risk Analysis*, *23*(4), 805–810. https://doi.org/10.1111/1539-6924.00357

Satterthwaite, D. (2008). *Climate change and urbanisation: Effects and implications for urban governance*. UN/POP/EGM-URB/2008/16, New York, USA.

Schuller, M. (2016). *Humanitarian aftershocks in Haiti*. Rutgers University Press.

Smith, N. (2005). There's no such thing as a natural disaster. *Understanding Katrina: Perspectives from the social sciences*. http://understandingkatrina.ssrc.org/Smith

Sun, L., & Faas, A. J. (2018). Social production of disasters and disaster social constructs: An exercise in disambiguation and reframing. *Disaster Prevention and Management*, *27*(5), 623–635. https://doi.org/10.1108/DPM-05-2018-0135

Turner, B. L., Kasperson, R. E., Matson, P. A., McCarthy, J. J., Corell, R. W., Christensen, L., Eckley, N., Kasperson, J. X., Luers, A., Martello, M. L., Polsky, C., Pulsipher, A., & Schiller, A. (2003). A framework for vulnerability analysis in sustainability science. *Proceedings of the National Academy of Sciences of the United States of America*, *100*(14), 8074–8079. https://doi.org/10.1073/pnas.1231335100

UNDRR. (2019). *Global assessment report*. United Nations. https://gar.unisdr.org/sites/default/files/reports/2019-05/full_gar_report.pdf

UNISDR. (2018). *Terminology*. https://www.unisdr.org/we/inform/terminology#letter-d

United Nations. (2014). *World urbanisation prospects: The 2014 revision, highlights, ST/ESA/SER.A/352*. UN Publications.

Von Meding, J., & Thai, H. (2017, August 29). In Vietnam poverty and poor development, not just floods, kill the most marginalised. *The Conversation*. https://theconversation.com/in-vietnam-poverty-and-poor-development-not-just-floods-kill-the-most-marginalised-82785.

Wamsler, C. (2004). Managing urban risk: Perceptions of housing and planning as a tool for reducing disaster risk. *Global Built Environmental Review*, *4*(2), 11–28.

White, G. F. (1945). *Human adjustment to floods*. Department of Geography Research Paper, University of Chicago, Chicago.

Wisner, B. (2010). Marginality. In P. Bobrowky (Ed.), *Springer encyclopedia of natural hazards*. Springer. Entry 226.

Wisner, B. (2012). Ch. 7. Violent conflict, natural hazard and disaster. In B. Wisner, J. C. Gaillard, & I. Kelman (Eds.), *The Routledge handbook of hazards and disaster risk reduction*. Routledge.

Wisner, B. (2013). Disaster risk and vulnerability reduction. In L. Sygna, K. O'Brien, & J. Wolf (Eds.), *A changing environment for human security* (pp. 257–276). Routledge.

Wisner, B. (2016). *Vulnerability as concept, model, metric, and tool*. Oxford Research Encyclopedia of Natural Hazard Science.

Wisner, B. (2018). Ch. 38. Core elements of natural hazard mitigation: Natural hazard mitigation and preparedness practices. In B. J. Gerber (Ed.), *Oxford encyclopedia of natural hazards governance*. Oxford University Press.

Wisner, B., Gaillard, J. C., & Kelman, I. (2012). Ch. 3. Framing disaster: Theories and stories seeking to understand Hazards, vulnerability and risk. In B. Wisner, J. C. Gaillard & I. Kelman (Eds.), *The Routledge handbook of hazards and disaster risk reduction*. Routledge.

CHAPTER 5

People's vulnerability

Fig 5.1
Bonds of love.
Source: photo courtesy of Jason von Meding

5.1 What is vulnerability and how do people become vulnerable?

'Vulnerability' is one of the most widely used terms in disaster studies – but it is also highly contested. Originating from the word 'wound', there are now many definitions of vulnerability adopted by various disciplines. In disaster studies, the most common definition is from UNDRR (2018): "the conditions determined by physical, social, economic and environmental factors or processes which increase the susceptibility of an individual, a community, assets or systems to the impacts of hazards". This definition, however, prompts an in-depth discussion on how these conditions come to be – and why some individuals and communities are more susceptible than others (Figure 5.1).

Bankoff (2019) provides a historic window into how the term became so prominent in disaster studies: in the 1970s, many Western programmes aimed at the Third World were criticised for focusing on the emergency rather than the reasons for the emergency and finding the ways of lifting people out of poverty. The term 'vulnerability' shifted the focus from hazards to the conditions that made communities unsafe and emphasised not only the exposure to hazards but, more importantly, people's capacity to recover from loss. The critique was unequivocal: colonial heritage, unequal power relationships, unfair developmental

DOI: 10.4324/9781315469614-8

policies as well as neoliberal interests make some communities and individuals less able to deal with disasters, leaving them at risk. Vulnerability, thus, offered a way of critiquing developmentalism, capitalism, consumerism, materialism – and all other isms that have become prominent in showing economic progress that was used to assess the state of various nations and that, at the same time, was precisely the root cause of many vulnerabilities.

Many developmental policies introduced after the Second World War were conceived as a way of increasing economic development of the Third World – as this was the only way to develop (or otherwise be seen as backward and poor). Many development projects (such as building dams, mines, tourist resources and other facilities that would enhance economic development) turned prime agricultural land or coastal areas into industrial and commercial areas without community consultations, resulting in displacement of local communities. Many lost not only their land, homes and livelihoods, but also their identity and roots (Bankoff, 2019).

Vulnerability is a complex system of characteristics related to human experience, geographical location, power, politics, money and so on. Disasters do not affect all people, communities and societies equally: the social groups who have been pushed to the margins of society disproportionately feel the impacts of disasters. This chapter attempts to unpack some of these complexities.

5.1.1 Vulnerability and structural violence

Structural violence – a concept that refers to the avoidable (but often invisible) political, economic, cultural, legal and other limitations "society places on groups of people that constrain them from achieving the quality of life that would have otherwise been possible" (Lee, 2016, p. 110) – has only recently started to be considered in the context on vulnerability. These limitations are embedded in social structures and are, therefore, often perceived to be ordinary difficulties that everyone encounters in the course of life, such as issues with access to healthcare, education and politics. Lee (2016) provides an excellent summary: "Structural violence directly illustrates a power system wherein social structures or institutions cause harm to people in a way that results in mal-development or deprivation. Because it is a *product of human decisions* and not natural occurrences, and because it is correctable and preventable through human agency, there is increasing advocacy that we call it violence, rather than simply social injustice or oppression" (p. 110, authors' emphasis). The harm may be unintentional: it often occurs through performing our regular duties as the structure defines them (Galtung, 1985). Health, racial and gender disparities, poverty, and denied access to education are all manifestations of structural violence. It is critical to bear this argument in mind as it demonstrates perfectly why understanding vulnerability – and its root causes (Box 5.1, Figure 5.2a and Figure 5.2b) – are important if we want to avoid disaster risk creation.

Box 5.1: Root causes of disasters

This two-part episode of *Disasters: Deconstructed* podcast is a conversation with Anthony Oliver-Smith, Emeritus Professor of Anthropology at the University of Florida, about root causes of disasters.

Fig 5.2a
QR code 1 for Box 5.1. In this first part (of two), the discussion focuses on what root causes are and why it is so important to look beneath the symptoms when disaster impacts are analysed, illustrating how disaster are anything but natural.

Fig 5.2b
QR code 2 for Box 5.1. The second part of the discussion focuses on how society constructs risk using the example of Florida, as well as language, politics and the media, in relation to root causes of disasters. Available at https://disastersdecon.podbean.com/e/s1e10-root-causes-part-1/ and https://disastersdecon.podbean.com/e/s1e11-root-causes-part-2/

Vulnerability is linked with the concepts of conflict and inequality, in the context of the common struggle over the capital; note, however, that capital is not just about the money. *Cultural capital* (acquired through, for instance, education) plays an important role in acquiring a status in the society. Deficits in *linguistic capital* may lead to misunderstandings of warnings or difficulties in applying for disaster relief aid or generally engaging with the disaster-related advice. *Social capital* represents both the number and value of connections one has within a network; it often lacks amongst the recent immigrants, as they lack a certain community connectedness known to be important in a disaster (Donner & Rodriguez, 2008).

Disasters magnify pre-existing social inequalities; as a result of a disaster, those who are already on the margins of society become even more vulnerable. However, as already noted in Chapter 4, vulnerability and poverty are not the same – although people living in poverty may be caught in protracted cycles of unemployment or underemployment, low productivity and low wages, all of which increases their vulnerability (UNDRR, 2019).

5.1.2. Vulnerability and human rights

Vulnerability plays a core role in human rights agenda. Human rights have emerged to recognise vulnerability: all humans feel pain and can suffer; they are dependent on one another to grow and mature, to become autonomous individuals, and to be cared for in illness and ageing; they are socially connected, as this provides social support and legal protection (ten Have, 2018). Vulnerability, thus, demands that humans build social and political institutions to provide collective security, and human rights demand that people learn to empathise with others and to think of others as equals (Hunt, 2007). Turner (2006) points out that human rights are universal principles because vulnerability is shared and, thus, constitutes a common humanity; it connects us at different levels: from the rights of individual human beings to social rights of citizens through social institutions and arrangements. However, humans can never be completely protected and made invulnerable.

It is critical that vulnerability and human rights are considered together in the context of a disaster (ten Have, 2018). Four categories of human rights are at stake in disasters – all of which are related to vulnerability:

- The right to the protection of life is the priority of disaster relief and of human efforts;
- The rights related to basic necessities, such as food, health, shelter and education, and the right to health;
- The rights related to more long-term economic and social needs (housing, land, property and livelihood); and
- The rights related to other civil and political protection needs (documentation, movement and freedom of expression).

Connecting vulnerability and human rights allows emphasising equality – and providing assistance on the basis of need. If any of these rights are ignored, vulnerabilities will be reinstated and enhanced – and the root causes of disasters recreated.

5.2 Vulnerability and exposure

In general, when attempting to understand vulnerability, it is important to consider exposure to risk. Vulnerable households are typically more exposed to risk and less protected from it (UNDRR, 2019): consider, for instance, informal settlements located in flood-prone areas – such exposure directly affects their socio-economic status and welfare. Attempts to reduce exposure – by, for instance, moving away from the sources of risk that may at the same time be the sources of income (e.g. coastal communities dependent on fisheries) – could increase the likelihood of falling into poverty.

Exposure can be structural, developmental or environmental (UNDRR, 2019). Structural exposure requires considering the likelihood of a hazard occurring as well as the nature of construction, building use, materials, population density and so forth. For instance, despite sound building codes and regulations introduced in the early 1990s, more than 98% of buildings in Nepal are constructed by informally employed workers (Figure 5.3). Consequently, most residential buildings are not designed with earthquake resistance in mind. Moreover, although a system of building permits exists in most municipalities, there is no

Fig 5.3
Reconstruction in Banepa after the 2015 earthquake.
Source: photo courtesy of Ksenia Chmutina

provision for checking the submitted plans against the strength criteria. This applies not only to newly built stock but also to the reconstruction of damaged buildings and the retrofit of existing building stock. Additionally, there is poor institutional and technical capacity within the local authorities for implementing strength-related provisions, even if they were introduced into the building permit process (Chmutina & Rose, 2018).

Considering structural exposure, however, is not sufficient: some hazards (such as droughts, epidemics, agricultural infestations) do not damage structures, but their direct and indirect economic cost could be devastating. It is, thus, important to take into account developmental and environmental exposure.

Risk is not static, and this is why it is also important to consider developmental exposure (i.e. how exposure links to growth). For instance, exposure can be changed – and thus, the risk increased – if a three-storey building can become five storeys over the course of a few weeks without sufficient implementation of building regulations or appropriate materials. In low-income countries, growing middle classes and expanded access to the global market are fuelling growth of assets. This has been occurring in the context of neoliberal policymaking, with urban areas rapidly developing due to the state's focus on enabling investments in construction through the provision of infrastructure, financial mechanisms and act of making land available for development but with little consideration for environmental concerns and disaster risks (Chmutina et al., 2017).

Environmental exposure considers systems that are difficult to quantify: for instance, in the last 20 years, approximately 20% of the productivity of the Earth's vegetated surface has been reduced due to climate change, biodiversity loss, poor management practices, overharvesting of resources and land-use changes. More than half of the world's ecosystem services are in decline (UNDRR, 2019). This has a major effect on risk reduction and the mitigation of environmental hazards. Such exposure is often created by developmental projects. For example, the devastating impact of Hurricane Mitch, in 2008, on communities living on the Casita's volcano slopes was a result of such a project. In the first half of the 20th century, the farmers were encouraged to support Nicaraguan agricultural exports: the farmers cleared trees on the slopes to facilitate cattle production and, later on, expanded cotton production in the valley. This led to accelerated deforestation rates and the increase in land ownership, with households being pushed towards living on the slopes. Whilst some redistribution reforms had taken place in the 1990s, the majority of the small farmers remained living on the slopes. Hurricane Mitch brought heavy rains that produced lahars of mud, rock and trees that sped down the slopes, destroying more than 1,500 houses and killing over 2,500 people (Bacon, 2012).

Remote location can also be considered here: of course, not all people living in the location that are hard or nearly impossible to reach are equally vulnerable; yet some physical (e.g. terrain) as well as political (e.g. restrictions of movement and conflicts) factors that prevent an access to a location may deepen the root causes of vulnerabilities. A combination of distance, challenging terrain and lack of transport can have a detrimental impact on a post-disaster response. For instance, in Nepal's mountainous regions, 25% of households have to travel more than an hour to reach a health facility; only 40% of births have a skilled birth attendant present, which significantly increases infant and post-neonatal mortality (Nepal Ministry of Health, 2016).

Marginalised indigenous and minority ethnic groups often live in remote locations where physical isolation and social exclusion can reinforce each other. This can also lead to inadequate housing and food insecurity; moreover, these communities have significantly less access to health services, frequently resulting in health crises (IFRC, 2018). Some communities living in remote locations can become vulnerable if their livelihood – for instance, agriculture – is destroyed by a natural hazard.

There are different dimensions to exposure, but all are closely linked with the idea of human ecological dimension, suggesting that the unique environment within which society, community and/or individual are located should be taken into account when exploring vulnerability (Donner & Rodriguez, 2008). As has been already noted in Chapter 4, people have always lived in hazard-prone areas. Vulnerability originates in a human experience manifested in a potential susceptibility to damage (Cardona, 2003); thus, a series of extreme (yet often permanent) conditions make some social groups (and individuals) particularly fragile – but at the same time, what constitutes vulnerability for one group, may provide livelihood for another. For instance, floods frequently affect poor people who live on riverbanks in Jakarta. But at the same time, flood is a necessary condition in creating a viable livelihood: floods create a cheap place to live in Central Jakarta and keep land-grabbers at bay (Hellman, 2015). Thus, the main challenge people living in these areas are facing is not the risk of being flooded but a risk of relocation that would destroy their livelihood.

5.3 Who's vulnerable?

In order to understand vulnerability, it is critical to understand what people are vulnerable to. Many approaches to measuring vulnerability have been introduced over the years; some are described in Chapter 4. Understanding of vulnerability has been made particularly clear in the Pressure and Release model (Blaikie et al., 1994): the model shows the extent to which a particular social order puts people at risk and that vulnerability is reproduced over time. The *root causes* reflect high level of historical distribution and exercise of power that marginalises certain groups; the *dynamic pressures* reflect contemporary issues, such as conflicts, foreign debt, epidemics, urbanisation and environmental degradation; and the *unsafe conditions* show hazardous living conditions, inadequate food and water sources, and dangerous livelihoods that some people are exposed to; all these showing how disasters are directly linked not only to a natural hazard but largely to the extent to which a society is put at risk.

Amartya Sen defined 'vulnerable populations' as those "liable to serious hardship" (Sen, 1999); the hardship is often exposed by disasters. Different disasters do not impact all communities and societies equally: for instance, women die more frequently than men in coastal storms and tsunamis; women also suffer domestic violence and other forms of gender violence and insecurity after disasters. In Nepal, after the 2015 earthquake, widows have had difficulty obtaining grants to rebuild houses because all documentation was in their husband's name (Jackson et al., 2016). Many more Nepali women have been trafficked into slavery due to resource shortages (Preiss & Shahi, 2017).

The elements of vulnerability are complex and intertwined; they change over a person's life cycle. Vulnerabilities may emerge, change, compound and persist over long periods, passing from generation to generation – and widening inequalities. Thus, each individual, household, social group or community is more and less vulnerable over the course of time (due to a change in season, life cycle, economic cycle or political cycle). They are also more and less vulnerable to a wide variety of everyday risks, such as crime, violence, ill health and unemployment, as well as to less-frequent large-scale events, such as civil war, chemical accident, earthquake, flood or epidemic. The interaction of everyday and large-scale threats in a temporal context of multifaceted change demands understanding of people's situations, not their category. It is impossible to assert that women or people living with AIDS or the elderly are a vulnerable group or should be considered at risk in an absolute sense – patterns of vulnerability are far too complex and dynamic to support such sweeping generalisations (Wisner, 2016).

Risk is not a core characteristic of the problems of those who are labelled as vulnerable, although in some cases, risk may have contributed to their hardship, as their opportunities to cope with those risks are limited (UNDRR, 2019). Risk perception is grounded in demographics, socio-economic, cultural and other factors (Gierlach et al., 2010). Women are significantly more likely to seek information about a disaster and to believe it is likely to occur; they are also more proactive in making their household safer and developing an emergency response plan (Bradshaw & Fordham, 2013). Some immigrant groups have heightened perceptions of risk due to disasters they had previously experienced (Fothergill et al., 1999). There is also a difference between races: Turner et al. (1980, cited in Fothergill et al., 1999) found that in the USA, black population were much more fatalistic about earthquakes, feeling that there was little or nothing one could do to protect against them; white males were the least worried about the risks of disasters. But perceptions of risk do not necessarily correlate with the actual exposure rates or response behaviour, and neither do they lead to reduction in vulnerabilities that are grounded in root causes. Often, vulnerabilities are layered on top of one another, creating further complexity. They affect preparedness, response and evacuation, as well as long-term post-disaster recovery processes.

There are many ways of categorising vulnerability. Cannon (2008) breaks vulnerability into five groups: livelihood strength and resilience, well-being and baseline status, self-protection, social protection, and governance. Alexander (2013) distinguishes six vulnerability types:

- Economic (people lack adequate occupation);
- Technological (caused by the riskiness of technology);
- Residual (caused by lack of modernisation);
- Delinquent (caused by corruption, negligence etc.);
- Newly generated (caused by changes in circumstances); and
- Total (i.e. life is generally precarious).

The World Disaster Report 2018 (IFRC, 2018) proposed an interesting grouping of those who are the most vulnerable because they fall through the cracks:

- Those out of sight (e.g. people whose birth is not registered or those no one wants to talk about – such as sexual or ethnic minorities);
- Those out of reach (e.g. for geographical or political reasons);
- Those left out of the loop (e.g. those who could get the support, but it is not offered in an appropriate way and, thus, can be used due to physical, cultural, social or political limitations);
- Those out of money; and
- Those out of scope (i.e. people whose needs do not fit into traditional interpretation of those in need in a post-disaster situation).

However, for the purpose of this overview, some categories of people that are considered to be vulnerable due to their frequent marginalisation founded in cultural, political, economic and social contexts are discussed in the following sections.

5.3.1 Gender and sexuality

Gender refers to social and cultural differences rather than biological ones: the differences between women and men reflected in their roles, responsibilities, access to resources, needs, perceptions and views – be that within the same household or within and between cultures – are socially and culturally constructed, and thus, evolve over time (Moser, 1993). Sex and gender are never automatically the primary social factor on the ground (Box 5.2, Figure 5.4a and Figure 5.4b), nor are they ever in isolation from other dimensions of life (Enarson & Chakrabarti, 2009).

Box 5.2: Gender and disasters

These two episodes for *Disasters: Deconstructed* podcast focus on gender.

Fig 5.4a
QR code 1 for Box 5.2. Available at https://disastersdecon.podbean.com/e/s4e14-gender-sexuality/

The interview with Dr Sarah Brown, Alison Sneddon and Dr Mirianna Budimir from Practical Action focuses and explores the following questions through the lens of intersectionality and storytelling: Why is it problematic when women and girls are treated as generically vulnerable? How can a binary narrative of gender be dangerous and create further marginalisation?

Fig 5.4b
QR code 2 for Box 5.2. Available at https://disastersdecon.podbean.com/e/s2e12-gender/

The interview with Cheryl Potgieter focuses on sexuality: What are we doing wrong when we discuss sexuality in scientific discourses? How can disaster studies bring sexuality into the frame in a transformative way? Dr Cheryl Potgieter is Research Professor at Durban University of Technology and Head of Gender Justice, Health and Human Development. She was previously Deputy Vice Chancellor at the University of KwaZulu-Natal.

Nevertheless, more women than men die in disasters, as women's vulnerability to disasters is generally higher due to gender-based inequalities that characterise societies throughout the world. These inequalities are often culturally engrained and limit women's access to important financial, political and other societal resources because of inequitable gender norms (the ways in which different societies define what it means to be masculine and feminine, including division of labour, roles, responsibilities and customs) (Shreve & Fordham, 2018). Bradshaw and Fordham (2013) list a number of vulnerabilities that increase for women and girls in an aftermath of a disaster: violence against women and girls; decline in sexual and reproductive health for young, unmarried and/or adolescent girls, as well as the health of those with non-binary gender identities; early or forced marriage; interruption to, or loss of, education; changes to social networks and family support; increased time burden due to gender norms and roles being promoted; and psycho-social impacts. Globally, compared to men, women and girls have less control over material assets, and typically, losses in women's assets (such as sewing machines or animals) go unrecorded (Bradshaw & Fordham, 2013). Gender plays a role in tenure insecurity: women are less likely than men to inherit land or property and to hold documentation in their own names (IFRC, 2018). Gender is also reflected in employment, as disasters may leave women more economically dependent on men, formal disaster assistance and/or state support (Enarson, 1998).

But not all women are vulnerable in the same way: class differences position them differently to withstand the material losses, and rebuild their homes and livelihoods, and engage in daily routines after a disaster. Women also do not simply rely on men or the state in post-disaster situations: many interact unexpectedly with men in non-traditional ways and places, using new tools for home reconstruction, negotiating with aid agencies or local governments, conducting search and rescue, or speaking out as emergent group leaders, neighbourhood activists, political leaders or emergency managers during relief and recovery (Enarson, 1998).

Men are also vulnerable through risk-taking, over-confidence, loss of a sense of control, reluctance to seek help, and failure to live up to expectations of them as protector during the disasters and provider in the aftermath (Dunn, 2016). Men and boys may be more likely to engage in risky behaviour due to prevailing social norms (e.g. the male role as protector of the family), which may lead to higher-than-expected male mortality rates (Ferris et al., 2013). Men often lack coping skills in the aftermath of a disaster and can be especially prone to alcohol abuse, stress and anger (MenEngage Alliance, 2016). There is also little recognition of men's diversity (according to their social class, ethnic group, sexuality) that affects the way they respond to risks (MenEngage Alliance, 2016).

To date, there is still little research (although it is growing (Figure 5.5)) on the impacts of disasters on adults, children and youth who are sexual and gender minorities (Box 5.3 and Figure 5.6). A lot

Fig 5.5
Youth who do not identify within the male/man-female/woman binary discussing disaster risk in Irosin, Philippines, in January 2010.
Source: photo courtesy of JC Gaillard

Box 5.3: Representation of LGBTQI in disasters

Fig 5.6
QR code for Box 5.3. This episode of *Disasters: Deconstructed* podcast features an interview with Darien Alexander Williams from MIT. The conversation focuses on the experience of risk and disaster impacts by LGBTQI+ communities, who are often rendered invisible by those in power. We also talk about the role of narrative/stories in the marginalisation of LGBTQI+ people and how we can do better.
Available at https://disastersdecon.podbean.com/e/s2e3-lgbtqi/

of evidence, however, exists showing that people belonging to a sexual and gender minority are frequently discriminated against, and their vulnerability to abuse is often exacerbated during times of crisis (International Alert, 2017). Increased vulnerability of LGBTQ+ people often occurs because of the social and political marginality: they are forced to hide their identity or are discriminated and vilified on the basis of sexual and gender identity, which leads to experiencing increased fear, stress, anxiety and depression (Gorman-Murray et al., 2017).

Evidence also suggests that transgender people experience more harassment than gay, lesbian and bisexual people. This makes them apprehensive to access emergency services and disaster recovery services, as they fear insensitivity, intolerance as well as scrutiny and out-of-placeness in public space. Another concern is the loss of bodily integrity in disaster contexts due to lack of access to hormone replacement therapy and medical support, as well as the ability to maintain a legible gendered appearance (Gorman-Murray et al., 2018).

In over 75 countries where same-sex activity is illegal (UN OHCHR, 2018), homophobia and transphobia not only contribute to violence but also inhibit LGBTQ+ survivors from filing complaints or seeking help (IFRC, 2018). Gaillard (2011; and Gaillard et al., 2017), for instance, shows that following the cyclone in 2011 in the Philippines, young *baklas* (gay or transgender women[1]) are often expected to perform difficult and unpleasant tasks, such as cleaning the house after recurrent flash floods, looking after young children and doing laundry, whilst also completing demanding tasks, such as fetching firewood and water in the deep floodwaters. Many *bakla* were also left to eat last when their households were struck by two powerful cyclones in late 2009. Moreover, their specific needs were not recognised in evacuation centres, where they typically lacked privacy and faced regular gender discrimination and sexual harassment.

5.3.2 Ethnicity, race, caste

Ethnicity, race and caste play important roles in increasing vulnerability. The role of ethnicity and race is particularly prominent in disaster research in the USA (Fothergill et al., 1999). Although the population is diversifying rapidly, the minority groups population is still facing significant barriers to education and higher rates of poverty. Chronic poverty, unemployment or underemployment, as well as discrimination and racism adversely impact the trust of minority groups in institutions that are designed to provide them with assistance both generally and in times of disasters. Ethnic minorities are also less likely to be insured and are, thus, more likely to face greater hardship dealing with health and economic costs, leading to slower recovery (Donner & Rodriguez, 2008). Racial and ethnic communities are also less likely to have had disaster risk awareness raising opportunities. Moreover, the way of receiving and communicating early warning also differed: some minorities in the USA do not speak English (which is used for early warnings) and, thus, large rely on social networks (Fothergill et al., 1999; Abukhalaf & von Meding, 2021).

Ethnicity and race are closely linked to poverty. This has been demonstrated in New Orleans after Hurricane Katrina (Figure 5.7). Historically black communities made up nearly 70% of the city's total population; at the same time, the city's poverty rate was nearly 16% greater than the national average for individual and family poverty rates. And 91% of black families in New Orleans lived in poverty prior to Katrina, and this created a disaster: more than 80% of the city was flooded, and over 1,500 people died. Only 181,400 people remained living in New Orleans after Katrina (compared to 484,674 prior to the hurricane), with a majority of remaining population being black people who lacked the means to evacuate or sustain themselves during long-term evacuation (Donner & Rodriguez, 2008).

Fig 5.7
Ladies and Men of Unity second line in post-Katrina New Orleans.
Source: photo courtesy of Wesley Cheek

Ethnicity and belonging to a caste are usually assumed to differ from each other by common kinship, language, religion, beliefs, traditions and customs, as well as territory, with most of these features being inherited from ancestors. Interactions between and amongst castes often establish socio-economic and political hierarchies and processes that define society (Gaillard, 2012). There is very little data that is disaggregated by ethnicity and caste, yet some case studies from Asia, Africa and Latin America exist to show how these determinants increase vulnerability (Box 5.4 and Figure 5.8). Bosher et al. (2007) observe that poverty is affected not by exclusion from society but by inclusion in a caste-based hierarchical society, as various resources and means of protection are unequally distributed based on a position in society. This is demonstrated through segregation of the location of houses in some parts of India: in the Ganges River Delta, villages are often built around central elevated grounds, with more powerful castes settled at the higher grounds and lower-power castes living in the low-lying lands (Wisner et al., 2004). Inter-caste relationships may also be very valuable (see Chapter 6).

Response and post-disaster recovery often highlight intended and unintended discriminatory practices that disadvantage some castes or ethnic groups. For instance, the Burakumin caste (a traditional untouchable caste in Japan), as well as many foreigners from Vietnam, Korea and the Philippines, were badly affected by the 1995 Kobe earthquake. Both groups lived in a district of Nagata that has been left out of many governmental rehabilitation programmes; moreover, the government suspended water supply, making the post-earthquake firefighting difficult (given that most of the houses in the district were traditional wooden houses). Most of the district's population was forced to camp outside whilst other affected people were provided shelter in public buildings (Wisner et al., 2004).

Box 5.4: Disasters, race and black feminism

Fig 5.8

QR code for Box 5.4. This episode of *Disasters: Deconstructed* podcast bring you an interview with Dr Fayola Jacobs, assistant professor at the Humphrey School of Public Affairs at the University of Minnesota. The conversation focuses on the black feminist lens that Fayola brings to the topics of disaster planning, environmental justice and urban planning.Available at https://disastersdecon.podbean.com/e/s4e2-black-feminism/

Reconstruction often unintentionally discriminates by introducing new building techniques that replace traditional ones, gradually destroying craftsmanship of the local population (and often increasing exposure). Resettlement also often results in loss of livelihood, moving many ethnics groups from traditional crops and leading to deterioration of communities.

5.3.3 Family size and structure

Family size and structure are intertwined with gender vulnerability, as they reflect households' power relations in kinship and marriage that can also create risky living conditions for women, especially in contexts of divorce, desertion, widowhood and single mothering (Enarson, 1998). Kinship, for instance, often manifests how food is distributed in a family and, thus, explains lack of nutrition amongst women and girls (Rivers, 1982).

Responsibility for children and other dependents also often falls on women's shoulders and has important implications for decisions about preparedness and evacuation. For instance, larger families tend not to evacuate or take longer to do so. This is often intertwined with the ethnicity: non-white families in the USA tend to be larger (Gladwin & Peacock, 2000).

Similarly, single mothers tend to be less educated and poorer than the general population, thus placing them at greater risk to disasters. Even though women in general may be more likely to perceive situations as risky, risks may become irrelevant or less pressing relative to the burdens of single motherhood (e.g. providing food, shelter and other basic necessities) to their families (Donner & Rodriguez, 2008).

5.3.4 Age

Age is another factor that should be considered, but it is not a stand-alone determinant of vulnerability. It often becomes a factor in vulnerability when combined with infirmity, but it also cannot be separated from gender, ethnicity or social status.

It is estimated that, by 2050, 22% of the world population will be aged 60 or over; it is, thus, important to consider roles, capacities and vulnerabilities of older people. Older people are more vulnerable, as they confront unique difficulties during disasters and, in particular, in periods of evacuation. For instance, appropriate food may not be available; mobility of the older people may be reduced, thus, increasing their dependency on others; focus of the health services is often on communicable diseases and injuries rather than age-related health issues (Barbelet et al., 2018). These and many other factors lead to the elderly being more likely to be injured or die as a consequence of hazard events. In the Philippines in 2013, people over 60 represented approximately 7% of the population but accounted for 38% of the fatalities caused by Typhoon Haiyan that same year. Similarly, in Nepal, 29% of people who died in the earthquake in 2015 were aged over 60, although older people represented only 8% of the population (UNDESA, 2016).

Older people also have particular requirements when it comes to the design of temporary shelter: ramps and handrails are often lacking; sleeping on cold or hard surfaces may aggravate chronic health conditions that otherwise can be manageable. Older people may also have difficulties in accessing water distribution points, as their needs are not often taken into account when designing the schemes (Barbelet et al., 2018).

Apart from physical abilities, psychological factors play an important role in increasing vulnerability: older persons are less likely to engage with personalised warnings and receive warning messages; the latter can be due to a significant loss of social ties often associated with the aging process and the lack of community interactions or involvement (Donner & Rodriguez, 2008). Due to reduced social cohesion and interaction, older people also lose their social status and power in the community; this, in combination with less frequent social interactions, often leads to isolation and can result in a depression or loneliness. Social isolation is often enhanced in a disaster situation, when some people are forced to move to new locations where they do not have existing social ties, thus causing further distress for older people (Barbelet et al., 2018).

Disasters affect children and adolescents in a different way (Box 5.5 and Figure 5.10), and it is particularly important to consider how vulnerability would affect their longer-term development. For instance, drought can cause water stress and food insecurity that in turn lead to malnutrition. Children can be affected by disasters even before they are born as care for pregnant mothers is often reduced; malnutrition and inadequate food consumption can have a detrimental effect on in utero development and maternal health, thus affecting the overall health of the baby (Diwakar et al., 2019).

Children are not a homogenous group (Figure 5.9): those who come from poor backgrounds and are socially, economically, culturally, politically, environmentally or otherwise marginalised are much more vulnerable to disasters, as their ability to prepare for and cope with

Fig 5.9
Children using fine fuel moistures to assess the flammability of the vegetation as a part of their bushfire DRR awareness training in Victoria, Australia. Source: Strathewen Primary School

Box 5.5: Children in disasters (interview with Marla Petal)

Fig 5.10
QR code for Box 5.5. In this episode of *Disasters: Deconstructed* podcast, Dr Marla Petal discusses some of the particular inequalities and injustices faced by children and why their voices are so important. We discuss the power of narrative and framing as we approach discussion of a group so often identified as vulnerable.

Dr Marla Patel is Principal Advisor for Urban Resilience and School Safety for Save the Children, and is a prominent and fearless advocate for children and youth around the world.

Available at https://disastersdecon.podbean.com/e/s2e13-children/

a disaster is limited. Children's mortality and morbidity as well as injury and illness are often higher during and after a disaster, and is impacted by their socio-cultural restrictions and physical abilities. Their dependence on adults may also lead to enhanced vulnerability if the adults themselves are ill or have died (Diwakar et al., 2019). Older children may experience direct emotional, behavioural or psychological impacts after a disaster and require support that differs from the one provided to adults: disasters leave children orphaned or separated from their family, while some children are sexually or psychologically abused (Peek, 2008).

Damaged infrastructure also has indirect impact on children if the household's water or energy supply as well as communication are disrupted. Disasters destroy school buildings, books and equipment, as well as disrupt healthcare services. The costs of recovery may also lead to reduced spending on education (particularly amongst girls), healthcare or food, all of which impacts longer-term development.

Adolescents' vulnerability can also be enhanced by their invisibility during disasters, as they do not fit into the child or adult categories used in various guidelines and policies. Evidence shows that adolescent girls are particularly vulnerable as they lack control over their bodies and access to sexual and reproductive healthcare and family planning resources; they often face early or forced marriage, rape and unwanted pregnancies; removal from school; and malnutrition during disasters (Van der Gaag, 2013). Adolescent girls often have to take on adult roles and responsibilities, including (often unwanted) motherhood, engagement in wage labour and reconstruction, as well as unpaid household duties, such as fetching water and food.

5.3.5 Disability

Over one billion people worldwide have some form of disability (WHO, 2020). It is important to note that the type and severity of disability vary significantly, with the most common examples as follows: difficulties in personal mobility, visual impairment, deafness, problems of communication and articulation of words, cognitive disorders, various medical problems that may require the use of life-supporting system, intolerance of various substances, psychiatric disorders, and so on (Alexander et al., 2012). Persons with disabilities are not inherently vulnerable to disasters – instead, considerable evidence shows they often act as an important resource for their families and communities, particularly during times of crisis. There are, however, factors associated with disability that can increase vulnerability to the impact of disasters; these include reduced mobility, diminished employment opportunities, chronic health conditions, discrimination and other factors that may put persons with disabilities more at risk during times of crisis (IFRC, 2018). During a disaster, people with disabilities may encounter physical barriers (Figure 5.11) or experience difficulties in communication that slow or prevent their response, or stop them from using the emergency facilities (Alexander et al., 2012).

There is a myriad of discrimination examples: in earthquakes, people in a wheelchair cannot hide under a desk or run down the stairs; deaf or visually impaired people may not be able to receive verbal orders or follow the emergency lighting; transportation and shelter are often inaccessible. Only 41 of 102 residents who lived in the home of people

Fig 5.11
Emergency shelters with disability access in Barbados.
Source: photo courtesy of Ksenia Chmutina

with disabilities in Galle (Sri Lanka) have survived the Boxing Day tsunami in 2004; the rate of mortality was so large because many residents could not leave or underestimated the time it would take to leave.

Disasters may lead to the infliction of new injuries and subsequent or further loss of mobility; reduced access to medical services for chronic non-communicable diseases; increased risk from infectious diseases; nutritional deterioration; or distress, depression and anxiety. The latter are particularly frequent when family of community support structure is destroyed.

Similarly to age, disability is not a stand-alone determinant of vulnerability: it comes from the combination of age, gender, ethnicity or social exclusion, as well as poverty. Persons with disabilities are more likely than their younger and non-disabled peers to experience poverty; this can be due to their particular needs (e.g. healthcare needs) as well as barriers in their environment that prevent them from accessing key services and opportunities, such as education, healthcare, employment, justice and social support (IFRC, 2018). This reinforces and increases the vulnerability of persons with disabilities, leaving them with fewer resources to withstand and recover from disasters, and pushing them further into poverty.

Health should also be given an important consideration and should be framed in gender, age and cultural context. A gender-blind approach to healthcare, particularly in post-disaster situation, means that women and girls in some regions cannot access healthcare, as many doctors are male.

5.3.6 Poverty

Marginalisation is a matter of poverty, as poor people lack or are deprived of access to resources and are socially and culturally excluded from dominant policies and DRR activities. But vulnerability and poverty are not the same: poverty is the relationship with others in society, which reproduces this state, whilst vulnerability implies causal relationships with both society and also the physical environment at particular times (Wisner et al., 2004).

Limited economic resources prevent people from having a choice about the location of the households, being able to afford preventive measures, or investing in education, healthcare or safe assets. In general, the poor suffer from disasters more than the more affluent. Whist the houses of both affluent and poor can be affected by flooding, more affluent people are more likely to have insurance and be able to rebuild or find an alternative shelter relatively fast. This is not the case for the poor who frequently have their entire stock of capital (home, clothing, tools for artisan handicraft production etc.) assembled at the site of the disaster. They have few (if any) cash reserves and are generally not considered creditworthy. Moreover, the rich can choose whether to live or not to live in a hazard-prone area (e.g. on a riverbank that provides a beautiful view); for the poor who live in hillside slums or on the edge of waste dumps, the choice is not as voluntary. The urban poor use their location as the base for organising livelihood activities and economic opportunities (e.g. casual labour, street trading, crafts, crime, prostitution), regardless of the hazard risk (Wisner et al., 2004).

Poverty influences risk perception, preparedness, warning communication, physical and psychological impacts, emergency response, recovery, and reconstruction (Fothergill & Peek, 2004). Chronically poor (i.e. those who have lived in poverty for many years, whose poverty is often transmitted to future generations, and who often lack skills and opportunities to escape poverty) can be particularly vulnerable, as disasters can potentially reverse years of developmental gains (Diwakar et al., 2019). For instance, lower-cost social housing is vulnerable to disasters, thus, leaving those who are likely to reside there (i.e. the poor) without shelter and assets after a disaster, therefore increasing their vulnerability further.

Vulnerability of the poor, of those living in slums on the periphery of cities and towns, in housing that is frequently poorly constructed and tenure is insecure, and in the physical environment unfit for human habitation, with low access to services and few employment opportunities, is generated by a complex range of factors, including conflict, urbanisation, chronic underdevelopment, high levels of criminal violence and political instability. Such daily struggle makes it extremely difficult to draw a line between acute and chronic vulnerability (Pantuliano et al., 2012). Chambers (1989) rightly points out that poor households are always having to make resource expenditure decisions that play off poverty and vulnerability, and poverty invariably wins over vulnerability, as such choice responds to the more immediate pressures of daily life.

5.3.7 Livelihood

Livelihood is a combination of natural, physical, financial and social capitals, which together determine the living gained by an individual or households (DFID, 2000). Whilst the central feature of a livelihood is a household, it is also important to consider its social relations and roles (Figure 5.12). Livelihoods rarely refer to a single activity. It includes complex, contextual, diverse and dynamic strategies developed by households to meet their needs (Chambers, 1989); diversity and dynamics are crucial to ensure that livelihood is sustainable (Gaillard et al., 2009). Establishing a livelihood is not just about meeting basic needs: over time, people build up and use assets (ranging from small belongings to a house), as well as develop relationships, skills and

Fig 5.12
Informal market at Pétionville, Haiti.
Source: photo courtesy of Gonzalo Lizarralde

communities. These tangible and intangible assets can be both means and ends: for instance, earning an income can be means to making a livelihood more robust, but it can also be an end; but they all play a pivotal role in reducing vulnerability (Sanderson, 2012).

Livelihoods are often earned in locations that combine opportunities with hazards: land near volcanoes provides rich soil for agriculture, and floodplains provide cheap flat land for businesses and housing. In order to access opportunities (as described in previous section), poor people can often only afford to live in informal settlements in unsafe ravines and on low-lying land within and around the cities where they have to work. Location and land play an important role in how vulnerable the livelihood is: for instance, after the 2004 Indian Ocean tsunami, the Indonesian government resettled many people who lost land on government-acquired land. However, land acquisition took place before the government had identified the people who required land and their particular needs; this resulted in some land not being appropriate for the livelihoods of the people who were to be relocated there (UN-Habitat, 2007). Pastoralists' and nomadic groups' movement from a place can be hampered by state boundaries and laws that govern hunting, fishing and cultivation, as well as restrict the right to encampment. Often, pastoralists find that they do not have a legal right to be on the land that they have utilised for generations – which restricts their livelihood options (Gilbert, 2007).

Land tenure is often overlooked in vulnerability context; yet vulnerability may be enhanced by the perceptions of insecure land tenure or in cases where insecure tenure results in the loss of land, especially when alternative livelihood and housing options are limited. The lost land can represent a lost livelihood, lead to a loss of identity (e.g. spiritual or otherwise), result in the displacement of community or in homelessness or inadequate housing (Reale & Handmer, 2011). The depletion in housing stock was one of the major impediments to recovery in New Orleans after the Hurricane Katrina. And 56% of all rental units were flooded and rendered uninhabitable; this created a housing deficiency of about 70,000 homes. A drop in a supply of rental stock meant that the rent prices increased by 46%. High accommodation costs became unaffordable for many people who could have moved back to New Orleans to rebuild or start employment. With short-term accommodation out of reach for thousands of people, the local economic activity and rebuilding necessary for recovery was made more difficult (Green et al., 2007).

Social assets that form livelihood are also extremely important. Skills, abilities, relationships as well as participation in decision-making processes can play a significant role in increasing or decreasing one's vulnerability. These informal safety nets often lie at the basis of the coping strategies that people adopt in response to livelihood shocks: they involve drawing on social networks (such as extended family, friends and neighbours, wealthy patrons) for help in times of need, with or without expectations of reciprocity. Commercialisation has gradually been destroying the vertical redistribution (i.e. transfers from wealthier patrons to poorer clients) that has been traditional in some regions; however, horizontal redistributive practices (i.e. transfers between people of similar economic and social status) remain widespread. Such practices, however, are highly irregular, particularly in times of a disaster; a drought, for instance, can eliminate food production surpluses across a community (Devereux, 2001). The role of social assets in reducing vulnerability will be further discussed in Chapter 6.

5.3.8 Migration, displacement and informal population

Migration is driven by a range of political, economic and social factors, including economic, security and welfare inequalities, poor governance, and environmental factors, as well as other motivations. It, therefore, defies easy categorisation between voluntary and involuntary, and refugee and economic migrant. Asylum seekers and refugees may resort to people smugglers, and may enter a country irregularly (e.g. entering the county with false documentation, entering without crossing an official border point, or residing in a country without official residence permit or visa) due to limited safe and legal channels to seek asylum (IFRC, 2018). Migrant flows are increasingly a mix of refugees, asylum seekers and other migrants, massed in groups and led by smugglers, giving rise to what is increasingly termed 'mixed migration' (Horwood & Reitano, 2016).

Displacement (i.e. removal of habitation and people to make land available for other purposes, without compensation or alternative shelter) deprives families of accustomed livelihoods, which can exacerbate their vulnerability (Lewis & Kelman, 2012). Once displaced, people have to learn about the new location, context, habits as well as to readjust or develop new coping capacities, which require understanding of the local context.

Displaced populations and migrants are often exposed to discriminatory practices, have interrupted or no access to formal education, health or justice systems, and health services. Such social status may lead to disparities in income, gender, ethnicity, household status and job type, which are difficult to overcome (UNDRR, 2019) – and which further enhance vulnerability. The violence and abuse faced by many migrants are compounded by limited access to services in countries of transit and arrival, as this is often contingent on citizenship or legal residency. The starkest example is the curtailment of access to healthcare, where all but emergency care is often off limits (despite increasing evidence that restrictions on primary healthcare are costlier for states and do not affect migration levels) (Ingleby & Petrova-Benedict, 2016).

Migrants' occupations and their gender often determine their experiences: for instance, many women from Bangladesh work as housemaids when coming to India and depend on their employers for wages and security. They are often unaware of their rights, have no written contracts, work extra hours and carry out additional duties for which they are rarely paid. If they fall sick, there is no guarantee their employers will send them to the doctor or pay for treatment, so ill health may mean returning to Bangladesh for healthcare (Samuels et al., 2012). Male migrants often work in catering or construction workers, or as watchmen and casual labourers, with many holding more than one job to make ends meet. They also face discrimination in the workplace, often through verbal abuse and denial of services (Samuels et al., 2012).

Migrants' vulnerability can be enhanced by a lack of economic, cultural and social capital. For example, many Hispanic immigrants in the USA may not (or will not) speak English, which continues to be problematic in tornado-prone areas where authorities may or will not issue warnings in Spanish. Difficulties associated with relocation and adaptation to a new country can also influence how disaster risks are perceived and responded to in a new region, as new residents are often focusing on establishing the normal life and, thus, looking for work, housing, schools and so forth, rather than focusing on disaster preparedness. As new immigrants lack strong social capital or community networks, overcoming these problems is extremely difficult. Moreover, migrants (especially undocumented) may be reluctant to seek help for fear of deportation (Donner & Rodriguez, 2008).

Displaced populations most often join the ranks of the urban poor. Moreover, they are frequently seen as an expense for the local government and also as a security threat. Displaced populations often face challenges that are embedded in the surrounding environment; this includes a lack of urban development in informal areas, poor quality services, lack of access to justice, scarce employment opportunities, poor transport as well as various (and often unexpected) threats. Violence against Rohingya communities in Rakhine State, Myanmar, led to more than 720,000 people (most of them are women and children) fleeing the homes since 2017. The displaced Rohingya population accounts for about one-third of the total population in Cox's Bazar, the area in Bangladesh that was already densely populated and facing severe development challenges. Here, the displaced Rohingya people are sheltered in makeshift settlements with minimal access to basic infrastructure and services, making them particularly vulnerable to cyclones, floods and landslides. Moreover, the quick establishment of makeshift shelters has caused deforestation, further increasing vulnerability to the effects of monsoon rains: in 2018, monsoon rains caused landslides and resulted in damages to 3,300 shelters and affected 28,000 refugees. The emergency relocation of refugees living in Cox's Bazar was made even more challenging by a lack of suitable available land (UNDRR, 2019).

The urban poor in general have little influence over how or whether their needs are addressed, and the displaced also often suffer from legal and social discrimination. In addition, newly arrived refugees in some contexts may be less adapted to their host country's climate, and they may face increased vulnerability to weather extremes during their adjustment period (UNDRR, 2019).

It is also important to note that, sometimes, it is difficult to distinguish between vulnerabilities of displaced and local population. This is particularly the case in urban settings, where the population density is high, with displaced and host communities living side by side in the same shelters, streets and neighbourhoods. It is, thus, critical to consider the relationship between displaced and host communities, including the potential impact of the influx of displaced populations on the resources of the local community and the authorities (Pantuliano et al., 2012).

If populations in protracted displacement continue to be neglected and marginalised, there is a risk of creating ghettos of frustrated people that may lead to civic conflict (Haysome, 2013): the phenomenon of urban violence is now increasingly seen as linked to the discrimination and marginalisation of certain groups, including displaced communities, in under-resourced or poorly serviced urban areas (Pantuliano et al., 2012). On the other hand, displaced populations, by and large, profess a commitment to making their lives in the new place – and thus, their skills and assets should be recognised (as discussed in Chapter 6).

5.4 Intersectionality

The unspoken intention of reducing vulnerability is to be inclusive. However, in practice, much of the implied inclusivity is lost because specific needs and interests are not recognised, sometimes as a conscious attempt to avoid possible bias and discrimination through the prioritising of one group over another; or sometimes as an inability to see beyond the experience and worldview of the dominant privilege (often white and male).

Over the decades, the efforts have been made to focus on poor, women, indigenous and many other marginalised groups; such mono-lens is useful in practical terms. However, this naming, labelling and framing are expressions of power relations (Gaillard & Fordham, 2018). Such categorisation of marginalised groups not only creates the ways in which people belonging to a particular category will be recognised and understood but also excludes those who are uncategorised (and thus, undeserving of attention) (Moncrieffe & Eyben, 2007), thus often concealing the social and historical contexts and struggles.

Vulnerability should be explored without de-politicising and invisibilising marginalised interests and experiences (Gaillard & Fordham, 2018, p. 77); this can be done through the lens of intersectionality. Intersectionality is a framework that allows taking into account people's overlapping identities and experiences in order to understand the complexity of discrimination and privileges they face. The idea is not novel: over two decades ago, Crenshaw (1989) argued that compounded biases and prejudices mean, for instance, that a woman of colour may be denied recognition for her situation because each category of concern (gender and race) has been regarded as a single issue and dismissed (i.e. there may be little evidence of discrimination against white women or against black men, but black women may experience considerable concealed levels of injustice). In other words, intersectionality allows recognising that intersecting marginalisation creates "an interlocking prison from which there is little escape" (Hancock, 2007, p. 65).

Exclusive categories of vulnerability (elderly, female, migrant, black) falsely represent intersecting identities and social relationships. Vulnerability does not derive from a single factor, such as household structure or race, but reflects historically and culturally specific patterns of relations in social institutions, culture and personal lives. Intersecting with economic, racial and other inequalities, these relationships create hazardous social conditions, placing different groups differently at risk when disastrous events unfold (Enarson, 1998). In other words, understanding vulnerabilities of particular groups of people is important, but it should be done in an integrated manner; it is also important to consider that labelling a particular group as vulnerable denies them agency and may instead reinforce cultural stereotypes (Gaillard & Fordham, 2018).

We should, however, be careful with how we use the term 'intersectionality' – this concept has only recently become prominent in academic and political debates, and yet it is already at risk of becoming an over-debated (and thus, often empty) and malleable term, similar to 'vulnerability' or 'resilience', amongst others. The conceptual open-endedness of intersectionality exposes it to the risk of reductionism. By failing to address power, justice and equity, using such concepts leads to missing opportunities for gaining a deeper understanding of the politics that shape all systems in which humans are involved. On the other hand, intersectionality encourages critical engagement with one's own assumptions in the interests of gender-responsive enquiry, which is reflexive, critical and accountable. Intersectionality encourages complexity, stimulates creativity and avoids premature closure (Davis, 2008, p. 79).

5.5 Concluding remarks

Vulnerability is a multifaceted phenomenon reflecting the range of social, cultural, demographic and economic conditions interacting in complex ways. A single dimension of the social structure is hardly ever responsible for vulnerability. Just like risks vary in their frequency, intensity and welfare impact, the sources of vulnerability are also diverse. It is, thus, critical to understand vulnerability through intersectionality (Box 5.6, Figure 5.13a and Figure 5.13b).

The categories of vulnerability discussed in this chapter are not exclusive; other marginalised groups include people with animals

Box 5.6: A threat or a resource?

In these episodes of *Disasters: Deconstructed* podcast, we focus on two groups of people that are often neglected in conversations about disasters.

Fig 5.13a
QR code 1 for Box 5.6. Carlee Purdum from the Hazard Reduction & Recovery Center at Texas A&M shares about her research on disasters and prisoners, and unpacks some of the narratives that shape our understanding of incarceration, particularly in the United States. How are prisoners particularly at risk in disasters, and how does the state exploit their status as modern-day slaves?Available at https://disastersdecon.podbean.com/e/s2e15-prisoners/

Dr Jamie Vickery from the Natural Hazards Centre, Boulder, Colorado, discusses how disasters affect people experiencing homelessness and some of the deeper issues at play that are too often overlooked. We cover emergency response, policy, trust and narratives, and consider how we might challenge a culture that normalises homelessness as a cautionary tale.

Fig 5.13b
QR code 2 for Box 5.6.

and pets, homeless people, people with BMI over 40, indigenous communities; the list can go on. Marginalised are the sections of society that regularly lack access to resources and the means of protection available to those with higher levels of socio-economic or political power. Often, the root causes of vulnerability are found in oppressive societal systems. Vulnerability is, thus, a part of people's everyday lives and involves their capacities to avoid, resist and recover from harm (this is discussed in Chapter 6).

Although vulnerability is not synonymous with poverty, people with lower incomes and poor access to education, healthcare, clean water and sanitation are more vulnerable to the impacts of any given natural hazard and less able to recover. Such vulnerability is also reflected in lower capacity, as their access to preparedness measures, early warning systems as well as insurance programmes are limited: they have less time to go to meetings, volunteer, participate politically; they have less income to invest in personal protection; they often do not feel a strong sense of citizenship and may be suspicious of risk education messages and risk reduction programmes.

Structural measures of a bigger scale that are supposed to protect large areas (e.g. flood dams) may well displace some of the vulnerable people to even more exposed locations. Thus, in order to understand why disasters occur and whom they would impact the most, it is critical to understand the underlying causes of vulnerability and address them in practice. Most of the vulnerability assessments focus on poverty, inequality, gender, education, health status, disability and environmental concerns. However, it is important to remember that personal characteristics can be linked to vulnerability but not define it. The solution to a problem of vulnerability is in political engagement and inclusion, rather than in reinforcing isolation.

It is also important to remember that vulnerability is a Western concept – and one has to be careful not to abuse it, as vulnerability can be seen as a concept employed to portray certain localities and groups of people as fundamentally unstable, unsafe and in need of intervention (Bankoff, 2001). The meaning of vulnerability has been shaped at a particular historic juncture by a particular historic perspective; this signifies that it can only really be understood from the perspective of the prevailing socio-economic system (Bankoff, 2019; Gaillard, 2019).

There should be most consideration of how and when (if at all) the term 'vulnerability' is used: people living with risk should not be seen as "those who need help" but instead as "powerful claimants with rights, rather than poor victims or passive recipients" (Heijmans, 2004, p. 127). Children, elderly people or those living with disabilities have valuable knowledge, experience and skill to contribute to social protection and risk reduction; the potential for building on such local knowledge and skill can be missed. Of course, local potentials should also not be romanticised. But exclusive focus on vulnerability (rather than capacity, as will be discussed in the next chapter) is misleading because it does not allow for the information about the ways that local people are proactive in protecting themselves, and it creates the perception of people as victims (Wisner, 2016).

We should also remember that vulnerability – when approached through the human rights lens – is something we all have in common. It should be treated as a positive phenomenon. Vulnerability gives us the basis for exchange and reciprocity – we cannot come into being, flourish and survive if our existence is not connected to the existence of others, which means that vulnerability is about solidarity and mutuality, the needs of groups and communities, not just those of individuals (ten Have, 2018). To paraphrase Isobel Wilkerson, the world that embraces and is built upon vulnerability would set everyone free.

Take-away messages

1 The concept of vulnerability reflects a complex system of characteristics related to human experience, geographical location, power, politics, money and so on. It allows to demonstrate why disasters do not affect all communities and societies equally, and why

the social groups who have been pushed to the margins of society disproportionately feel the impacts of disasters.

2 The elements of vulnerability are complex and intertwined; they change over a person's life cycle. But they are all grounded in oppression and marginalisation that exclude some individuals/groups of individuals based on their ethnicity, sexuality, age, gender and many other personal characteristics as well as access to resources.
3 Vulnerable groups are often categorised – however, such categorisation not only treats a group as homogenous (which is never the case) but also excludes those who are uncategorised. Thus, exclusive categories of vulnerability (elderly, female, migrant, black) falsely represent intersecting identities and social relationships.
4 Intersectionality is a framework that allows taking into account people's overlapping identities and experiences in order to understand the complexity of discrimination and privileges they face.
5 Understanding vulnerabilities of particular groups of people is important, but it should be done in an integrated manner; it is also important to consider that labelling a particular group as vulnerable denies them agency and may instead reinforce cultural stereotypes.

To learn more about the topic discussed in this chapter, listen to the *Disasters: Deconstructed* episode interview with Dr Ksenia Chmutina, Dr Jason K. von Meding and Dr Darien Alexander Williams (Figure 5.14).

Fig 5.14
QR code for Chapter 5.

Note

1 I.e. those who identify as women but were classified as men when they were born.

Further suggested reading

Bankoff, G. (2001). Rendering the world unsafe: "vulnerability" as Western discourse. *Disasters*, *25*(1), 19–35. https://doi.org/10.1111/1467-7717.00159

Butler, J. (2009). *Frames of War: When is life Grievable?* Verso.

Butler, J. (2016). Rethinking vulnerability and resistance. In J. Butler, Z. Gambetti & L. Sabsay (Eds.), *Vulnerability in resistance*. Duke University Press.

Chambers, R. (1989). Editorial Introduction: Vulnerability, coping and policy [IDS bulletin]. *IDS Bulletin*, *20*(2), 1–7. https://doi.org/10.1111/j.1759-5436.1989.mp20002001.x

Davis, K. (2008). Intersectionality as buzzword: A sociology of science perspective on what makes a feminist theory successful. *Feminist Theory*, *9*(1), 67–85. https://doi.org/10.1177/1464700108086364

Enarson, E., & Pease, B. (Eds.). (2016). *Men, masculinities and disasters*. Routledge.

Fothergill, A., & Peek, L. A. (2004). Poverty and disasters in the United States: A review of recent sociological findings. *Natural Hazards*, *32*(1), 89–110. https://doi.org/10.1023/B:NHAZ.0000026792.76181.d9

Moncrieffe, J., & Eyben, R. (2007). *The power of labelling: How people are categorized and why it matters*. Earthscan Publications.

Peek, L. (2008). Children and disasters: Understanding vulnerability, developing capacities and promoting resilience. *Children, Youth and Environments*, *18*(1), 8.

Von Meding, J. (2021). Transformation comes through vulnerability: Contesting resilience praxis. *The Arrow*, *8*(1). https://arrow-journal.org/reframing-vulnerability-as-a-condition-of-potential/

References

Abukhalaf, A. H. I., & von Meding, J. (2021). Psycholinguistics and emergency communication: A qualitative descriptive study. *International Journal of Disaster Risk Reduction*, *55*. https://doi.org/10.1016/j.ijdrr.2021.102061

Alexander, D. (2013). Vulnerability. In K. Penuel, M. Statler, & R. Hagen (Eds.), *Encyclopedia of crisis management* (pp. 980–983). SAGE.

Alexander, D., Gaillard, J. C., & Wisner, B. (2012). Ch. 34. Disability and disaster. In B. Wisner, J. C. Gaillard, & I. Kelman (Eds.), *The Routledge handbook of hazards and disaster risk reduction* (pp. 413–423). Routledge.

Bacon, C. M. (2012). Ch. 14. Disaster risk and sustainable development. In B. Wisner, J. C. Gaillard, & I. Kelman (Eds.), *The Routledge handbook of hazards and disaster risk reduction* (pp. 156–167). Routledge.

Barbelet, V., Samuels, F., & Plank, G. (2018). *The role and vulnerabilities of older people in drought in East Africa. ODI*. HPG Commissioned report.

Bankoff, G. (2001). Rendering the world unsafe: "Vulnerability" as Western discourse. *Disasters*, *25*(1), 19–35. https://doi.org/10.1111/1467-7717.00159

Bankoff, G. (2019). Remaking the world in our own image: Vulnerability, resilience and adaptation as historical discourses. *Disasters*, *43*(2), 221–239. https://doi.org/10.1111/disa.12312

Blaikie, P., Cannon, T., Davis, I., & Wisner, B. (1994). *At risk: Natural hazards, people's vulnerability and disasters*. Routledge.

Bosher, L., Penning-Rowsell, E., & Tapsell, S. (2007). Resource accessibility and vulnerability in Andhra Pradesh: Caste and non-caste influences. *Development and Change*, *38*(4), 615–640. https://doi.org/10.1111/j.1467-7660.2007.00426.x

Bradshaw, S., & Fordham, M. (2013). *Women, girls and disasters. A review for DFID*. https://assets.publishing.service.gov.uk/government/uploads/system/uploads/attachment_data/file/236656/women-girls-disasters.pdf

Cannon, T. (2008). *Reducing people's vulnerability to natural hazards: Communities and resilience*. Research Papers. World Institute for Development Economics Research, *2008/34*.

Cardona, O. D. (2003). Ch. 3. The need for rethinking the concepts of vulnerability and risk from a holistic perspective A necessary review and criticism for effective risk management. In G. Bankoff, G. Frerks, & D. Hilhost (Eds.), *Mapping vulnerability: Disasters, development and people*. Earthscan Publications.

Chambers, R. (1989). Editorial Introduction: Vulnerability, coping and policy [IDS bulletin]. *IDS Bulletin*, *20*(2), 1–7. https://doi.org/10.1111/j.1759-5436.1989.mp20002001.x

Chmutina, K., & Rose, J. (2018). Building resilience: Knowledge, experience and perceptions among informal construction stakeholders. *International Journal of Disaster Risk Reduction*, *28*, 158–164. https://doi.org/10.1016/j.ijdrr.2018.02.039

Chmutina, K., Von Meding, J., Gaillard, J. C., & Bosher, L. (2017). *Why natural disasters aren't all that natural*. OpenDemocracy. Retrieved September 14 2017, from https://www.opendemocracy.net/ksenia-chmutina-jason-von-meding-jc-gaillard-lee-bosher/why-natural-disasters-arent-all-that-natural.

Crenshaw, K. (1989). Demarginalizing the intersection of race and sex: A black feminist critique of antidiscrimination doctrine, feminist theory and antiracist politics. *University of Chicago Legal Forum*, *1989*(1), 139–167.

Davis, K. (2008). Intersectionality as buzzword: A sociology of science perspective on what makes a feminist theory successful. *Feminist Theory*, *9*(1), 67–85. https://doi.org/10.1177/1464700108086364

Devereux, S. (2001). Livelihood insecurity and social protection: A re-emerging issue in rural development. *Development Policy Review*, *19*(4), 507–519. https://doi.org/10.1111/1467-7679.00148

DFID. (2000). *Sustainable livelihoods guidance sheets*. Department for International Development. http://www.livelihoods.org/info/info_guidancesheets.html

Diwakar, V., Lovell, E., Opitz-Stapleton, S., Shepherd, A., & Twigg, J. (2019). *Child poverty, disasters and climate change*. Overseas Development Institute.

Donner, W., & Rodriguez, H. (2008). Population composition, migration and inequality: The influence of demographic changes on disaster risk and vulnerability. *Social Forces*, *87*(2), 1089–1114. https://doi.org/10.1353/sof.0.0141

Dunn, L. (2016). Integrating men and masculinities in Caribbean disaster risk management. In E. Enarson & B. Pease (Eds.), *Men masculinities and disasters* (p. 209–218). Routledge.

Enarson, E. (1998). Through women's eyes: A gendered research agenda for disaster social science. *Disasters*, *22*(2), 157–173. https://doi.org/10.1111/1467-7717.00083

Enarson, E., & Chakrabarti, P. G. D. (2009). *Women, gender and disaster global issues and initiatives*. SAGE Publications.

Ferris, E., Petz, D., & Stark, C. (2013). *The year of recurring disasters: A review of natural disasters in 2012*. Retrieved April 11, 2016, from https://www.brookings.edu/research/the-year-of-recurring-disasters-a-review-of-natural-disasters-in-2012/

Fothergill, A., Maestas, E. G. M., & Darlington, J. D. (1999). Race, ethnicity and disasters in the United States: A review of the literature. *Disasters*, *23*(2), 156–173. https://doi.org/10.1111/1467-7717.00111

Fothergill, A., & Peek, L. A. (2004). Poverty and disasters in the United States: A review of recent sociological findings. *Natural Hazards*, *32*(1), 89–110. https://doi.org/10.1023/B:NHAZ.0000026792.76181.d9

Gaillard, J. C. (2011). *People's response to disasters: Vulnerability, capacities and resilience in Philippine context*. Pampanga, Centre for Kapampangan Studies.

Gaillard, J. C. (2012). Ch. 38. Caste, ethnicity, religious affiliation. In B. Wisner, J. C. Gaillard, & I. Kelman (Eds.), *The Routledge handbook of hazards and disaster risk reduction* (pp. 459–469). Routledge.

Gaillard, J. C. (2019). Disaster studies inside out. *Disasters*, *43*(Suppl. 1), S7–S17. https://doi.org/10.1111/disa.12323

Gaillard, J. C., & Fordham, M. (2018). Silent, silenced and less-heard voices in disaster risk reduction: Challenges and opportunities towards inclusion. *Australian Journal of Emergency Management*. Monograph #3.

Gaillard, J. C., Gorman-Murray, A., & Fordham, M. (2017). Sexual and gender minorities in disaster. *Gender, Place and Culture*, *24*(1), 18–26. https://doi.org/10.1080/0966369X.2016.1263438

Gaillard, J. C., Maceda, E. A., Stasiak, E., Le Berre, I., & Espaldon, M. V. O. (2009). Sustainable livelihoods and people's vulnerability in the face of coastal hazards. *Journal of Coastal Conservation*, *13*(2–3), 119–129. https://doi.org/10.1007/s11852-009-0054-y

Galtung, J. (1985). Twenty-five years of peace research: Ten challenges and some responses. *Journal of Peace Research*, *22*(2), 141–158. https://doi.org/10.1177/002234338502200205

Gierlach, E., Belsher, B. E., & Beutler, L. E. (2010). Cross-cultural differences in risk perceptions of disasters. *Risk Analysis*, *30*(10), 1539–1549. https://doi.org/10.1111/j.1539-6924.2010.01451.x

Gilbert, J. (2007). Nomadic territories: A human rights approach to nomadic peoples' land rights. *Human Rights Law Review*, *7*(4), 681–716. https://doi.org/10.1093/hrlr/ngm030

Gladwin, H., & Peacock, W. G. (2000). Warning and evacuation: A night for hard houses. In W. G. Peacock, B. H. Morrow, & H. Gladwin (Eds.), *Hurricane Andrew* (pp. 52–72). International Hurricane Center.

Gorman-Murray, A., McKinnon, S., Dominey-Howes, D., Nash, C. J., & Bolton, R. (2018). Listening and learning: Giving voice to trans experiences of disasters. *Gender, Place and Culture*, *25*(2), 166–187. https://doi.org/10.1080/0966369X.2017.1334632

Gorman-Murray, A., Morris, S., Keppel, J., McKinnon, S., & Dominey-Howes, D. (2017). Problems and possibilities on the margins: LGBT experiences in the 2011 Queensland floods. *Gender, Place and Culture*, *24*(1), 37–51. https://doi.org/10.1080/0966369X.2015.1136806

Green, R., Bates, L. K., & Smyth, A. (2007). Impediments to recovery in New Orleans' Upper and Lower Ninth Ward: One year after Hurricane Katrina. *Disasters*, *31*(4), 311–335. https://doi.org/10.1111/j.1467-7717.2007.01011.x

Hancock, A. M. (2007). When multiplication doesn't equal quick addition: Examining intersectionality as a research paradigm. *Perspectives on Politics*, *5*(1), 63–79. https://doi.org/10.1017/S1537592707070065

Haysome, S. (2013). *Sanctuary in the city? Urban displacement and vulnerability*. Overseas Development Institute [Final report].

Heijmans, A. (2004). From vulnerability to empowerment. In G. Bankoff, G. Frerks, & D. Hilhorst (Eds.), *Mapping vulnerability: Disasters, development and people* (pp. 115–127). Earthscan Publications.

Hellman, J. (2015). Living with floods and coping with vulnerability. *Disaster Prevention and Management*, *24*(4), 468–483. https://doi.org/10.1108/DPM-04-2014-0061

Horwood, C., & Reitano, T. (2016). *A perfect storm? Forces shaping modern migration and displacement*. RMMS Discussion Paper.

Hunt, L. (2007). *Inventing human rights. A history*. W.W. Norton & Company.

IFRC. (2018). *Leaving no one behind*. World Disasters Report.

Ingleby, D., & Petrova-Benedict, R. (2016). *Recommendations on access to health services for migrants in an irregular situation: An expert consensus*. https://eody.gov.gr/wp-content/uploads/2019/12/Συστάσεις-Διεθνούς-Οργανισμού-Μετανάστευσης.pdf

International Alert. (2017). *When merely existing is a risk. Sexual and gender minorities in conflict, displacement and peace building*. https://www.international-alert.org/publications/when-merely-existing-is-a-risk/

Jackson, R., Fitzpatrick, D., & Singh, P. M. (2016). *Building Back Right—Ensuring equality in land rights and reconstruction in Nepal*. https://oxfamilibrary.openrepository.com/bitstream/

handle/10546/606028/bp-building-back-right-nepal-210416-en.pdf?sequence=1&isAllowed=y. Oxfam International.

Lee, B. X. (2016). Causes and cures VII: Structural violence. *Aggression and Violent Behavior*, *28*, 109–114. https://doi.org/10.1016/j.avb.2016.05.003

Lewis, J., & Kelman, I. (2012, June 21). *The good, the bad and the ugly: Disaster risk reduction (DRR) versus disaster risk creation (DRC)* (1st ed.). PLOS Currents: Disasters. https://doi.org/10.1371/4f8d4eaec6af8

MenEngage Alliance (2016). *Men, masculinities and climate change*. A Discussion Paper. http://menengage.org/men-masculinities-and-climate-change-a-discussion-paper/

Moncrieffe, J., & Eyben, R. (2007). *The power of labelling: How people are categorized and why it matters*. Earthscan Publications.

Moser, C. (1993). *Gender planning and development: Theory, practice and training*. Routledge.

Nepal Ministry of Health. (2016). *Demographic and health survey 2016*. https://www.dhsprogram.com/pubs/pdf/fr336/fr336.pdf

Pantuliano, S., Metcalfe, V., Haysom, S., & Davey, E. (2012). Urban vulnerability and displacement: A review of current issues. *Disasters*, *36*(Suppl. 1), S1–22. https://doi.org/10.1111/j.1467-7717.2012.01282.x

Peek, L. (2008). Children and disasters: Understanding vulnerability, developing capacities and promoting resilience. *Children, Youth and Environments*, *18*(1), 8.

Preiss, D., & Shahi, P. (2017). Two years after the devastating earthquake, Nepal's women have become easy prey for traffickers. *Time Magazine*. http://time.com/4442805/nepal-earthquake-anniversary-poverty-women-trafficking/

Reale, A., & Handmer, J. (2011). Land tenure, disasters and vulnerability. *Disasters*, *35*(1), 160–182. https://doi.org/10.1111/j.1467-7717.2010.01198.x

Rivers, J. P. W. (1982). Women and children last: An essay on sex discrimination in disasters. *Disasters*, *6*(4), 256–267. https://doi.org/10.1111/j.1467-7717.1982.tb00548.x

Samuels, F., Wagle, S., Sultana, T., Sultana, M. M., Kaur, N., & Chatterjee, S. (2012). *Stories of harassment, violence and discrimination: Migrant experiences between India, Nepal and Bangladesh*. ODI Project Briefing #70.

Sanderson, D. (2012). Livelihood protection and support for disasters. Ch. 58 in B. Wisner, J. C. Gaillard, & I. Kelman (Eds.), *The Routledge handbook of hazards and disaster risk reduction* (pp. 697–710). Routledge.

Sen, A. (1999). *Development as freedom*. Oxford University Press.

Shreve, C., & Fordham, M. (2018). *Child-centred research-into-action brief: Gender and disasters: Considering children*. GADRRRES.

Ten Have, H. (2018). Ch. 11. Disasters, vulnerability and human rights. In D. P. O'Mathúna et al. (Eds.), *Advancing global bioethics* (pp. 157–174). https://doi.org/10.1007/978-3-319-92722-0_11

Turner, B. S. (2006). *Vulnerability and human rights*. Penn State University Press.
UN Department of Economic and Social Affairs. (2016). *ODA allocation and other trends in development cooperation in LDCs and vulnerable contexts*. Development Cooperation Forum Policy Briefs No. 13.
UNDRR. (2018). *Terminology*. https://www.unisdr.org/we/inform/terminology#letter-d
UNDRR. (2019). *Global assessment report, 2019*. United Nations.
UN-Habitat. (2007). *Enhancing urban safety and security*. Earthscan Publications.
UN OHCHR (2018). *Combatting discrimination based on sexual orientation and gender identity*. Last accessed June 10, 2018.
Van der Gaag, N. (2013). Because I am a girl: The state of the world's girls 2013. In *Double Jeopardy: Adolescent Girls and Disasters* (pp. 1–224). Plan International.
Wisner, B. (2016). Vulnerability as concept, model, metric, and tool. In *Oxford research encyclopaedia of natural hazard science*. https://doi.org/10.1093/acrefore/9780199389407.013.25.
Wisner, B., Blaikie, P., Cannon, T., & Davis, I. (2004). *At risk* (2nd ed.). Routledge.
World Health Organization. (2020). *Disability and health: Key facts*. WHO. https://www.who.int/news-room/fact-sheets/detail/disability-and-health.

CHAPTER 6

People's capacities

Fig 6.1
Life goes on.
Source: photo courtesy of Ksenia Chmutina

6.1 What is capacity?

Since the beginning of humanity, people have lived in hazard-prone locations that – due to their inherent characteristics – could support and provide for livelihoods: volcanic slopes provide excellent agricultural land, rivers give access to trade and so on. However, it is not the location

DOI: 10.4324/9781315469614-9

that necessarily determines people's capacities; instead, capacities are endogenous to the community of people who share and combine them in dealing with the same hazards (Figure 6.1).

Capacity is probably one of the most overlooked concepts in disaster studies (although a popular one amongst practitioners). Whilst many appreciate the role of social capital, skills, traditional knowledge or alternative income-generating activities in increasing resilience to natural hazards, the focus of the research tends to inevitably move towards vulnerability (which is often seen as an opposite end of the spectrum) or resilience (which is sometimes seen as the same idea).

Capacity as a concept became prominent in the 1980s (largely thanks to Anderson and Woodrow (1989) who offered guidelines for developing coping strategies in a disaster) as a way to demonstrate that labelling people as vulnerable (i.e. victims or helpless) underplays their skills, knowledge and resources that are reliable, familiar and accessible, and thus, are the first go-to resources when facing a disaster. This also showed that local/indigenous people should lead capacity development because they understand their context best. Such approach has challenged the traditional technocratic approach to DRR that has been prominent then (and remains prominent now) and promoted the idea that people, including most marginalised, should be at the forefront of development because they are knowledgeable and resourceful (Freire, 1970).

The UNISDR (2018) defines 'capacity' as "the combination of all the strengths, attributes and resources available within an organization, community or society to manage and reduce disaster risks and strengthen resilience". This definition emphasises that capacity is not just about the availability of resources and highlights that the potential of communities themselves to play a central role in dealing with disaster risk should not be underestimated. In other words, capacity demonstrates how individuals and communities make gradual adjustments to cope with environmental changes caused by natural hazards without modifying the fundamentals of their social organisation (Gaillard, 2007).

It is important to note that there is a difference between capacity and capacities: the former refers to an ability to do something, whereas the latter should be understood as a set of knowledge, skills and resources. Capacities reflect the anticipation of future changes in the environment (i.e. adaptive capacity) and the experience of past events (founded on a retrospective nature of coping capacity).

Capacities are both an individual and a collective set of diverse knowledge, skills and resources people can claim, access and resort to in dealing with hazards. Everyone possesses a unique set of knowledge that, when combined, becomes diverse. Capacities allow active prevention to avoid an occurrence of hazards, to foster preparedness in facing impending hazards as well as to respond to disasters, and to cope with and recover from the impacts of disasters (Gaillard et al., 2019).

Capacity development has been identified as one of the main ways of substantially reducing disaster losses, as through it, society can foster change and enhance resilient to risks from natural hazards (Hagelsteen & Burke, 2016). But often, the opportunity to harness the powerful capacities of local community is missed through ignorance or reluctance by key decision makers to give up power. There is often a lack of analysis of the relevant risks and initial capacities. The division of roles, responsibilities and ownership is sometimes understood differently by different partners, leading to confusion. External experts, instead of harnessing capacities, apply ready-made solutions and leave before any institutional memory is created. Harnessing capacities for reducing the risk of disaster requires people's genuine participation in assessing and enhancing their existing knowledge, skills and resources (Gaillard et al., 2019). It also implies a shift in power – and gradual empowerment, the process "by which people, organisations, and communities gain mastery over their lives" (Rappaport, 1984, p. 3). Ignoring established systems, strategies and capacities can only result in the creation of parallel structures and processes (Hagelsteen & Becker, 2014).

6.1.1 Coping capacity and adaptive capacity

Sometimes capacity is referred to in terms of coping or adaptation, and occasionally, these are used interchangeably – which may cause some confusion. The differentiation is not clear-cut and is very context specific – but understanding of both is important for enhancing resilience and transformation (Wamsler & Brink, 2014).

In general terms, in the context of disaster risk (and in particular, climate change), coping is a response to an experienced impact; it is the ability to survive amidst an adverse environment. Risk, impact and capacity to cope evolve throughout a person's life cycle and vary amongst social groups. *Coping capacities* are usually affected by and evolve because of physical and temporal characteristics of a natural hazard. For example, communities that live in a flood-prone area could develop different coping strategies, depending on the intensity and the duration of the inundation. Coping capacities are, thus, retrospective by nature, as they are based on the experience of past events (Figure 6.2); but they are flexible and respond to changes in economic, social, political and institutional conditions over time (Smit & Wandel, 2006). For instance, resource depletion may gradually reduce a coping capacity of a community.

Adaptive capacity is the ability to adapt to environmental change (e.g. Smit & Wandel, 2006); it is the process of adjusting to change (both experienced and expected), which is longer term. These changes may be small or large, punctual or long-lasting. Adaptive capacity has a prospective dimension, as it builds around the ability to anticipate future changes in the environment. In recent years, adaptive capacity has been

Fig 6.2
Peace Boat volunteers distribute curry to survivors of the 11 March tsunami in Ishinomaki, Japan, March 2011. Source: photo courtesy of Wesley Cheek

largely used in the context of climate change, somewhat synonymously with adaptability, and it has been argued that "increasing adaptive capacity improves the opportunity of systems to manage varying ranges and magnitudes of climate impacts, while allowing for flexibility to rework approaches if deemed at a later date to be on an undesirable trajectory" (Engle, 2011, p. 647).

Very often, people rely upon and develop both coping and adaptive capacities. Coping capacities, for instance, are important during a disaster, as local people are invariably the first responders (Quarantelli & Dynes, 1972). This is particularly the case during small and frequent events that do not qualify as disasters by the outsiders. Often, these capacities are the only available resources for marginalised people, whose rights and needs are overlooked by those with more power (Gaillard et al., 2019). Many people who live in remote – and thus, largely neglected – areas that are regularly affected by hazards rebuild their lives on their own: they clear debris, repair houses and engage in diverse income-generating activities to sustain their daily needs, without relying on outside help.

However, in the long run, coping capacities – since usability and effectiveness may be limited or even gradually decrease because of values, processes and power relations in society (Adger et al., 2009) – should be transformed into adaptive capacities aimed at the transformation of the structure, functioning and organisation of a system (Berman et al., 2012). There is still very little understanding how, in practical terms, adaptive capacity can be built on coping capacity. It is also important to note that whilst here we discuss coping and adaptive capacity, as well as vulnerability and resilience separately, these concepts are intertwined and overlapping (as well as contested and politically weighted).

6.2 Characteristics of capacity

Capacities are dependent on four factors that change significantly in time and space (Gaillard, 2007):

The *nature of a hazard* implies the frequency of recurrence as well as level of magnitude of the hazard that affects a community. This shapes the long-term consequences for societies: annual floods brought by monsoon seasons are, arguably, easier to prepare for than an earthquake that may happen unexpectedly. The extent of damages also plays an important role. For the communities whose livelihoods get completely wiped out and relocation is the only option, developing adaptive capacities are existential, as often, relocation results in living in a close proximity to another group, whose culture maybe completely different. Hazards, such as hurricanes and floods, on the other hand, allow post-disaster reoccupation; here, coping capacities are important.

The *intrinsic socio-cultural condition* of a particular group that faces a hazard are the conditions that explain pre-disaster level of acculturation, the relationships between the affected groups and their neighbours, the diversity of livelihood, the cultural attachment to a place, the size of the community, its demographics as well as the willingness of traditional leaders to change. For instance, the 1991 eruption of Mount Pinatubo Volcano in the Philippines forced the Aeta Negrito communities residing on the slopes to move to the foothills, which were already occupied by lowland ethnic groups. The relocation resulted in intensification of economic, social and political interactions, leading to integration of foreign socio-economic and cultural elements that were traditional for the mountain community into the lives of the community living at the foothill. The mountain community impacted changes in the settlement pattern, religion, language, medicinal treatments, diet and farming activities. At the same time, the community that lived at the foothills of the volcano and had already established contacts with other lowland groups and American servicepeople on duty at the neighbouring airbase, has brought to the mountain community Western socio-economic references (Gaillard, 2006).

Geographical setting is directly linked to the two factors discussed earlier and is closely linked to the magnitude and the extent of damage brought by a hazard. Geographical location determines the amount of space available and, therefore, the extent to which relocation would result in an encroachment on other groups. For instance, the 1961–1962 eruption of the volcano of Tristan da Cunha left no other alternative to the inhabitants of this small island but to relocate to England, a place with very different Western habits that consequently deeply penetrated their traditional way of life (Lewis et al., 1972).

The *post-disaster governance* also plays a critical role in how capacities can be enhanced or decreased. Some post-disaster policies (particularly those where relocation is involved) have hidden agendas aimed at pushing the cultural change. The Mexican government, for example, tried to use the resettlement policy and the associated social services programme to "civilise" the Indians following the eruption of Paricutín Volcano (Nolan, 1979).

These four factors are closely intertwined yet extremely diverse (and they change constantly through time and space). Such diversity emphasises that enhancing capacity is only possible through a local contextual consideration of a challenge (rather than through a transfer of technology from one country to another).

Capacities are often characterised using the same typology of resources (also referred to as assets or capital) used for assessing vulnerability (see Chapter 4):

- Human resources provide strength, knowledge and skills to face hazards; these include local knowledge, local healthcare, literacy and numeracy, experiences lived through past events (positive and negative);
- Physical resources help establishing safe housing and infrastructure; these include vernacular architectural, traditional water management, irrigation, homeland tenure;
- Natural resources ensure that there is enough food and water to cope with shortages; these include biodiversity, hazard-resistant crops, wild fruits and animals, water springs, hazard-resistant crops, seed banks;
- Political resources provide decision-making power and ability; these include leadership, transparent and flexible decision-making;
- Economic resources ensure that there is enough money to cope with crises and losses; these include local market, micro-credits, micro-insurance; and
- Social resources are the foundation for solidarity and include social networks, kinship ties (Wisner et al., 2012).

All these resources are employed by various people in different – and often creative – ways in order to prevent, mitigate, cope or recover from the impacts of natural hazards. These resources can be tangible and intangible, local and transnational, accessible to all or to just a few – thus, identifying them can be problematic; moreover, they are intertwined and overlapping. For instance, floodwalls that are built using traditional material (Figure 6.3) are usually considered as physical resources; but in reality, they reflect a combination of a deep understanding of the natural environment and crafts kills learnt from the elders.

Capacities usually include a whole range of resources (rather than one particular resource), but each type of resource requires specific knowledge and skills (Gaillard et al., 2019). Diversity of resources is

Fig 6.3
Traditional floodwalls in Japan.
Source: photo courtesy of Ksenia Chmutina

also important: usually, societies that rely on a unique livelihood have less capacity. For example, communities relying exclusively on the natural resources available in their immediate environment are much more vulnerable if these resources are partially or totally destroyed by a hazard. Having a spectrum of resources that help sustain a livelihood allows people relying less on external support. Intangible resources should also consider power relations, as they often determine the levels of capacity.

6.3 Capacity and vulnerability

Although often seen as two ends of the same spectrum, capacity and vulnerability are not the opposite of each other: reducing vulnerability does not always enhance capacity and the other way around (Davis et al., 2004). The majority of people – even most marginalised – have some capacities (although these are often underappreciated), and no one should be labelled as a helpless victim (which the interpretation of vulnerability often leads to).

Whilst not always, capacity is often grounded in resources that are local and endogenous to the community that is facing the hazard. Vulnerability, on the other hand, is often (but not always) about the structural constraints on access (such as unequal redistribution of wealth, political systems and market forces). In other words, people have more control over capacities than vulnerabilities, as they often have little or control over external factors that create vulnerability. In practical terms, it is easier to develop and enhance capacity than to reduce vulnerability (Wisner et al., 2012).

Portraying people as vulnerable, however, allows justifying an intervention and attracting resources, and this is one of the reasons why DRR efforts focus on addressing the weakness instead of emphasising the strength. However, in reality, what these efforts are doing is enhancing capacity; they do not change the root causes of disasters – that is, the cultural, political and social systems (that have already been discussed in detail in Chapters 4 and 5) – and thus, vulnerability still remains.

6.4 Capacities at a household/community level

Capacities that people display when facing a disaster are seldom unexpected. At a household level, capacities are often endogenous to communities, meaning that people have more control over them; they are built on everyday skills, resources and strengths that people use in their daily life. Some of these are local (e.g. traditional architecture), others are transnational (e.g. money sent back home by a migrant worker); some are individual (e.g. ability to repair), others are collective (i.e. pulled together through social networks). It is very difficult (if not impossible) for an outsider to recognise these resources – but for the locals, they are an organic part of their daily life. Ultimately, the more knowledge, skills and resources people have, the more capacities they can resort to. The diversity of capacities is, thus, important: should one or several of capacities be affected, there are others that are resistant enough.

Although capacities are often shared and pooled between people, knowledge, skills and resources are never evenly distributed across individuals living in the same place or sharing kinship. The possession of information is a central element of power (Ostrom et al., 1993): people are often excluded from participating in decision-making because of their exclusion from education and formal knowledge (Freire, 1985). Local knowledge possessed by, for example, a chief of a village may be a source of power and prestige kept away from most people by a few individuals entitled to hold such knowledge (Wisner et al., 2014). Capacities also differ according to age, gender, ethnicity and physical ability. The ability to claim and access collective capacities, thus, depends on power relations amongst individuals.

The levels of capacities are not independent: the capacity of a household to cope with risks to an extent depends on the enabling environment of the community, which in turn is reflective of the resources and processes of the region (Smit & Wandel, 2006). Thus, the way society is organised has huge implications for capacities. Family units, for instance, have very different meanings for different communities: in the West, these are small, and capacities would be aimed at protecting these small units and accumulating resources for them (i.e. quite individualistic); whereas in many Asian countries, the family unit is seen as a much larger social network based on sharing culture. All these factors have to be considered to truly understand what capacities already exist and what capacities need enhancing.

6.4.1 Traditional knowledge and practice

Knowledge is neither unitary nor universal, and a whole range of knowledge – and its implementation – is required to understand hazards, risks and vulnerability to be able to enhance the right capacities. Nevertheless, knowledge helps constructing meanings for places, and various cultures have developed their knowledge and wisdom over a long period, inhibiting or spurring on desire to address disasters.

Traditional practice can broadly be divided into four types based on their origins and use (Program for Strengthening Household Access to Resources, 2014):

- Traditional practice is an established or inherent way of acting in the community; it is widely accepted and integrated with broader cultural traditions. Although it gradually evolves, there are certain ways in a community that ensure its continuity and reproduction.
- Experience-based practice is empirical and evolves as a consequence of practical engagement in everyday life, irrespective of the origin of the knowledge. Community develops it through experimentation and experiences, and then transmits through demonstration and replication.
- Communally-trusted practice is based on informally applying intimate understanding of the environment. Its efficacy and dependability are not validated by scientific methods; nevertheless, the community continues to rely on these beliefs to get desired results from their application.
- Reliant on locally available resources practice includes materials and tools (that are often a by-product of household materials) as well as skills required that are readily available within the community.

Traditional knowledge informs capacities that allow avoiding an occurrence of a hazard (Box 6.1). For instance, multiple accounts of traditional bushfire management practices across Latin America, Africa and Australia are recorded – all emphasising the reliance on people's very fine understanding of their environment, of factors at the origin of fire, and of the negative and positive impacts of fire for their livelihoods (Gaillard et al., 2019). Traditional houses associated with specific building techniques are usually resistant to the destructive effects of earthquakes. Such measures are usually much cheaper than "innovative" technologies and easier to implement.

Traditional stories relating to cloud formation, wind direction, rain, drought or the impact of weather of crops are common in Rajasthan, India; they help various indigenous communities adapt to the local environment. For instance, the unusual sounds or behaviour of wild

Box 6.1: Early warning systems based on local knowledge in Malawi

The early warning indicators of flooding in Malawi are often based on various phenomena that local people observe in the surrounding environment. Here are some examples:

- Elderly community members feeling pain in certain body parts before the occurrence of heavy rainfall;
- Villagers unable to sleep due to increased temperatures;
- Number of ants (*nyerere*) increase in the villages;
- Animals (e.g. hippos or *mvuu*, crocodiles) migrating from the rivers to the fields and villages;
- Birds producing specific sounds (e.g. trumpet bird or *n'gombe n'gombe*);
- Tamarind tree (*bwemba*) producing an increased number of flowers;
- Production of fruits of the mango tree increase;
- Bamboo (*bende*) growing next to riverbanks increase;
- Halo around the moon (*chikwa*);
- Occurrence of Orion star (*nthanda*);
- Sounds of waters in the rivers increasing;
- Colours of waters getting dirty and muddy; and
- Rate of water level increase.

Source: Šakic Trogrlic et al. (2019)

animals or insects, or changes in water flow or colour are used to assess the probability of flood. The traditional texts also contain information about DRM; *Atharvaveda*, for example, discusses drought mitigation strategies, and *Arthashastra* (fourth century BC) describes mitigation measures in case there is a famine due to a drought (Pareek & Trivedi, 2011).

The indigenous knowledge plays an important role for some communities during the Indian Ocean tsunami in December 2004. For example, the Moken community in Thailand knew the signs, such as unusual behaviour of animals, birds and low tide, as indications for a tsunami from their traditional stories. This allowed the community to move away from the sea towards protective areas (Arunotai, 2008). Some indigenous communities in Bangladesh believe that the cyclone is coming when crabs are climbing on houses, vata fish is coming near the river ghat or heron is flying in flocks. They also use various traditional techniques to minimise the impact of a cyclone; these include keeping jute stalks on platforms covered with plastic sheets, laying a row of bricks around plinths or keeping tools (ploughs and yokes) on raised platforms (Program for Strengthening Household Access to Resources, 2014).

Through preserving traditional knowledge, the people of Niuafo'ou in Tonga, who were relocated to the neighbouring islands of Nukualofa and Eua because of the 1946 volcanic eruption, were successful in re-establishing their community with respect to their ethnic traditions and architecture. The main Niuafo'ou territorial landmarks (churches, schools, stores) were conscientiously rebuilt (despite some hostility from the local population). This meant that when some of the relocated people were permitted to go back to their native island, they are able to quickly rebuild their villages abandoned for more than a decade.

This, however, does not mean that people do not change their ways and that capacities are rigid – as has been noted in Section 6.1, capacity development is about gradual adjustments that are informed by what is known. Observing unusual animal behaviour, for instance, may activate an early warning system, which involves both traditional devices, such as gongs, as well as modern technology, such as mobile phones. A good example of gradual adjustment is an island of Tikopia in the Solomon Archipelago: its population was severely affected by typhoons in 1952 and 1953 that led to a famine (Boehm, 1996). As the resources of the island were largely destroyed, the Tikopia people collectively had to adjust their traditional way of life to meet their daily needs; as a result, they temporarily abandoned fallow periods, redefined agricultural rights, introduced stricter punishments for crime and even adjourned wedding ceremonies. Such changes had been agreed by the whole community during daily public assemblies. Spontaneity and creativity contribute to the constant enrichment and evolution of capacities, as they allow increasing knowledge, enhancing skills and diversifying resources (Gaillard et al., 2019).

However, it is also important to note that some traditional knowledge actually increases exposure and vulnerability. For instance, the Javanese community living near Mount Merapi Volcano believes that the village they live in and the land they cultivate are also their ancestors, and is, thus, controlled by divine forces. During the eruptions, this community often refuses to evacuate their village, or when they evacuate – they always return (Lavigne et al., 2008).

6.4.2 Traditional practices and rituals

Culture shapes the meaning and perceived impact of a disaster; community members' behaviours before, during and after a disaster; and efforts to cope at individual and community levels. The impulse to perform rituals as a response to good or bad events in life is probably as old as humankind; thus, it is critical to explore existing cultural practices and traditions as part of people's capacities.

A ritual (in religious or other contexts) is expressed symbolically, as a performative act, which people carry out in order to make or find meaning. This functional experience of a ritual is often individual but highly culturally dependent, making one feel as a part of a community. Meaning is important for re-establishing strength and well-being in a situation when a normal functioning is broken apart (Danbolt & Stifoss-Hanssen, 2017).

Cultural representations, such as rituals, embody widely accessible ideas and practices that provide meaning to the social life of people, both pre- and post-disaster (Figure 6.4). In Lalitpur, Nepal, some rituals are built around a belief that a disaster may occur – and thus, generate awareness, provide early warning for their (possible) occurrence and facilitate appropriate actions to cope with the misfortune.

Traditions can also help in a post-disaster recovery process (Box 6.2 and Figure 6.5). Rituals and festivals act as an occasion to get together and socialise, thus helping to cope with stress. These rituals are also familiar – people grow up with them – helping people to feel normal again (Bhandari et al., 2011). Evidence exists that traditional religious practices, including rituals, arts and festivals, have provided psychological support to the survivors of the 2011 earthquake and tsunami in Japan (Miichi, 2016).

Cultural norms, however, can also create barriers for capacity development, as often, deeply rooted discriminatory socio-cultural values and traditions exclude some community groups from participation, thus creating further vulnerabilities. Such norms often form a foundation of rules of behaviour that govern belief systems as well as an organisational structure, and therefore, guide how each individual must act – or, crucially, not act – in certain situations (Jones & Boyd, 2011). For example, the local informal rules dictate to the lower-caste Hindu women in parts of Western Nepal not only their daily activities (e.g. household duties, unequal access to education, inability to participate in village meetings and politics) but also the appropriate behavioural norms that are afforded to the lower castes (e.g. restrictions in the ability to own land and in employment, and access to key resources; compliance with Hindu rituals, values, and beliefs; abiding by caste structures; dietary

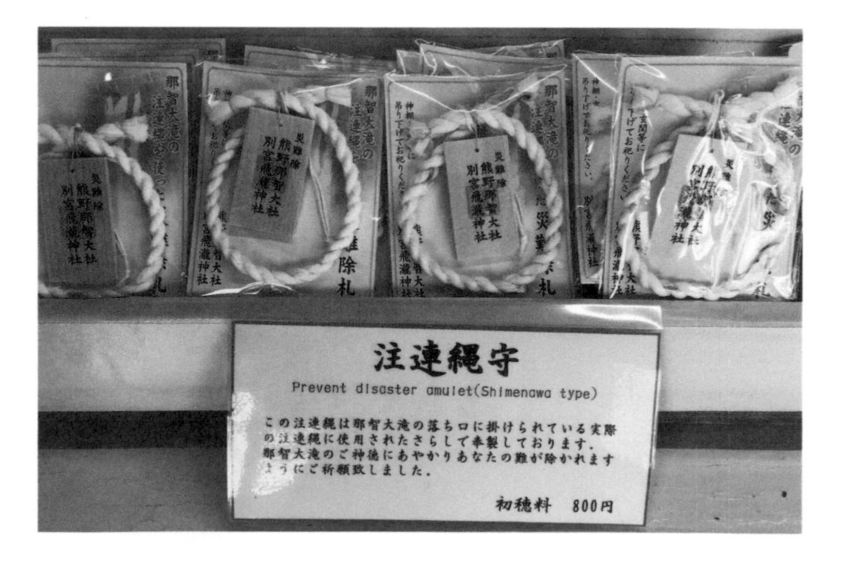

Fig 6.4
Disaster-preventing amulet.
Source: photo courtesy of Lee Bosher

Box 6.2: Culture and capacities

Fig 6.5
QR code for Box 6.2. In this episode of *Disasters: Deconstructed* podcast, Aparna Tandon, Senior Programme Leader at the International Centre for the Study of the Preservation and Restoration of Cultural Property (ICCROM), discusses the importance of preserving cultural heritage through disasters and the tensions that arise. Available at https://disastersdecon.podbean.com/e/s3e9-culture-capacities-good-bad-ugly/

restrictions). Whilst marginalised groups have developed their own capacities, these often lead to maladaptation in the long run.

6.4.3 Interpersonal relationships and social networks

Interpersonal relationships and social networks play key roles in establishing collective capacities: here, individual knowledge and skills get shared amongst different people. These networks and kinship ties are support systems people resort to in dealing with hazards through pooling, sharing, offering, swapping, pawning, buying and loaning resources. These are often referred to as social capital.

Social capital is what gives us the sense of community and is usually based on four elements (Gil-Rivas & Kilmer, 2016; McMillan & Chavis, 1986):

- Membership: a personal sense of belonging to and investment in the community that provides a sense of emotional safety and identification;
- Influence: the effects of members on each other and on the whole of the group, and the reciprocal power the group has on individual members;
- Integration and fulfilment of needs: the degree of shared their values and goals, and the extent to which the community allows to fulfil these goals; and
- Shared emotional connection: frequency and quality of interaction between the members and shared rituals.

Interpersonal relationships can be divided into two categories: bonding and bridging. Bonding ties are shared between co-identifying individuals (e.g. ethnic or religious groups). Bridging ties are the relationships of exchange, where people have shared interests or goals but may have contrasting social identity (Woolcock & Narayan, 2000). Strong bonding ties are often prominent during a disaster; for instance, 85% of the 1985 Mexico City earthquake survivors were rescued by their friends or neighbours. These ties are also critical in a recovery process: turning to close-knit groups reduces the exposure of group members to perceived external risks (Pelling, 2003). Social relationships are also important in a pre- and post-disaster setting: social connections and the actions of community members can facilitate preparedness and increase the likelihood of evacuation and joint social action (Aldrich, 2012). Māori can call upon their network of kin and cover the food and sleeping needs of tens to hundreds of evacuees in a couple of hours through *marae* (Māori traditional settlement compounds and spiritual homes) (Hudson & Hughes, 2007).

Social networks are not just local networks. These are also endogenous resources, access to which is ensured through social networks (e.g. money sent back home by a migrant worker), as well as practice networks. Social ties with a variety of formal (e.g. schools, work, national organisation) and informal (e.g. neighbourhood) groups can facilitate access to economic and informational resources, and increase capacity to influence decision makers and policymakers. Often place-based networks and practice-based networks overlap, and this sometimes can lead to conflicts of interests.

The core components of social networks are trust and reciprocity. 'Trust' can be defined as "the expectation that arises within a community of regular, honest and cooperative behaviour, based on commonly shared norms on the part of other members of that community" (Fukuyama, 1995, p. 26). This implies that trust is not an action but a notion that exist through the command of social norms; trust is built on actions but also on credentials or reputation (e.g. trust in what an expert says) and is critical in mobilising a community.

Reciprocity is a social attribute through which trust is enacted, be that in transferring information or resources. Reciprocity manifests in a number of ways: it can be a routine exchange of gifts (e.g. between neighbours), or an individual might help another without expecting anything in return. At the same time, it can be used as an informal system where participants with a wide array of social contacts are able to shortcut formal procedures for accessing goods and information (Pelling & High, 2005). Both trust and reciprocity are, therefore, important to consider in the context of capacity, as it can play an important role in shifting the power balance.

6.5 Capacity at an institutional level

Capacity at the institutional level refer to a broader system within which individuals and organisations function. Here, institutions comprise the formal legal rules, as well as informal social norms and cultural beliefs, through which they can enable and maintain certain practices, as well as exclude certain actors or constrain certain practices. Establishing capacities at an institutional level if often referred to as a way of establishing an enabling environment.

Capacity at individual, household and community levels affect and are affected by capacity at an institutional level. The interaction between individuals within a community, and at the same time, interaction of communities and individuals with policies and processes that are determined by external agents (individuals and organisations), impact how we respond to hazards (Berman et al., 2012). Governments, civil society and the private sector, therefore, have an opportunity and obligation to work together to commit to a safer future (Figure 6.6), and therefore, their capacity for engagement can be assessed across all sectors (e.g. finance, planning) and levels (e.g. small and medium enterprise, farmers) (UNISDR, 2018).

Institutions dictate the functioning of markets, local governance, common-pool resources, and land tenure and access, all of which are important for the ability of a community to respond through short-term coping, as well as to adapt over the longer term. Capacity at an institutional (e.g. local government) level should incorporate the following features (Jones et al., 2010):

- Asset based: availability of key assets allowing to respond to changing circumstances;
- Institutions and entitlements: existence of an appropriate and evolving institutional environment that supports fair access and entitlement to key assets and capital;
- Knowledge and information: the ability to collect, analyse and disseminate knowledge and information in support of adaption activities;
- Innovation: an enabling environment for fostering innovation, experimentation and the ability to explore niche solutions in order to take advantage of new opportunities; and

Fig 6.6
Water canon at Ninna-ji Temple in Kyoto, Japan. Source: photo courtesy of Wesley Cheek

- Flexible, forward-looking decision-making and governance: governance structures that are able to anticipate, incorporate and respond to changes, taking into account future planning.

It is argued that transformation from coping to adaptive capacity should be driven at an institutional level; this, however, is not easy to achieve (Berman et al., 2012). Transformation is necessary, as present coping activities may unintentionally affect the future adaptive capacity of a system and lead to maladaptation (O'Brien et al., 2008). Similarly, pursuing long-term goals may distract from the immediate needs and, thus, increase vulnerability. For example, charcoal is an important income source during the drought period as well as a key fuel source for many rural communities in Uganda – yet policies that are designed to decrease deforestation (and thus, limit the production of charcoal) can undermine the very coping strategies that these communities use in times of drought.

Moreover, capacity development often comes down to resources and commitments to ensure that these resources are effectively utilised. For low- and middle-income countries, international aid could help – but the contrast is striking when compared with what is spent by donors and development banks on relief (humanitarian assistance), including post-disaster reconstruction, as opposed to capacity development (O'Brien et al., 2006). Thus, at an institutional level, capacities are best developed through mainstreaming capacity development activities into governance processes. This requires relatively low amounts of personnel and resources but can achieve high levels of impact (IFRC, 2015).

6.6 Assessing capacity

Capacity assessment is not as common as vulnerability assessment – although it very often relies on similar typologies. The assessment usually focuses on reviewing individuals', community's or organisation's capacity against desired goals through understanding existing strengthens that should be enhanced and maintained, and gaps that require further action (UNISDR, 2018). In other words, capacity assessment should be about the analysis of risks and the analysis of capacities that are currently available to manage them. Four key components that feature in most of capacity assessments are institutional arrangements, leadership, knowledge and accountability; whilst not every assessment covers all four components, they should, nevertheless, be considered when defining the scope of an assessment.

6.6.1 Vulnerability-Capacity Assessments (VCA)

VCA is a common tool used by NGOs in DRR practice. It allows looking at vulnerabilities and capacities at the same time, presenting opportunities for strengthening – in terms of the impact of a disaster on physical, social and attitudinal element of life.

A typical VCA would usually come as a two-by-three matrix constructed of these elements as the basis for conversations with local people as participants in planning recovery and reduction of risk to the next hazard. Such matrix allows disaggregation by gender, economic class, age, ethnicity, religion or any other characteristic. The objectives of a VCA are typically to support a community exposed to various hazards in identifying the hazards that affected them, categorising their vulnerabilities and mapping out their capacities, in order to engage in developing a DRR plan and disseminate the results amongst the relevant stakeholders. The assessment relies on a combination of qualitative and quantitative methods, including secondary data (e.g. hazard maps), focus group discussions involving a variety of participatory tools (e.g. timelines, seasonal calendars and transect walks) and discussions with local stakeholders (Wisner, 2016).

There are myriads of guidelines for creating and carrying out a VCA – and there are plenty of examples showing that a more localised application of VCA can be an essential part of a longer-term process of building confidence in local people to make their needs and their ideas heard by decision makers. For instance, a VCA conducted by IFRC in Yemen in 2005 in 2 districts badly affected by flash floods exposed that over the past 15 years, more people have been killed in road accidents than as a result of flooding. A road safety programme designed to reduce such accidents, especially near schools, was therefore initiated, which has been much appreciated by the local population.

However, there is also a lot of critique of VCAs, as, after all, it is a management tool based on the pre-defined concepts (whether they make sense locally or not), and quantitative and/or demographic indicators that help ticking the box. Critics of the VCA argue that people's response to disasters cannot always be understood through standard criteria and methodologies designed by outsiders. An interpretation of an outsider would always reflect their own knowledge, assumptions and values, thus "creating false 'stories' that fit her or his expectations" (Bhatt, 1998, p. 71). This is especially problematic when terminology used in VCAs (and other tools) does not translate well and, thus, is not meaningful in a local language (Chmutina et al., 2020).

Assessing and strengthening capacities should, therefore, be considered as a community-driven process. An effective VCA should emphasise the process through which people realise by themselves the extent, strength and diversity of their individual and collective capacities – and how these can be maximised when facing hazards.

6.7 Harnessing capacities

Harnessing capacities is about strengthening people's strategies that help facing a hazard. Whilst the aim is generally shared by different organisations, the terminology that used to achieve it is not. Some focus on 'capacity building', whereas others refer to 'capacity development'.

These are often used interchangeably, but there are subtle differences. 'Building' suggest erecting a new structure, starting from a clean slate – based on a preconceived design – and implies that capacity is something that can be 'built' by those who have created that preconceived design (often an outside). Such an approach fails to incorporate existing structures and norms that have been discussed earlier in this chapter. 'Development', on the other hand, implies that capacities already exist and need growing (Hagelsteen & Becker, 2013). Here, 'harnessing', 'developing' and 'enhancing' are used to highlight that capacities need strengthening rather than reinventing.

Capacity enhancement often comes with a price: labelled as 'capacity building' or 'capacity development', the activities delivered as a result of capacity assessments entail a transfer of knowledge and endowment of external resources, frequently with no or little consideration for local people's knowledge, skills and resources. This results in those who are building local capacity actually focusing on building our capacity to reduce their risks or our capacity to build their capacity to reduce risks, instead of strengthening people's own capacities to reduce their own risks (Bhatt & Pandya, 2021).

As noted earlier, in practical terms, it is easier to develop and enhance capacity than to reduce vulnerability. Activities that enhance capacities are usually focused on individual/household or community level, as the key aim of such activities is to strengthen people's strategies that help facing a hazard. These strategies range widely, from developing a system of signals for early warning, to creating evaluation routes and meeting points, to building or strengthening shelters and preparing resources that would be needed in case of a disasters. However, for capacity development programmes to be effective, they should include the following seven elements: terminology (i.e. removing ambiguity in the terminology used regarding DRR); local context; ownership; capacity assessment; roles and responsibilities; mix of activities; and monitoring, evaluation and learning (Hagelsteen & Becker, 2014).

If the outcome of these strategies is to be effective, they have to be based on what is available locally and foster creativity and should focus on people's ability to cope with hazards organically. The strategies have to be informed by the local context, including general political, social, cultural, economic, physical and environmental factors, and by the understanding of the relationships and dependencies between individuals or organisations (Hagelsteen & Becker, 2014). These strategies cannot be normative, imposing the normative DRR measures. Enhancing and harnessing capacities is, therefore, not an individual task. This means that strengthening capacities involves addressing power relations amongst local people to facilitate claim and access to locally available knowledge, skills and resources amongst individuals (Gaillard et al., 2019). Enhancing capacity also requires challenging the norms. Local knowledge, for instance, can be a symbol of privilege; having knowledge is closely connected to having power. If a chief of a village does not share the knowledge with the villagers, getting access to this knowledge would mean challenging the power of the chief.

There is no silver bullet to empowerment – every such approach should be context driven, as it requires a process through which people realise by themselves the extent, strength and diversity of their individual and collective knowledges and perspectives. However, the fundamental ideas of such approaches can be informed by well-known principles of situated learning, offered by Paulo Freire, as a way to reflect on consciousness, generativity and shared objectives. These ideas can, thus, be framed around five underlying principles of Freire's ontological call (Suzina & Tufte, 2020):

- *Dialogue*, as a central mechanism of change, is communication that allows us to learn and deconstruct knowledges in a way that enables a deeper understanding of the context. The dialogue is never finished and always open.
- *Humility* recognises that every person is knowledgeable and allows us to become aware (and critical) of oppressive structures. Knowledge is constantly constructed and reconstructed, but authentic word, in Freire's terminology, requires permanent action and reflection that takes into account the multiple realities and the perspectives.
- *Empathy* recognises that there are different points of departure that make it harder for some to speak about or even employ their knowledges. It allows to deal with contradictions and disagreements, both in reality and in in epistemology and ontology, in contextualised ways.
- *Love* is a model of development based on collective ties and interconnections, including all human and non-human beings. Love would teach how to connect local solutions to global problems, and to understand the relationships and connections of ecologies and societies, engaging with historical legacies and practicing courage and solidarity.
- *Hope* is the principle and the rule for achieving a critical view and a permanent search for change towards a more just society. It allows envisioning possibilities that traditional knowledges could lead to, and mobilisations for and with communities and individuals who would benefit from these knowledges.

Considered in the context of DRR and empowerment of indigenous and other communities, Freire's principles can, thus, challenge the top-down decisions and approaches, and instead lead to creating a more open, plural, inclusive and democratic ones.

6.7.1 The role of participation

People's genuine participation is essential for harnessing capacities, as effective capacity harnessing can only occur through assessing and enhancing existing knowledge, skills and resources (Wisner, 2016). People who face disasters are the people who know and understand the nature, extent, strength and diversity as well as the limitations of their own capacities.

Inevitably, capacity and participation are closely intertwined – and the idea of participation is politically loaded. Participation is about challenging power relations and recognising the power of the locals who have skills and knowledge to make an informed decision by themselves. Enhancing capacities, therefore, requires a transfer of power to benefit local people, whilst recognising their knowledge, skills and resourcefulness. Participation needs to be bottom-up (i.e. local population is included in decision-making from the earlier opportunity and throughout the process), be inclusive (i.e. the group is not dominated by an interest group or a particular demographic group (e.g. men or houseowners)) and have representative leadership.

Inclusivity is extremely hard to ensure, particularly in those societies without a history of public involvement in decision-making or in deeply divided communities (e.g. migrant communities and slums). As noted earlier, communities are not homogenous: the information asymmetries and unequal distributions do not only exist between a community and other political actors but also within the community (Pelling, 1998). Political neglect and social discrimination are often evident when capacities of certain marginalised groups are ignored. For instance, *warias* (individuals who are biologically male but adopt distinctly feminine features and identity, and who have very low social status) were largely invisible during the 2010 Mount Merapi eruption in Central Java Indonesia. Nevertheless, they played an important role in recovery by providing free haircuts and make-up services to the people in evacuation centres, and thus helped to re-establish the social, emotional and physical well-being of people (Balgos et al., 2012).

Establishment of leadership can also pose challenges, as it has to be both deeply rooted in the local community to ensure representation and accountability, as well as have substantial linkages with external institutions to gain resources for achieving the community goals. Community leaders, thus, occupy a strategic position: they are the gatekeepers to social power and information in the local community. Because of religious and cultural norms, leadership positions can be less accessible to women, the young and old, or minority ethnic and low-caste groups. This again can lead to the emergence of a dominant leader group (Pelling, 1998).

It is important, therefore, to acknowledge the political dimension of capacity to be able to mobilise people and communities. Capacity building activities can help communities to feel more connected and empowered; this, however, does not mean that by mobilising a community, the vulnerability of that community would be decreased. Participation also helps to grow the sentence of ownership, commitment and primarily responsibility.

Box 6.3: Learning from communities

Fig 6.7
QR code for Box 6.3. In this *Disasters: Deconstructed* episode, Mihir Bhatt, Director of the All India Disaster Mitigation Institute, focuses on how we can learn so much from those experiencing risk, and how we need to centre these people and their capacities rather than always bringing external knowledge through teaching and training – "too much schooling, not enough active learning". Available at https://disastersdecon.podbean.com/e/s3e14-learning-from-communities/

A desire to work in partnership is not enough to ensure ownership: it requires genuine commitment of time, structured activities, creativity and flexibility from the actors involved (IFRC, 2015). Taking ownership is always voluntary; this cannot be imposed on someone by someone – and these capacity development strategies should be driven from the inside, with external partners supporting the activities but not leading them (Hagelsteen & Becker, 2013).

Harnessing capacities requires a careful approach to participation not only through the appropriate and carefully thought through tools and methods but also through proper consideration of behaviours and attitudes that should be designed to empower local people to share, analyse, and enhance their knowledge, skills, and resources in order to plan, implement, monitor, and assess their own initiatives for DRR, rather than to blindly participate in what others have offered. Participation should be built on trust and should offer room for innovation, building upon people's inherent spontaneity and creativity (Gaillard et al., 2019). It should be built on a commitment to balance power relations between those currently holding most power and people who face hazards and disasters (Box 6.3 and Figure 6.7).

6.7.2 Community-based disaster risk reduction

Traditionally, top-down, interventionist and technocratic approaches, often driven by outside experts, were predominant in DRM. However, in recent decades, increasing emphasis has been placed on community-based approaches (Box 6.4 and Figure 6.8).

Community-based disaster risk reduction (CBDRR) (also known as community-based disaster preparedness (CBDP)) are based on the idea that the inherent capacities of communities get examined after every disaster. Whilst people can control their own capacities, they cannot control external conditions (Wisner et al., 2014). Thus, capacities should be strengthened – and community level (which, in case of CBDRR, is interpreted as the population living with specific territorial bounds) is

Box 6.4: Games as a participatory activity

Humans have played games for centuries, and in recent years, there has been an increasing focus on the use of games for participatory activities – mainly because games encourage greater engagement in people. Games are increasingly used by international aid agencies for increasing disaster risk awareness, risk assessment and decision-making. The International Federation of the Red Cross and Red Crescent Societies have over 30 online games surrounding hazards, climate change and climate risks available on their website, whilst PreventionWeb lists over 45 games for DRR on their website. Games are appealing because they provide the following feelings amongst those who are involved in playing:

- Urgent optimism – the desire to act and the belief in achieving success;

Fig 6.8
Games as a participatory activity.

- Social fabric – the ability to trust and form stronger social bonds through game playing;
- Blissful productivity – the belief that the task they are engaging with is meaningful, hence the dedication towards the game task itself; and
- Epic meaning – the strong attachment to a meaningful and awe-inspiring story that they are personally involved in and striving to make their mark on it.

Game design and game elements confer such power to people that it can transform their relation with services, products, policies or even everyday tasks that can be monitored tracked and modelled within a gamespace. Games enable the participants to test theories and ideas with a freedom to fail without real-world consequences, with full transparency and real-time feedback – and therefore, learn from mistakes and errors. Game design can be such that individuals compete against each other and the game, or that teams must work together to complete the game. Finally, games provide a sense of control with the gamer as the decision maker (Figure 6.8).

Figure 6.8 shows community members in Georgia identifying the relationship between cultural heritage, vulnerabilities and capacities through a game.

Source: author; the text is a section of the paper by Rose and Chmutina (2020)

deemed to be the most appropriate. CBDRR does not simply promote local knowledge and skills; instead, it emphasises how the insights of communities (including not only local resources but also social capital) and the outsiders can support each other, if based on a successful dialogue (Rolsted & Raju, 2019). Reinforcing the ideas from the bottom-up, CBDRR empowers communities with self-developed and culturally, socially and economically acceptable ways of coping with hazards. CBDRR, thus, enhances endogenous resources, which prevent people from resorting to exogenous (harder to access) resources, and this often creates a cycle of dependency (Gaillard, 2010).

The successful outcomes of CBDRR depend on many factors, including the procedures and funding arrangements of supporting organisations, the worldviews and bias of the various partners, their relative negotiating power as well as the socio-economic and political context in which the programme operates. The established social structures must be considered carefully; within the same co-located community, various other cross-cutting communities (e.g. migrants, homeless, women) co-exist and may display competing interests and priorities (Allen, 2006). Different community members, thus, have different degrees of access to community institutions and resources, depending on social status or social capital (that can, for instance, be provided by family networks). CBDRR projects that ignore social heterogeneity are unlikely to be acceptable to intended beneficiaries (Tobin, 1999).

The CBDRR approaches recognise that, generally, it is impossible to prevent people from settling in hazardous areas: these locations are often a part of their livelihood and provide resources on a daily basis, as in the case of fertile floodplains and coastal zones with fisheries, and sometimes are imposed on people's external circumstances. The focusing on livelihoods, thus, allows enhancing not only people's capacities to face hazards but also people's ability to sustain their daily needs. CBDRR helps to enable local communities to live with risk on an everyday basis, favouring the integration of DRR into development policy and planning (Gaillard, 2010).

The CBDRR programmes usually involve activities focusing on anticipation (e.g. awareness raising of risk, education, and participation and implementation of risk assessments), coping (e.g. training in first aid, securing home, learning to swim), preparedness (e.g. establishing early warning systems, designing evacuation strategies, stockpiling emergency equipment) and recovery (e.g. alternative means of income, social protection). These activities should not be just about building of technical capacities but also be combined with the promotion of leadership and other managerial capacities (such as budgeting, implementation and evaluation).

Strengthening capacities at a community level does not exclude governments and other (for example, research) organisations from action. Local decision makers are encouraged to foster participatory approaches to support and sustain community actions (Kafle & Murshed, 2006). In Japan, for instance, local governments serve as a driver for the organisation of community-based organisations in charge of emergency management of the neighbourhoods. These *Jushibo* proved efficient in mobilising people and resources, and have become popular since the 1995 Kobe earthquake. Moreover, a directive approach that can only be delivered from the top (i.e. by the government) is necessary to enforce law and regulations that would actually reduce vulnerabilities.

However, the CBDRR projects may be seen as a panacea, without realising that, as the world is changing, capacities also need to evolve – and sometimes, the pace of change of the external environment is too fast. Moreover, some programmes state that CBDRR help communities to reduce vulnerability; whilst, in theory, these approaches may be used to mobilise communities to resist unsustainable changes in development and livelihood practices (Allen, 2006); in reality, enhanced capacities are rarely able to make changes to socio-economic systems, as discussed in Section 6.3. The key weakness of CBDRR is its relative lack of resources and decision-making: local actors and institutions usually lack legislative and regulatory powers (Lavell, 1994). And unless such projects are "as part of a wider and deeper process of [developmental] change" (Eade, 1997, p. 25), they do not reduce vulnerabilities.

The CBDRR approaches can potentially make a significant and long-lasting contribution to strengthening capacities, but there is a danger that they may also place greater responsibility on the shoulders of local people without necessarily proportionately increasing their capacities (Allen, 2006). Any capacity enhancing activity must involve a monitoring and evaluation stage that allows measuring the progress and results to determine whether the project has caused any actual change towards the overall objective, as well as to learn from the experience. The problem, however, is that most evaluations are short term: they are carried out at the end of a project (that is driven by budgetary time cycles) and usually miss out on capturing long-term consequences (Hagelsteen & Becker, 2014). This has to be changed if CBDRR approaches are to have legacy through empowering people not to merely cope and adapt but also to shape social institutions and contribute to decision-making.

6.8 Concluding remarks

Capacity is a difficult concept, and there is very little theoretical debate around it. When discussing capacity, we often focusing on the lack of capacity (i.e. what people cannot do as opposed to what they can do) – and this patronises the work of many organisations as well as people's knowledges and skills.

We need to remember that capacity and vulnerability are not on the same spectrum: developing capacity does not reduce vulnerability, as vulnerability is engrained in political and social systems of oppression. Realistically, DRR practitioners and researchers are very rarely in a position to actually address these structural issues. This, however, does not mean that vulnerabilities should be neglected when building capacity; both should be addressed in tandem.

Capacities differ: often, those who are seen as the most vulnerable can make up for their lack of access to economic and political resources by relying upon strong social and human resources. Thus, it is important to understand that capacities are people- and context-specific. The recognition of diverse capacities from diverse range of communities gives primary importance to utilising local resources and emphasising the overarching contribution of local communities in facing natural hazards.

It is also important to remember that having capacities does not imply that people can rely on their knowledge, skills and resources whenever they need to. Structural barriers and temporary impediments, ranging from state ideologies and political decisions, to physical impairments, technological failures and environmental constraints, are the challenges that may stop people from using the capacities (Wisner, 2016). A government's decision to block access to social media may weaken people's social networks and ability to seek support in time of hardship; a power outage and communication breakdown may prevent people from seeking support from friends and relatives.

Neither should people's capacities be overestimated: whilst local knowledge can help dealing with frequent flooding, it is not enough to build a large dike to prevent a long-term flooding. External and scientific knowledge can also help with dealing with future events, such as climate change, the full impacts of which are still not clearly understood – but already experienced. It is, thus, important to consider how local capacities can be intertwined with external support in an integrated manner (Gaillard et al., 2019).

Any efforts aimed at enhancing capacities must be context-specific and locally appropriate. But at the same time, capacities are best enhanced through sharing between local people and outside actors who may provide appropriate support in case of genuine needs. Such sharing must be built on trust and mutual understanding.

The willingness to enhance one's capacity is a moral one; however, it is often built on Western ideas of inclusion and participation. In this context, enhancing capacities may mean challenging the established traditional norms. This creates an ethical dilemma: is it okay for an outsider to challenge the caste norms in India by empowering lower castes, for instance? Is it okay to intervene when the situation seems unjust from a Western perspective? Are inclusions and participation ethical? These questions are not new: the issue of respecting culture and – at the same time – helping the most marginalised is a classic humanitarian dilemma; but the answers are still not clear (Box 6.5, Figure 6.9a and Figure 6.9b).

Box 6.5: Power, prestige and forgotten values

Fig 6.9a
QR code 1 for Box 6.5. In 2019, a group of disaster scholars published a disaster studies "Manifesto: Power, Prestige and Forgotten Values". In this two-part episode, many academics and practitioners share their views of the manifesto. Together with JC Gaillard, in the first part, we consider whether current terminology and concepts are appropriate for working in a non-Anglophone context.

Fig 6.9b
QR code 2 for Box 6.5.
Available at https://disastersdecon.podbean.com/e/s3e11-manifesto-part-1/ and https://disastersdecon.podbean.com/e/s3e12-manifesto-part-2/

We also discuss whether Western research methodologies serve the needs of those who research and those who are researched. In the second part, we discuss the relationship between researchers around the world and reflect on the importance of considerations of power in research and practice.

Take-away messages

1 The concept of capacities demonstrates how individuals and communities make gradual adjustments to cope with environmental changes caused by natural hazards without modifying the fundamentals of their social organisation.
2 Capacities evolve, as they are dependent on four factors that change significantly in time and space: the nature of the hazard, the pre-disaster socio-cultural context, the geographical setting and the rehabilitation policy set up by the authorities.
3 Capacity and vulnerability are not the opposites each other: reducing vulnerability does not always enhance capacity and other way around.
4 Although capacities are often shared and pooled between people, knowledge, skills and resources are never evenly distributed across individuals living in the same place or sharing kinship. Capacities differ according to age, gender, ethnicity and physical ability. The ability to claim and access collective capacities, thus, depends on power relations amongst individuals.
5 Capacity assessments should focus on institutional arrangements, leadership, knowledge and accountability. Any efforts aimed at enhancing capacities must be context-specific and locally appropriate.

To learn more about the topic discussed in this chapter, listen to the *Disasters: Deconstructed* episode with JC Gaillard (Figures 6.10a and 6.10b).

Fig 6.10a and 6.10b
QR codes for Chapter 6.

Further suggested reading

Aldrich, D. P. (2012). *Building resilience: Social capital in post-disaster recovery*. University of Chicago Press.
Freire, P. (1970). *Pedagogy of the oppressed*. Bloomsbury Publishing.
Gaillard, J. C., Cadag, J. R. D., & Rampengan, M. M. F. (2019). People's capacities in facing hazards and disasters: An overview. *Natural Hazards*, *95*(3), 863–876. https://doi.org/10.1007/s11069-018-3519-1
Hagelsteen, M., & Becker, P. (2013). Challenging disparities in capacity development for disaster risk reduction. *International Journal of Disaster Risk Reduction*, *3*, 4–13. https://doi.org/10.1016/j.ijdrr.2012.11.001

Pelling, M., & High, C. (2005). Understanding adaptation: What can social capital offer assessments of adaptive capacity? *Global Environmental Change*, *15*(4), 308–319. https://doi.org/10.1016/j.gloenvcha.2005.02.001

Quarantelli, E. L., & Dynes, R. R. (1972). When disaster strikes: It isn't much like what you've heard and read about. *Psychology Today*, *5*(9), 66–70.

References

Adger, W. N., Dessai, S., Goulden, M., Hulme, M., Lorenzoni, I., Nelson, D. R., Naess, L. O., Wolf, J., & Wreford, A. (2009). Are there social limits to adaptation to climate change? *Climatic Change*, *93*(3–4), 335–354. https://doi.org/10.1007/s10584-008-9520-z

Aldrich, D. P. (2012). *Building resilience: Social capital in post-disaster recovery*. University of Chicago Press.

Allen, K. M. (2006). Community-based disaster preparedness and climate adaptation: Local capacity building in the Philippines. *Disasters*, *30*(1), 81–101.

Anderson, M. B., & Woodrow, P. J. (1989). *Rising from the ashes*. Lynne Rienner Publisher.

Arunotai, N. (2008). Saved by an old legend and a keen observation: The case of Moken sea nomads in Thailand. In R. Shaw, N. Uy, & J. Baumwoll (Eds.), *Indigenous knowledge for disaster risk reduction: Good practices and lessons learnt from the Asia-Pacific region. UNISDR* (pp. 73–78). Asia and Pacific.

Balgos, B., Gaillard, J. C., & Sanz, K. (2012). The warias of Indonesia in disaster risk reduction: The case of the 2010 Mt Merapi eruption in Indonesia. *Gender and Development*, *20*(2), 337–348. https://doi.org/10.1080/13552074.2012.687218

Berman, R., Quinn, C., & Paavola, J. (2012). The role of institutions in the transformation of coping capacity to sustainable adaptive capacity. *Environmental Development*, *2*, 86–100. https://doi.org/10.1016/j.envdev.2012.03.017

Bhandari, R. B., Okado, N., & Knottnerus, J. D. (2011). Urban ritual events and coping with disaster risk a case study of Lalitpur, Nepal. *Journal of Applied Social Science*, *5*(2), 13–32.

Bhatt, M. R. (1998). Can vulnerability be understood? In J. Twigg & M. R. Bhatt (Eds.), *Understanding vulnerability: South Asian perspectives* (pp. 68–77). Intermediate Technology Publications.

Bhatt, M., & Pandya, M. (2021), Rethinking capacity development for disaster risk reduction: Lessons from bottom up. *Disaster Prevention and Management*, *30*(3), 259–260.

Boehm, C., Antweiler, C., Eibl-Eibesfeldt, I., Kent, S., Knauft, B. M., Mithen, S., Richerson, P. J., & Wilson, D. S. (1996). Emergency decisions, cultural-selection mechanics, and group selection. *Current Anthropology*, *37*(5), 763–793. https://doi.org/10.1086/204561

Chmutina, K., Sadler, N., von Meding, J., & Abukhalaf, A. H. I. (2020). Lost (and Found?) in translation: Key terminology in disaster studies. *Disaster Prevention and Management*, *30*(2), 149–162. https://doi.org/10.1108/DPM-07-2020-0232

Danbolt, L. J., & Stifoss-Hanssen, H. (2017). Ritual and recovery: Traditions in disaster ritualizing. *Dialog*, *56*(4), 352–360. https://doi.org/10.1111/dial.12355

Davis, I., Haghebeart, B., & Peppiatt, D. (2004). Social vulnerability and capacity analysis. In *Discussion paper and workshop report*. ProVention Consortium.

Eade, D. (1997). *Capacity-building: An approach to people-centred development*. Oxfam.

Engle, N. L. (2011). Adaptive capacity and its assessment. *Global Environmental Change*, *21*(2), 647–656. https://doi.org/10.1016/j.gloenvcha.2011.01.019

Freire, P. (1970). *Pedagogy of the oppressed*. Bloomsbury Publishing.

Freire, P. (1985). *The politics of education: Culture, power and liberation*. Macmillan Press.

Fukuyama, F. (1995). *Trust: The social virtues and the creation of prosperity*. Free Press.

Gaillard, J. C. (2006). Traditional societies in the face of natural hazards: The 1991 Mt Pinatubo eruption and the Aetas of the Philippines. *International Journal of Mass Emergencies and Disasters*, *24*(1), 5–43.

Gaillard, J. C. (2007). Resilience of traditional societies in facing natural hazards. *Disaster Prevention and Management*, *16*(4), 522–544. https://doi.org/10.1108/09653560710817011

Gaillard, J. C. (2010). Vulnerability, capacity and resilience: Perspectives for climate and development policy. *Journal of International Development*, *22*(2), 218–232. https://doi.org/10.1002/jid.1675

Gaillard, J. C., Cadag, J. R. D., & Rampengan, M. M. F. (2019). People's capacities in facing hazards and disasters: An overview. *Natural Hazards*, *95*(3), 863–876. https://doi.org/10.1007/s11069-018-3519-1

Gil-Rivas, V., & Kilmer, R. P. (2016). Building Community Capacity and Fostering Disaster Resilience. *Journal of Clinical Psychology*, *72*(12), 1318–1332. https://doi.org/10.1002/jclp.22281

Hagelsteen, M., & Becker, P. (2013). Challenging disparities in capacity development for disaster risk reduction. *International Journal of Disaster Risk Reduction*, *3*, 4–13. https://doi.org/10.1016/j.ijdrr.2012.11.001

Hagelsteen, M., & Becker, P. (2014, August 26–30). Forwarding a challenging task: Seven elements for capacity development for disaster risk reduction. *Proceedings of the 4th International Disaster and Risk Conference IDRC*, Davos, Switzerland.

Hagelsteen, M., & Burke, J. (2016). Practical aspects of capacity development in the context of disaster risk reduction. *International Journal of Disaster Risk Reduction*, *16*, 43–52. https://doi.org/10.1016/j.ijdrr.2016.01.010

Hudson, J., & Hughes, E. (2007). The role of marae and Maori communities in post-disaster recovery: A case study. Glucosamine (N-acetyl)-6-sulfatase science report, *2007/15*, GNS Science, Lower Hutt.

IFRC. (2015). *World disaster report*. https://ifrc-media.org/interactive/wp-content/uploads/2015/09/1293600-World-Disasters-Report-2015_en.pdf. IFRC.

Jones, L., & Boyd, E. (2011). Exploring social barriers to adaptation: Insights from Western Nepal. *Global Environmental Change*, *21*(4), 1262–1274. https://doi.org/10.1016/j.gloenvcha.2011.06.002

Jones, L., Ludi, E., & Levine, S. (2010). *Towards a characterisation of adaptive capacity: A framework for analysing adaptive capacity at the local level* [Background paper]. Overseas Development Institute.

Kafle, S. K., & Murshed, Z. (2006). *Community-based disaster risk management for local authorities*. Asian Disaster Preparedness Center.

Lavell, A. (1994). Prevention and mitigation of disasters in Central America: Vulnerability to disasters at the local level. In A. Varley (Ed.), *Disasters, development and environment* (pp. 49–63). John Wiley & Sons.

Lavigne, F., De Coster, B., Juvin, N., Flohic, F., Gaillard, J. C., Texier, P., Morin, J., & Sartohadi, J. (2008). People's behaviour in the face of volcanic hazards: Perspectives from Javanese communities, Indonesia. *Journal of Volcanology and Geothermal Research*, *172*(3–4), 273–287. https://doi.org/10.1016/j.jvolgeores.2007.12.013

Lewis, H. E., Roberts, D. F., & Edwards, A. W. F. (1972). Biological problems, and opportunities, of isolation among the islanders of Tristan da Cunha. In D. V. Glass & R. Revelle (Ed.), *Population and social change* (pp. 383–417). E. Arnold.

McMillan, D. W., & Chavis, D. M. (1986). Sense of community: A definition and theory. *Journal of Community Psychology*, *14*(1), 6–23. https://doi.org/10.1002/1520-6629(198601)14:1<6::AID-JCOP2290140103>3.0.CO;2-I

Miichi, K. (2016). Playful relief: Folk performing arts in Japan after the 2011 tsunami. *Asian Ethnology*, *75*(1), 139–162. https://doi.org/10.18874/ae.75.1.06

Nolan, M. L. (1979). Impact of Paricutı'n on five communities. In P. D. Sheets & D. K. Grayson (Eds.), *Volcanic activity and human ecology* (pp. 293–335). Academic Press.

O'Brien, G., O'Keefe, P., Meena, H., Rose, J., & Wilson, L. (2008). Climate adaptation from a poverty perspective. *Climate Policy*, *8*(2), 194–201. https://doi.org/10.3763/cpol.2007.0430

O'Brien, G., O'Keefe, P., Rose, J., & Wisner, B. (2006). Climate change and disaster management. *Disasters*, *30*(1), 64–80. https://doi.org/10.1111/j.1467-9523.2006.00307.x

Ostrom, E., Schroeder, L., & Wynne, S. (1993). *Institutional incentives and sustainable development: Infrastructure in perspective*. Westview Press.

Pareek, A., & Trivedi, P. C. (2011). Cultural values and indigenous knowledge of climate change and disaster predictions in Rajasthan, India. *Indian Journal of Traditional Knowledge*, *10*(1), 183–189.

Pelling, M. (1998). Participation, social capital and vulnerability to Urban flooding in Guyana. *Journal of International Development*, *10*(4), 469–486. https://doi.org/10.1002/(SICI)1099-1328(199806)10:4<469::AID-JID539>3.0.CO;2-4

Pelling, M. (2003). *The vulnerability of cities: Natural disaster and social resilience*. Earthscan Publications.

Pelling, M., & High, C. (2005). Understanding adaptation: What can social capital offer assessments of adaptive capacity? *Global Environmental Change*, *15*(4), 308–319. https://doi.org/10.1016/j.gloenvcha.2005.02.001

Program for Strengthening Household Access to Resources. (2014). *Local wisdom: Indigenous practices for mitigating disaster loss*. https://www.preventionweb.net/publications/view/37377. United States Agency for International Development.

Quarantelli, E. L., & Dynes, R. R. (1972). When disaster strikes: It isn't much like what you've heard and read about. *Psychology Today*, *5*(9), 66–70.

Rappaport, J. (1984). Studies in empowerment: Introduction to the issue. *Journal of Prevention and Intervention in the Community*, *3*(2–3), 1–7.

Rolsted, M., & Raju, E. (2019). *Addressing capacities of local communities in a changing context in Nepal*. Concept paper. The United Nations report on global assessment report on disaster risk Reduction 2019. United Nations.

Rose, J., & Chmutina, K. (2020). Learning to learn: Developing disaster risk reduction skills among the informal construction workers. *Disasters*, *45*(3), 627–646

Šakic Trogrlic, R., Wright, G. B., Duncan, M. J., van den Homberg, M. J. C., Adeloye, A. J., Mwale, F. D., & Mwafulirwa, J. (2019). Characterising local knowledge across the flood risk management cycle: A case study of Southern Malawi. *Sustainability*, *11*(6), 1–23. https://doi.org/10.3390/su110⟩

Smit, B., & Wandel, J. (2006). Adaptation, adaptive capacity and vulnerability. *Global Environmental Change*, *16*(3), 282–292. https://doi.org/10.1016/j.gloenvcha.2006.03.008

Suzina, A. C., & Tufte, T. (2020). Freire's vision of development and social change: Past experiences, present challenges and perspectives for the future. *International Communication Gazette*, *82*(5), 411–424. https://doi.org/10.1177/1748048520943692

Tobin, G. A. (1999). Sustainability and community resilience: The holy grail of hazards planning? *Environmental Hazards*, *1*(1), 13–25. https://doi.org/10.3763/ehaz.1999.0103

UNISDR. (2018). *Terminology*. https://www.unisdr.org/we/inform/terminology#letter-d

Wamsler, C., & Brink, E. (2014). Moving beyond short-term coping and adaptation. *Environment and Urbanization*, *26*(1), 86–111. https://doi.org/10.1177/0956247813516061

Wisner, B. (2016). Vulnerability as concept, model, metric, and tool. In S. L. Cutter (Ed.), *Oxford research encyclopedia: Natural hazard science*. Oxford University Press.

Wisner, B., Gaillard, J. C., & Kelman, K. (2012). Framing disasters. Ch. 3. in B. Wisner, J. C. Gaillard & I. Kelman (Eds.), *The Routledge handbook of hazards and disaster risk reduction* (pp. 18–33). Routledge.

Wisner, B., Gaillard, J. C., & Kelman, K. (2014). Hazard, vulnerability, capacity, risk and participation. In A. LopezCarresi, M. Fordham, B. Wisner, I. Kelman, & J. C. Gaillard (Eds.), *Disaster management: International lessons in risk reduction, response and recovery* (pp. 13–22). Earthscan Publications.

Woolcock, M., & Narayan, D. (2000). Social Capital: Implications for Development Theory, Research, and Policy. *World Bank Research Observer*, *15*(2), 225–249. https://doi.org/10.1093/wbro/15.2.225

PART III

Natural and socio-natural hazards

CHAPTER 7

Endogenous processes

Earthquakes, volcanoes and tsunamis

7.1 Introduction: the Earth's internal engine and plate tectonic

Earthquakes, volcanic eruptions and, in most cases, tsunamis as well, are triggered as the release of the Earth internal energy at the surface (Figure 7.1). The first question we may want to start with is where does the internal energy come from and how does it manifest itself at the surface of the planet? At the planetary scale, the mechanism of plate tectonic is driven by the endogenous energy of the Earth – the resulting heat flow – and gravity. Deep inside the Earth, the thermogenesis (heat generation) is the product of the decay of radioactive isotopes ^{235}U, ^{238}U, ^{206}Pb and so forth, and of the friction between the core and the mantle. More occasionally (at the human scale), impacts with other celestial bodies can liberate similar forms of energy. The core has been calculated to be about 4,000 degrees C, and the gradient towards the surface to be, on average, 0.62 degrees C/km, although the latter is irregular through the Earth interior. This energy gradient results in convection in the mantle, acting like water in a hot kettle: the hot mantle cooling whilst rising towards the surface and then plunging back.

However, those mechanisms are not well understood yet. The widely accepted theory states that it is this convection mechanism acting as a traction carpet that moves the oceanic and continental plates at the surface of the Earth. Furthermore, the ascent of magma in the mantle creates a thermal bulge in the lithospheric slabs, and the plates are believed to also slide by gravity. Today, readers will recognise this theory as plate tectonic and accept it without any hesitation, but this idea only dates back from the second half of the 20th century.

During the 18th and 19th century, Europeans were thinking that earthquakes were the products of volcanic activity, and mostly so because observations had been made in Italy in the vicinity of several active volcanoes. It is only after entering the 20th century that

DOI: 10.4324/9781315469614-11

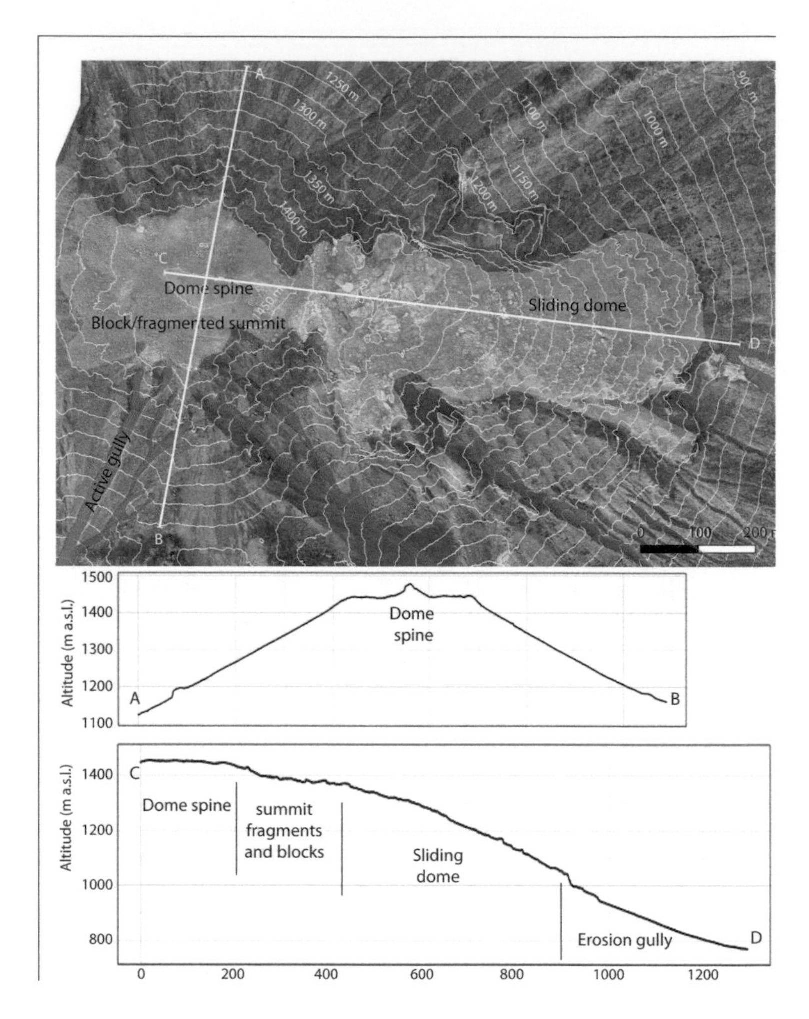

Fig 7.1
Topography of Unzen Volcano in 2016 (Kyushu Island, South Japan). The 1990–1995 lava dome is still sliding today, with an increasing potential for a full collapse of the dome, especially in the event of an earthquake.
Source: author created

earthquakes started to be linked to faults and movements of the Earth surface. This conceptual shift can be linked to two individuals. First, the geologist John Milne, who at the end of the 19th century, during a voyage in Japan, evidenced that earthquakes were occurring commonly on mainland Japan but not necessarily in the vicinity of volcanoes. He was even surprised to hear from local inhabitant of the volcanic island of Izu Ooshima (100 km south of Tokyo City) that earthquakes seldom occurred on the island, despite the volcano being active. Second, from this observation, the British geologist Charles Davison made the next step, stating at the turn of the 20th century that earthquakes were most certainly the product of the formation of faults. Our contemporary idea linking fault and earthquake was born. Relating faults and earthquakes was one step forward, but the theory of continental drift and plate tectonic was still a long way away. Vertical movement of the Earth was still the common explanation of the deformation of the Earth surface, except for De Saussure in 1796 and Hall in 1815, who proposed that the horizontal forces were at the origin of Earth surface deformation.

Although it was only to be adopted in the 1960s, the idea of continental drift is not fundamentally novel. As early as the 16th century, a Dutch cartographer noticed the match between the African coast and the South American coast, and he suggested that the block Europe and Africa must have been torn apart was from the American block. In the mid-19th century, a French cartographer even published a map showing continental drift. Those ideas could not be proven nor supported by further evidences at that time, and they were forgotten for a while until Alfred Wegener, a German meteorologist and astronomer, theorised that all land must have been part of a super-continent that he named Pangea. Although Wegener published his ideas in 1915, it is only in the 1960s that earth sciences made sufficient progress to start accepting them. And in the 1970s, continental drift was accepted as the main idea for plate tectonic and earthquake explanations. This sudden change of direction amongst most geologists came from the results of the Atlantic Ocean bathymetric and sampling survey. It was found that the Atlantic Ocean mid-ocean ridge had volcanic rocks much younger than anticipated and with virtually no sediment cover, when at that time, it was thought that sediments from land must have accumulated at the bottom of the ocean. It is then, in 1960, that Harry Hess explained the birth of the ocean floor from the oceanic ridge. This rather grinding experience, which defied the arcane less than 50 years ago, is now the commonly adopted theory to explain oceanic and continental crust behaviour as well as tectonic earthquakes, and explain the origin of the endogenous hazards (this is a sound reminder that all crazy ideas are worth testing).

Tectonic plates

The Earth is, therefore, divided in a series of plates of different sizes that can be divergent (where new material is created), convergent or also slide parallel to one another (those are named transform zones or transcurrent horizontal slips), like it is the case over New Zealand (Figure 7.2). The plates' relative motion occurs at different velocities. The Pacific Plate is one of the fastest moving, with a rhythm of 60 to 100 $mm{\cdot}yr^{-1}$, with the highest velocities recorded on the segment to the north of Japan. India, for instance, moves north into the Eurasian Plate at velocities between 20 and 40 $mm{\cdot}yr^{-1}$, and this velocity is similar to the one at the opening of the Atlantic mid-ocean ridge. Based on present theories, the relative movements of one plate over another is then the origin of tectonic earthquakes and also a portion of volcanism.

Endogenous hazards are associated with the relative motion of the tectonic plates; it is, therefore, the boundaries that interest us, and plate boundaries cover about 15% of the Earth surface, with boundary width that can extend hundreds of kilometres, creating complex

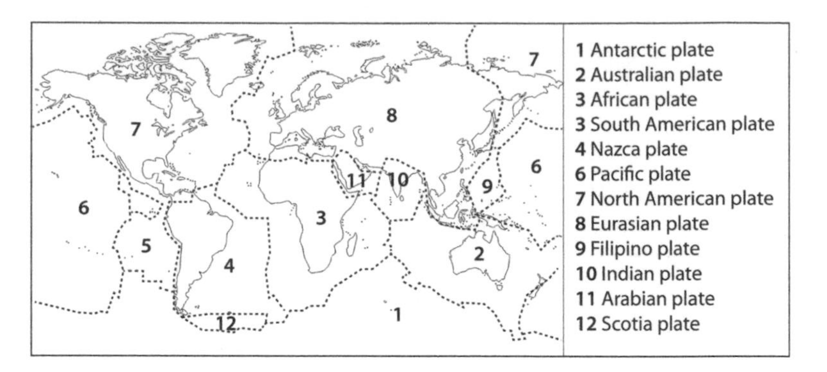

Fig 7.2
The major tectonic plates of the world. The limits of the major plates control the distribution of earthquakes and active volcanoes dominated by explosive activity.
Source: author created

geomorphologic systems, like the Andes, where a variety of hazards related to the plate boundaries occur (Duarte & Schellart, 2016). Contemporary concepts of the limits of rigid tectonic plates start with Wilson's contribution in 1965 (those ideas are, thus, all very recent, considering the history of scientific ideas), where he wrote about "mobile belts, which may take the form of mountains, mid-ocean ridges and major faults", recognising the geomorphologic features associated with plate boundary activity. As plates are in movement, some boundaries will be divergent; those are places where new surface is being created, whilst other boundaries will be converging; those are the places where material is being recycled with either collision or subduction – the second one also including obduction processes (i.e. ocean crust is being scrapped off the plunging oceanic plate).

The rock and sediment cycles, and their distribution

The processes of planetary-scale plate tectonic are the backdrop to several regional and local cycles and smaller-scale environmental processes. At the regional and the local scales, the rock, sedimentary and water cycles all add to the complexity to the distribution of seismic hazards. In other words, it is not only the distance to the epicentre that controls seismic phenomena at the surface.

The rock and sediment cycles are the cyclic pathways that both rocks and sediments take at the surface and inside the Earth. The rock material generated at the surface, either by uplift (like in mountain chains, such as the Himalaya or the Andes) or by volcanic activity (like Hawaiian volcanoes), is being faulted, eroded, broken into material of different size, depending on the process involved (explosive volcanoes can produce very fine ash, whereas a lava flow will create a slab-like deposit that can extend across very large areas), and this material is then transported back to zones that scientists call sinks, like the bottom of the ocean. The sediment is then reintegrated into the deep-rock cycle, by either subduction or reintegration by melting during volcanic activity. This cycle interacts with the water cycle controlled mostly by the atmospheric processes during erosion and deposition, and with living organisms that

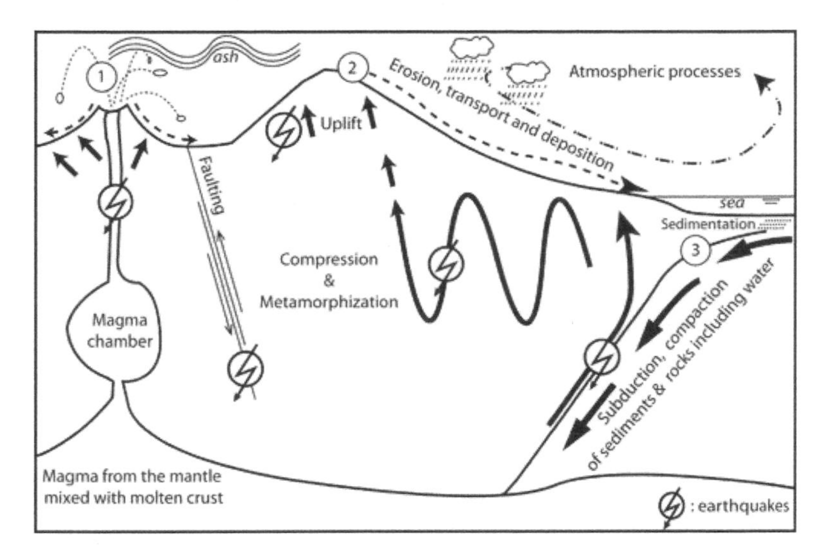

Fig 7.3
The rock and sediment cycles. The production of material includes (1) volcanic activity, which first brings material from the entrails of the Earth to its surface, and then distributes them at the surface of the planet through processes such as lava flows, pyroclastic flows, local ballistics and long-range ashfalls. The Earth internal energy is also at the origin of (2) uplift and faulting processes, which in combination with atmospheric processes help eroded material to travel back into sediment sinks (like the seafloor), where (3) subduction will combine and melt the material back into the Earth interior.
Source: author created

create carbonate rock formations from their skeletons and so forth. Those cycles are the backbones to understanding the distribution of geological processes and related hazards at the surface of the Earth and differentiate the hazards associated with endogenic processes (Figure 7.3).

7.2 Earthquake hazards

The first hazard any of us would link to endogenous processes are certainly earthquakes, but they are by far those we understand the less. This is due to their long return period (compared to human scale) and the depth at which they occur, which have been so far beyond reach to human technologies, and the only knowledge we have is through a kind of remote-sensing technology: seismology. Direct measurements are not possible because earthquakes usually occur at depth several tens of kilometres, with few shallower ones. At the surface, an earthquake is characterised by ground shaking due to the sudden release of energy in the Earth's crust. This energy can originate from either natural (mostly volcanism and tectonism) or anthropogenic sources (e.g. fracking), and it is this phenomenon that seismology measures.

Reiter (1990) defines seismic hazard as "the potential for dangerous, earthquake-related natural phenomena such as ground shaking, fault rupture, or soil liquefaction".

If you compare a dataset of "earthquakes and surface acceleration" or "earthquakes and damages and death-tolls", you can clearly determine that the variables are not linked by any form of linear relationship. Same earthquake, different hazards? Why is it that the same magnitude earthquakes can generate such different outcomes? Earthquakes in Dasht-e-Bayaz in Iran took the lives of 12,100 in 1968, 22,778 in 1976 in Guatemala, 20,500 in Iran in 1978, 25,000 in Spitak (Armenia) in 1988,

17,500 in Kocaeli (Turkey) in 1999, 13,800 in India (2001), 31,000 in Bam (Iran) in 2003 and 86,000 in Kashmir Pakistan and India in 2005 (So, 2016). The main factor that controls those outcomes is undeniably attributed to the societies that inhabit the place the earthquake strikes, but this fact must not hide the great variability in ground acceleration and shaking characteristics, notably. The surface paradigm is essential to acknowledge because seismology and the field of earthquake sciences have been historically interested in finding the source of earthquakes and understanding the mechanisms at great depth. It is only in the 1970s that the world of hazard and disaster risk research went to focus on people – a necessity. This shift of the focal to the societies has produced a bipolarity in seismic hazards' research, where the importance of the surface and subsurface is only being recently seriously integrated in seismic models. Although there have been certainly fewer interests for the complexity generated at the surface or the near surface, this research field is presently expanding, with, for instance, research on the Canterbury earthquake sequence in New Zealand, where engineers and geologists and geographers teamed up to generate more accurate pictures of the earthquake hazards, showing that the surface morphology and the near subsurface played a central role in controlling the severity and distribution of hazards.

It is arguably the way we build our environment that is the source of earthquake casualties. One of the mottoes amongst seismic hazard specialists is that "earthquakes don't kill people, buildings do" (Box 7.1). I would even extend it to ill-development, as we have seen with the 2010 Haiti earthquake. The event occurred on 12 January 2010. It was a Mw 7.0 earthquake located only 13 km deep that hit the republic of Haiti located at the Western tip of Hispaniola, which resulted in an estimated 250,000 deaths, 300,000 injuries and 1.2 million people left homeless, out of the 5 million people who were living in the earthquake-impacted area. In other words, almost 35% of the population was directly affected by the earthquake, with about 110,000 buildings destroyed. The inadequacy of infrastructures and the built environment resulted in the disaster, even if the area was known to be seismically active. In the vicinity of the Hispaniola Island, a Mw 8.1 earthquake triggered a tsunami in 1946, and another major earthquake occurred underneath Port-au-Prince City in 1770. Despite of it, historical reasons lead to a less-than adequate development. In the aftermath of the 2010 main earthquake, 59 > Mw 4.5 earthquakes occurred up to the end of February, and yet to date, another major earthquake is still expected as the stress accumulated in the North-eastern Caribbean did not seem to have been released during the 2010 earthquake (Calais et al., 2010). In 2010, the lifelines were particularly impacted. Out of the 900 healthcare facilities that were known prior to the earthquake, only 10% survived the earthquake. The water system of the city of Port-au-Prince was relying on chlorinated sources located in the mountains around the city, but the water was not potable and not reliable anymore after the

Box 7.1: Earthquakes do not kill people, buildings do

Starting on 4 September 2010 with the Mw 7.1 Darfield earthquake, the Canterbury region of the South Island of New Zealand was rattled by several thousands of aftershocks > Mw 6.0, including the 6.3 Christchurch earthquake of 22 February 2011, which left the city scarred. The stone buildings particularly suffered in the centre of Christchurch City, where the author was living at the time (Figure 7.4). Although river stopbanks rose, liquefaction occurred extensively and rockfalls even left car-sized holes in houses; it was not the natural phenomena that took the lives of individuals but masonry failing and falling on people, and the collapse of poorly engineered buildings (the multistorey building, which collapsed on the Japanese school children who came to study English, had been designed by an engineer who had forged his engineering degree). It was not earthquakes that took people's lives in Christchurch but buildings that were not suited for an earthquake-prone country.

Fig 7.4 Collapsed 19th-century brick-and-mortar buildings in Central Christchurch as a result of the 22 February 2011 earthquake. Source: photo courtesy of Christopher Gomez (2011)

earthquake. The electric grid faced the same fate. The road network, which was not optimised, suffered heavy damages on the coast due to liquefaction, and in the city of Port-au-Prince, the outdated network was not brought back to capacity for a long period. The American Society of Civil Engineers monograph (Curtis, 2012) on the Haiti earthquake highlighted that (1) the dominance of unreinforced masonry structures in buildings caused numerous lifeline damages; (2) the unstable excavation need reinforcing in critical areas where lifeline structures are; (3) the lack of equipment and supplies delayed the restoration of the island;

(4) the lack of clean water increased the risk of death; (5) the network operated very rapidly after the earthquake because Haitians used cell phones instead of landlines.

Earthquake mechanisms

Faulting mechanisms – the great majority of natural earthquakes is triggered by tectonic plate motion and volcanic processes – such as magma rise underneath a volcano that cracks the surrounding rocks during its ascension. For both processes, it is the sudden release of energy that becomes a series of waves travelling through the rock and the sediment medium. The waves' amplitude, frequency and velocity depend on the local characteristics of the rock and the amount of energy released. For tectonic earthquakes, the sudden release of energy originates from the accumulation of strain, notably when two slabs of rocks move towards one other. By this process, the elastic strain energy accumulates, and it is then released at the slipping interface. The accumulation and release process is the main process driving tectonic earthquakes. The interface of rupture notably materialises itself as a fault, and its geometry is defined by the azimuth angle (P), which is the angle of the trace and the north; the dip angle (*t*), or the angle at which the fault plane plunges in the ground; and the slip or rake angle, which is the angle between the direction of relative displacement and the horizontal (Figure 7.5).

Depending on how the two (or more) slabs move against one another, the fault will then be described as a lateral strike-slip fault (left or right), normal fault or reverse fault (Figure 7.5). Those movements are seldom

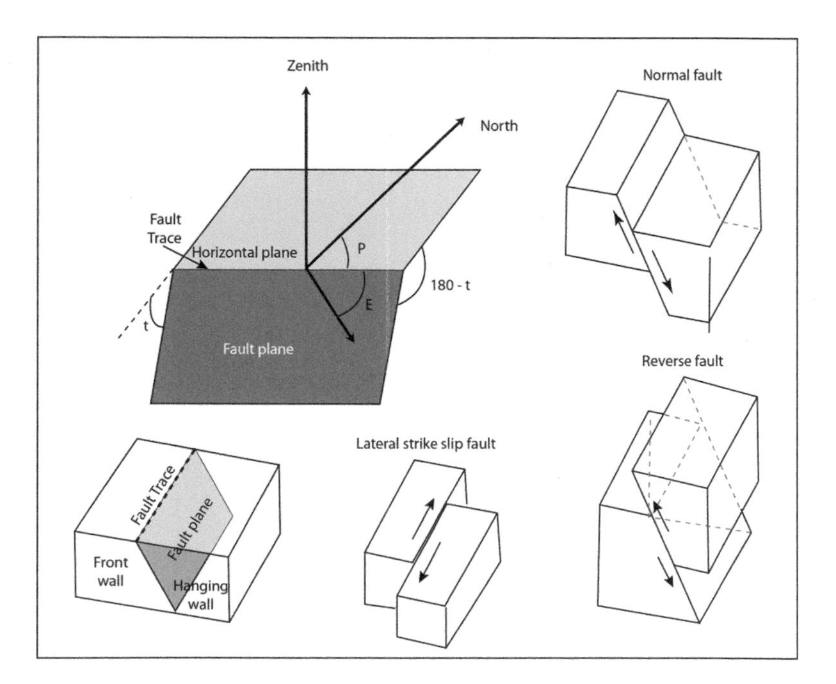

Fig 7.5
Fault descriptors and principal movement types (P is the azimuth angle, *t* is the dip angle and E is the slip or rake angle). Source: author created

strictly one or another, and they will combine into different types of complex movements; for example, a reverse fault can see the rise of the hanging wall over the front wall, whilst at the same time, a lateral strike-slip can also dislocate the blocks in such a way that they are not aligned anymore.

The shaking itself and the rupture are now supposed to be related to the nature of the material and the presence of water in subduction zones. According to Stefansson (2011), in Iceland, earthquakes are believed to be triggered by highly pressurised fluid injection from the mantle – about 100 km underneath Iceland; this fluid is conjectured to be water that rises from the separation of basaltic fluid from the upwelling mantle. The water migrates upwards, and it accumulates in highly pressurised pockets. The earthquake cycle is, therefore, believed to start with, first, the accumulation of fluid below the brittle part of the lithosphere, where faults are. Although the time it takes for the fluid to accumulate is difficult to predict, evidences show that it is much slower than the ~ 140 years' strain accumulation resulting from plate motion. As the accumulated water rises through the fault system, it generates micro-tremors, and it also modifies the condition of the fault, creating heterogeneity along the sliding plans. As the slab movement increases strain, it also increases the heterogeneity, until this weakening process lets a large portion of the fault slip, in turn generating large earthquakes. During the earthquake generation and its immediate aftermath, the fluid continues to rise, lubricating the fault and generating further micro-tremors (Stefansson, 2011).

Although theories and models are progressing, earthquake mechanisms and Earth interior processes remain one of the Holy Grails of contemporary earth science, notably because it is impossible to go and directly measure 100 km underneath the surface.

It is worth noting that, at the surface, the processes can be absolutely invisible. In such case, one says that such an earthquake does not break the surface. Often, the various rock and sedimentary layers deform and absorb the deformation, dissipating the movement necessary to rupture the surface. Moreover, even if an earthquake breaks the surface, very often it is condemned to disappear due to the erosion and deposition processes. If you take the case of the Canterbury Plains of the South Island of New Zealand, for example, the Quaternary sediments generated from the glaciers in the New Zealand Alps have coated the coastal plain and hidden fault lines and obvious signs of seismic activity underneath Christchurch City. This is one of the reasons why the earthquake that occurred just underneath Christchurch City during the 2011 Canterbury earthquake sequence surprised numerous scientists, as there was little visible evidence of faults in the landscape underneath the city. But in this case, it is once again more a historical reason and the late coming of Europeans with buildings not designed to withstand strong earthquakes, which turned the earthquake into a disaster.

Earthquakes in floodplains and at the coast: liquefaction

As mentioned in the introduction, all earthquakes are not equal, and depending on where they occur and the characteristics of the geomorphology, their impacts will be very different. Notably, Quaternary sediment deposits tend to amplify earthquake effects – the sediments that have been travelling from mountains, volcanoes and all topographic highs towards the sea during the Quaternary period have settled in floodplains, in coastal areas, around lakes and in all topographic sinks (low locations where sediments can settle). As the sediments haven't had the time to harden and properly consolidate, they tend to amplify seismic waves, like a jelly on a plate, increasing surface ruptures and deformation (Figure 7.6). In coastal and floodplain areas, the presence of unconsolidated sediments mixed with water leads to several phenomena.

Liquefaction – under cyclic seismic loading, liquefaction occurs in saturated sands and silts. It is the result of excess pore water pressure triggering the loss of soil shear strength and the expulsion of water mixed with clay, silt and sand particles towards the surface. As the particles settle after liquefaction, ground subsidence often occurs. They create features known as sand boils (also named sand boiling, sand blows, mud spouts or sand volcanoes), which can be either a cluster or one ingle cone-like feature (like a volcano) or like fissure eruption. The size of the sand boil volcanoes and fissure features have been reported to be on average between a few centimetres to a few tens of centimetres in height, and with a perimeter or a length of a few tens of centimetres to several metres. During the 2011 Christchurch earthquake in New Zealand, several of those features coalesced to expend over several hundred metres' length. Although liquefaction does not represent a direct threat to human lives, it is often the cause of building and infrastructure failures, either above ground (damage to bridge piles, poles, cracks in roads and buildings, uneven settlements of buildings etc.) or underground (tunnels and pipes failures etc.).

Fig 7.6
Photos of Kumamoto earthquake. Ground deformation and ground breaking by fault movement: (A) surface breakage of the ground in Mashiki village, (B) road surface collapse in south Aso area, and (C) ground deformation and building collapse in Aso Prefecture. Source: open data of the Geospatial Authority of Japan, http://www.gsi.go.jp/BOUSAI/H27-kumamoto-earthquake-index.html

Liquefaction does not occur in every substratum. It does not occur in bedrock, in gravels nor in coarse sands. It takes a water-saturated loose silt, fine sand and clay soil to trigger liquefaction. Consequently, young soils less than a few hundred years old are easily liquefiable compared to older indurated material. Youd and Perkins (1978) defined soils less than 500 years old with a high liquefaction susceptibility, soils less than 10,000 years as moderately liquefiable and soils of 10,000 to 1.8 million years as having a liquefaction susceptibility of low to very low. In New Zealand, the Canterbury earthquake sequence started with the Darfield earthquake in September 2010, and a year later, more than 100 M_w 4 earthquakes had occurred, with two $M_w > 6$ events and one $M_w > 7$ event. This resulted in extensive liquefaction in Christchurch City. As the city lies near the sea and has been built on a mix of river, estuarine and marine sediments (mostly less than 6,000 years old), both the surface acceleration and liquefaction were severe. Liquefaction occurred for most of the events with $M_w > 4.5$, and it has been observed that $M_w > 5.5$ (Quigley et al., 2013) even triggered flood-like events, when pushing water out of the ground. To predict the potential for liquefaction, engineers usually perform various in situ tests, such as the standard penetration test, cone penetration test, Becker penetration test or indirect measurement using the shear wave velocity from an artificial source. These testing methods have shown to be accurate to determine liquefaction potential compared to ground acceleration from strong motion stations during the Canterbury earthquake sequence of 2010–2011 (Wotherspoon et al., 2014). One approach to estimate the potential for liquefaction is to measure the maximum shear stress during an earthquake as a function of depth and the surface maximum acceleration so that the maximum shear stress is as follows:

$$\tau_{max} = \frac{\gamma * z}{g} a_{max}$$

expressing the ratio of the averaged unit weight of the material multiplied by the depth divided by the gravitational acceleration, and then multiplied by the last term, being the maximum acceleration at the surface.

Coastal earthquake: an accelerated experience of sea level rise

One of the effects of an earthquake that is acutely felt near the coast is land subsidence. It is of particular importance because the 0 level of the ocean or the sea virtually rises by the amount of subsidence. Furthermore, the effects are not linear, and subsidence is not only a general lowering of the land; it is also a local deformation eventually creating depressions or shallow pits, where water can accumulate more easily or where groundwater can reach the surface more easily.

In the city of Banda Aceh (Sumatra, Indonesia) and along the coastal plain to the south of the city, the 2004 earthquake that triggered the

Boxing Day tsunami also generated coastal plain subsidence of about one metre, and this process must have repeated itself several times over the coast history, notably in Banda Aceh City, because the 13th-century Muslim cemetery is presently underwater at the bottom of the port of Banda Aceh. In turn, it means that any coastal engineering structure becomes lower and needs reworking. Likewise, in the aftermath of the 2011 Tōhoku earthquake and tsunami – the *Higashi Nihon Daishinsai* – a metre of subsidence was also observed locally. Similar measurements were also made in Christchurch after the 2011 earthquakes, for which LiDAR (light detection and ranging) pre- and post-earthquake revealed spatially varied patterns of subsidence. The maximum subsidence occurred around the rivers, where the banks slid and the surrounding land subsided further, due to lateral expansion. River paleochannels and their edges also experienced the same subsidence patterns, and sand boils also emerged at those locations. Because Christchurch City is constructed on estuarine and river floodplain material, which are mostly less than 6,000 years old, the sediment deformed in numerous places, creating local depressions, where rainwater accumulated later on. In places that locally subsided, the underground storm water system also overflew to the streets. Because of Christchurch's position on an estuary, and because of the extensive local subsidence, the city suddenly experienced the equivalent of some of the effects of sea level rise. Rainfalls that would traditionally not trigger inundations were having important impacts on the city. Because the sea level was virtually higher, the river slope profile and energy profile were thus reduced, and water, instead of being evacuated at sea, was piling up in the waterways. The spillways connecting the storm water system were also being clogged because of their lowered position closer to the river level (due to subsidence). Furthermore, the water that accumulated in the rivers was also further stopping the flow in the storm water pipes, which in turn flooded on land, especially from the depressions created by the earthquake. In many aspects, this earthquake was the equivalent to the fate awaiting most of coastal cities worldwide.

Coseismic landslides and mass movements

Mass movements and landslides can occur as the resultant of a combination of processes, including heavy rainfall, slope downcutting and earthquakes, in which case they are named coseismic landslides (for more general information on landslide processes, please refer to the chapter on exogenous hazards). They differ from other landslides because they are mainly driven by the modification of the balance of inertial and driving forces due to the ground acceleration and the associated stress, and as the shaking can continue as they flow, their reach can be much longer than water-triggered landslides in the same soil and topographic conditions. The shear strength is then modified by the earthquake magnitude, the distance to the epicentre and the orientation of the slope to the seismic waves. The effects are plural, and they can be one of or a combination of the direct acceleration of a mass above a slip surface, the liquefaction of the material on the slope, the change in groundwater and the groundwater pressure.

To destabilise the slope, sufficient energy is necessary for the shear stress to overcome the shear strength. One way to realise this condition is through the seismic acceleration. It has been shown that landslides can be triggered from $M_w > 4$, with a dominance of coherent landslides at $M_w > 4.5$ and non-coherent landslides at $Mw > 5.5$ (Rodríguez et al., 1999). Soil and wood debris mass movements will in turn preferentially happen at $M_w > 6.5$ (Murphy, 2015). Although the energy released by the earthquake – $\log E = 1.5\ M_w + 11.8$ – is attenuated based on the distance and the type of media that the seismic elastic waves travel through, high topography will usually have the reverse effect, and it will tend to amplify the seismic waves by resonance. In the aftermath of the Christchurch earthquake in 2011, people in the town of Littleton to the south of Christchurch mentioned how they could see the earthquake wave ramming up the slopes and dislocating blocks, which were popping successively. Towards the ridges and the top of the hills and mountains, the shaking is then amplified, eventually fracturing and moving large blocks. Research has demonstrated that shaking increased towards hilltops and mountain ridges, explaining partly the distribution of coseismic landslides. Indeed, coseismic landslides can start in the uppermost part of slopes, whereas other rainfall/water-triggered landslides will traditionally start slightly below the ridges, as rainfall needs to concentrate and as groundwater will not concentrate in the hilltops and mountaintops. Moreover, it has also been shown that slopes with an angle > 17 degrees seem to play an increasing role in topographic amplification, but that the topographic effects for lower-angle slopes can be negligible. This research results in the inclusion of seismic coefficient into the slope stability analysis, known as a horizontal seismic coefficient, from 0.05 to 0.5 for catastrophic earthquakes (Terzaghi, 1950).

Earthquake measure and hazard assessment

Because scientists do not comprehend fully the processes that lead to earthquakes yet, seismic monitoring is based on surface monitoring. The latter is the prime method of choice for both earthquake and volcanic activity monitoring, notably, because the sensors can be deployed for a low cost, and they can record seismic activity several kilometres away from the source, and they provide information even when ground deformation does not occur or when the water physical and geochemical properties remain unchanged. The idea of measuring earthquakes from their surface manifestation is not new. One of the first known measurement device was made in China in about AD 132, but the contemporary understanding of waves travelling through a medium only arose in the 19th century, and the first scale to measure their intensity, the Richter Ms scale, was only established in 1935. This scale records the logarithm on base 10 of the maximum wave amplitude, in such a way that the released energy (E) is measured as $\log_{10}(E) = 4.8 + 1.5M_s$.

Another popular scale based on how humans feel the tremor and the level of impacts on the built environment is the Modified Mercali Scale of Intensity. It is a subjective scale as it integrates the appreciation of

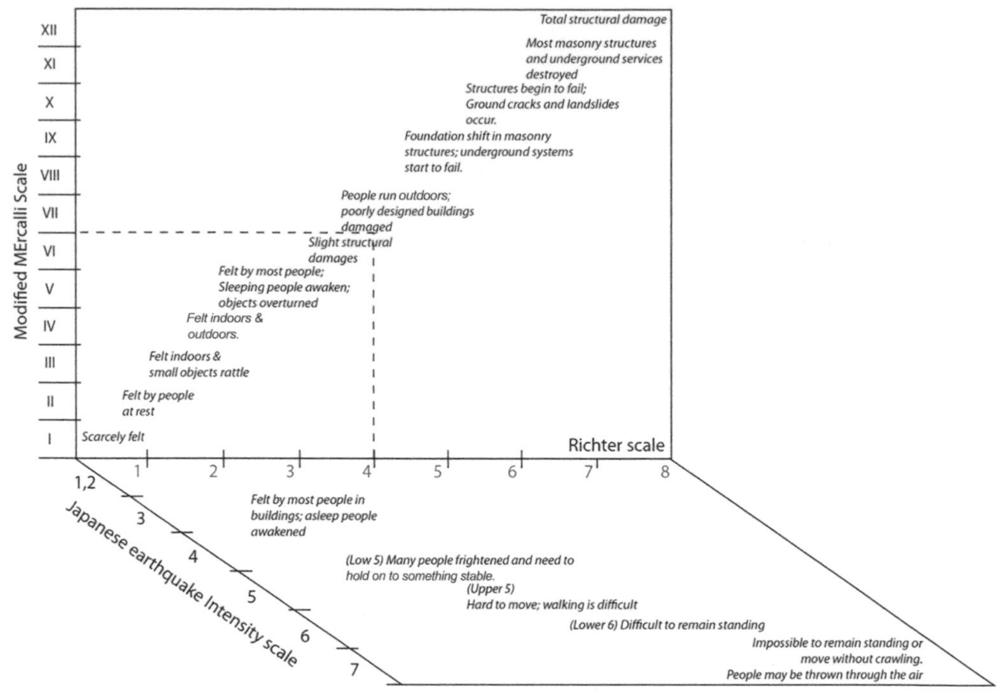

Fig 7.7
Modified Mercali Scale, Richter Scale and the Japanese Intensity Scale of earthquakes in relation to one another, with some of the common descriptors. Source: author created

parameters that are specific to individuals and a community (Figure 7.7), although the scale was built as an objective tool.

Those scales (Figure 7.7) are, however, not well suited for calculations of impacts on buildings and infrastructures; therefore, engineers and scientists often measure the PGA (Peak Ground Acceleration), PGV (Peak Ground Velocity) and PGD (Peak Ground Deformation). The PGA is the maximum acceleration that occurs during an earthquake event, and it is calculated in $m \cdot s^{-2}$; the PGV is the maximum velocity reached by the ground, and it is measured in $m \cdot s^{-1}$. Finally, the PGD is a measure, after the earthquake, of the maximum change that the ground surface has experienced.

Examples of seismic hazard assessment in Japan, New Zealand and Europe

Seismic hazards in Japan – scientific efforts: in the aftermath of the 17 January 1995 *Hanshin Awaji-Daishinsai* earthquake in Kobe, the government established the Headquarters for Earthquake Research Promotion, and the National Research Institute for Earth Science and Disaster Prevention (NIED) developed a programme to map seismic hazards across Japan, providing maps updated yearly of earthquake probability (www.bosai.go.jp/e/).

Insurance: based on the documents of the Japanese Ministry of Finance (www.mof.go.jp), the earthquake insurance system covers residential buildings and household goods from the direct earthquake damages as well as the impacts of indirect effects, such as earthquake fires, and volcanic- and tsunami-related events. To avoid insurance companies being overwhelmed by surges of claims during an earthquake, the government is providing reinsurance, with a limit decided by the Diet every fiscal year and with a maximum limit of ¥11.1268 trillion (about US$98 billion). This payout level is calculated from what is considered as the worst-case scenario, defined by the Great Kanto earthquake. So far, both the Tōhoku event of 2011 and the Kobe event of 1995 have fallen within the set limits. At the individual level, limits have been set to ¥50 million (US$440,000) for a building and ¥10 million (US$90,000) for household goods.

Seismic hazards in New Zealand (Aotearoa): the country has been named the Shaky Isles because of the frequency of high-magnitude shallow earthquakes. Out of the 16 $M_w > 6$ earthquakes between 2007 and 2017, 87.5% are less than 60 km deep, and 31% are very shallow earthquakes < 10 km deep (Table 7.1).

Policy: New Zealand is a country of 4.6 million people in 2016, and consequently, a small economy in size (US$185 billion in 2016), with an important primary industry sector, representing 8% of the GDP but 50% of the international exports in 2012. The manufacturing and the services, respectively, are 12% and 71% of the GDP. Consequently, the funds available for hazards and disaster management are proportional in scale. In New Zealand, the hazards and disaster risk resilience framework is organised around the Civil Defence Emergency Management Act of 17 October 2002 (http://www.legislation.govt.nz/act/public/2002/0033/48.0/DLM149789.html). It

Table 7.1 Recurrence of earthquakes of $M_w > 6$ in New Zealand during the last decade

Year	*Name/Location*	*Magnitude (Mw)*	*Depth*
2017	South Tasman Sea E.	6.6	10 km
2016	Kaikoura E.	7.8	15 km
2016	Te Araroa E.	7.1	55 km
2016	Macauley Island E.	6.2	366 km
2015	Marlborough E.	6.1	52 km
2014	Gisborne E.	6.7	22 km
2014	Eketahuna E.	6.1	34 km
2013	Lake Grassmere E.	6.5	8 km
2013	Seddon E.	6.5	17 km
2012	Opunake E.	6.3	254 km
2011	Christchurch E.	6.0	7 km
2011	Christchurch E.	6.1	6 km
2010	Darfield E.	7.0	11 km
2009	Fiordland E.	7.8	12 km
2007	Gisborne E.	6.6	44 km
2007	Fiordland E.	6.7	5 km

proposes the 4 Rs integrated approach, with reduction, readiness, response and recovery. It is the central document of the Ministry of Civil Defence and Emergency Management (MCDEM), which is a unit of the Department of Internal Affairs. The Civil Defence Emergency Management Act is a document of decentralised power, with a national controller and a national recovery manager, who manages different Civil Defence Emergency Management Groups at the local and regional level. Those groups have to follow different rules set by the document. This document does not operate in a vacuum and works in coordination with the Resource Management Act (1991) and the Building Act (2004). The latter ensures the integrity and safety of building structures, as well as their compliance to the building code, which defines how buildings must behave during an earthquake.

Seismic hazard assessment in Europe – in countries like Indonesia, Japan or New Zealand, seismic hazards are part of the everyday life, in such a way that governments at different levels are expected to undertake significant steps and actions, but in European countries where the seismic hazard is more diffused over time, it is unreasonable to ask every homeowner to retrofit their dwelling. Nevertheless, owners and operators of critical infrastructures, such as nuclear power plants, cannot ignore earthquake hazards, especially in France, where the majority of them are built along the Saone River and Rhone River, both rivers following major fault lines of the European Alps.

If individual dwellings do not follow the same strict rules as in Japan or in New Zealand, critical infrastructures, such as power plants have attracted the attention of politics and scientists in Europe. SIGMA (http://www.projet-sigma.com/) is a Franco-Italian scientific project created and funded by a consortium of electrical and nuclear power plant companies (the French power company EDF, the French nuclear fuel prospection and waste management AREVA, the Italian power company ENEL and the CEA, and the French Atomic Energy Commission), with the aims to homogenise seismic hazards calculation, improve the existing sciences and engineering relying on it, and share the research with the European community. This project started in January 2011 made even more sense when two months later, in March 2011, the Great Eastern Japan earthquake and tsunami triggered the explosion and meltdown of nuclear reactors in Fukushima Prefecture.

In 2016, at the end of the project, the research has produced a database of ground motion for Europe named RESORCE (http://www.resorce-portal.eu/), providing a first-of-its-kind integrated database for Europe. The research has also demonstrated weaknesses in the present state of knowledge. For instance, France, the first provider of nuclear electric energy in Europe, does not have a good understanding of the fault's characteristics across the country yet and further statistical and deterministic research on seismic source characterisation are most certainly necessary.

7.3 Volcanic hazards

The shaking of the Earth is not the sole manifestation of the Earth thermodynamic energy. By convection, the magma sometimes finds its way to the surface to create volcanoes and volcanic eruptions, which have both captivated the imagination of humankind and brought havoc.

Volcanoes are the manifestation at the Earth surface of endogenous mechanisms that lead to magma rising through the oceanic or the continental crust, forming pathways to the Earth surface, which deform the latter and also lead to the effusion and ejection of volcanic material. Volcanic hazards originate from the accumulation of magma in a magma chamber, which then ascends to the surface by a volcanic conduit. Then "for magma to erupt, the chamber pressure has to exceed the sum of frictional and magma-static pressure losses, as well as losses associated with opening of the conduit", writes Gonnermann and Manga (2013, p. 56). Depending on the location, the time it takes for the magma to rise and the different types of rock layers that the magma goes through, a variety of volcanic activity and hazards occur.

Different volcanoes = different eruptive styles

Depending on the location where a volcano grows – either through the oceanic crust or near plate boundaries, where it can be characterised by a mixture of material coming from the continent and tectonic plate material – the eruptive style, morphology and morphometric signatures of the volcanoes will differ (Box 7.2). It creates a scale of material that range from mafic (basalt) to felsic rocks (rhyolite), with intermediate andesite. The distinction is mostly based on the proportion of SiO_2, which range from 45% for mafic rocks to over 65% for felsic rocks, and it will control the type of eruption a volcano will have. More specifically, we can differentiate effusive eruption from explosive eruptions. The second type needs magma fragmentation, triggered by a recipe of critical bubble fraction, stress and strain rate as well as potential energy to expel the material out of the vent. Importantly, an explosion is believed to occur when the overpressure in the melt allows for the rupture of the vesicles' walls. Spieler et al. (2004) defines this critical overpressure limit ($\Delta \rho_f$) as follows:

$$\Delta \rho_f = 10^6 \, Pa / \varphi_b$$

where φ_b, is the volume fraction of bubbles, often reported to be > 0.75 for critical cases. Below this critical threshold, the probability of an explosive activity is reduced and effusive activity might become dominant.

Shield volcanoes are the largest in size, and they are common in Iceland or Hawaii, for instance. They are mostly basaltic and have generally non-explosive eruptions. They are constructed mostly from lava

Box 7.2: Izu-Ooshima in Japan: material from the 1986 eruption, tourist shelter and monitoring system

Izu-Ooshima (or Izu-Oshima) is a volcanic island located approximately 100 km south of Tokyo City, and 30 km south of the Izu Peninsula. The island hosts a 3–4 km diameter caldera, with the Mihayarama cone in its centre (764 m a.s.l.). On 15 November 1986, the latest eruption started from Miharayama Volcano with an effusion of lava that reached and spread on the caldera floor. This first stage was then followed by the opening of a 1 km-long fissure, from which a lava fountain spewing material as high as 1,600 m, according to Japanese scientists. This eruption led to a mass evacuation of the island on November 1986, with 12,000 transported to Tokyo on the main island. In the aftermath of the eruption, this event holds a dual reality, as it attracts tourists and school visits, helping the local economy, whilst it is still being monitored for potential eruption, and shelters are a reminder of the danger (Figure 7.8). However, the presence of andesite at the bottom of the 1986 fissure suggests that the next eruption could be different in nature. As the island moves north with the oceanic plate towards the main islands of Japan, the volcanism will eventually transform to become more explosive. And this is one of the main challenges in volcanic hazard monitoring; despite apparent cycles, volcanoes also change as they age and move, creating different types of hazards. This brings us to the question: how reliable are predictions based on the statistical analysis of past data?

Fig 7.8
Inside the caldera of Izu-Ooshima, Japan. The 1986 eruption fallout material: (A) Dr Patrick Wassmer examining the sediment layers, (B) a view from the caldera of Mount Miharayama with one of the vents at the top, (C) a hand sample of the vesiculated lava, (D) a bomb shelter built within the caldera with Dr Danang Sri Hadmoko for scale and (E) one of the laser reflectors used by the Japanese authority to monitor the deformation of the volcano.
Source: author created

flows, and they can also produce highly vesiculated tephras. The material can be recognised in the landscape from its typical black colour.

Composite volcanoes can be recognised in the landscape from their almost-perfect conic shapes, in the like of Mount Fuji in Japan, Mount

Merapi in Indonesia, Volcan de Colima in Mexico or Mount Rainier in the USA. Because the magma that creates composite volcanoes is more viscous than in shield volcanoes (more differentiated), the eruptive activity combines explosions and lava flows. Accompanying the explosions, pyroclastic flows are one of the main sources of hazards during an eruptive event. At the summit of composite volcanoes, or more rarely, on their own, volcanic domes are structures that result from very viscous magma, which can either explode or slide by gravity, like it is the case for Unzen Volcano in Japan.

Finally, volcanoes can also generate larger structures, with the collapse of the floor surrounded by steep walls. These structures are named caldera and are the origin of the most devastating eruptions. The typical depression is due to the collapse of a volcanic floor, often over areas of several tens of square kilometres, surrounded by subvertical walls, like the Aso Caldera in Japan. Following the floor collapse, some of the calderas become filled with water, creating caldera lakes, like at Toba Volcano in Indonesia. Associated with calderas, large explosive and complex eruptions occur. Very often, the central part of the caldera sees the rise of smaller volcanic structures, like Mount Bromo in the Tengger Caldera (Indonesia)

Effusive volcanism: lava flow hazards

Lava appears at the surface in lava lakes, and it becomes a hazard when it starts flowing from the vent, either in a conduit or at the surface. The temperature of lava is usually between 1000 degrees C to 1150 degrees C for basaltic lava, whilst more silicic lava can be slightly cooler, at 850 to 1050 degrees. Its flow dynamics can be modelled using an equation often used in mass movements:

$$\tau = h\rho g \sin(\theta)$$

This equation relates stress or deformation ability of the lava (τ), with the depth of the flow (h), density of the lava (ρ), gravity *(g)* and slope angle (θ). The deformation of the fluid depending on the velocity will then vary depending on the type of fluid; lava being often modelled as a visco-plastic material or, in other words, a fluid where the shear stress and the yield strength of the fluid is not governed by a straight line passing by the origin of the two variables. One of the earliest empirical relation that relates the fluid velocity to the lava density is the relation by Jeffreys (1925, as reported by Harris, 2013):

$$\eta = \frac{h^2 g\rho \sin(\theta)}{Bu}$$

where the nomenclature is the same as the previous equation, with η as the fluid viscosity, B is an empirical parameter depending on the flowing cross-section shape, and u is the velocity of the flow. From this empirical equation, it is easy to see that with increasing viscosity, the flow velocity will decrease and present velocities much lower than debris flows or water floods. This viscosity depends on the lava composition, temperature, dissolved gas content and percentage of crystal-per-melt ratio, and very importantly, for field assessment, these differences can be distinguished from the roughness of the lava surface. Basaltic lavas travel on land at a speed of a few metres per hours, although extreme cases have been documented at 65 km/h. As the lava flow tends to cool down on the surface first, they create a crust, which often hide the flow occurring below the surface, forming pipes in the aftermath of eruptions. The very same pipes can then be reused in future eruption, easing underground flowage of lava. In 2006, the Karthala Volcano on the Grand Comoros Island, located in the Indian Ocean near Madagascar Island, entered in eruption. The eruption started with tremors and the rise of lava in the central crater at the summit of the caldera. At that time, the local government was worrying about underground lava flow that could flow into existing pipes underneath the airport. Compared to explosive and dome-collapse phenomena, effusive activity can be seen as a slow-onset hazard, which is primarily a threat to buildings and infrastructures, and trapped livestock, rather than a direct threat to human lives.

Lava flows can then be differentiated between ‘a ‘a, pahoehoe and block lava. ‘A ‘a lava has a rough surface with angular fragments, pahoehoe is smooth, whilst block lava has large fragments and is typical of very sticky lava. Lava are commonly associated with basaltic volcanoes, and 90% of effusive eruptions occur on those volcanoes, but not only (e.g. Semeru Volcano in Indonesia, where pyroclastic flow deposits and lahars are controlled by a lava flow base in the Curah Lengkong Valley), and they can also be found with island arc andesitic and rhyolitic volcanoes as well. Effusive volcanism seldom poses a threat to human lives, as its relatively low speed leaves sufficient time to evacuate, but the infrastructure and land impacts are often irreversible at human timescale. As the lava recovers the pre-eruption surface, soils are lost, vegetation is burnt and infrastructures are destroyed or buried. As lava flow mostly occur at oceanic islands, where there is less opportunity to differentiate the magma, it then becomes a space issue. In 2006, the Karthala Volcano, located on the Grande Comore Island enters in eruption, and a lava lake appears in the crater, and the crater floor rose by several hundred metres. Although the eruption did not cause any casualties, the government had concerns over the potential need to evacuate the island should the lava start to flow outside the crater. Indeed, limited space and resources on the island did not need for the eruption to be a direct threat to be the source of hazards. Furthermore, the eruption coincided with the presidential election, and it led to all sorts of conjectures over what nature wanted to happen in the elections. Although lava flows are difficult to stop, attempts have been made to control them and limit their speed. In 1992, in Italy, the Civil

Defence Force dropped blocks of concrete in Etna Volcano, and various diversion-like stopbanks were built to channel lava flows. If the concrete blocks in the crater hardly had any effect, flow diversion structures met some success. Furthermore, chemical control has been proposed, by notably injecting carbonate rocks that makes the lava more viscous and, thus, slow it down. In 1973, for the Heimaey eruption in Iceland, seawater was pumped and poured on the lava flow to slow it down. Those methods have, however, relatively limited success, and one can't state that human engineering is able to stop lava flows as yet.

Pyroclastic flows

Pyroclastic activity is arguably the most common manifestation of explosive volcanism, when tephras are propelled from a vent into the atmosphere. The explosions result from the decompression of the magma in the volcanic conduit, which then transforms fluids into gas. Then the gas released in the viscous magma increases the internal pressure, and when it can't escape, it can generate explosion through the pressure differential. When the gas pressure is not sufficient to create a direct explosion, a volcanic dome can grow at the summit of the volcano, like at the summit of Merapi Volcano (Indonesia) during the 2010 eruption or at the summit of Shiveluch Volcano (Russia). The dome can then either collapse by gravity or explode. One of the by-products of this activity is the generation of pyroclastic flows, also named pyroclastic density currents.

Pyroclastic flows and the associated surges are rapidly moving mix of gas and clasts at temperature reaching several hundred degrees, with often a sole layer hugging the ground and an elutriated ash cloud travelling on top. The latter can move independently from the sole layer, as scientists and mass media learned the hard way during the last eruption of Unzen Volcano in Japan (1990–1995). Because they travelled following the topography at speeds that can reach 200 km/h, escaping them is almost impossible, and prior evacuation is essential. Both for scientific and hazard research purposes, the propagation, dynamics and deposition have been studied using mostly field deposits (Branney & Kokelaar, 2002).

Pyroclastic flow hazards are the rapid (a few seconds) burial of valleys and entire areas. During the 2006 and 2010 eruptions of Merapi Volcano, the village of Kali Adem was both times recovered and destroyed by pyroclastic flow deposits. Furthermore, the dynamic pressure of the flow can also flatten enclosed buildings. As the flow can transport large clasts in a mixture of gas and tephras, the blocks can also impact and flatten structures. Even away from the main flow, the elutriated ash cloud transports ash-saturated gas at high temperatures, which can suffocate and burn from the inside anyone breathing the mixture.

Volcanoes, like the Merapi Volcano (Box 7.3) or Semeru Volcano in Indonesia, have their valleys barred by large check dams built to stop

Box 7.3: Merapi Volcano

Merapi Volcano, located in Central Java, Indonesia, is approximately 35 km north of the capital of Central Java, Yogyakarta. The volcano is one of the most active in the world, and it has been generating eruptive events once every five to six years during the last several decades. In 2010, however, a sub-Plinian eruption generated the largest eruption of the last 1,000 years. Villages on the volcanic apron were evacuated for months at a time, and the Kali Adem village that was already affected by the 2006 eruption was swept once more by pyroclastic flows (Figure 7.9).

For the latest eruption of Merapi Volcano (2010), and rich of the too-many previous experiences, the Indonesian government displaced an estimated 400,000 people from Merapi Volcano. In 2010, the magnitude of the event surpassed the previous smaller events, and eruption hazards and disaster risk maps became rapidly obsolete with the increasing magnitude of the eruption, putting forward the weaknesses of our hazard approaches based on statistical analysis of past events. Nevertheless, the government adaptation was rapid, and inhabitants rated very high their confidence in the authorities and scientists, which used different media of communications to reach everyone, and adapt the pace and scale of evacuations. This positive

Fig 7.9

Merapi Volcano (meaning the "mountain of fire" in Bahasa) from 1994 to 2010. Note: (A) pyroclastic flow at Merapi Volcano (photo courtesy of Professor Franck Lavigne); (B) the summit of Mount Merapi in 2006 in the background, with pyroclastic flow deposits near the village of Kali Adem; (C) the summit of Mount Merapi in 2010 with the new dome growing in the open vent; (D) aerial photograph of the Kali Adem village (meaning "cold-valley" in Javanese), with the locations of photographs B, E and F, in white, is the extent of the ignimbrite (pyroclastic flow deposit); (E) pyroclastic surge deposit that burnt the forest; and (F) mining of the pyroclastic flow deposit and the subsequent lahar for road pavement and the concrete block industry.

Source: author created

outcome was somehow contrasted, as 50% to 70% of people did at least one trip back to their homes in the closed-off zone (Mei et al., 2013).

Once the first hazard phase linked to the eruption itself finished, a new form of hazard took the lead, as the pyroclastic material then became the source of several lahars brought by the rainy season (Hadmoko et al., 2018). This is another difficulty working with pyroclastic and other forms of tephra hazards: when it is finished, it is not the end but the start of another set of processes and hazards that can last for several years. In Indonesia, people working in the pyroclastic flow and lahar deposits mining for roadwork and concrete block constructions are then at the forefront of lahar hazards, where economic realms compete with the danger of being in a valley enclosed by a few tens of metres of high walls.

sediments from lahars and *banjir*, but when pyroclastic flows fill up the dams and flow over them, they can act as ramp propelling the pyroclastic flow outside the valley and then worsening the disaster risk. Structures for hazard management can, thus, have the opposite effect on other hazards, like lahar check dam effects on pyroclastic flows.

Water remobilisation: lahars

Lahar (an Indonesian word) is a mixture of water and sediment that flows on or from the slope of a volcano. It flows rapidly, and in term of content, it differentiates itself from streamflows with a typical sediment concentration > 60% of the volume and > 80% of its mass. Lahar can be considered as the equivalent of debris flow that is found in mountain areas. Because of the high-sediment concentration, the flow can carry large sized blocks, and it is not uncommon to see a small sedan-sized block being transported by a flow that is less than a metre deep. For example, the villagers near the Curah Lengkong Valley at Semeru Volcano, in Indonesia, know not to cross a lahar flow even when it is just ankle deep because they know it can sweep an adult away. Lahars are commonly associated with andesitic and dacitic volcanoes, therefore, most common in Indonesian volcanoes, in the Philippines, Japan, the Aleutians, the South American Andes and so forth.

The material feeding lahars range in scale from large-scale sector collapses, like at Mount St Helens in 1980; to pyroclastic flows and vulcanian explosions, like during the major eruption of Merapi in 2010; and also smaller-scale vulcanian explosions. At Semeru Volcano, in East Java, Indonesia, a vulcanian eruption occurs on average once every 5 to 15 minutes, producing ash and small clast deposits at the summit. This material is regularly remobilised by rainfalls generating lahars several times a year, especially during the rainy season. Finally, lahars can also be associated with processes in between eruptions. In such case, the material is supplied through existing stocks, like at Unzen Volcano, where, in 2018, lahars still travel the Tansandani and Gokurakudani valleys on the east flank of the volcano.

Although they are less common than their lithic counterpart, lahars can also occur in pumiceous material, emplaced by ignimbrite. Because the material is vesiculated and presents a density that is originally lower than the one of water, the material will, therefore, float or be very buoyant, until it is abraded and transformed into smaller fine fractions or until the pumiceous material is filled with water. The low density generates flows that can sweep valleys for hundreds of kilometres, like the megafloods triggered from Numazawa Volcano in Japan.

Check dams or sabo(u) dams (originally *sabou* in Japanese and often referred as *sabo* in the international literature) are types of constructions along a valley aimed to slow down or stop the course of a debris flow or a lahar, or to separate large clasts from the rest of the material flowing. Although their proliferation, notably in East Asia, is to be attributed to the Japanese influence, they originated from 19th-century France, Italy and Austria, notably, where very often they were simple walls or steps across a valley. Today, there are several types of sabou dams that are often used in combination with complementary functions, such as slit dams, which are large metallic tube structures designed to stop the largest rocks and wood debris. Very often, different types of sabou dams are associated together, depending on their functions. At Unzen Volcano, for instance, a series of steps first reduce the velocity of the sediments, then the largest blocks are stopped by large slit structures, then check dams trap the main part of the flow, and then downstream sets of sediment pockets riddle the valley floor to stop any fine sediments that would have escaped from the set of dams upstream.

Those dams used to be very popular in an era of population expansion and global economic wealth, with at its heart the notion that human beings could control nature. Today, they are way less popular due to their negative aspects, stopping the flow of sediments, storing sediments in large amounts to eventually collapse, notwithstanding the aesthetic aspect in the landscape, and breaking the natural chain of sediment and ecosystem continuity.

Volcanic tele-hazards: atmospheric transport and ashfalls

The spatial distribution of smaller grain-sized fractions – tephras to volcanic ashes – are usually not limited to the volcano or to its vicinities. It can travel from local to several times around the globe, depending on the altitude of the eruptive column, the season, the position of the volcano and the morphometric characteristics of the ashes. Because of the wide spatial range and the potential pervasive nature of the material, ashfall hazards affect agriculture, the industry, the transport sector, the health of most living organisms, water quality and chemistry, and so forth (Wilson et al., 2015).

Ash clouds in the atmosphere are made of fragmented material of micron to sub-micron size, and they are rarely superior to 0.5 mm in size. They are made of glass shards and acid droplets, gaining their low pH from the volcanic material. As volcanic ash is typically microscopic glass shards, it can, therefore, impair vision if people rub their eyes; it has direct breathing and pulmonary impacts; and it is at the source of all sorts of allergies and skin problems. During the 1990–1995 eruption of Mount Unzen in Japan, the population in the city of Shimabara suffered from the direct impact of ashfalls, and air-conditioning systems with filters had to be installed in the schools to limit the health issues.

There are three characteristics of volcanic ash that makes them a hazard for the built environment and a number of the electro-mechanical machines human beings rely on the following: (1) the microscopic size of ash allows it to enter through common industrial filters, (2) its composition turns to glass at high temperature, like in a jet engine, and (3) its electrostatic quality allows it to be stacked together, eventually clogging systems. It has been estimated that the costs of ash encounters for the period 1980–1998 was superior to US$250 million in damages to engines. The main damages are the potential complete frosting of the cockpit window, the blockage of instruments that use any form of vent for their sensors and, of course, the entry into the reactor, where they melt into glass that deposit on the internal blades, eventually shutting the engine. The concerns in the aviation industry around hazards to engine only really began with the jet engine and with the 1980 Mount St Helens eruption in the USA, where a commercial jet engine lost two of its engines and experienced serious damages to the other two whilst flying through a volcanic ash cloud. As international flights have been increasing since the 1970s, the number of incidents has only increased, leading to near catastrophe, like when in 1982, the British Airways Boeing 747 that flew through a volcanic ash cloud from the Galungung Volcano in Indonesia lost all its four engines and fell by more than 7,000 m before successfully restarting the engine and making an emergency landing at the Jakarta airport in Indonesia. Despite of the gravity of these incidents, planes are not well equipped to detect ash clouds, as the particles are too small and dispersed to be detected by traditional radars. However, real safety improvements have come from collaboration with meteorological agencies, remote sensors and geologists helping predict the trajectory and spread of volcanic ash clouds. The real-time seismic monitoring of volcanoes also helps prevent unexpected ash dispersal and volcanic eruptions.

Volcanic hazards without eruptions: sector collapses

Volcanic sector or flank collapses are the destabilisation of a large portion of a volcanic structure, which fails either by gravity or from the internal pressure, or a combination of both. A sector collapse leading to a debris avalanche was first observed at the end of the 19th century at Bandai-san Volcano in Japan, where it is believed that hydrothermal alteration of the material is at the origin of the destabilisation of the slope. This idea, however, did not immediately percolate through the Western world, and the West really woke up to such disaster potentials with the bulging and sector collapse of Mount St Helens in the USA in

1980. At the geological timescale, however, such events are common processes, and it is estimated that about 400 Quaternary volcanoes have experienced a sector collapse. They have been identified around the world; for instance, in Indonesia (MacLeod, 1989), Japan (Ui et al., 1986), Hawaii (Moore et al., 1989) and Central America (Siebert et al., 2006).

Sector collapses are produced through the explosion and/or the gravitational collapse of a large part of a volcanic structure following lateral bulging, destabilising the structure (like at Mount St Helens), or following hydrothermal alteration (like at Mount Bandai). It results in the very rapid onset of a mass of debris that leave small mounts of blocks and rocks called hummocks. Because of the consequent size of a sector collapse, the landscape around the volcano is modified across tens of square kilometres, with potential lake impoundments (at Mount Bandai) and the creation of islands near the seashore (at Mount Mayuyama of the Unzen complex). Because of the scarcity of events since the beginning of contemporary sciences, predicting those movements, anticipating them and doing proper hazard analysis is still haphazard. Siebert et al. (2006) write that debris avalanche and sector collapse research is "still in its infancy", even if such events are potentially the most destructive that can occur in the vicinity of a volcano.

Freak events: Plinian eruptions and volcanic winters

Looking further afield, sector collapses are not the worst-case scenarios, and volcanoes are known to be able to create freak events known as Plinian eruptions, which have the potential to modify the climate of the planet for a long period and eventually wipe part of, if not entire, civilisations from the face of the planet. Volcanoes that are known to have created such events are the caldera (see earlier section) volcanoes and those named supervolcanoes, like the Yellowstone National Park in the USA, which can cover 75% of the USA with ash covers and tephras. Supervolcanoes are large volcanoes with at least an eruption of > 7 Volcanic Explosivity Index (see later). The main issue with such volcanoes is that even under the sea, where very little is known or observed, those volcanoes still hold the same destructive power, whereas smaller volcanic eruptions can be supressed by the water pressure. One such volcano is the Kikai Caldera to the South of Kyushu Island in Japan. This volcano is known to have exploded and created pyroclastic flow deposits that covered the totality of Kyushu Islands and generated ash deposits more than 500 km away in the vicinity of Osaka City.

Because those events are very rare, they are not often included in volcanic hazard maps and emergency plans. In Japan, Tsuji et al., (2017) have investigated the possible impact of a Plinian eruption from Kuju Volcano (Kyushu Island). They have shown that the Ikata nuclear power plant located on the nearby island of Shikoku and downwind of the vent of Kuju Volcano could be covered by as much as 10 cm of volcanic ash. Such research and suggestions remain, however, very scarce because the cost-benefit analysis over such long return period event is often ignored, but like the 2011 tsunami that crippled the Fukushima power plants, low return period events do not mean that they don't occur.

Measuring and monitoring volcanoes

One of the challenges in volcanic monitoring are the eruption rhythms that show that out of the almost 600 volcanoes that have erupted in the historical period, long periods of slumber characterise those structures, which are, therefore, not monitored regularly nor sufficiently (most of the instrumentation is composed of seismometers and geodetic survey equipment, and only countries like Japan or the USA have a network of continuous GNSS instruments monitor volcanoes). Scarpa and Tilling (1996) provided a table of the monitoring systems on the various volcanoes by regions, and only little change has happened since then. The most common scale to measure the size of an eruption is the Volcanic Explosivity Index (VEI), which was first proposed by Chris Newhall and Stephen Self in the early 1980s. The VEI scale range from 0 to 7, from non-explosive eruptions (VEI 0) to very large explosive eruptions (VEI 7). It corresponds to the volume of erupted tephras from 0.00001 km^3 (the summit explosions at the summit of Mount Semeru every 5 to 15 minutes), to more than 100 km^3 (the 1815 eruption of Tambora > 100 km^3; the eruption 600,000 years ago of Yellowstone ~ 1,000 km^3).

One of the most popular ways to monitor a volcano is to use the seismic signal that volcanoes generate when the magma rises and move beneath and inside the volcano. Active volcanoes have been differentiated based on their seismic signals, with the A-type being the high-frequency seismic signals generated by shear failure and slip, and with a typical frequency range of 5–15 Hz; B-type or the low-frequency ones are supposed to be linked to fluid pressurisation and have typical frequencies between 1 and 5 Hz; and explosion and volcanic tremors. This classification was first proposed by Minakami (1974). There are two principal types of data generated for volcanic monitoring: (1) RSAM, which stands for Real-time Seismic-Amplitude Measurement, represents the overall signal over a period of 10 minutes and records the signal regardless of what produces it – it can record wind, for instance – and (2) SSAM, which stands for Seismic Spectral-Amplitude Measurement, shows the relative signal size in different frequency bands. Tremor sources can be tracked with seismic sensors, using array processing methods. The sensors can be used to determine the shallow structure of a volcano from the wave field characteristics of the tremors, and they can be used to track the spatio-temporal evolution of the source. There are two distinct approaches to do so; the first one considers the tremor wave field as a stationary stochastic wave field, from which the structure beneath the antenna can be calculated. The second approach used a deterministic stationary wave field and frequency-slowness power spectra computed to determine the ray parameters. This second method relies on the spatio-temporal properties of a tremor source from a spatial Fourier decomposition of the tremor signals recorded by the seismic antenna. For instance, in 1986, Izu-Oshima produced a spectacular basalt fissure eruption on 21 November, following seven months of increased seismicity and four months of tremor. The largest precursory event was M = 3.1. Banded tremors of 30-minute durations with 2-hour intervals appeared from 19 July to late October and then changed into continuous tremors, coinciding with a 3-fold increase in the rate of energy release. Just after the eruption started, two large earthquakes occurred with a Mw = 5.1 and Mw = 6.0.

Three-dimension high-resolution travel-time tomography can also be calculated through the inversion of first-arrival times from man-made tremors and natural ones (earthquakes). It is a method that relies on criss-crossing raypaths to separate the integrated effects of the slowness on travel times and then derive an image of the velocity structure. Seismic monitoring is, therefore, a challenge to quantitatively interpret because of the complexity of both the methods and the media (mix of rock, gas, liquid and voids) investigated.

Another way to monitor volcanic activity is concerned with surface deformation at and around the volcano, even if the ground deformation can originate from movements several kilometres underneath the surface due to magma movement. Using the Mogi model proposed in 1958, it is possible to then interpret the surface deformation in terms of change at depth, using, notably, Hooke's law of stress-strain relation for an elastic body. The monitoring techniques can be used for vertical variations: (1) geometric levelling, which consists in the measure of height change in between known monuments using Invar rods, uses successive measures to infer displacement; (2) trigonometric levelling can also be carried out using an electronic distance meter complemented by one or two theodolites, and although it is not as precise as the geometric levelling, it is faster and needs fewer operators; (3) tiltmeters – originally using mechanical pendulum, the bubble tiltmeter consists of a spirit level in which the movement of a bubble is recorded electronically, and the electric output transformed into a signal. The strain between points is also measured using strainmeters, which use cables less than 50 m long, usually put in perpendicular positions with their end fixed to the ground, in order to measure the strain between two points. Finally, the multiplication of satellites used for positioning systems now allow millimetre monitoring of ground deformation on volcanoes and slowly changing surfaces.

There are numerous other monitoring methods – using fumarole gas content, the chemistry of waters, microgravity to monitor changes deep underneath the surface and so forth – methods that are beyond the scope of this chapter.

7.4 Tsunamis

It has been estimated that about US$4 billion a year and about 60,000 individuals are exposed to tsunami hazards (UNISDR, 2015). From a scientific (Western science) perspective, a tsunami is a long-amplitude wave generated by the displacement of the water column, due to earthquakes, volcanic activities, slides and meteoritic impacts. These waves can travel across deep sea at the speed of a modern jet (~ 700–800 km/h), and the accumulated energy turns into waves that can grow to height of 20–30 m at the coast, creating the hazard we are concerned with. It is to be noted that the 20 to 30 m high waves observed in 2004 and 2011 are not the maximum wave height that our contemporary oceans can create. Indeed, meteoritic impacts can result in several hundred-metre-high waves, like for the case of the Eltanin impact at the southern tip of South America at the beginning of the Quaternary period.

If not for meteoritic impact that can reach any waterbodies with similar probabilities (although this is not completely accurate due to our galaxy rotation, alignment and expansion directions), the coastlines located at the tectonic plate boundaries or the coastline of volcanic landmasses are more prone to tsunami hazards, in terms of return period (only tele-tsunami of a given amplitude will reach distant coastlines, whereas small tsunamis will have impacts near the source) and in terms of amplitude and time span between a triggering event and the arrival of the tsunami: indeed, there is enough time to prepare for evacuation on the coast of New Zealand when a tsunami triggered in South America crosses the South Pacific, but the tsunami can be virtually instantaneous in the coastal cities of Chile.

As tsunami propagates through the ocean, scientists explain the phenomenon using the long wave (or shallow wave) assumptions because the ocean is never deeper than 5 km. Most of these models assume that the horizontal distance is at least four times the depth and that viscous effects can be ignored. The computational fluid dynamics is then based on the Navier-Stokes equations:

$$\nabla * u = 0$$

$$\rho\left(\frac{\partial u}{\partial t} + u * \nabla u\right) = -\nabla p + \mu * \nabla^2 u + g$$

where u is the velocity, p is the pressure, $\frac{\partial u}{\partial t}$ is the time derivative and $u * \nabla u$ is the spatial component. To these equations, boundary conditions are applied to define the sea-floor displacement, its bathymetric characteristics, on top of which the water surface displacement also needs description.

Away from the coast, when the water column is > 50 m, the wave's speed and the wavelength can be approximated using simple trigonometric functions:

$$v = (g * d)^{0.5} \quad \& \quad L = v * T$$

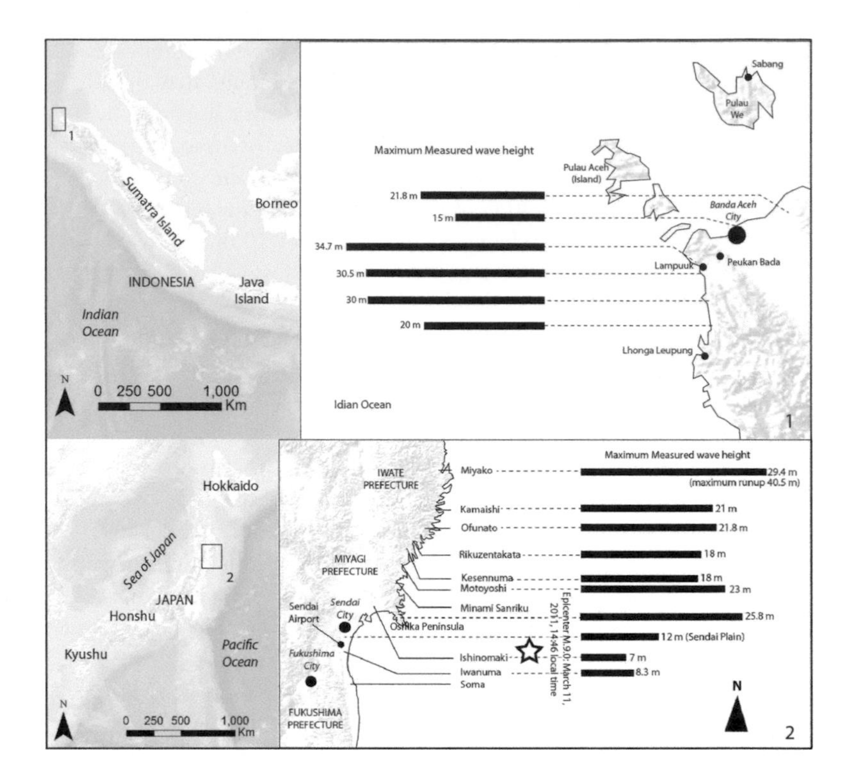

Fig 7.10
Tsunami inundation heights at selected locations in (1) Northern Sumatra and (2) Tōhoku region.
Source: author created

where v is the velocity of the wave, g is the gravity and d the depth of the water column. The wavelength L is then a function of the velocity (v) and the wave period.

Once the tsunami waves approach the coast, the vertical difference between the trough and the crest increases because of energy conservation. For hazard purposes, the waves are measured at the coast using two data: (1) from the mean sea level, which is the runup, and (2) from the ground directly underneath, which is the wave height. Therefore, the wave height tends towards 0, when the runup tends to increase with the topography. During the 2004 tsunami, south of Banda Aceh, the wave had a wave height of 35 m near the mosque of Lampuuk. In 2011, the wave heights of the tsunami in East Japan was around 10 m at its maximum, but in Miyako City, the runup was 39 m (Figure 7.10).

Generation mechanisms

As a tsunami requires the movement of the entire column of water, or at least a large portion of it, the generation mechanisms are linked to a rapid change of the floor underneath the waterbody, due to either or a combination of the following phenomena: (1) an earthquake, (2) a large-scale volcanic eruption, (3) a large-scale submarine mass movement, (4) aerial mass movements entering the sea and, finally, (5) objects coming from space, such as meteorite and other stellar objects entering the sea. Although they have never been directly observed, evidences, such as impact craters at the bottom of the ocean, are reasons to believe that such events aren't rare over geological time. Although we don't discuss it in the present book, it is worth knowing that human beings can also create tsunamis if blasting atomic bombs in the deep ocean, such as the Mariana Trench, and it could be very well used as a military weapon.

Impacts at the coast

A tsunami is much more than its direct impacts on human beings and their activity at the coast; it also modifies the environmental dynamics for eventually a very long period, generating, in term, secondary hazards. It is, for instance, difficult to go and use a water well after a tsunami wave brings debris and salts in the soil and the water. It is also difficult to continue growing crops when the soil has been salinised or when the soil itself has been eroded. Those environmental changes operate at different scales.

The immediate or very-short-term impacts of tsunamis are the direct loss of lives and livelihood, as well as all the destruction of engineered structures. The short-term impacts (~ 1 to ~ 5 years) include floods, salinisation and debris spread over land. Furthermore, because the water is brackish, even remaining vegetation can perish due to the change of soil chemistry, like it was the case for the plantation at the bay of Lampuuk in the aftermath of the 2004 tsunami in Sumatra or, as it was observed, for some of the littoral pine trees in Japan after 2011.

Then the mid-term impacts include the coastal and river system reconfiguration, and sediment spread, with impacts in the range of > 5 to 50 years. Modifications along this time range include the environmental destruction of sand dunes chopped off to the ground in the Lampuuk area (south of Banda Aceh), the redistribution of sediments and erosion of riverbanks, and so forth. Those environmental impacts have a deep impact on the communities because of the land configuration change and the legal relation of people to their environment. There are then the longer-term impacts, like the destruction of coastal cliffs or the full destruction of coastal islands.

Tsunami warning systems

In their tsunami warning review paper, Bernard and Titov (2015) have explained that Japan developed their first understanding of tsunami generation after the 1896 Sanriku giant tsunami, which claimed the lives of 22,000. Then the 1933 earthquake and tsunami in the Sanriku only took the lives of 3,000 because the Japanese had developed a warning campaign, stating, "if the Earth shakes violently, take cover on high ground". The warning system has been progressively improved. The next system is established in 1949 in Hawaii, as a response to the deadly 1946 Alaska tsunami. Afterwards, Russian and French systems were developed to, respectively, cater for Kamchatka and the islands of Tahiti and Papeete. Japan was the first to develop warning systems, and it is still one of the world leaders, with, notably, the DONET network (Dense

Oceanfloor Network System for Earthquakes and Tsunamis) developed by JAMSTEC (Japan Agency for Marine-Earth Sciences and Technology). The system is developed across the Nankai Trough to protect the cities of Osaka and Nagoya, which would be the first major urban and economic centres to be ever impacted during the contemporary period. The DONET network is cabled with seismometers and pressure sensors at depth between 1,900 and 4,300 m depth.

7.5 Concluding remarks

The Earth is a hot planet; the internal energy is fuelling plate tectonics and volcanism, which in turn can displace large waterbodies at the surface of the planet that generate tsunamis. Because of the interaction with the topography and the near-surface sedimentary formations, earthquake shaking can then be amplified, and volcano magmas can be enriched to change effusive volcanism into an explosive one. The complexity of the surface cover and the realities of communities, in turn, contribute to generate differentiated hazards. Furthermore, endogenous hazards can (1) affect other hazards by modifying the framework where they occur (for instance, earthquake can reshape the morphology of valleys, modifying the relationship between rainfall characteristics and floods), and (2) other hazards can also modify endogenous hazards (for instance, important rainfall saturating soils will modify the surface acceleration during an earthquake and eventually contribute to liquefaction). It is, therefore, essential not to consider hazards one by one in a vacuum but as a complex system. Unfortunately, the way we are educated, with disciplinary divisions at school from secondary to the end of University, and then very often, the divisions between different sections in the company and administrations we work in, are all counterproductive when it comes to finding appropriate solutions. Even the field of hazards and disaster risk, which has moved from hard sciences and engineering to a more social sciences approach, have evolved, but it has replaced walls by others, and fully integrated approaches are still rare.

Take-away messages

1 Earthquake hazards have arisen because communities have developed infrastructures and constructions that were not adapted to a seismic environment. The direct threat from an earthquake is very limited, except when it triggers secondary hazards, like rockfalls.
2 There are different types of volcanoes, and depending on their origin (intraplate or at the boundary of divergent or convergent tectonic plates), they will display very different volcanic activity, turning into different types of hazards.
3 Tsunamis are mostly the results of earthquakes or other forms of mass movements that can displace a tall column of water all at once.

4 Tsunami warning systems have been developed by economically wealthy nations, and they are, therefore, not always well-placed for other nations of the world to forecast a coming tsunami.

To learn more about the topic discussed in this chapter, listen to the *Disasters: Deconstructed* interview with Dr Jazmin Scarlett (Figure 7.11).

Fig 7.11
QR code for Chapter 7.

Further suggested reading

Lee, W. H. L., Kanamori, H., Jennings, P. C., & Kisslinger, C. *International handbook of earthquake and engineering seismology*. Academic Press 937 p.

Mackie, S., Cashman, K., Ricketts, H., Rust, A., & Watson, M. (2016). *Volcanic ash. Hazard observation*. Elsevier Publishing.

McGuire, R. K. (2004). *Seismic Hazard and risk analysis*. Earthquake Engineering Research Institute, 119 p.

Papale, P., & Schroder, J. F. (2014). *Volcanic hazards, risks and disaster*. Elsevier Publishing.

Wyss, M., & Schroder, J. F. (2014). *Earthquake Hazard, risk, and disasters*. Academic Press, 606 p.

References

Bernard, E., & Titov, V. (2015). Evolution of tsunami warning systems and products. *Philosophical Transactions. Series A, Mathematical, Physical, and Engineering Sciences*, *373*(2053). https://doi.org/10.1098/rsta.2014.0371

Branney, M. J., & Kokelaar, B. P. (2002). *Pyroclastic density currents and the sedimentation of ignimbrites* (p. 27). Memoire of the Geological Society, 143 p.

Calais, E., Freed, A., Mattioli, G., Amelung, F., Jónsson, S., Jansma, P., Hong, S. H., Dixon, T., Prépetit, C., & Momplaisir, R. (2010). Transpressional rupture of an unmapped fault during the 2010 Haiti earthquake. *Nature Geoscience, 3*(11), 794–799. https://doi.org/10.1038/ngeo992

Curtis, L. E. (2012). *HAITI earthquake BOX—Haiti Mw 7.0 earthquake of January 12 2010 lifeline performance*. American Society of Civil Engineers (Ed.) [Monograph p. 35].

Duarte, J., & Schellart, W. (2016). *Plate boundaries and natural hazards*. American Geophysical Union Publishing.

Gonnermann, H. M., & Manga, M. (2013). Dynamics of magma ascent in the volcanic conduit. In S. A. Fagents, T. K. P. Gregg & R. M. C. Lopes (Eds.), *Modeling volcanic processes, the physics and mathematics of volcanism* (pp. 55–84). Cambridge University Press.

Hadmoko, D. S., de Belizal, E., Mutaqin, B. W., Dipayana, G. A., Marfai, M. A., Lavigne, F., Sartohadi, J., Worosuprojo, S., Starheim, C. S. A., & Gomez, C. (2018). Post-erutpive lahars at Kali Putih following the 2010 eruption of Merapi volcano, Indonesia: Occurrences and impacts. *Natural Hazards*, *94*, 419–444.

Harris, A. J. L. (2013). Lava flows. In S. A. Fagents, T. K. P. Gregg & R. M. C. Lopes (Eds.), *Modeling volcanic processes, the physics and mathematics of volcanism* (pp. 85–106). Cambridge University Press.

Jeffreys, H. (1925). The flow of water in an inclined channel of rectangular section. *Philosophical Magazine, 49*, 793–807. https://doi.org/10.1080/14786442508634662

MacLeod, N. (1989). Sector-failure eruptions in Indonesian volcanoes. *Geology of Indonesia*, *12*, 563–601.

Mei, E. T., Lavigne, F., Picquout, A., De Belizal, E., Brunstein, D., Grancher, D., Sartohadi, J., Cholik, N., & Vidal, C. (2013). Lessons learned from the 2010 evacuations at Merapi volcano. *Journal of Volcanology and Geotehrmal Research*, *261*, 348–365.

Minakami, T. (1974). Seismology of volcanoes in Japan. In Civetta et al. (Eds.), *Physical volcanology developments in solid earth geophysics*, 6, 1–27. https://doi.org/10.1016/B978-0-444-41141-9.50007-3

Moore, J. G., Clague, D. A., Holcomb, R. T., Lipman, P. W., Normark, W. R., & Torresan, M. E. (1989). Prodigious submarine landslides on the Hawaiian Ridge. *Journal of Geophysical Research, 94*(B12), 465–484. https://doi.org/10.1029/JB094iB12p17465

Murphy, B. (2015). Coseismic landslides. In Schroder & Davies (Eds.), *Landslides hazards, risks and disasters* (pp. 91–129). Elsevier.

Quigley, M. C., Bastin, S., & Bradley, B. A. (2013). Recurrent liquefaction in Christchurch, New Zealand, during the Canterbury earthquake sequence. *Geology, 41*(4), 419–422. https://doi.org/10.1130/G33944.1

Reiter, L. (1990). *Earthquake Hazard analysis*. Columbia University Press.

Rodríguez, C. E., Bommer, J. J., & Chandler, R. J. (1999). Earthquake-induced landslides: 1980–1997. *Soil Dynamics and Earthquake Engineering, 18*(5), 325–346. https://doi.org/10.1016/S0267-7261(99)00012-3

Scarpa, R., & Tilling, R. I. (Eds.). (1996). *Monitoring and mitigation of volcano hazards*. Springer.

Siebert, L., Alvarado, G. E., Vallance, J. W., & van Wyk de Vries, B. (2006). Large volume volcanic edifice failures in Central America and associated hazards. In B. Rose, E. Carr, & V. Patino (Eds.), *Volcanic hazards in Central America*. Geological Society of America Special Paper 412 (pp. 1–26). Geological Society of America Publisher.

So, E. (2016). *Estimating fatality rates for earthquake loss models*. Springer-Briefs in Earth Sciences. Springer Verlag, 71 pp.

Spieler, O., Kennedy, B., Kueppers, U., Dingwell, D. B., Scheu, B., & Taddeucci, J. (2004). The fragmentation threshold of pyroclastic rocks. *Earth and Planetary Science Letters, 226*(1–2), 139–148. https://doi.org/10.1016/j.epsl.2004.07.016

Stefansson, R. (2011). *Advances in earthquake prediction—Research and risk mitigation*. Springer Praxis Book, 245 p.

Terzaghi, K. (1950). *Mechanisms of landslides. Engineering geology*. Geological Society of America.

Tsuji, T., Ikeda, M., Kishimoto, H., Fujita, K., Nishizaka, N., & Onishi, K. (2017). Tephra fallout Hazard assessment for VEI5 Plinian eruption at Kuju volcano, Japan, using TEPHRA2. *IOP Conference Series: Earth and Environmental Science, 71*(012002), 12 p. https://doi.org/10.1088/1755-1315/71/1/012002

Ui, T., Yamamoto, H., & Suzuki-Kamata, K. (1986). Characterization of debris avalanche deposits in Japan. *Journal of Volcanology and Geothermal Research, 29*(1–4), 231–243. https://doi.org/10.1016/0377-0273(86)90046-6

UNISDR. (2015). *Making development sustainable: The future of disaster risk management. Global assessment report on disaster risk reduction*. United Nations Office for Disaster Risk Reduction.

Wilson, J. T. (1965). A new class of faults and their bearing on continental drift. *Nature, 207*(4995), 343–347. https://doi.org/10.1038/207343a0

Wilson, M. T., Jenkins, S., & Stewart, C. (2015). Impacts from volcanic ash fall. In Papale et al. (Eds.), *Volcanic hazards, risks and disasters* (pp. 47–86). Elsevier.

Wotherspoon, L., Orense, R. P., Green, R., Bradley, B., Cox, B., & Wood, C. (2014). Analysis of liquefaction characteristics at Christchurch strong motion stations. In R. P. Orense, I. Towhata, & N. Chouw (Eds.), *Soil liquefaction during recent large-scale earthquakes* (pp. 33–43). Routledge.

Youd, T. L., & Perkins, D. M. (1978). Mapping liquefaction-induced ground failure potential. *Journal of the Geotechnical Engineering Division, 104*(4), 433–446. https://doi.org/10.1061/AJGEB6.0000612

CHAPTER 8

Gravity-driven natural exogenous processes

Fig 8.1
At Merapi Volcano, Central Java, Indonesia, lahars are not perceived solely as a hazard but also a resource, opening the door to a post-eruption economy. The team of workers on this photograph is taking a break in the middle of the pyroclastic flow and lahar-ridden valley.
Source: photo courtesy of Christopher Gomez (2017)

8.1 Introduction

This chapter deals with gravity-driven exogenous hazards that mix water and engineering soils and rocks. It first provides an anatomy of those processes, what they are made of and how they move across the land. It then turns towards a differentiation of the events based on whether they occur in engineering rock or in engineering soils. Particular attention is given to the family of debris flow and their volcanic counterpart because of the important place they have amongst the gravity-driven hazards and the scientific challenges they still pause. The chapter then turns towards the triggering processes of these hazards; and how scientists, engineers and practitioners define levels of safety in the face of gravity-driven natural exogenous hazards; and what subsurface, surface and corollary data are usually collected for hazard mitigation.

Competing with the endogenous forces that build mountains up and construct volcanic structures, the erosion is a set of processes, which,

DOI: 10.4324/9781315469614-12

by and large, smooths the Earth surface. It is mostly driven by water on and in the ground, and the atmospheric processes and gravity. It is an easy task to look at a landslide and classify such event as a hazard, but they are first and foremost environmental processes, and they only become hazards when they are of a given size, recurrence and timing. Imagine a small landslide in your garden planter when the aged wood planks give way under the pounding of a heavy rainfall. You effectively have a landslide in your garden, but nothing could relate it to a hazard. However, if we were the size of an ant in your garden, well, we may consider this flower pot movement a hazardous phenomenon.

Recently, an essential step has been taken in the hazard and disaster risk space, with the cornerstone statement that there is no such a thing as a "natural disaster" (Kelman, 2020). This concept can go one step further by also adding that natural hazards are a human construct, not out of our direct involvement in the hazards but through the very fact that our perception of a hazard is shaped by our human reality (lifespan, size, lifestyle, cultural background, socio-economic realm). Most Europeans or North Americans do not look at the weather with anxiety, wondering whether it is going to rain or shine because this may be the start of the next drought hazard (food security issues etc.), but in other parts of the world, this may be the case. If natural hazards are not a human construct, how can you explain that, depending on where you are and your socio-cultural background, a single event can be seen very differently?

For the present chapter on gravity-driven natural exogenous hazards, we are investigating the events that we, as humans, can consider as hazards. And those processes become hazards because they are a manifestation of energy that can exceed what individuals, communities and the built environment can withstand. In other publications, they have been traditionally grouped together into the hydrometeorological hazards (cf. Chapter 9), but processes involving mostly water against those involving engineering soils and rocks and water (the object of this chapter) were used as a boundary in the present work. Compared to waterborne events, the physics and mechanics of these events are less clearly understood, basic physics processes still have a question mark, and for those reasons as well, I thought it was opportune time to separate the two. Consequently, the present chapter is interested with gravity-dominated mass movements (GDM2) in rocks and in sediments, and all the other hydrometeorological hazards are dealt with in Chapter 9.

More precisely, in this chapter, we will investigate the movements driven by gravity, from the collapse of entire mountains to more modest rockfalls, from a range of soil and rock movements triggered without water as a lubricant to debris flows that are a mixture at the limit with waterborne processes, such as river floods.

Mass movement hazards can unfortunately take the lives of many within a very short time span: on 16 December 1920, in China, loess landslides took the lives of 180,000 people; in Colombia, at the city of Armero, downstream of the Nevado Del Ruiz Volcano, around 25,000 people perished in a lahar on 13 November 1985. On 31 May 1970, still in South America, 18,000 perished at Yungay, from a rock avalanche mixed with snow and ice, which swallowed the community. In Tajikistan (1949), 12,000 casualties were reported at Khait from another rock avalanche triggered by an earthquake and so forth.

Whether one argues that the casualties are due to population concentration or to the phenomenon itself (or most probably, a combination of the two), the commonality of events, with > 10,000 casualties, reveal the important role that gravity-driven hazards have in the disaster risk realm, even if a large majority of the events is of smaller scale.

8.2 Gravity-driven rock and sediment movements as a hazard

Drawing – one or several – limits between processes depending on their water content, amount of sediment, shear size, whether they occur underwater or on land, is based on some level of understanding of the mechanics and physics of those processes, but it is, above all, an experience of the mind. At this stage of our understanding, different authors may draw the line between one type of event in different ways.

For the purpose of structuring this handbook, we have fenced the content of the present chapter around mixtures of rocks, sediments, debris and water moving mostly under the action of gravity.

These events can range from small roadside slides to a full mountain being set in movement, like it had been the case for the Flims rockslide that occurred in Eastern Switzerland, where the top half of the mountain slid downslope and into the valley. In gravity events – even of smaller scale – the size of the debris can range from a few millimetres to objects several kilometres in size (Table 8.1), and it is not uncommon to find outsized clasts and blocks that can be several tens of metres in scale. For instance, at Mount Unzen in Japan, some parts of the dome that collapsed on the apron underneath during the 1990–1995 eruption can be more than 10 m across and was mixed with a supporting matrix of pyroclasts, where the smaller particles can be as small as a few tens of microns.

But in mixtures of water and debris, even the smallest material fraction plays an important role in the events. Indeed, the calculation of the impacts of blocks and other large-scale objects on a structure due to their momentum in a flow is an essential part of hazard assessment, but the small fraction often plays the role of a lubricant, influencing the runout of these events. For instance, in large rock avalanches, like at Flims, the material crushing and fragmentation has been proposed by Professor Tim Davies to be one of the important drivers explaining why such a large mass can travel kilometres even with a small slope gradient.

Class	*Size range (mm)*	*Class*	*Size range (mm)*
Outsized clasts	> 4,000	Fine gravel	8–2
Very large boulders	4,000–2,000	Coarse sand	2–0.5
Large boulders	2,000–1,000	Medium sand	0.5–0.25
Small and medium boulders	1,000–250	Fine sand	0.25–0.062
Cobbles	250–64	Silt	0.062–0.004
Coarse gravel	64–16	Clay	0.004–0.001
Medium gravel	16–8	Fine clay	< 0.001

Table 8.1
Simplified sediment grade scale by class and in millimetres
Source: author created

Scientists are, therefore, classifying GDM2 events based on their horizontal reach or distance travelled compared to the vertical gradient (the distance travelled vertically). This relationship is often expressed as the Fahrböschung. It is usually calculated as the angle linking the highest point to the lowest one reached by the mass movement. The heavier the mass movement is, the lower the Fahrböschung angle; meaning that large mass movements can reach long distances even without an important vertical gradient. For instance, Hungr and Evans (1996) reported that the Blackhawk mass movement (USA) of 283 10^6 m^3 has a Fahrböschung angle of only 6.3 degrees, when the Dusty Creek mass movement (USA), which is only 7.10^6 m^3 has a 21 degree angle.

Velocity and energy

The GDM2 processes are also a hazard (1) because they move rapidly (at a speed that can overcome a human on foot), and they do so from a location where they were a fix mass, into a deposition area, invading it (like a cliff that suddenly gives way and sweeps the land underneath), and (2) because of the movement, the surface changes, creating deformation and breakage at the surface of the moving mass (opening of a crack in the ground, impacting buildings' foundations, for instance). If we think of it as a landslide, it is the difference between the hazard of being in the way of the moving mass and being on top of the moving mass.

For the first reason considered, it is the combination of velocity and volume and density (i.e. the momentum) that impacts and modifies the natural and built environment and, thus, causes casualties (Figure 8.2); this can, therefore, be associated with rapid-onset hazards. For slower-onset hazards (second case), it is not so much the momentum of the translated material but the deformation and transformation of their surface and subsurface, which is the source of hazards. Such hazards are more commonly an issue for the built environment and infrastructures, which will fail under the forced deformation of the soil and rocks they are built on. Consequently, the types of impacts vary based on the velocity of the gravity-driven mass movements, which can display

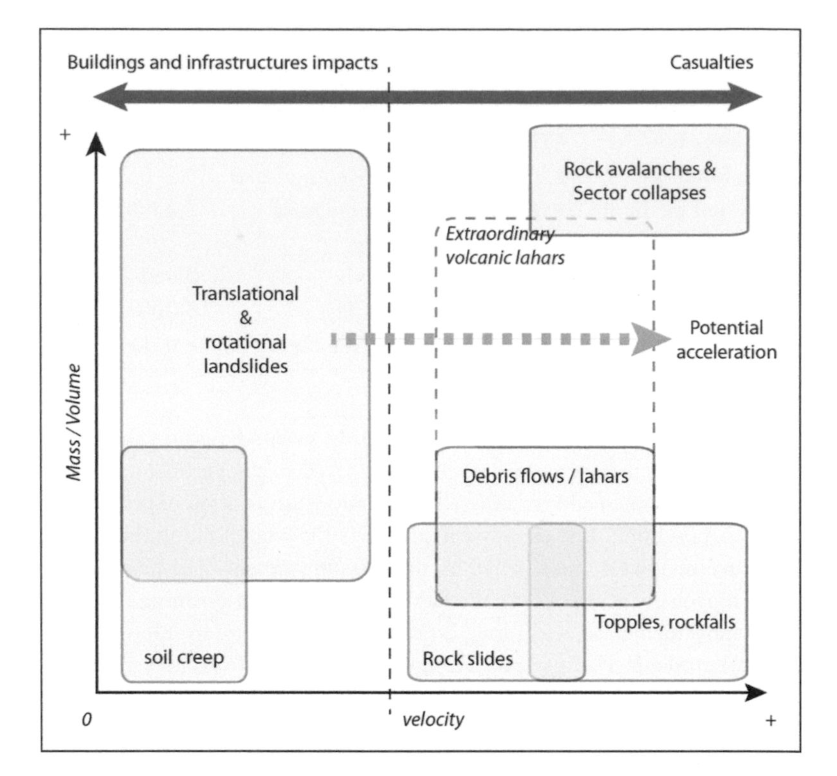

Fig 8.2
Relation between velocity * mass/volume of gravity-driven and granular movements. The combination of the two variables provide the momentum, which is an essential factor linking casualties to hazards. Source: author created

velocities varying from a couple of metres per second to events as slow as a few millimetres per decades. The common distinction based on velocities distinguishes extremely rapid landslides, 3 m/s; very rapid landslides, above 0.3 m/min; rapid landslides, above 1.5 m/day; moderate velocity landslides, above 1.5 m/month; slow landslides, above 1.5 m/year; very slow landslides, 0.06 m/year; and extremely slow landslides, below 0.06 m/year.

Waxing of mass movement: the role of water in gravity-driven hazards

With earthquake-induced acceleration, slope saturation by water is one of the primary causes of mass movements (Table 8.2) in soils (by opposition to bedrock). The water can either come from intense rainfall, snowmelts or modification of the groundwater levels. The water moves vertically and horizontally within the soil, and it will usually concentrate at the sliding surface, which is often a more impermeable subsurface layer located below a more permeable soil layer, in such a way that the water becomes trapped, accumulates and flows at the limit between the two layers, and in turn, help waxing a sliding surface in between the two layers. It is then that the excess of pore water pressure triggers the sliding. External manifestation of this phenomenon is often the appearance of unusual moisture at the toe of the sliding surface or the presence of water sources on the slope.

Table 8.2
Simplified classification of mass movements
Source: after Varnes (1978)

Types of movements	*Material*		
	Cohesive rock	*Unconsolidated material*	
		Coarse	*Fine*
Falls	Rockfall	Debris fall	Earth fall
Topples	Rock topple	Debris topple	Earth topple
Slides	Rock slump	Debris slump	Earth slump
Lateral spread	Rockslide	Debris block slide	Earth block slide
Flows	Rock flow	Debris flow	Mud flows
Complex combination of several principal types of movements together			

Similar observation can be made when a mass movement occurs as a series of sliding blocks contingent to one another, with either faults or cracks in between them, so that the vertically uplifted rock blocks act as barriers that drive the groundwater up to the surface. Once again, the water at the interfaces act as lubricant between the blocks, and whilst the blocks are moving against one another, they grind a proportion of the original material into fine particles out of the main blocks. The mixing of the fine particles with water create a paste that helps block movement.

Water also has a role in increasing the mass of the sliding unit by replacing weightless air by water in the soil. This increase of mass then affects the stability of the slope, but as water does not offer any shear resistance, the more water is in the soil pores, the more it is likely to slide. Moreover, research in Japan and in Taiwan have also shown that rainfall water over large areas can act as a blanket whilst entering the soil and then force gas to move in the soil, creating excess gas pore pressure, which in turn destabilises the slope. The role of water in mass movement is, however, not restricted to groundwater; it can also play a role in eroding the toe of the sliding section, for instance, destabilising the slope from the outside.

Furthermore, the role of water then dramatically changes when the sediment concentration drops below ~ 80% in mass; the slide changes into a mixture of debris and water known as debris flow.

The landslide differentiation is based on the mechanics, the type, the velocity of failures as well as the ratio between water and debris. Such classification is convenient for hazard researchers and practitioners because it fits the distinctions that can be made between different types of impacts. In the present case, we separate mass movements that occur favourably in rocks and mass movements occurring in soils (as engineers define 'soils'), and then divide the different types of landslides based on their mechanical properties. Soils can be separated from rocks based on the intact strength boundary of 1 MPa unconfined compressive strength, value above which the material can be considered as a rock.

8.3 Landslides classification

Landslide classifications have emerged in the first half of the 20th century, with the Ladd classification (1935) that differentiated these types of movements: flows (mud flows of clay or mud flows of volcanic ash), slope readjustments in soils of clay to sand sizes, undermined strata with horizontal elements in movement and structural slide (along bedding planes, joint planes, fault planes, schistose planes). One of Ladd's contemporaries, Sharpe, made further distinction in landslides, based on the temperature, involving ice against liquid water, and between dry and wet landslides. The classification of Sharpe (1938) was segmented as follows: (1) ice-involved movement, such as rock glacier creep, solifluction and debris avalanches; (2) those involving water and ice, with rock creep, talus creep, soil creep, slump, debris slide; (3) those involving water, such as solifluction, earthflow, mud flow and debris avalanche; and finally, (4) movements that can occur in any area regardless of the presence of ice or liquid water, such as rockslide, rockfall and subsidence. A still-popular classification today is the one proposed by Varnes in 1978 (Table 8.2).

Another complementary approach to landslide hazards and mechanism appraisal is based on the final state, once a landslide has occurred. One of these methods is the Newmark displacement method (Newmark, 1965), which has been used to quantitatively measure the displacement of translational landslides, notably. It defines the permanent displacement of the translated mass using a set of geometric descriptors. From the Newmark displacement method, numerous variations have been proposed in the last 60 years (e.g. Rathje & Saigili, 2006). And one could argue that such a method using the geometric changes of the landslide has been particularly successful with the development of GIS. In Mexico, for instance, this tool has allowed a stochastic simulation of landslide return periods of 150–500 years for urbanism and management purposes (Niño et al., 2014). For hazards and DRM, the spatial and temporal statistical analysis are not limited to scientific research, and they have dominated numerous professional sectors, such as the insurance industry.

Mass movements in rock and bedrock

Bedrock mass movements are an essential erosion process that occurs frequently at a small scale (small rockfalls, scree movements) or by infrequent large events, such as debris avalanches and large bedrock deformations (Box 8.1). They can be characterised from the unconfined compressive strength and from their structure (density, orientation and size of fractures in the rock etc.), as they often appear as discontinuous series of blocks separated by fractures, faults and tectonic folds – contrasting with soils that can be thought to be continuous material (the approach is, therefore, different).

If we were to draw a profile through all the events in rocks and bedrocks from the simplest to the most complex, we could start with falls. One of the smallest and simplest gravity-driven hazards is the fall.

Box 8.1: Rock movements and rockfalls from the dome of Unzen Volcano

Rock movements and subsequent rockfalls can occur in a variety of environment, including volcanoes, although other processes are often the centre of the attention. At Mount Unzen, on Kyushu Island, Japan, for instance, the volcanic dome made of several lobes is slowly sliding down the volcano, even several decades after the eruption stopped. Although the dome may be collapsing all together, present activity is characterised by localised rockfalls. These processes can be measured using successive LiDAR (Light Distance And Ranging) datasets of the topography (Figure 8.3).

Note: for the period 2004–2016, dome blocks have moved up to 7 m horizontally – in collapse and erosion zones. At LOC1, metre-scale cracks in blocks open with movement between 2004 and 2016. The observed horizontal movement is between 0.8 and 1.3 m. At LOC2, the subvertical face of the dome collapsed (D), and the erosional gully (E) deepens, whilst the gully (F) is being filled by material deposited and coming from upper parts of the dome. ND areas are new deposits showing volcaniclastic material. At LOC3, a part of the dome in G shows that the block has rotated whilst moving, and per the zoom given for the zone H, the dome is being excavated from the sediments whilst new cracks have appeared. A portion of the dome appears to have collapsed as well on the north face (the half to the east).

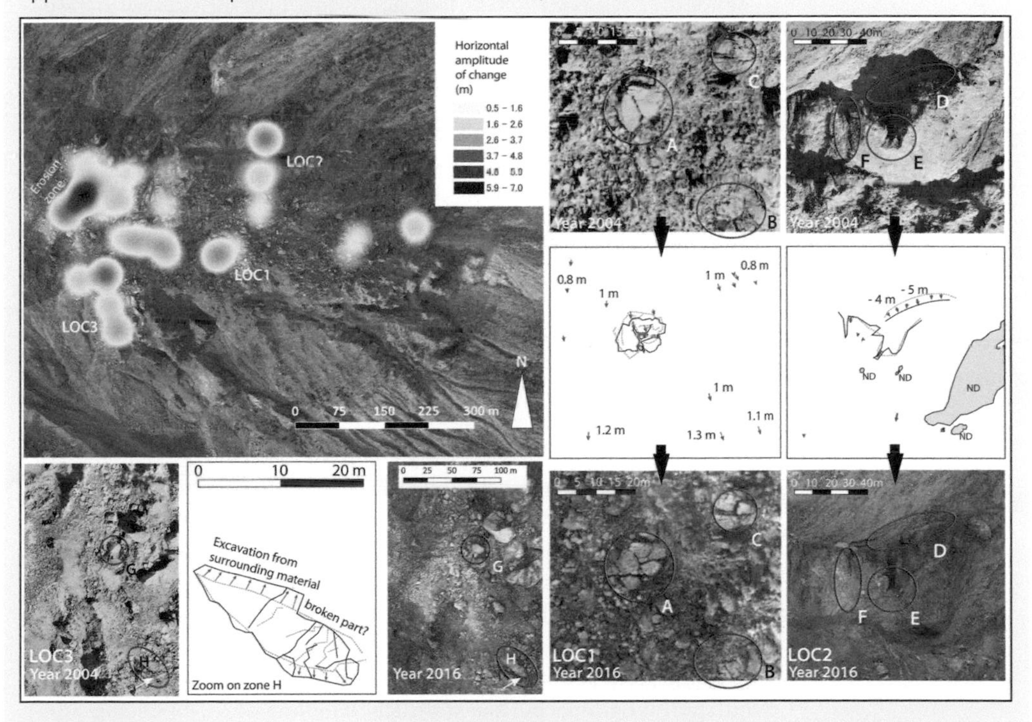

Fig 8.3
Planform changes observed at the dome surface using aerial photographs, with three focus zones (LOC1, LOC2 and LOC3) between the year 2004 and 2016.
Source: author created

It is simple because the amount of material involved is often limited to one or a small group of blocks, and the physics based on Newton's second law alone allow some simple estimation of velocity, impacts and potential runout. At the other end of the complexity and size scale of mass movements in bedrocks, there are the rock avalanches, which are virtually the largest types of rapid mass movements known on Earth.

Falls: rockfalls and topples – falls are driven solely by gravity and involve the failure of a mass of material, which is most often involving rock masses of a relatively small volume. This type of mass movement is represented by the two sub-categories of rockfalls and topples, which are mostly associated with subvertical geomorphology and tectonic triggers (e.g. a block from a horst brought up by tectonic activity), or in locations where subvertical slope toes are being eroded (i.e. coastal cliffs or eroded glacial valleys). Typically, these movements are abrupt, are downward and occur in rock, soil or both, usually with various bio-debris composed of the eventual vegetation located close to the surface. They occur commonly along coastal cliffs, like in the southern coast of England (United Kingdom). For those cliffs, the main causative factor is the undercutting by the action of sea waves, which generate unbalance and collapse by gravity. As those events tend to happen very suddenly, it is difficult to predict when a fall will occur.

Once the rock is set free, it falls, following different patterns that are related to the slope: free falling (90 degrees), bouncing (~ 70 degrees), rolling and sliding (< 45 degrees), and rolling. The mass falling either breaks on impact or deforms, in the case of soils. The highest velocity is associated when the rockfall approaches free fall, with some elements eventually bouncing against the subvertical slope where it occurs. When the rockfall doesn't tend to free fall on subvertical slopes, it will bounce and roll, losing energy until it stops, and it has been traditionally defined as $\tan \phi = \tan \phi_0 + k * d/(2*R)$ (Kirkby & Statham, 1975), where ϕ is the dynamic angle of friction (degrees), ϕ_0 is the angle of internal friction (degrees between 20.3 and 33.8), *k* is a constant (0.17–0.26), *d* is mean diameter of scree on the slope surface (m) and R is the radius of the rock (m). This method can be haphazard for hazard assessment, as the *k* factor can vary for one single slope. Two empirical models using the angle between the top of the rockfall triggering zone and the maximum travelled distance are the Fahrböschung (Heim, 1932) and the minimum shadow angle (Evans & Hungr, 1993).

Because of the limitations of empirical methods and the difficulties to physically model rockfall loss and transfer of energy during bouncing, and because most of the models can't incorporate the shape of the blocks either, statistical methods are often preferred for hazard analysis. One of the early popular models is the CRSP (Colorado Rockfall Simulation Program), but the statistical handling of the bouncing directions and energy can result in a lot of variation in the results. Another popular 2D model is the RockFall model, which also allows for a probabilistic triggering zone. More recent models are 3D, like STONE, Hy-STONE, Rockyfor3D, Rockfall Analyst and RAMMS Rockfall model, which is probably the most accomplished system taking into account the 3D shape of the rocks and a non-smooth contact dynamic (Leine et al., 2014).

The topple is a gravity movement similar to a rockfall except that it has a specific forward rotation out of the slope where it originated from. It will occur anywhere where a failure plane parallel to the subvertical slope is possible. Although the movement is slightly slower than the rockfall, it is also in the order of a second to a few seconds, depending on the scale of the topple.

Corrective and mitigation measures for rockfalls and topples are mostly composed of rock curtains and other slope covers that can be made of concrete, a biofilm or a combination, including anchors in the sane rock. When space allows, the protective measures hugging the slope are combined with retaining walls, which can be concrete walls or metallic restraining nets hung from metallic poles, and which are meant to capture or slow bouncing rocks. Around those physical barriers, sand pools can also be installed to absorb the rockfall energy and limit the bouncing. Finally, soft engineering, such as warning signs and no-stopping signs, can be installed on the roads.

Debris/rock avalanche – if rockfalls and topples are some of the most localised events, dry bedrock mass movements also include the largest of all: debris/rock avalanches. There are events the size of a hill, half a mountain that suddenly starts to move. Such large and extremely rapid open-slope flow occurs when an unstable slope collapse.

Because rock avalanches are rare events at the timescale of modern sciences, and because the place where they will start is difficult to appreciate, scientists only have proxies from witnesses, notably about the velocity of rock avalanches, which are supposed to move at velocities from 100 km/h to approximately 350 km/h. If the body travels at those averaged velocities, single blocks can be propelled at much higher velocity, like at Nevados Huascaran, where the maximum estimated velocity for a single block reached 800 km/h. More conservative values are most certainly more common, and seismic recording have provided mean velocities between 100 km/h and 210 km/h (Sosio et al., 2008). Some of the best measurements of velocities come from the nuclear test sites of the ex-Soviet Union, where at Novaya Zemlia in the Barents Sea, the nuclear test triggered rock avalanches on the slopes located above.

The largest rock avalanches are around several tens of cubic kilometres, like the Langtang rock avalanche in Nepal that could be as much as 15 giga cubic metres, the Baga Bogd in Mongolia that is 5 giga cubic metres or the Green Lake landslide that is 2.7 giga cubic metres. Amongst those giants, more localised events occur more regularly. For instance, in 1929, the volume of the Falling Mountain rock avalanche at Arthur's Pass National Park, New Zealand, was estimated to be 55×10^6 m^3. The probability distribution of the rock avalanche by volume has been estimated to follow a power law, where the power has been estimated to be between 0.7 and 0.88 for historical and events during the last 10,000 years in different areas of the world.

Rock avalanche starts from either a stable rocky surface or a slowly creeping mass moving towards maturity. The first phase of the movement can almost be one of a ballistic on steep slope, as it was noticed during the nuclear-test triggered rock avalanches in the USSR and as it was depicted in Flims, in Switzerland, where the local inhabitants described how they could see the church spire of the neighbouring village underneath the travelling rock avalanche, as though it were flying. Whether this airborne phase occurs or not, the material hugs the ground, it fractures and it rapidly disintegrates either partially or fully. The recognition of this process has given rise to the most robust explanation of the extraordinary long runout, which partly relies on the fragmentation of the rock avalanches during movement (Davies et al., 1999). The overburden stress and the basal shear stress fractures part of, if not all, the moving mass. As the sole layer is being fragmented, it acts as a granular carpet, waxing the movement. As the rock avalanche slows down, it spreads and thins. Finally, the rock avalanche comes to a stop.

To better understand this process, numerous simulations have taken place. Early rock avalanches simulations have been performed with the Lagrangian DAN numerical model (Hungr, 1995) developed to simulate rapid flow slides and debris avalanches. Davies and McSaveney (1999) demonstrated that the model was applicable to small avalanches > 1 million m^3 and was representing relatively accurately the runup of the 1929 Falling Mountain rock avalanche using a field-derived friction with an angle of 27 degrees (Davies & McSaveney, 2002). Conceptually, one of the simplest ways to consider a rock avalanche is to think of a rigid slab moving downslope, where the driving forces oppose the driving force:

$$\mathrm{M}\frac{dU}{dt} = Mg\sin S - R$$

where U is the mass velocity, t is the time, M is the mass, g is the gravitational acceleration, S is the local slope and R is the resistive force, which can be computed as the Coulomb friction force. The concept behind this equation is the one dominating landslide modelling (i.e. the assessment of the forces resisting movement and the forces driving movement). Although conceptually simple, the application to a real event is often much more difficult due to the complexity and the unknown variables in a real setting.

At this stage, there are no solutions humans can apply to mitigate rock avalanche hazards (unless of very small size), and evacuation, when the movement is predictable, has been the only viable solution (Box 8.2).

Box 8.2: The Frank rockslide: a contemporary rock avalanche in Canada

The Frank slide is a rock avalanche that occurred on 29 April 1903, just before dawn, in the province of Alberta, Canada. The rock avalanche took the lives of 83 and further injured 23. Approximated 8 million tons of rocks collapsed from the Turtle Mountain at the back of the Frank Village, blocking the Crowsnest River and crossing on the other side of the valley, notably blocking the local coal mine (Figure 8.4).

The trigger of the slide is still a subject of debate, as the toe of the slope is located along a fault line, and because the Crowsnest River is also eroding the bottom of the slope. Mining activity was another weakening process of the slope's toe, with notably the creation of a mining chamber that was 40 m long, 75 to 120 m high and 5 m in width. It has also been suggested that the weather may have been a contributing trigger as well because the day before the slide occurred, the temperature were recorded to have reach 40 degrees C in the region, whilst local sources mentioned temperatures that may have been as low as −18 degrees C on the day of the collapse. Water that filled the cracks on the previous day would have frozen and helped destabilise the slope. Amongst those different theories, the famous Dr Terzaghi would have stated that "the slope failed when it was ripe for failure". Indeed, as the slope had been steepening during the Holocene in a material that was already weathered and fractured by the building of the Rocky Mountains, the mining, the river undercutting and the weather conditions may have just been the extra-push that was needed to trigger the rock avalanche.

Source: Charrière et al. (2016)

Fig 8.4
3D aerial view of the Frank slide using the Google Earth engine and display application.
Source: author created

Mass movements in soils

Mass movements in soils can be shallow events only concerning the first few metres of the subsurface (type dominated by soil flows or solifluction, earth creep and earth flows, and spreads), or they can be deep-seated landslides, with a sliding plane located several tens of metres deep and more (Figure 8.5). Large events, such as sector collapses and deep-seated landslides, can destroy a whole valley and human settlements in just a matter of seconds, like it happened with Typhoon Morakot in Taiwan in 2009. The Morakot Typhoon took 673 lives and cost an estimated of US$3 billion, initiating the rapid movement of deep-seated landslides that turned into rock avalanches and, in a few minutes, covered all the village of Shaorin under several tens of metres of material.

Translational landslide – it is a landslide translating on a planar rupture surface, which moves downward in a relatively homogeneous way. It can be triggered by seismic activity or intense rainfall water accumulating at the limit between the sliding mass and the substratum underneath, which is usually more impermeable. Such landslides are common along rivers because the erosion of the slope's toe can destabilise and help the triggering of such mass movement. The translation velocity can range from slow landslides with velocities less than a metre per year to rapid movement occurring within seconds to minutes.

Rotational landslide – it is a landslide for which the surface of rupture is curved upward along a spoon-shaped surface on which the landslide rotates whilst sliding. It is associated with slopes ranging from 20 to 40 degrees. The velocity can be very slow, from 0.3 m/year to several metres per months. Compared to translational landslide, a rotational landslide can experience different speeds during its lifetime, with phases

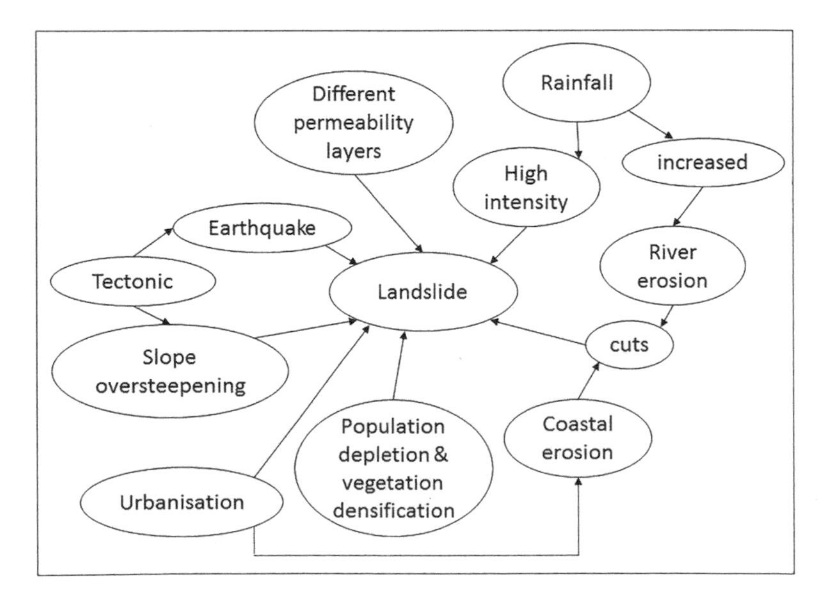

Fig 8.5
Schema crudely dividing the main triggers of landslides in soils. The arrows indicate one event triggering or helping/ pushing another.
Source: author created

of acceleration and speed decrease. Rotational landslides occur in soils and not in rocks, as they are associated with a deformation of the media most often due to wetting of the soil, after events such as intense or sustained (or both) rainfalls, rapid snowmelt and/or earthquakes, when the material is already saturated.

Earthflow and creep – earthflows are movements in fine-grained soil, characterised by a loss of structure and a movement measurable at the scale of m/s. On the other hand, a creep is slightly slower and allows the soil to conserve elements of its original structure. Earthflow is rather a viscous flow, whereas a creep is closer to a plastic flow. Creep usually generates characteristics micro-landforms, typical steps and ridges perpendicular to the slope direction. In the suburbs of Barcelona (Spain), for instance, the urbanised area of El Papiol is sliding slowly through an earth creep. The creeping slope is located above the Argiles Creek with a 70 m vertical altitude difference. The hazards have damaged the road by fracturing the surface, displacing the foundations and the walls. Crosetto et al. (2018) have measured vertical and horizontal displacements of concrete slabs by, respectively, 5 and 6 cm in the field, and satellite monitoring has revealed a slope displacement velocity between –7.5 to 7.5 mm/y (vertical and horizontal), which cumulated over the period 2007–2012, resulting in deformation up to 150 mm, which is sufficient to fracture and destabilise structures.

Spreads – spreads, such as lateral spreads, usually happen on gentle slope to sub-horizontal surfaces. Spreads are the product of the extension of the material that cause cracks to appear. As the material spreads apart, blocks usually slide and subside at the same time as they move. Lateral spreading commonly occurs during earthquakes along rivers; as the riverbanks can slide inward into the river, it generates spread around it.

Debris flows – closer to the boundary with waterborne hazards and other flood phenomenon, but still belonging to the category of gravity-driven mass movements, one will find the broad and complex family of debris flows. They are a highly unpredictable hazard, potentially carrying large debris, able to create important damages kilometres away from their source. Within the landslide's family, debris flows are the most mobile hazards, and they are gravity-driven events with mixtures of solid and fluid in varying proportions. Because of their high mobility, impacted infrastructures range from buildings, roads, to railroads; they can bury infrastructures, destroy river management infrastructures, forests and agricultural fields all together (Figure 8.6).

Debris flows and their volcanic cousins, lahars, are given their own sub-section in this chapter because they are the first causes of casualties with volcanic activity in the last two centuries and also because the realm of climate change is only likely to see their frequency and magnitude increase, becoming an increasing issue for mountain foot and plain settlement located in proximity of sediment-rich slopes. In Japan (except for the northern island of Hokkaido maybe), this setup is true

Fig 8.6
Debris flow impacts in the aftermath of heavy rainfalls in Kobe (Japan).
Source: author created

for the entire country, and recent and repeated debris flow disasters in Hiroshima (2015, 2018), Kobe (2018), Kagoshima Prefecture (2019) and Asakura (2017) are only the most recent examples in a too-long list. Finally, debris flows are still a mystery in terms of mechanics and physics, and despite several efforts to create physically true models, they have, so far, all fell short of explaining and reproducing those events.

In the broadest sense, debris flows have been defined as mixtures of debris and water, eventually including bio-debris. The dry material involved in debris flows are the debris, which are coarse clasts and at least a fraction of 20% or more of gravel, as well as earth that are defined as unsorted clay-rich colluvium derived from parent material (Hungr et al., 2001).

Further distinction can be made from the sediment concentration between debris flow (sediment > 80% weight, 60% volume) and hyperconcentrated flows (40% to 80% weight, 20% to 60% in volume), with typical values per volume varying between 30% and 70% (one will note that different rock densities will also have an impact on this ratio and have an effect on the flow, especially with pumices and material with a specific density inferior to 1). Distinction has also been made between flows carrying large clasts and those made of solely finer material, with specific terminologies like *coulee de boue* in French and *dosharyu* in Japanese. In China, a different classification exists. It is based on the viscosity and the composition of the debris flow, with three types: the viscous debris flows, the sub-viscous debris flows and the dilute debris flows. From these three categories, scientists have then made the distinction between those that are stony debris flows, lacking fine grains, and mud flows that lack coarser grains.

In Anglo-Saxon literature, Jakob and Hungr (2005) have proposed a differentiation based on different types of flows: (1) debris flow is "a very rapid to extremely rapid flow of saturated non-clastic debris in a steep channel. Plasticity index is less than 5% in sand and finer fractions"; (2) mud flow is "a very rapid to extremely rapid flow of saturated plastic debris in a channel, involving significantly greater water content relative to the source material (plasticity index >5%)"; (3) debris flood is "a very rapid, surging flow of water, heavily charged with debris, in a steep channel"; and (4) debris avalanche is "a very rapid to extremely rapid shallow flow of partially or fully saturated debris on a steep slope, without confinement in an established channel".

Note: in four days, more than 910 mm of rain fell in the mountains behind Kobe City, with local peaks recorded at 47 mm per hour. It triggered numerous landslides and debris flows in the weathered granitic mountains. The start of the debris flows was often associated with a road cut (A), and it transported drifted wood recruited in the gullies (B). The impacts in Kobe City destroyed house walls (C), and the mud also penetrated from the windows (D). The mud splash shows that the mud-water did exceed one-metre height (D) (photos courtesy of Christopher Gomez, 2018).

The frequency-magnitude of debris flows

The frequency magnitude of debris flows is controlled by two inlets: sediments and debris availability, and the water input (most often, rainwater). In other words, you need a reservoir of material ready to be remobilised in a catchment. When this reservoir of material mostly come from material created by the last glaciation, as it is often the case in the European Alps, for instance, a discrete reservoir can be emptied over time. In other words, even if you have intense and large-volume rainfalls, if the reservoir of sediments is not sufficient, then a debris flow won't be triggered. Reservoirs can be replenished by other processes, bringing material within a channel, but it is certainly one of the elements that has left scientists hungry for further research. And it seems that the dynamics of these sediment reservoirs refill controls long cycles.

In the European Alps, where debris flows have been identified to be the most important hazard causing important damages to infrastructures and transport corridors, the frequency and the magnitude of debris flows have been shown to correlate positively. For valleys in the Alps of Central Europe, debris flows of volume $< 10^5$ m^3 have been reported to occur on a return period of one to ten years, whilst debris flow between 10^5 and 10^6 magnitudes tend to occur with a return period of a hundred to a thousand years. Studies have also shown that although this general relation holds, it is very much site dependent, and within one single region, the frequency-magnitude of debris flows change drastically from one valley to another.

The conditions of occurrence also modify the frequency-magnitude of debris flows. Out of 988 events, Riley et al. (2013) demonstrated that the 264 events that occurred in the aftermath of forest fires had a slightly smaller volume (median volume of 1,579 m^3 and maximum volume around 864,000 m^3), whilst non-fire-related debris flows had a 2,000 m^3 median volume and a maximum volume of 100,000,000 m^3. The burnt surface and vegetation create a more impermeable layer that has been, supposedly, at the origin of more runoff and, thus, debris flows.

On volcanoes, debris flows or lahars are linked to their time occurrence with an eruption. Large-scale lahars occur in the aftermath of eruption, and their frequency and magnitude decrease over time, until it reaches a different equilibrium, when bursts of events can occur. At Unzen Volcano, in Japan, the lahars that swept the Mizunashigawa basin ceased at the beginning of the years 2000 when the stock of material available reduced and the slope of the river joined an equilibrium level. Although this process also occurred in other smaller valleys and gullies, such as the Gokurakudani and the Tansandani gullies, there has been a regain of activity after 2008 in these two gullies. The reason comes from the erosion of the volcaniclastic material produced during the 1990–1995 eruption to a point when the pre-eruption topography is close enough to the surface to increase the amount of water available for lahar triggering. As long as the gully was not dug vertically sufficiently, the groundwater remained groundwater, but when the gully removed enough material, the groundwater could turn easily to surface water, triggering small lahars and eroding further the gully walls, which, in turn, makes more material available for lahars.

In some sense, the lahars functioned as volcanic debris flows in the aftermath of the eruption to then reach a new equilibrium with events that are closer to a traditional mountain debris flow, with valley clogging due to lateral material collapsing in the gully, and increased pore water pressure due to the rise of the groundwater. Such process is very different from what has been recognised in the immediate aftermath of eruptions, like at Merapi Volcano in Indonesia.

The spatial characteristics of debris flows

Debris flows start in what is named as the initiation zone; it then flows into the transport zone and stops in the deposition zone. Those terms are interrelated with the behaviour of the dry load that forms the debris flow.

The initiation zone – in mountain areas, debris flows typically starts on slopes of angles ranging from 20 to 45 degrees because it needs sufficient energy for the soils to be mobilised, but slopes that are too steep are not sufficiently prone to the development of soils nor to the accumulation of material necessary to start debris flows. The material can be mobilised by rockslide, slope failure, debris avalanche, topples, falls and deep-seated slides, or progressive mobilisation of the material from one or several locations.

Landslides-triggered debris flows – when a landslide mass enters a channel (i.e. a gully, a torrent or a stream), the initial movement has already unconsolidated part of, if not all, the material, easing the incorporation of water. Furthermore, the landslide adds an undrained load to the channel bed, which then liquefies more easily.

On top of the mechanic effects, the entry of material in a drained channel will generate pressure on the deposit due to the upstream water accumulation, eventually turning into sudden failure.

The transport zone – a debris flow is typically comprised of a front that is rich in debris, a body that has variable densities and, at the end, the tail of the flow, which tends to be more fluidised. In the channel, debris flow typically moves as a series of surges. The surges can be rich in debris, with more fluidised periods in between, which are named intersurge flows. The surges can also be fluidised events that travel as a wave in the flow, usually at slightly higher velocities than the rest of the body.

Although the right mixture of solid and fluid on slopes that are sufficiently steep is essential for the triggering of debris flows, it rarely controls the volume of the latter. The incorporation of bed material constitutes a large portion of the deposits. In Hong Kong, a 400 m^3 landslide turned into a 500-fold debris flow of 20,000 m^3 by this process. Takahashi (1991) first proposed to model this process by extending the infinite slope stability theory and by assuming a steady-state seepage parallel to the slope, with the pore fluid pressure as hydrostatic. Equating the shear failure due to the pressure added by the debris flow overriding the bed, with the shear strength of material, he calculated the erosion depth of the bed. Hungr et al. (2005) argued that during a surge, it is unlikely that seepage conditions are in a steady state, and instead, they proposed a relation transferring the bulk weight of the debris flow to pore water by undrained loading.

The total volume of debris flows has also been estimated using empirical algorithms. One approach has been to link the total volume of the debris flow to the yield rate, in such a way that the total volume of a debris flow is equal to the volume of the initiation process, plus the volume coming from tributaries and side processes, plus the yield rate from the bed, being a function of the length and width of the channel.

The deposition zone – it is unclear where the erosion zone and the deposition zone departs from one another, most probably because of the variety of factors involved. Numerous studies have shown that debris flows tend to deposit below 10 to 20 degrees angle, but volcanic debris flows (lahars) also commonly present erosion below this value, on slopes as low as 1 degree.

A special case: volcanic debris flows – lahars

Lahars are the volcanic equivalent of debris flows in the mountains. They usually contain a debris flow phase and a hyperconcentrated flow phase that leave deposits of a very different nature so that they can be recognised from their deposit, for hazard identification purposes. Debris flow phase: the debris flow phase usually comes first as a plug of material being pushed in front of the lahar, and where water plays a secondary role. After the flows, the corresponding deposit can typically lobate arcs of large blocks, which represent the limit of a front that stopped. Debris flow phases during a flow can repeat themselves several times and can be separated by other flow phases. The debris flow phase on volcanoes is typically poorer in fines, and lahars are in general less cohesive than their mountain counterparts. Hyperconcentrated flow phase: these flows have been traditionally defined by its boundaries, with Newtonian fluviatile flows on one end and debris flow at the other. In terms of sediment concentration, they have been presented as limited by 20–60% of sediment concentration in volume or 40–80% of the weight. An effort to characterised hyperconcentrated flows in international literature has been proposed by Pierson (2005), although Japanese scientists had already made similar work several decades earlier. When lahars do not include a debris flow phase and are made only of a hyperconcentrated flow that becomes diluted, Indonesians named them *banjir*, but this terminology has been less exported than the word 'lahar'.

Lahar is a difficult hazard to deal with because evacuation and hazard prevention linked to eruptive processes are relatively short-lived, and although several lahars can be syn-eruptive, the majority of them are triggered by rainfalls in the years following the end of an eruption. At Unzen Volcano, in Japan, the last eruption ended in 1995, and in 2018, lahars were still being recorded in the gullies of the volcano (Box 8.3).

Box 8.3: Lahar gully erosion and material feeding at Mount Unzen (Japan)

Lahars are mixtures of debris and water flowing on and from the slopes of volcanoes, and are non-Newtonian flows (i.e. the relation between shear stress and flow deformation is not linear). They are often classified as volcanic debris flows, but researchers agree that the nature of the material and the way they form are often different from their mountain relatives. As for a lot of processes that we name 'hazards', we don't understand much about their physics, their triggering processes and their period of return. This is most certainly why they are seen as hazards. They tend to occur mostly on stratovolcanoes, and their activity can be evaluated from the long radial gullies that can be seen on volcanoes, like the Merapi or the Semeru in Indonesia, and Mount Unzen or Sakurajima in Japan. At Mount Unzen, several gullies start from the dome at the summit and convey lahars downstream. Those gullies have steep subvertical walls, which tend to collapse, bringing remobilisable material to the bottom of the gully, which in turn feeds the next lahar (Figure 8.7).

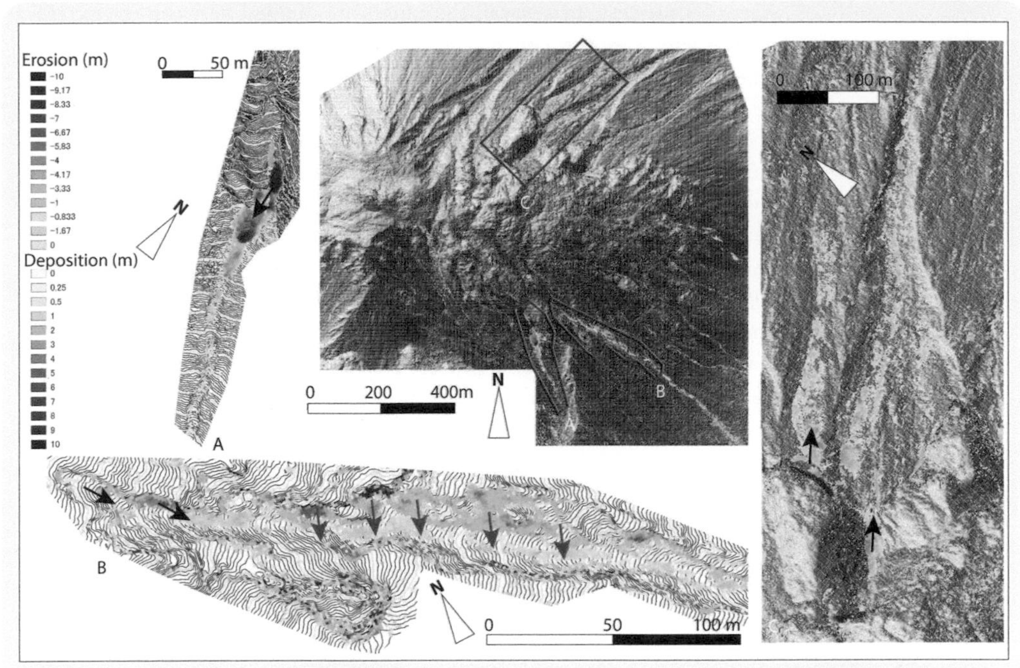

Fig 8.7
Gully erosion at Unzen Volcano in Japan. Gullies A, B and C are all connected to the summit dome, and they are partly fed by collapses from the dome, whilst lahar and other processes erode the gullies.
Source: author created

Note: in the gully at location B, the progressive wall collapse is also one of the reasons for the change in the morphology but also the transfer of wall material to the gully bottom, which can then be remobilised by lahars.

As the volcano has been cordoned off and the reconstruction of some portions of the Shimabara City that were destroyed are still off limits, those events do not cause any casualties, but they continue to be a cost to the community. In Indonesia, at Merapi or Semeru Volcano of Java Island, where the demographic pressure is very important, the attraction of fertile land, the possibility to resell the pyroclastic material as construction material and the use of the eruption as a tourist attraction are community revival opportunities, similar restrictions can't be applied; and the population, therefore, comes back in areas where lahar-related disaster risk are significant.

8.4 Triggering mechanisms and causes of gravity-driven movements and flows

The mechanisms that lead to the triggering of gravity-driven movements and flows are still poorly understood because of the lack of observations and, often, the impossibility to make direct measurements at the very start of a process. In general, the causes are divided between the external causes, which mostly control an increase in the shearing stress, and the internal causes, which result from the decrease of the shearing resistance.

Posteriori causes are all the events and processes that push the trigger of a landslide, and they contrast with the priori causes, which are the intrinsic characteristics of a slope, a location, making it prone to mass movement.

Geometrical changes – the change in the slope geometry, such as undercutting, erosion, stream incision, excavation and road construction, all contribute to the destabilisation of the slope. The processes of unloading (erosion, incision) at the bottom of a slope, as well as the loading (construction and addition of material, vegetation overgrowth) impact the slope stability balance. For instance, it has been argued that the shallow landslide hazards that occurred in Asakura (North Kyushu, Japan) in July 2017, have been amplified by the plantation trees that have not been cut down due to low prices of timber in recent years, increasing the mass on the slopes.

Shocks and vibrations – earthquakes and other vibrations are either contributing factors or direct triggering factors, like during the 1999 Chichi earthquake in Taiwan, which generated several thousands of landslides in the central part of the country. Earthquake acts by either increasing the loading due to the extra-acceleration, then decreasing shear strength.

Changes in water levels – the changes in water levels have their origin in rainfalls, which in turn impact surface waters and groundwater. This increase also increases water pore pressure and also the overall mass of a sliding lump. The drawdown of the water level, for instance, of a lake or a reservoir at the foot of a mass movement is also an issue, as it accelerates seepage through the slope and, thus, increases mass movement hazard.

Internal causes – internal causes are as varied as the processes that are acting on rocks and sediments at the surface of the Earth.

Material formation: depending on whether the material is in situ soil originating mostly from hard bedrock, or whether the material originates from deposited sediments from water or wind transport into an aerial location or a deeper sink in water, the material characteristics will then differ and influence the mass movement hazard. Furthermore, material once deposited in the sea can then be brought back up in the mountains to create rocks, like argillite, which are metamorphised clay sediments from the seabed. In South Taiwan, at the deep-seated landslide location of Zhulin or Maolin, the bedrock is made of bedded poorly consolidated argillite that can be crushed by hand, opposing very little resistance to reshaping and buckling, but creating clay layers when crushed, which are the ideal material for creating failure planes.

Material weathering – before a lump of material is set to be a potential mass movement, it goes through numerous transformations, which can be chemical weathering, like it is often the case under tropical regions, or freeze-saw processes, when temperature oscillate around the

0 degree Celsius mark. Material weathering can, over a long period, help in maturing subsurface material to trigger mass movements.

8.5 Mass movement soil mechanics for safety assessment

Soil characteristics and mass movement

Consequently, the triggering of mass movement is mostly tributary of the physical characteristics of the soils and rocks, combined with a temporally limited trigger. This is the reason why engineers in soil mechanics measure and calculate elements, such as the mass, density, porosity, effective stress and plasticity, to better understand how those elements react to a trigger. Knowing about those various parameters is essential for hazard specialist to be able to dialogue effectively with engineers and make sense of the data they produce.

Phase relationships of soils – soils are considered to be made of a combination of solid particles, referred to as the soil skeleton, and voids, which can be filled by some liquid, the pore water and/or the pore air. From these, the void ratio (traditionally noted e) defines the volume of voids/volume of solids. The porosity (n) is defined as the volume of voids divided by the total volume. Then the saturation ratio can be obtained, and it is defined as the volume of water divided by the volume of voids, and it is usually given the letter S. If a soil is fully saturated, then $S = 1$, and its values vary between 0 and 1. In the same way the volume of air, named air content, can also be calculated.

Unit weight (Υ) – it is the overall mass density of the soil, that is, the bulk density, multiplied by the gravitational constant (on Earth) $g = 9.81\ m/s^2$, and the unit weight have kN/m^3 as unity. Using the data of phase relationship of soils, it is then possible to calculate the dry unit weight and the saturated unit weight of the soil.

Effective stress – in soils and rocks, applied forces are resisted by both the particle assemblage and by fluids in the voids. The particle assemblage resists both the normal and the shear forces, but the water located in the pores only resist the normal forces. As the deformation resistance comes from the contacts between particles, it consequently does not include the water in between the particles. This essential concept in mass movement and slope stability research has been first proposed by the geotechnical engineer Karl Terzaghi (Box 8.4), who is considered as the father of contemporary geotechnical engineering. Effective stress is expressed as follows:

$$\sigma' = \sigma - u \tag{1}$$

This equation, although relatively simple, is essential, as it defines contemporary soil mechanics and its application to hazards research. Even a century onward, the work of Professor Karl von Terzaghi is still seen as the start of contemporary geotechnical engineering and slope stability analysis.

Box 8.4: Karl von Terzaghi, the father of contemporary soil mechanics

Karl von Terzaghi was born in 1883, in what is today the Czech Republic. He started his academic career in Austria, and travelled to Russia and the USA. After a post-WWI position in Istanbul, where his academic writing and teaching shifted from French to English, his work on soil mechanics gave him a position at the prestigious Massachusetts Institute of Technology in the USA. Towards the end of his career, he left MIT to return to Vienna in Austria.

Karl von Terzaghi is famous in the field of landslide engineering and soil mechanics for his work on effective stress, a relation that he proposed in 1925. In this relation, he demonstrates that the strength of a soil structure to shearing is controlled by the soil skeleton only, meaning that when extra mass is added on a sloped soil unit, the mass increased due to the water only contributes to the destabilising forces but not to the stabilising forces. Consequently, increasing rainfalls destabilises the slopes.

Slope strength and slope stability

To initiate landslide movement or avoid it, two principle concepts are at play: the shear stress and the shear strength of a soil. The stress (σ) can be defined as the force (F) per total surface area (A), and the total stress is, thus, the sum of all the forces per surface area:

$$\sigma = \frac{F}{A} \tag{2}$$

Going one step further, effective stress can be calculated from the stress. It corresponds to the forces that are transmitted by the particle contact divided by the total surface area, excluding any surrounding fluid. It is usually computed as the stress minus the water pressure.

The second principle is the shear strength, which is the maximum stress that a soil can withstand before failing. This relation between the shear strength and the shear stress is controlled by the Mohr-Coulomb relationship:

$$s = c + \sigma_f . \tan \varnothing \tag{3}$$

where s is the shear strength, c is the effective stress cohesion, σ_f is the effective stress at failure and $\varnothing$ is the internal friction angle. For both drained and undrained soils (regardless of the water content), this relation is built from the effective stress.

At equilibrium, the shear strength of the soil has to be higher than the shear stress. Failure occurs when either the shear strength reduces or the shear stress increases (or both change together) in such a way that the shear strength of the soil becomes lower than the shear stress.

The first and most common cause for a decrease in shear strength of a soil is an increase in pore pressure, either through rainfalls or a modification of the groundwater level and how the seepage occurs. The second process that reduces shear strength is the opening of vertical gaps and cracks in the soil. This phenomenon is commonly observed near the scarp of landslides, where extension processes occur. Moreover, the loss of tensile strength of the soil at the crack location decreases the resistance to movement of the outer soil segment. The third is the weathering of material, its leaching and the bedrock alteration, which are all processes that also contribute towards a modification of the chemical composition and organisation of the soil grains. Thus, they also impact the balance between the shear strength and the shear stress.

The shear stress can also increase when the load changes, like an increase due to buildings with shallow foundations, excess in vegetation and increase in the volume of water contained on and in the soil. Changes in the slope geometry, such as excavation or other processes occurring on or around the slope (e.g. slope erosion at the toe by a river), also increase the shear stress due to the steepening of the slope toe. This is a problem when building a road on a slope, for instance. In volcanic and tectonically active areas, the steepening of the slope can also occur with the rapid increase in the altitude of a volcano or the rapid movement of a fault, for instance. Combined with those relatively slow processes, the instantaneous seismic ground acceleration is a common trigger for various types of mass movements.

These characteristics do a balancing act between forces that help material downward and forces that help material stay on the slope, and it is, thus, this approach that is being used to estimate slope stability.

Estimating the stability of slopes: the factor of safety

The factor of safety is a measure of the ratio between the stress and the strength of the material, defining whether it is going to be prone to movement. A safety factor < 0 means that the slope is stable, at 0 means it is just in equilibrium and > 0 signifies that the slope is unstable. It is possible to formulate the factor of safety (F) with a ratio using the shear strength (s) and (τ) the equilibrium stress as follows:

$$F = \frac{s}{\tau} \text{ or } \tau = \frac{F}{s} \qquad (4,5)$$

Rewriting the shear strength using the Mohr-Coulomb equation, one can then calculate the factor of safety as follow:

$$\tau = \frac{C + \sigma \tan\phi}{F} \qquad (6)$$

Using this concept, and in order to model results, as all data aren't always available, there exist a series of procedures developed by engineers in order to evaluate the stability of slopes. One of the most popular methods is the Infinite Slope Procedure, which assumes that a sliding is occurring on an infinitely extending slope, and that the sliding is occurring along a plane parallel to the slope (Taylor, 1948). This method was developed to ignore the forces acting at each end of a sliding block as they balance each other. The model, therefore, sums the forces triggering movement against those resisting as follows:

$$F = \frac{c + \gamma z \cos^2\beta \tan\phi}{\gamma z \cos\beta \sin\beta} \tag{7}$$

where F is the factor of safety, c is the cohesion factor of the soil, γ is the total unit weight of the soil, β is the slope inclination angle, z is the vertical depth of the shear plane and ϕ is the friction angle of the soil.

This method works well for shallow movements that are parallel to the slope and where the seepage is also parallel to the slope, but a lot of deeper landslides have a curved slip surface, for instance. Therefore, there exist numerous procedures to solve those sliding problems: the Logarithmic Spiral Procedure, the Procedure of Slices, the Bishop and Simplified Bishop procedure, and so forth. Duncan and Wright (2005) propose a practical presentation of those methods.

8.6 The usual suspects: data used for evaluating landslide hazards

Landslide analysis to evaluate associated hazards tends to focus on one of two scales: the local scale, which is intensive and dominated by geotechnical engineering approaches, and regional approaches, which are based on statistical, mixed with simplified geotechnical approaches, to fit the regional scale (Keefer & Larsen, 2007). At the local scale, a single site can be instrumented, and soil and water sample can be analysed to obtain data on soil shear strength, water pressure and so forth, with different physical-based analysis.

At the regional scale, it is not possible to instrument every single site, but similar datasets, such as groundwater, geology, geomorphology, rainfall patterns are used. The inference made and data combination can be seen as an extrapolation of observations and laws developed from local site research. For further reading on those issues, the reader is invited to consult, notably, Sassa (1999).

At the local scale, and without the use of any complex equipment, the presence of features that indicate landslide movements are springs, seeps, and wet or saturated ground in previously dry area on the slopes or at the toe of a slope.

Further evidences showing at the surface that movement is happening are ground cracks, in the soil rock or the snow cover, as well as in the built environment, like in concrete slabs of houses, or like destabilised and pulled-away walls and features from buildings. As the cracks are showing movements from extension, there are also zones of compressions, which are characterised by unusual bulks in the ground, building basements. Finally, on top of compression and extension, translational movements will be characterised by the tilting of trees, telephone poles and various signs, such as windows and doorframe that are tilted with the doors and windows not closing properly anymore, with various cracking sounds and popping noises in buildings and from tree roots (you may want to make a quick escape when hearing those noises, as they are usually the sign of imminent failure that can suddenly accelerate).

Topography – identifying deep-seated landslides is essential because their onset can be very slow at the thousands to tens of thousands of years' scale. It is, therefore, necessary to deploy high-precision instrumentation to identify them (Box 8.5). However, before proceeding

Box 8.5: Total station and scanners – harnessing laser technologies for high-resolution topography

Considerable progress has been made with the harnessing of laser technology to measure the topography and change of landslides, debris flows and other hazards involving the movement of the ground, either through slow or rapid onset. Laser technology can be either carried on an airplane, where a scanning head using a prism and a moving mirror is sending back and forth a laser swath, which, combined with the airplane movement, can generate long strips of data over several kilometres. The data collected is the return from the laser at the ground or against the vegetation, and once this return is combined with a precise position and orientation data combining the onboard GNSS and fixed GNSS on the ground, it can provide centimetre-scale precision data of the topography. The same technology has also been brought to the ground with laser scanners and hybrid solutions, combining a total station that precisely measures single points and laser scanning (Figure 8.8).

Fig 8.8
Leica Total Station for topographic survey integrating a short-range scanning capacity. (A) Instrument belonging to Canterbury University set up in the Alps of New Zealand and (B) same instrument on a debris flow apron seen from face. Source: author created

with instrumentation, deep-seated landslides can often be recognised in the landscape from the typical landform that they can create. First, for a given slope, a landslide is often characterised by a topographic drop at the crown, where the material started to slide. Furthermore, as sliding planes sometimes form partially, the movement of a portion of the slope whilst other portions remain stable create compression and distension in the material leading to large-scale deformations, such as non-tectonic faulting and folding of the material, named buckling. Whilst the faulting process will generate series of blocks that can be connected to one another by different sediment aprons and accumulations, the buckling will create a lump on the slope with counter-slope directions. This phenomenon is not limited to the first few metres of the subsurface, but in deep-seated landslides, it can refold and reshape the lithology down to several hundreds of metres under the slope surface.

Such a process can be measured and quantified from topographic data, revealing portions of slopes that may be prone to landslide. Furthermore, the measure of the density of the rills and surface channels is often another marker of landslide potentials and activity, with less channels and channels that are less developed over landslide, compared to other mountain slopes that are more stable.

At the time this chapter is being written, the latest advances in the field include (1) the use of InSaR (Interferometry Satellite Radar), showing the deformation of a slope between repeated satellite radar measures of a slope; and (2) topographic data acquisition that can occur at a high resolution using aerial LiDAR. LiDAR is a remote-sensing method using laser illumination of the target and a record of the reflected light wave parameters and time. It has been traditionally mounted on small airplanes, but a fixed laser on the ground also exists (TLS: terrestrial laser scanners), and smaller versions have been adapted to be mounted on UAVs (unmanned aerial vehicles).

Subsurface – the topography is often a reliable proxy of what happens underground, but as the sliding plane is out of sight, various measures of the subsurface are necessary. The most common method includes the drilling of boreholes in a landslide, until one finds the sliding plane (typically a clay-rich layer), and from these boreholes, a log of the stratigraphy, and a punctual or continuous record of the groundwater head and pressure, helps monitoring the landslide activity. Furthermore, numerous instruments looking at the deformation of the borehole with the sliding movement are also used to assess the potential for a hazardous event.

On top of the physical measurements, which are often costly to put in place, scientists and practitioners have used proxy methods, based on geophysical methods. Those methods are the electric resistivity tomography, Ground Penetrating Radar (Box 8.6) investigation and seismic reflectivity. These three methods all involve sending into the ground electric, electromagnetic and seismic waves to recreate the subsurface structure from the return intensity and velocity of the sent

Box 8.6: Ground Penetrating Radar – investigating the subsurface

One of the popular methods for measuring the ground surface structure and define the presence or not of blocks, groundwater and other elements of interest to assess the structure and the stability of a slope is GPR (Ground Penetrating Radar). The method relies on Maxwell's equations of electromagnetism and the dielectric characteristics of the subsurface. What can be seen from the radar data depends on the subsurface geochemical and sedimentological characteristics (clays or ferromagnetic minerals tend to hamper the electromagnetic signal). Their influence can be seen from the velocity of the radar in the ground as well as the reflections from the punctual objects in the ground. The instrument itself is composed of an antenna, the electronics to create the signal needed and a coding wheel or a time trigger. All the system is then linked to a control unit, which is often rugged, as it has to operate in sometimes difficult environments (Figure 8.9).

Radar velocity

In vacuum, radar velocity is the speed of light (c), and the speed then varies mostly due to the electric resistivity (ξ_r) and the relative magnetic permeability (μ_r). As a rule of thumb, one can consider that the dielectric permittivity increases with water content:

$$V = \frac{c}{\left[(\mu_r \xi_r)^{\frac{1}{2}} \right]}$$

Except for highly magnetic material. For dry sand, $V = \left(3x10^8 \text{m/s}\right) / \left[(1*4)^{1/2}\right]$ $= 0.15$ m/ns.

Fig 8.9
Ground Penetrating Radar used in volcanic setting. (A) GPR with the antenna and the control unit connected by the blue cable (the GPR is made by GSSI in the USA), and (B) Ramac GPR with a 500 MhZ antenna at Semeru Volcano in Indonesia operated by the author as the volcano is spewing material.
Source: author created

The radar reflection coefficient, R_c

It is a function of the ratio of the amplitude of reflected waves (W_r) to incident waves (W_i):

$$R_c = \frac{W_r}{W_i} = \frac{\left(\xi_2^{\frac{1}{2}} - \xi_1^{\frac{1}{2}}\right)}{\left(\xi_2^{\frac{1}{2}} + \xi_1^{\frac{1}{2}}\right)} = \frac{V_2 - V_1}{V_2 + V_1}$$

signal. Those methods involve dragging antennas and laying cables on the ground surface, but they are not invasive as such.

From those methods, the scientists and practitioners have an idea of the internal structures, and the material that constitutes the slope, in such a way that they can then be plugged in slope safety equations or other types of models used to assess hazard potentials.

Groundwater movement and sources – as mentioned earlier, boreholes drilled in a landslide are often used to monitor the groundwater, from which pore water pressure can be calculated as well as any increase in the water pressure. Such a measure is completed by a monitoring of sources on the surface of the landslide and the discharge of water from each source, to monitor any activity increase, clogging, modification in the water path. This path can notably be surveyed using stable isotopes in the water, from which information on the landslide activity can be deduced, as it has been demonstrated by Dr Hotta Norifumi in Japan. He explains that the ratio and the number of selected isotopes (different depending on each site) provide information on the level of activity of a given slope.

Vegetation monitoring and tree activity – as a mass movement translates, whether it is deep-seated, or whether it is a shallower event, the mass movement will deform underneath the surface (e.g. buckling), affecting the vegetation with lower access to water, or by rupturing tree roots and so forth, in such a way that the photosynthetic activity of the vegetation and the evapotranspiration may be modified by the landslide. Thus, recent research has been investigating modification in vegetation activity using thermal cameras hooked to UAVs to determine whether an area over another is showing signs of impacts on the vegetation.

Tree trunks can also be used as a proxy of the movement of a landslide, even if it is not moving at present, by investigating the morphology of a tree trunk, whether it is bended or not, whether it had to grow more roots to counterbalance its structure on the slope. Such work has been developed in Japan and New Zealand by Professor Christopher Gomez, through the analysis of the British and the Japanese white oak trees, pines and Japanese cypress tree bark grooves. Their irregularities and opening and closing upslope and downslope are proxies of the effort a tree has made to remain stable during the years it has been growing. Similar results can be obtained from tree rings, but when possible, tree bark has the advantage of being non-invasive.

8.7 Concluding remarks

Hazards generated from gravity-driven mass movement happening at the surface of the Earth are complex combinations of slope, internal structure, soil and rock geo-mechanical characteristics and water concentration and distribution within a slope; and the interaction of this defined lump of material with the surrounding external processes modify further the conditions of triggering, flowage and deposition.

If long-term monitoring can provide reliable answers on the amount of water in a slope, for instance, or if high-resolution imagery and topography can provide very accurate measures of the surface, the internal structure and lithological formations that have developed over geological periods are still almost impossible to define with sufficient details, and scientists and practitioners are arguing that it is one of the main sources of uncertainty that stops us from being able to predict those mass movements (amongst other reasons).

Furthermore, if numerous physical theories exist, they still remain an approximation of processes needing further investigation. Indeed, although the science has made significant progress since the beginning of the 20th century and the dawn of contemporary slope stability and geotechnical engineering with Professor Terzaghi, those advances require consequent amount of field data and processing, which are not available at the regional scale. Major advances will most certainly arise from the remote-sensing monitoring of surface proxies of groundwater, surface water, impact on vegetation, geomorphology and so forth. For instance, multiplying automated UAV flights with LiDAR data to investigate progressive ground deformation, combined with the analysis of anomalies in the water sources location and the vegetative activity using infrared and near-infrared data, combined then with airborne electromagnetic data to investigate subsurface water movement and geological discontinuities, and so forth, would all contribute to provide data, which are presently only available at selected sites at the regional scale (Gomez et al., 2021). Landslide investigation is, therefore, a very active field of study that still needs improvement.

Furthermore, as climate change is predicted to bring a more intense and higher volume of rainfalls, it is most likely that the field of landslide research has to make a major contribution to climate change impacts mitigation. Moreover, this near future is already happening in certain parts of the world: in Japan, the 2017 heavy rainfall event that triggered about 1,500 landslide events has been proven to be the result of exceptionally warm sea temperature that are imputed to human-induced climate change.

Take-away messages

1 Hazards involving a mixture of rocks, sediments and water are differentiated based on their velocity, water content and location where they happen;
2 Soil mechanics, under the influence of Karl von Terzaghi, has developed geometric tools to estimate slope stability; and
3 Because the physics of those mixtures of rocks, sediments and water are not well understood, scientists and practitioners rely on empirical and statistical approaches using slope values, rainfall data, seismic activity and land-cover data to estimate the potential for mass movement triggering.

To learn more about the topic discussed in this chapter, listen to the *Disasters: Deconstructed* interview with Dr Christopher Gomez (Figure 8.10).

Fig 8.10
QR code for Chapter 8.

Further suggested reading

Conforth, D. (2005). *Landslides in practice: Investigation, analysis and remedial preventative options in soils*. Wiley, 642 p.
Davies, T. R. H. (2015). *Landslide hazards, risks, and disasters*. Hazards and Disaster Series. Elsevier Academic Press, 475 p.
Glade, T., Anderson, M. G., & Crozier, M. J. (2005). *Landslide hazard and risk*. Wiley 834 p.
Pradhan, S. P., Vishal, V., & Singh, T. N. (2019). *Landslides: Theory, practice and modelling*. Springer, 322 p.

References

Charrière, M., Humair, F., Froese, C., Jaboyedoff, M., Pedrazzini, A., & Longchamp, C. (2016). From the source area to the deposit: Collapse, fragmentation, and propagation of the Frank Slide. *Geological Society of America Bulletin*, *128*, 332–352. https://doi.org/10.1130/B31243.1
Crosetto, M., Copons, R., Cuevas-González, M., Devanthéry, N., & Monserrat, O. (2018). Monitoring soil creep landsliding in an urban area using persistent scatterer interferometry (El Papiol, Catalonia, Spain). *Landslides, 15*(7), 1317–1329. https://doi.org/10.1007/s10346-018-0965-5
Davies, T. R., & McSaveney, M. J. (1999). Runout of dry granular

avalanches. *Canadian Geotechnical Journal, 36*(2), 313–320. https://doi.org/10.1139/t98-108

Davies, T. R., & McSaveney, M. J. (2002). Dynamic simulation of the motion of fragmenting rock avalanches. *Canadian Geotechnical Journal, 39*(4), 789–798. https://doi.org/10.1139/t02-035

Davies, T. R., McSaveney, M. J., & Hodgson, K. A. (1999). A fragmentation-spreading model for long-runout rock avalanches. *Canadian Geotechnical Journal, 36*(6), 1096–1110. https://doi.org/10.1139/t99-067

Duncan, J. M., & Wright, S. G. (2005). *Soil strength and slope stability*. Wiley.

Evans, S. G., & Hungr, O. (1993). The assessment of rockfall hazard at the base of talus slope. *Canadian Geotechnical Journal, 30*, 620–636.

Gomez, C., Allouis, T., Lissak, C., Hotta, N., Shinohara, Y., Hadmoko, D. S., Vilimek, V., Wassmer, P., Lavigne, F., Setiawan, A., Sartohadi, J., Saputra, A., & Rahardianto, T. (2021). High-resolution point-cloud for landslides in the 21st century: From data acquisition to new processing concepts. In Z. Arbanas, P. T. Bobrowsky, K. Konagai, K. Sassa, K. Takara et al. (Eds.). *ICL contribution to landslide disaster risk reduction* (pp. 199–213). Springer. https://doi.org/10.1007/978-3-030-60713-5_22

Heim, A. (1932). Bergsturz und Menschenleben. *Beiblatt zur Vierteljahrschrift der Naturforschenden Gesellschaft in Zurich*, 77, 218p.

Hungr, O. (1995). A model for the runout analysis of a rapid flow slides, debris flows, and avalanches. *Canadian Geotechnical Journal, 34*, 610–623.

Hungr, O., & Evans, S. G. (1996). Rock avalanche run out prediction using a dynamic model. In Senneset (Ed.), *Landslides, proceeding of the internal symposium in Trondheim, 1996*. Balkema.

Hungr, O., Evans, S. G., Bovis, M. J., & Hutchinson, J. N. (2001). A Review of the classification of landslides of the flow type. *Environmental and Engineering Geoscience, 7*(3), 221–238. https://doi.org/10.2113/gseegeosci.7.3.221

Hungr, O., McDougall, S., & Bovis, M. (2005). Entrainment of material by debris flows. In M. Jakob & O. Hungr (Eds.), *Debris flow hazards and related phenomena. Praxis*. Springer.

Jakob, M., & Hungr, O. (2005). *Debris-flow hazards and related phenomena*. Springer and Praxis Publishers.

Keefer, D. K., & Larsen, M. C. (2007). Geology. Assessing landslide hazards. *Science, 316*(5828), 1136–1138. https://doi.org/10.1126/science.1143308

Kelman, I. (2020). *Disaster by choice*. Oxford University Press.

Kirkby, M. J., & Statham, I. (1975). Surface stone movement and scree formation. *Journal of Geology, 83*(3), 349–362. https://doi.org/10.1086/628097

Leine, R. I., Schweizer, A., Christen, M., Glover, J., Bartelt, P., & Gerber, W. (2014). Simulation of rockfall trajectories with consideration of rock shape. *Multibody System Dynamics, 32*(2), 241–271. https://doi.org/10.1007/s11044-013-9393-4

Newmark, N. M. (1965). Effects of earthquakes on dams and embankments. *Géotechnique, 15*(2), 139–160. https://doi.org/10.1680/geot.1965.15.2.139

Niño, M., Jaimes, M. A., & Reinoso, E. (2014). Seismic-event-based methodology to obtain earthquake-induced translational landslide regional hazard maps. *Natural Hazards, 73*(3), 1697–1713. https://doi.org/10.1007/s11069-014-1163-y

Pierson, T. C. (2005). Hyperconcentrated flow—Transitional process between water flow and debris flow. In M. Jakob & O. Hungr (Eds.), *Debris flow hazards and related phenomena. Praxis* (pp. 159–202). Springer.

Rathje, E., & Saigili, G. (2006). A vector hazard approach for Newmark sliding block analysis. In *Proceedings, New Zealand workshop on geotechnical earthquake engineering workshop* (pp. 205–216). University of Canterbury.

Riley, K. L., Bendick, R., Hyde, K. D., & Gabet, E. J. (2013). Frequency–magnitude distribution of debris flows compiled from global data, and comparison with post-fire debris flows in the western U.S. *Geomorphology, 191*, 118–128. https://doi.org/10.1016/j.geomorph.2013.03.008

Sassa, K. (Ed.). (1999). *Landslides of the world—Japanese Landslide Society*. Kyoto University Press.

Sharpe, C. F. S. (1938). *Landslides and related phenomena*. Pageant.

Sosio, R., Crosta, G. B., & Hungr, O. (2008). Complete dynamic modeling calibration for the Thurwieser rock avalanche (Italian Central Alps). *Engineering Geology, 100*(1–2), 11–26. https://doi.org/10.1016/j.enggeo.2008.02.012

Takahashi, T. (1991). *Debris flow*. A. A. Balkema Publishers.

Taylor, D. W. (1948). *Fundamentals of soil mechanics*. Wiley.

CHAPTER 9

Climatological and hydrometeorological hazards

9.1 Introduction

This chapter explores the different types of climatological and hydrometeorological hazards, at different timescales, and notably so, under the influences of anthropogenic activity and anthropogenic climate change (Figure 9.1). The chapter begins with an overview of the environmental mechanisms related to the atmosphere, water in the atmosphere and water in the watershed so that the reader can gain a better understanding of the systemic articulations between the different components of the climatological and hydrometeorological hazards.

This chapter on hazards follows Chapter 7, which first presented hazards with the energy coming from within the Earth, then Chapter 8 which described the interface with sediments and water, and the types of hazards created at the surface of the Earth. This chapter leaves the hard surface to look at water, the atmosphere and how the movements and distribution (too much, too little or transported with very high energy) of water are also source of hazards.

This theme is presented with a deliberate emphasis on water, as it will arguably be the most precious commodity of the 21st century, which could lead to mass migrations and potential warfare. Furthermore, it was calculated that scarcity or too much water accounted for 90% of the 1,000 most disastrous disasters between 1990 and 2006 (Adikari & Yoshitani, 2009). Indeed, freshwater in the atmosphere, on land and underground is an irreplaceable ingredient to life on Earth and to human survival, but too little of it triggers droughts, and impacts agriculture and, eventually, our very own survival; on the contrary, too much of it generates floods and inundations. The human relation to water is, therefore, a difficult balancing act linking resource and hazard management. Furthermore, this complex relationship threatens to radically change under the impact of anthropogenic climate change.

Finally, because of the ubiquity of those ingredients on Earth, it is virtually impossible to stay away from the associated hazards, as you could do from staying away from an erupting volcano, for instance.

DOI: 10.4324/9781315469614-13

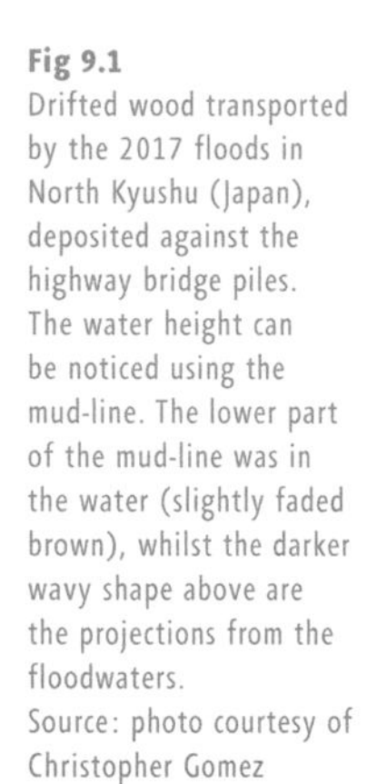

Fig 9.1
Drifted wood transported by the 2017 floods in North Kyushu (Japan), deposited against the highway bridge piles. The water height can be noticed using the mud-line. The lower part of the mud-line was in the water (slightly faded brown), whilst the darker wavy shape above are the projections from the floodwaters.
Source: photo courtesy of Christopher Gomez

Water is an integrant part of the fabric of human activity and so are the related hazards.

9.2 From atmospheric energy to water resource

Besides the internal energy generated from within the Earth, the rest of the energy available on Earth comes from solar radiation, in turn fuelling the atmosphere thermodynamics system (Curry & Webster, 1999). Out of the solar energy reaching the Earth, 25% is absorbed by the atmosphere and 45% is absorbed by the Earth surface; the rest is reflected by the top of the atmosphere or the Earth surface. We have all experienced the cold dry night when there is no cloud cover, compared to cloudy ones that trap the solar energy received during the day. This is the greenhouse effect. This thin gas layer that mantles our planet is made mostly of nitrogen and oxygen, and a plethora of other minor gases, as well as water vapour, which defines the relative humidity (percent of moisture a volume of air can contain). The distribution of those different elements varies with the conditions on the ground (forest, desert, oceans etc.) with which the atmosphere exchange gas and water, and it also varies with the dynamic conditions of the atmosphere. If the entire atmospheric layer around the Earth is considered to be 140 km thick on average, the part that interests us for hazard purposes is actually about the first 10 km near the ground, namely the troposphere that is limited at the top by the tropopause. It is within this layer that most of the clouds form and evolve, and where the weather is occurring. Within this layer, solar radiation is the main source of energy, and it controls how temperatures are distributed from tropical highs to cold polar temperatures, and how water circulates around the globe in the water cycle. A secondary gradient is the vertical gradient, with temperature lowering on average by 0.6 degrees C per 100 m altitude. This layer can, thus, be understood as a system of connection for water in the atmosphere to the water on and in the ground, linking different pockets of water, the biggest one being the oceans. Indeed, about 96.5% of the water resources is ocean water; less than 2.5% is freshwater, from which glaciers count for 68.7%, groundwater 30.1% and surface freshwater is only 12% of the total resource. Despite an abundance of water on the Earth, usable water is, therefore, relatively limited, and access issues or modification of the cycle (e.g. through climate change) can lead to significant hazards. It is even sometimes a political weapon between different states. For instance, the Mekong River that starts flowing through China is being controlled by numerous dams, which in turn limits the discharge in all the downstream countries (Thailand, Laos, Cambodia and Vietnam). At present, the number of hydroelectric dams planned by the countries other than China along the Mekong River exceeds a hundred, and there are very serious concerns that those projects would dry the delta and ruin the rice culture in Vietnam, which is presently feeding the region. In the present chapter, we do not concentrate on the various political aspects of water nor the warfare around its access, we will rather investigate the role of its presence, excess and scarcity as a source of hazard, but the reader will be aware that the political dimension of water resource and hazards, more often than not, exceed the natural realm.

9.3 Water in the atmosphere and acid rainfall hazards

The total amount of water in the atmosphere is called the precipitable water. Despite being spread over atmospheric layers several kilometres thick, if condensed, it would result in a layer of only 25 cm covering the entire Earth. Its distribution is, however, far from regular, and the tropical region concentrates the largest proportion of precipitable water, whereas the poles are characterised by dryer air conditions. The precipitable water, for a given column of the atmosphere, is an essential value for numerous exogenous hazards, such as debris flows and landslides and, of course, all the water-related hazards. It can be calculated as follows:

$$W_{prec} = \int_0^{\rho} q \, dp/g$$

where W_{prec} is the precipitable water, ρ is the surface pressure, p is the total pressure of the air, g is the gravity and q is the specific humidity, the latter being the ratio between the density of the water vapour and the addition of the water vapour to the dry air density. From the water present in the atmosphere, one must then ask the question of whether this water present in the atmosphere will be transformed into precipitation and under which condition the precipitation will occur. To make such estimate, one can consider the saturation vapour pressure, which is a condition of the temperature and which has been formulated by Goff-Gratch as the following log relation:

$$\log(e) = -7.90298\frac{T_{sat}}{T-1} + 5.02808\log\left(\frac{T_{sat}}{T}\right) - 1.3816*10^{7}\left(10^{11.344\left(\frac{1-T}{T_{Sat}}\right)} - 1\right) + 8.1328*10^{-3}\left(10^{-3.19149\left(\frac{T_{sat}}{T-1}\right)} - 1\right) + \log(e_{sat})$$

where e is the saturation vapour pressure, T is the temperature, and T_{sat} is the steam-point temperature (373.16 K) and e_{sat} is the saturation vapour pressure at the steam-point temperature. As the atmosphere is not stable, and the air tends to be advective and convective, either from dynamic or thermal processes, the saturation vapour pressure conditions change over time and space. The vertical evolution of a partly saturated air parcel follows the dry adiabatic lapse rate, whereas saturated air parcels follow the saturated adiabatic. The water in the atmosphere is colourless and does not affect solar radiation through the troposphere; however, it has a greenhouse effect, as water drops can store heat. When forming clouds as fine droplets of water or ice, the water has a further effect, as it reflects about 20% of the incoming solar radiation.

The process of rainfall nucleation is complex, and so is the droplet formation, which can also turn the explosion of drops in miniature showers (Tomkinson, 2009) and so forth. Raindrops tend to form around a nucleus of dust present in the atmosphere, with which it chemically interacts. This chemical reaction between the nucleus and the surrounding water can become detrimental, when the chemical composition of the nucleus leads to the acidification of the water. It can be a natural process around volcanic ash nucleus, for instance, but it is particularly acute with atmospheric pollution particles. As natural and, most especially, anthropogenic activity injects large amount of sulphates and nitrates in the atmosphere, the raindrops or the fog it contaminates brings the acid contaminant to the Earth surface: the rainfall itself, regardless of its intensity or duration, becomes a hazard then.

The acidity of the rainwater is measured with the pH logarithmic scale ranging from 1 (acid) to 14 (alkaline), 7 being neutral. Non-contaminated rainwater has a pH of ~ 4.5 to ~ 5.5, but after sulphates and nitrates contamination, the pH can drop to values between 1 and 3. Subsequently, when the water reaches the surface, it results in the burning of the vegetation, in skin rashes and burns, as well as respiratory issues. Because the acidic components travel with the winds, the pollution created in one location, one country, can travel afar and generate issues in remote locations. In the 1990s, the atmospheric pollution generated in the United Kingdom was reaching Norway and Sweden, notably burning the forest. Similar issues happened at the border between France and Germany, along the Rhine River, in the vicinity of a European industrial corridor. Acid rain also strongly impacts the waterways and waterbodies, being harmful to the wildlife. The acid rain, percolating through the soils and coming in contact with clay soil, releases the aluminium contained in the clay, in turn creating aluminium pollution in the waterways and lakes.

For instance, internal combustion engines of cars create nitric oxide (NO), and the combustion of fossil fuel produces SO_2, the second being several orders of magnitude more important – climate change speech is often very quick to blame individuals and modes of transports, but large industrial groups have a far deeper impact.

Indeed, the SO_2 combines with atmospheric O_2 to create SO_3, which in turn recombines with H_2O to create sulphuric acid (H_2SO_4). The acid then dissociates in water to give a H^+ ion and a HSO_{4-} ion, and in turn, the HSO_{4-} transforms into $SO_4^{2-} + H^+$ so that the pH is dramatically reduced to harmful levels in water droplets.

$$SO_2 \left(+\text{ atmospheric } 0_2\right) \rightarrow SO_3 \left(+\text{ atmospheric } H_2O\right)$$
$$\rightarrow H_2SO_4 \rightarrow HSO_4^- + H^+$$

$$HSO_4^- \rightarrow SO_4^{2-} + H^+$$

The United States Environmental Protection Agency (https://www.epa.gov/acidrain/effects-acid-rain) defines the effects of acid rain on wildlife, plants and trees with pH tolerance relatively limited. They show that snails and clams need a pH > 6, crayfish, bass > 5, which is unfortunately higher than the values < 4.2 recorded in the 1960s through year 2000, in the north-eastern part of the United States of America, for instance. It influences the environment and the food chain, and it is also a hazard to human activity and health, with agriculture and forestry being on the front line (Sandra, 1984). Zhang and Chang (2012) defined indicators of socio-economic vulnerability to acid rain in China, as it is a hazard to development. Although this reality is often downplayed, and most probably because of the economic power that the polluters hold, acid rain is arguably the sole chemical hazard from the atmosphere; all the other hazards being mechanical due to movement and energy manifestation, such as tornadoes and typhoons.

9.4 Typhoons, hurricanes and tropical cyclones

Cranking up the energy in the atmosphere means increased kinetic energy manifesting itself through windborne, and sometimes accompanied by precipitation, hazards, like it is for typhoons, hurricanes and cyclones. The three are intense atmospheric vortices, which have gained different names, depending on the locations where they occur around the world, but which all describe a similar process. East Asia names them 'typhoons', whilst Northern American defines them as 'hurricanes'. The last term, 'cyclone' is kept for events in the Indian Ocean, and it uses the French wording. Regardless of their names, scientists argue that they are one of the most destructive natural hazards (Emmanuel, 2005). For the USA alone, the NOAA (National Oceanic and Atmospheric Administration) calculated that Hurricane Katrina (2005) costed US$161 billion; Hurricane Harvey in 2017, US$125 billion; and the same year, Hurricane Maria costed US$90 billion. For the period 1986–2015, the NOAA calculated that, after adjustment for inflation, the insured losses alone (not the real losses, as everybody does not have an insurance and there are values that can't be captured by insurances) reached a staggering US$515.4 billion.

If hurricanes that strike Central and Northern America mostly start in the vicinity of Cape Verde Islands, at sea from West Africa, in East Asia, the seawater at the east of the Philippines is the main factory of typhoons. The latter have particularly devastating effects in the low-lying islands and volcanic arc islands of the Philippines, Japan, Taiwan and also mainland China. In this region of the world alone, it is estimated that 15% of the world population lives in the shadow of typhoons, placing some of the fastest-growing populations and economies on the front line.

Scientists define a typhoon/hurricane/cyclone when the threshold of 33 m/s is reached, often upgrading a tropical storm or tropical depression into a typhoon (or hurricane or cyclone). Another condition is the typical rotating wind systems around a central eye. When reaching maturity, they can be between 20 to several hundreds of kilometres in size, and the pressure deficit can reach values in the ~ 10%, locally raising the sea level. The life cycle of a typhoon is usually between a few days to several weeks, and its movement is partly controlled by the surrounding air masses that can ease, or not, its progression over the sea. The large typhoon that hit Japan during the week of 11 to 17 August 2019, progressed very slowly because high-pressure systems were slowing its advance towards Japan. It resulted in rainfalls that lasted longer, generating cumulated precipitations of 700 mm to 800 mm in 24 hours, with a peak rainfall exceeding 100 mm per hour. Before the event, models had even predicted that up to 1,000 mm of cumulated rainfall was locally possible with this typhoon. Typhoons, hurricanes and cyclones all start over the ocean, when the sea surface temperature is at least 26 degrees for about at least 1 m depth. Therefore, they occur in the summer or at the end of the summer in a belt centred on the equator. This is the reason why numerous typhoons start in the warm seas ashore the Philippines or hurricanes in the warm waters of the Gulf of Mexico, for instance. As the storms are low-pressure cells, the wind effects on waves are combined with a local rise in the sea level, which amplifies the flooding effects when reaching the shore.

Table 9.1
The Saffir/Simpson scale for hurricanes, typhoons and cyclones

Scale	*Damage*	*Lowest pressure (mb)*	*Windspeed (m/s)*	*Sea surge (m)*
1	Minimal	> 980	33–42	1.2–1.6
2	Moderate	965–979	43–49	1.7–2.5
3	Extensive	645–964	50–58	2.6–3.8
4	Extreme	620–944	59–69	3.9–5.5
5	Total/catastrophic	< 920	> 69	> 5.5

The greatest impacts to infrastructure are traditionally due to the winds, but most casualties are often associated with water and drowning, as it has been repeatedly experienced in India and Bangladesh (Chittibabu et al., 2004) or in New Orleans, for instance.

Typhoons, hurricanes and cyclones have been the deadliest hazards (not that the loss of lives are caused by hazards, rather than cultural, economic, political and other realms) in front of all the others natural phenomena recognised as hazards. In 1970, 300,000 people perished due to the effects of tropical cyclones; in 1922, about 60,000 lost their lives in China; and in India, in 1935, an estimated 60,000 people perished. The scale of hazards is read on the Saffir/Simpson scale (Table 9.1), which emphasises the fact that typhoons are not strictly winds but also heavy rainfalls, and eventually, coastal and coastal waterways floods.

9.5 Tornadoes and twisters

Tornadoes or twisters are another type of rotating wind, but their dynamic is different, and their scale is usually in the order of magnitude of several tens to hundreds of metres only. They can reach velocities close to 500 km/h, and their fast-spinning columns usually reach down to the ground from the base of a thunderstorm cloud. They occur when there is enough wind shear in the lower atmosphere and always ahead of cold fronts and low-pressure atmospheric systems.

Tornadoes are measured on the Fujita scale of tornado intensity (Table 9.2). One of the deadliest tornados on record in the United States is the Worcester tornado of June 1953, which had more than 90 casualties and injured 1,300, for a total cost of US$52 million. In the US, tornadoes are most commonly created by summer thunderstorms that develop from the states north of the Gulf of Mexico, across the central states to the Canadian border. Out of the 100,000 thunderstorms occurring each year in the USA, about 1,000 trigger tornadoes. Tornadoes can also have other dynamic origins, such as the swirling around mega-structures, like Taranaki Volcano in New Zealand: on 5 July 2007, winds wrapped around Taranaki Volcano and generated tornadoes that damaged 50 houses, cut electricity to 7,000 customers and impacted the water supply for 10,000 homes. The magnitude of those tornadoes is more modest than those in the USA, displaying a

Table 9.2
Fujita intensity scale of tornadoes

Category	*Damage*	*Wind speed (m/s)*	*Impacts*
F1	Light	18–32	Trees and signs damaged
F2	Moderate	33–50	Roofs damaged, cars overturned
F3	Severe	51–70	Large trees uprooted, flying debris
F4	Devastating	71–92	Masonry building damaged, cars airborne
F5	Disastrous	93–142	Wood-frame buildings lifted up, car airborne more than 100 m in the air
F6	Disastrous	> 142	No evidence yet

top speed at 200 km/h, which is slightly more than half of the maxima recorded in the USA (Katurji & Zawar-Reza, 2011).

9.6 Floods from precipitations on land

If not near the sea, human settlements have developed near rivers and lakes for centuries, to harness the transport potentials and the energy that can be generated, but at the same time, these advantages can turn into hazards when the waterbody starts to behave in unexpected manners or in expected manners but exceeding capacities. Consequently, floods become one of the dominant hazards in riverine and coastal settings, where human settlements are concentrated. Floods are complex phenomena that are driven and dominated by water, but which also involve the transport of sediments and bio-debris (e.g. uprooted trees), creating a wide array of hazards. Floods can be defined as an event occurrence of a water level increasing above expected values, overtopping natural or artificial banks. It is usually infrequent – so that humans settled near enough – but in a recurrent manner. From a human being point of view, flood hazards can also be defined as a flood condition when people are likely to be swept or drowned – according to the UK environmental agency – and when damage to the build environment and infrastructures are likely to happen. Floods can occur from torrents, streams, rivers, coastal areas of lakes and the sea. Floods, in front of any other type of hazards, are the most frequent disasters in Asia, involving about 2.2 billion people during the period 1975–2000, and they are one of the most frequent anywhere else in the world (Noji & Lee, 2005). Historically, humans have not been able to stay away from the vicinity of rivers and waterbodies for well-being and economic reasons, fostering a love-hate relationship. Necessary to sustain life and a major vector of human economic and cultural exchanges – river transport, hydroelectricity and so forth – the 3% of freshwater on the planet is, at the same time, the source of flood hazards. Indeed,

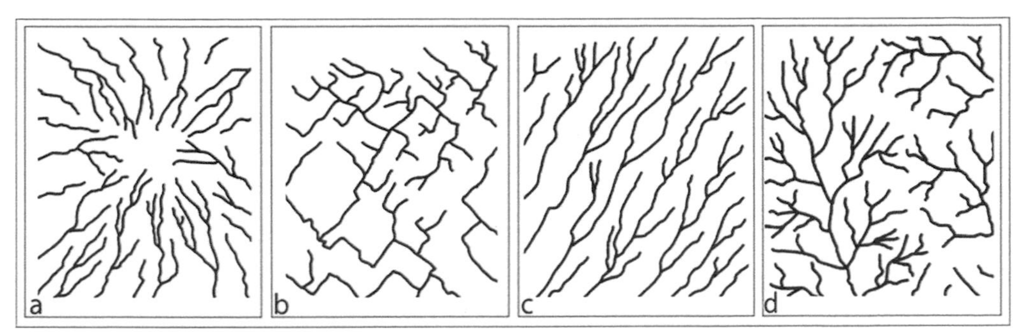

Fig 9.2
Examples of river networks: (a) radial network, (b) rectangular network, (c) parallel network and (d) dendritic network.
Source: author created

"Flooding is a natural process that sustains ecosystems around the globe, but at times floods can have devastating effects on society and the environment" (Schumann et al., 2016).

To understand the variety and complexity of floods on land, one must look at the shape of the catchment area that concentrates the water, and then the characteristics of the floodplains and the channels, from which indications of the characteristics of past and extreme floods can also be gathered, as extreme events leave geomorphological traces in the landscape.

To define the rainfall concentration land units, three terms are commonly used: watershed, basin and catchment area (although British English makes a difference in the terminology between precipitated water and water contributed from groundwater + directly precipitated water). Regardless of the terminology one chooses, the 'catchment' is one of the essential spatial units used in hydrology and in water hazard management. In the present case, and for simplicity, we consider it as the area for which all the precipitation falling over the defined surface converges towards one single exit: the mouth of the watershed. The catchment is usually characterised by one main channel with several tributaries. The channel will have a varying gradient through the catchment, and the channel slope and the watershed slope will contribute towards the modalities of convergence of surface water towards the main channel. This variability is accentuated by the multitude of river network types (Figure 9.2), and it is around these networks that floods are most likely to happen, as the water concentrates in channels, rivers and lakes, and to a lesser extent, in engineered waterways, such as storm water pipes (Ashley et al., 2007).

Depending on the land cover, the topography and the geology, the water will converge at variable speeds, modifying the modality of flooding. Indeed, the physiographic characteristics of catchments are an essential control on the shape of the flood hydrograph (Figure 9.3). Although catchments are often drawn as elongated oval shape objects, their shape vary from almost circular features to very linear features. In general, the closer to a circle a catchment is, the faster the water will concentrate, creating a fast-rising peak in the hydrograph, whereas elongated catchments will produce slow and progressively rising peak flows.

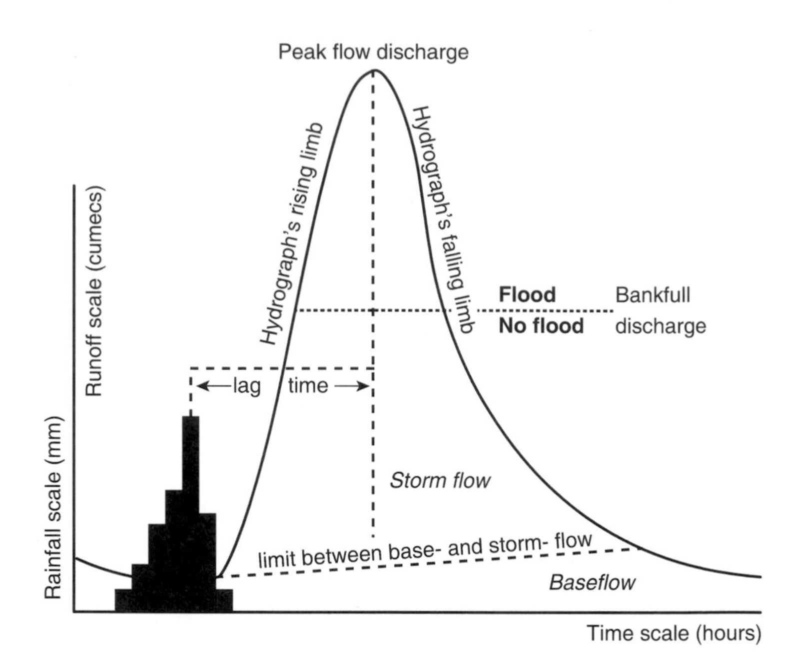

Fig 9.3
Flood hydrograph. The flood hydrograph is a tool for hydrologists and flood hazard managers to link the rainfalls (here represented as a black bar graph) and the peak flow discharge at a location in the basin. Those graphics provide an idea of the time needed before a flood peak arrives at a location, and it also provides an idea of the amplitude of the hydrograph and the potential flood (when bankfull discharge is exceeded). The arrival of the flood is named on the hydrograph as the rising limb, and the recession phase the falling limb.
Source: author created

On top of the morphology of the catchment, its surface cover plays an important role in controlling the amplitude of the peak flow discharge and the lag time between the peak rainfall and the peak discharge (Figure 9.3). Artificial impervious surfaces in cities tend to supply the stream water rapidly, whereas soil and vegetation will retain some of the water, capping or retarding the hydrograph peak flow.

Finally, as rainfall does not pour evenly over a catchment and as the rainfall pattern can range from widespread low-intensity to high-intensity localised rainfalls, the resulting flood can be slow rising or, on the contrary, a flash flood, like it is the case in the Canterbury Plains of New Zealand, depending on whether rainfalls pour in the mountains in the headwaters or whether they pour on smaller sub-catchments (Box 9.1). Furthermore, the location and altitude and season control the temperature, transforming the precipitations to snow and ice, in which case, the precipitation does not always directly contribute to the flood hydrograph. In such case, the water can be stored for a later delivery. All those parameters result in a relative complexity in predicting flood stages, even for basins that are well instrumented, due to the roles of multiple parameters and antecedent conditions. This is the reason why engineers, planners and scientists use models, like the SWAT model (https://swat.tamu.edu/), which combines rainfall, subsurface and surface information to model catchment processes.

Box 9.1: River flooding in Canterbury (New Zealand)

The Canterbury Plains extends for 100 km, linking the Southern Alps mountain range to the east coast of New Zealand, and are prone to river flooding (Figure 9.4). The river network drains sand and gravels mostly accumulated during and at the end of the last glaciation, with deformation due to tectonic activity. The flooding in the Canterbury Plains can be divided between the river flooding on the major rivers, such as the Waimakariri River, or on the minor coastal streams, which do not extend up in the mountain basins. The first type of floods occurring on the major rivers, like the Waitaki River, the Rangi Tata River, the Rakaia River, the Waimakariri River and the Hurunui River (Figure 9.5), occurs due to snow melt and/or heavy rainfalls in the head of the catchment, providing more than 12 hours of warning time. The second type of flood on the secondary network is more related to the localised rainfall. On 28 January 1975, localised rainfall of 150 mm fell in 24 hours, following a wet period of cumulated rainfalls of 300 mm that lasted 20 days. This combination resulted in the Kowhai Stream flash flood, which entered the village of Blanswood along the stream, taking the lives of four and spreading floodwaters 1.5 to 3 m deep in and around the village. More important flooding affected the Canterbury Plains, with the South Canterbury flood on 13 March 1986 (Figure 9.4). Wet conditions that lasted a month and high-intensity rainfall in the upper catchment near the Southern Alps divided, creating extended flooding and damage, notably in Pleasant Point and Geraldine, causing damages worth NZ$120 million at the time. To

Fig 9.4
The 1986 flood in Canterbury, South Island of New Zealand.
Source: author created

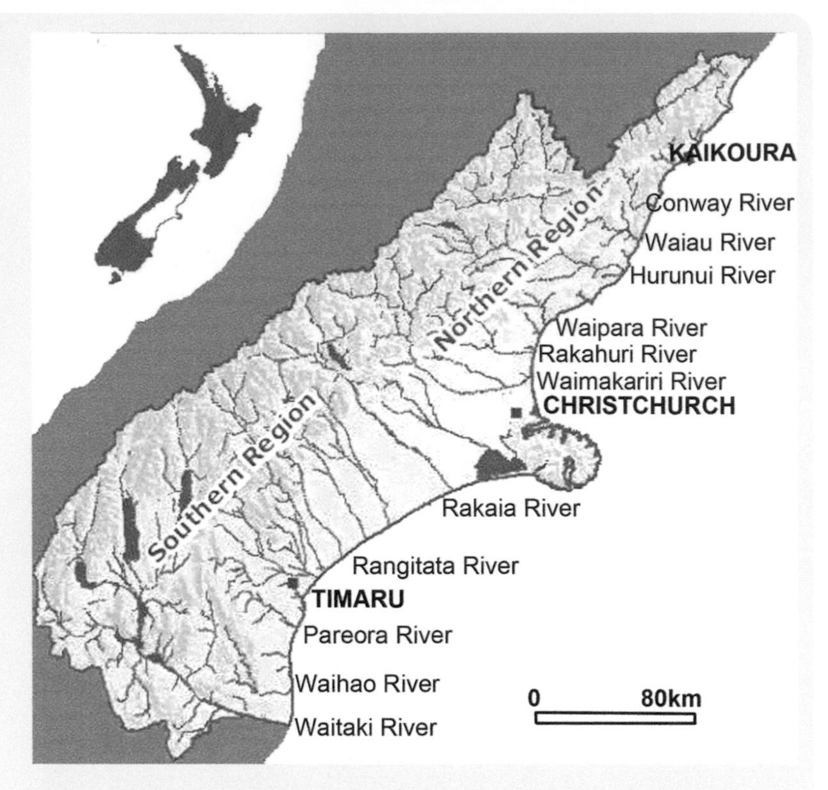

Fig 9.5
The Canterbury Plains, on the South Island of New Zealand, with the major natural drainage and the three important regional cities.
Source: modified from the Canterbury Regional Council

respond to this major issue, a large set of stopbanks were built along the major rivers 20 years ago, and a new law enforced a 100 m no-development zone from the stopbanks, completed with flood warning and forecasting systems, managed by E-Can (Environment Canterbury).

9.7 Surface water and open-channel flood assessment parameters

Surface waters are the resultant of precipitated waters and resurging groundwater either due to internal dynamic or to topographic characteristics, with regard to the groundwater table. In urban setting, the level of water surface can also be related to saturation of the storm water network that is usually calculated to use the road network as a secondary waterway.

After interception by tree canopies, trunks, building roofs and other elements at the surface of the Earth, and absorption by the soil, the excess water first fills in the topographic concavities, and then it starts to flow over land. Those flows that can occur as sheet flows on the slopes gather into rills and gullies through self-organisation patterns before exerting sufficient shear stress to the substrate to erode it and generate favourable water pathways. Those pathways grow into rivers that can be several kilometres wide, like the Amazon or the Congo

Rivers. Open-channel flow can be either steady or unsteady, with steady uniform and non-uniform flows. Uniform steady flow is extremely rare and improbable in nature, but it is often assumed for engineering calculation purposes. Those calculations are essential to estimate hazards related to a given rainfall and a discharge/water-level relation. Even if numerous computation solutions exist to estimate the relation between rainfalls and floods, simple estimations for hazards and risk analysis can be done using the Chezy formula and the Manning or Manning-Strickler formulae. The Chezy formula was developed by a French scientist of the same name in 1755:

$$V = C\sqrt{R.S}$$

where V is the velocity in the channel direction, and C is the roughness coefficient:

$$C = \sqrt{8.\frac{g}{f}}$$

where *f* is the Darcy's friction factor and *g* the gravity acceleration, and R is the hydraulic radius (cross-section area divided by the wetted perimeter).

The second equation is the Manning or Manning-Strickler equation, that was first presented in 1890 and that also relate the velocity to channel structure and the slope:

$$V = \frac{1}{n}.R^{\frac{2}{3}}.\sqrt{S}$$

where S is the energy slope of the river and *n* is the roughness coefficient (there are several tables that exist depending on the Manning roughness coefficient). Based on those equations, one can calculate whether a channel will be able to convey the water away from a cross section without flowing out of the channel and generating flooding, and it can be used easily on calibrated rivers and cross sections (Figure 9.6).

The geometry of streams and rivers is, thus, essential to understanding flood hazards. In natural setting, it is the resultant of the interplay between water, rock, and sediment characteristics and size, all related by the energy developed by the flowing water, which in turn contribute to material erosion, transport and deposition. Emerging from a balance between the driving and the resisting forces channelled flows, one can generally observe a negative downstream gradient of energy. Streams tend to have a higher competence on steep slopes of mountainous areas where they can carry large blocks in the headwaters. Downstream, closer to the base level, slopes tend to be less steep. Consequently, transported sediments are of smaller size, with the stream flowing on a lower velocity gradient or slope, which is the vertical change for a given horizontal

Fig 9.6
Embankments in Kesennuma City (Tōhoku District, Iwate Prefecture, Japan). Note: surface water is fast flowing, and as one of the interfaces is with the atmosphere, it can also carry and transport debris, making them particularly hazardous. For this reason, important hard engineering protections are often developed. In the case of Kesennuma, the stopbanks on the photographs and the newly raised bridges act both as protections for river floods and tsunamis. Source: photo courtesy of Christopher Gomez (December 2018)

distance. Therefore, depending on where the flood occurs, transported bedload impacts can range from being insignificant to having enough momentum to bring down walls and buildings.

Following oscillations of the energy cells in cross sections, river channels tend to create a variety of planform patterns (shape of the river as seen from an aerial view) through long-term slow processes and by rapid short-lived large-scale movements during flood events, which tends to be more terraforming than low-energy processes. The type of planform patterns of a river is dependent on the slope of the riverbed, the discharge, the flow regime, and the amount and mean size of sediment transported. Numerous rivers and streams display meandering planforms, like the Mississippi River, which migrates outward, cutting steep cut-banks. Although the channel tends to create increasingly elongated meanders, during floods, the river can take the shortest pathway, cutting off meanders. This process generates abandoned channels and oxbow lakes in the floodplain. Along rivers, it is, therefore, not solely the floodwater that generates hazards; it is also the bank erosion that modify the geometry of the channel and, therefore, threaten the structures in the vicinity of the channel. Rivers with steeper gradient that transport gravels and larger clasts tend to be braided rivers. Braided rivers consist of at least two channels that are adjacent and are interconnected. Each channel is called a braid, which tends to migrate seasonally. Braided rivers are shallow waterways that spread widely laterally. They are mostly found in mountain areas or on short floodplain downstream of mountains, like the Waimakariri River or the Rakaia River on the South Island of New Zealand, or the Tenryugawa in Shizuoka Prefecture in Japan. Hazards are, thus, related to water inundations and gravel transport. Finally, channels can also be classified as straight channels. There is no such thing as a perfectly straight channel in nature, but it is a distinction made for all the channels that have a lateral sinusoidal amplitude inferior to the longitudinal amplitude of 1/5 for one cycle. They can be commonly found in mountain torrent and for lower longitudinal gradient rivers shaped from bedrock fractures where they have been confined.

Consequently, there are two types of hazards that arise from rivers: (1) the one due to the movement of water and sediment inundating land, with the water flowing into settlements and communities when breaking the channel banks, for instance; and (2) the hazards due to the geomorphological dynamic of the river. As the river erodes its banks, constructions and infrastructures too close to the river can be washed away during floods, and in turn, the transported material will then have further impact downstream on communities. A good example of this effect is the recruitment of drifted wood from the banks being eroded, which then can go and impact bridge piles and so forth.

This phenomenon has been increasing, notably because in the mountains of Europe, for instance, there has seen an expansion of riparian forest in the last 100 years (Kondolf et al., 2007), increasing the wood recruitment and transport during flood events, eventually clogging bridges and other infrastructures crossing river channels.

During the heavy rainfalls of July 2017, in South Japan, in the region of Kita-Kyushu (Japan) near Asakura City, several hundreds of landslides and debris flows were triggered from forested mountain slopes delivering drifted wood from the catchment slopes and the riverbanks to then enter the main river, system, floodplain, destroying houses and bridges (Box 9.2) and eventually reaching the Ariake Sea downstream.

Box 9.2: Drifted wood and debris in floods: the role of catchments' land cover

River floods are in part controlled by the physiography and the land cover of the watershed, and the interactions between the floodwater and the land cover also determine the type and volume of debris transported by the flood. One type of debris transported by rivers is the drifted wood. Drifted wood debris are an important hazard associated with floods because the wood tends to float at the surface of the floodwater. In contrast with sediments that can be deposited on fans at slope breaks or that can't travel rapidly downstream in the absence of steep slopes, drifted wood tends to stay afloat, even with low slope gradients. Therefore, they can impact the built environments and infrastructures downstream, far away from the wood debris sources.

This issue appeared in the aftermath of the July 2017 heavy rainfall in Northern Kyushu Island in Japan, when more than 400 mm of rainfall fell over the Asakura area (Figure 9.7). As the forestry industry has seen a price

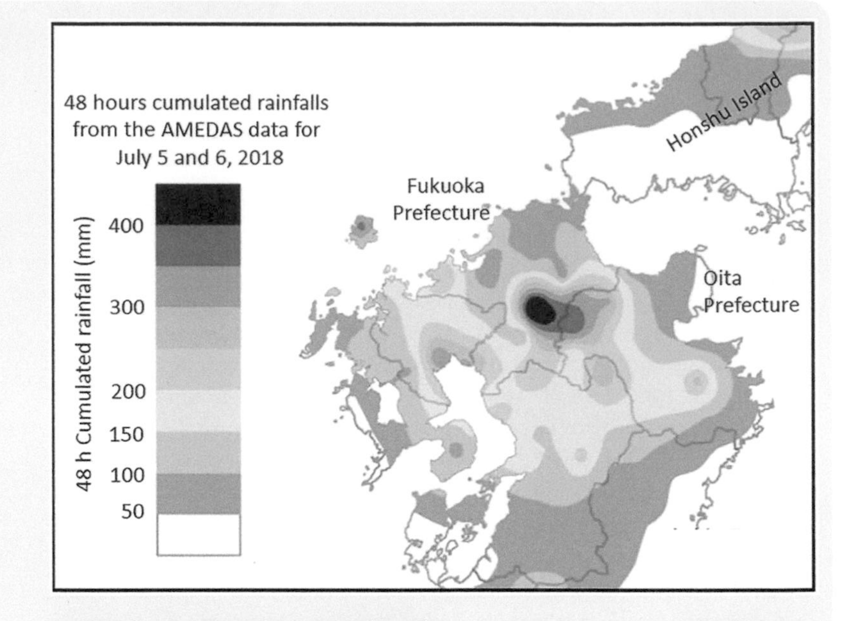

Fig 9.7
Distribution of the cumulated rainfalls over 48 hours on 5 and 6 July 2017, showing the spatial hotspot in the Asakura area, at the border between Fukuoka and Oita Prefectures.
Source: author created from JMA data

Fig 9.8
Log jams in rivers in the Asakura City area. Note: (A) drifted wood debris runup against the riverbanks in a river bend, (B) drifted wood trapped against a small bridge near the headwaters, (C) drifted wood trapped against the highway bridge pile and (D) bio-debris entangled in the still-standing bamboos. The bottom sediment layer is a 30 cm thick sedimentary debris deposit.
Source: author created

drop in timber, the logging has slowed down significantly in Japan, leaving very large trees on the slopes. Furthermore, the fast ageing population in Japan and the increasing concentration of population in urban centres have left portions of the country, which used to be cultivated, going back to the wild or being turned into forest plantations. There is consequently more forest with larger timber, with less hands for maintenance, in a climate change setting with increased rainfalls.

The mountainous river basins to the north of Asakura City were, therefore, a remarkable reservoir of potential timber in 2017. This timber was turned into wood debris due to the heavy rainfalls, and it destroyed buildings located along the streams. Part of it was floated down to the Ariake Sea, when it did not stop on gravel point bars or debris flow fans, or when it wasn't trapped on the riverbanks (Figure 9.8 A), on small and large bridges (Figure 9.8B and C), or on the floodplain natural combs: trees and vegetation (Figure 9.8 D).

This phenomenon has also an anthropogenic aspect to it. Indeed, in Japan, the drifted wood issue has been growing with the ageing population that have left once cultivated slopes to forestry plantations. Furthermore, as in recent years, the price of timber dropped, and timber cuts have been limited. Consequently, the July flood recruited a large number of mature trees, of large size, severely impacting bridges and other infrastructures.

9.8 Flood hazards

Evaluating river flood hazards

Hazard assessment is the first step in flood risk assessment (some of the other traditional steps being exposure, vulnerability and risk assessments). Hazard assessment should provide information on the velocity and the depth of the flow – most probably, maximum data – the presence of debris in the flow and the spatio-temporal parameters of the flood (frequency, rate of rise, seasonality etc.). The estimation of these parameters can be approached through empirical methods and modelling, based on surveys and simulations, as well as work from a fluid mechanics perspective, using a combination of computer modelling and laboratory experiments.

A popular empirical method of flood assessment is based on the geomorphological analysis of the landscape, as floods are hydrological and geomorphological events as well. It is based on the investigation of past flood traces in the landscape and their analysis. As floods occur, they leave sediments and create typical geomorphological forms that can be dated and from which energy and flood pathways can be determined. One of the recent tools that has been developed towards this purpose is the Anisotropy of Magnetic Susceptibility, which is provided from the fabric of sand grain information on the orientation and the energy of past flood events. The geomorphologic method is, therefore, a powerful way to investigate past floods, even when numerical and human records aren't available.

Another way to consider flood hazards assessment is from the perspective of the impacts' measure. One of the empirical equations developed by Ramsbottom et al. (2003) links flood parameters to the effects on human beings:

$$HR = d\left(v + [0.5\sim2]\right) + DF$$

where HR is the flood hazard rating, d is the water depth (m), v is the velocity of the flow (m/s) and DF is the debris factor (m^2/s), and for which the degree of flood hazard is considered as low for HR < 0.75, moderate between 0.75 and 1.5 (i.e. dangerous for elderly and children), significant between 1.5 to 2.5, and > 2.5 is dangerous for anyone. There are numerous such empirical formula in the literature, and it is important to determine the one that is the most appropriated before applying them.

Flood hazards in mountain areas

Today, mountains cover about 25% of the land surface, with higher proportions in Asia (52%) than in North America, Europe and Africa (25%, 22% and 20%). Like for rivers and streams on floodplains, mountain streams tend to generally be incised in the landscape, leaving numerous terraces levels, where people have favourably settled. In turn, population have modified mountain streams, used channel regulation, done sediment mining in the channels and constructed dams, weirs and also sediment check dams (*sabo* dams). The main difference with floodplains is the slope gradients conducting the water and also the sediments more rapidly. Furthermore, as the slope gradient is higher, larger-sized sediments travel with the floods, which are often loose material resulting from the last glaciation, and from the dislocation and fracturation of material by earthquakes and other terragenic processes pushing the mountain upwards.

Consequently, the response of mountain catchments to rainfalls or lake (natural or man-made) outburst floods is controlled by the temperature (low temperature due to altitude gradients, notably, controls the water phase: liquid, solid), precipitation, lithology and soil, as well as the vegetation. Rainfall-triggered floods in mountain catchments can produce extreme events, translating into very fast-paced flowing water in the landscape, due to the proximity to the catchment divide and the high slope gradients. Furthermore, the topography creates clusters of small basins, which, under 1000 km^2, becomes very prone to flash floods, which are characteristics of mountain flood hazards, in combination with debris flows (Borga et al., 2013).

This picture, however, varies strongly with the season. Indeed, a large amount of precipitation that falls in mountains and mountain basins remains trapped for a season to thousands of years in the seasonal snow and the glaciers, creating time delay in the release of the water, making the prediction of hazards often haphazard (Box 9.3).

Flood hazards on alluvial fans

The connection between the mountain areas and the floodplains is made by conic-shaped features, named alluvial fans (and sometimes debris flow cones/fans when the system is fed predominantly by debris flow). Because of their intermediate nature between mountain and floodplains, their flood hazards have characteristics slightly different from the two geomorphologic units they connect.

Box 9.3: The August 2005 flood in Switzerland

Switzerland is no stranger to floods and flash floods. On 27 July 1999, 18 people who went canyoning in Switzerland lost their lives in a flash flood in the River Saxeten.

Switzerland is located in the heart of the European Alps, in between France, Italy and Germany; and 30% of its territory is located above 1000 m a.s.l. The country is a collection of valleys and intramountain basins with steep slopes and thin soils that make the country prone to flash floods. The rainfalls occur either through localised convective events or orographic advective rainfalls. Convective rainfalls in Switzerland have been measured to rate up to 100 mm/h, with an increasing trend.

In August 2005, a large flood occurred in the Prealpine Foreland, flooding the city of Bern and taking the lives of six people. The disaster did cost 3 billion Swiss Francs (about US$3.19 billion), and 75% was registered on private properties and 25% on infrastructures.

The flood was triggered by prolonged and intense rainfalls. The western and south-eastern part of the country recorded only ~ 20 mm of rainfall in 48 hours, but the centre of the country between the Sarine River to the eastern border recorded values of 220 mm over two days, with local maxima exceeding 300 mm. This resulted in the highest ever recorded lakes and river level, exceeding the channel capacity.

Sources: Stoffel et al. (2016a, 2016b) and the Swiss Federal Office for Water and Geology and the news channel Swissinfo.ch

The alluvial fans are triangular fan-shaped geomorphologic features located at the junction between mountainous high-gradient streams and the lower-angle plain where they develop. Their presence is the sign of the deposition of sediments due to rapid energy loss and due to avulsion and flooding processes that construct the fan shape, either through the lateral movement of channelled flow or through sheet flows that can develop over all the fan. They are consequently particularly prone to flooding, most especially in semi-arid and arid environments, because the low recurrence of events makes scientific observations problematic as well as effective soft engineering campaigns. Moreover, floods on alluvial fans can follow the main channel, but they can also affect larger areas through sheet flow floods. As the flow erodes, flooding points vary during a single event and avulsion also occurs.

In the USA, the FEMA relates flooding on alluvial fans to high-intensity, short-duration storms and thunderstorms that tend to occur in the United States summer. Because of the relatively limited understanding of the hydraulic of alluvial fans and because of the rapid onset of the triggering rainfalls, floods can happen with little to no warning and follow different patterns depending on the location on the alluvial fan (FEMA, 1989).

In the face of fan flooding hazards, scientists have developed numerous formulas to provide potential estimates of the hazard's spatial extension and amplitude. The USGS (United States Geological Survey) provides the following equation:

$$y = \varphi Q^{0.4}$$

$$T = \beta Q^{0.4}$$

$$u = \frac{Q}{Ty}$$

where y is the depth of flow, with $\varphi = 0.07$, and Q is the flow rate. Then the width of the channel is T, with $\beta = 9.41$, and u is the velocity of the flow.

The estimation of flood hazards on alluvial fans has been increasingly handled using numerical simulations of flows, like the Flo-2D model (https://www.flo-2d.com/), MIKE FLOOD (https://www.mikepoweredbydhi.com/products/mike-flood) or HEC suite created by the US Corps of Engineers (http://www.hec.usace.army.mil/software/default.aspx). Simpler approaches using flow routines in 2D using Lisflood-FP2d, like Caesar-Lisflood (https://sourceforge.net/projects/caesar-lisflood/), also offer open-source solutions for flood modelling. Such modelling is usually combined with geological fieldwork, in order to assess the past activity of the fans and the energy that is developed over its surface, before management tools are proposed. According to FEMA, the whole fan protection system includes the use of levees, use of channels engineering, creation of water detention basins to dissipate the peak flow amplitude and arrival time, and finally, use of different debris basins, fences and deflectors to limit the impacts of the sediments transported by the water. At the same time, in a 1965 document, FEMA also suggests a combination with local structures, such as debris fences, street design and alignment, to convey flows, elevation and armoured fills, as well as house constructions on pillars or raised grounds.

Flood hazards over the floodplain

Basins and floodplains are characterised by river longitudinal profile slope that is close or virtually equal to the slope of the riverbed. Those rivers are the major conveyor belts of water and sediments outside mountainous areas, and their water levels are controlled by the headland rainfalls, the groundwater levels and, specifically, the change in the base level (either a lake or the sea) and the availability of sediments. If the river is starved in sediments, it will erode its bed, but if the river has a surplus of sediments, it will create natural levees and generate a bed rising above the floodplain, like it is the case in China with the Yangtse River. Hazards on those rivers can be particularly severe because there is no topographic break to limit or stop the flood once the river breaks its banks or breaches a natural or artificial levee (Figure 9.9).

Fig 9.9
Transect across the city of New Orleans explaining the weak position of the entire city once the city levees and floodwall are breached. There is no stopping the flood.
Source: adapted from information from the New Orleans City Hall

Box 9.4: The July 2018 heavy rainfall and flood event in Mabichou (Okayama Prefecture, Japan)

The months of June and July in 2018 can be characterised by an important amount of antecedent precipitations on all West Japan, from Kyoto to Nagasaki, due to the combination of low-pressure systems trapping typhoon tracks along the west coast of Japan, pumping important amount of rainfall on land (Figure 9.7). It generated peaks of monthly rainfalls that exceeded 1,800 mm in Kouchi Prefecture (Shikoku Island) and exceeded 500 mm for virtually the entire western part of Honshu Island. The Chubu area also recorded rainfalls of 1,200 mm for the period of 28 June to 8 July, or an average of 120 mm per day every day for ten days. Kyushu Island also experienced regional values exceeding 650 mm in Shimabara peninsula and 900 mm in Kitakyushu. In Okayama Prefecture, near Mabichou City, rainfall exceeded 300 mm in three days, with peaks above 20 mm/hour. Although it is not an unusual amount of rainfall for Okayama Prefecture, the rivers broke their banks at ten locations, inundating 1,200 ha of land. The event resulted in the death of 200 across West Japan, and it comes second after the worst rainfall flood and sediment disaster event that impacted Nagasaki City in 1982, taking the lives of 300 individuals.

The July 2018 disaster is not so much the resultant of severe rainfalls on a single day but rather the resultant of exceptional cumulated rainfalls. For this event, the city of Hiroshima recorded 444 mm of rainfall during the 72 hours before 7 July, when its last maximum was 340 mm in 1985, in Matsuyama City, the 72 hour cumulated rainfalls reached 360 mm when the last record was 329 mm in 1979. In Kobe, the 72 hour cumulated rainfall reached 435 mm when the record was only 300 mm in 2015.

Sources: *Asahi Shimbun*, 12 July 2018; information and data gathered from the Geospatial Authority of Japan (www.gsi.go.jp) and the Japan Meteorological Agency (www.jma.go.jp)

Riverbanks are some of the most engineered protection against floods, but although those protections were very attractive until the end of the 1970s, their popularity has decreased, and the related issues with those structures have progressively increased. Recently, economic downturns are some of the reasons about half of the levees and sluice gates in the UK are considered to be in a good state, and in Japan (Box 9.4), economic issues, combined with an ageing population, have left numerous structures with too little maintenance. Infrastructures that are built to keep the water within the waterways are then often combined with flood abatement strategies with bypasses and retention ponds, like the one found on the Sacramento River North of Sacramento City, near "Jenny Lind Bend". This structure was designed by the USA Corps of Engineers to protect the cities of Sacramento and the San Francisco Bay area. Finally, river improvements are also carried out by widening or by dredging sediments at the bottom of waterways and then increase their capacity.

9.9 The interplay between river and coastal processes in generating river floods

Climate change impacts on river flood

It is widely recognised that flood hazards will increase in frequency and amplitude with the anthropogenic global warming. The global temperature increase of 2 degrees C will increase flood-prone population from 10% to 50%, depending on the scenarios (Arnell & Gosling, 2016). At 4 degrees C global warming, it has been estimated that 70% of the world population would face an increase of 500% in flood risk (Alfieri et al., 2016).

At the world scale, different scenarios and models of climate change for year 2050 provide varying results, but overall, Northern America, and the belt extending from India through China, Russia, Japan and all the countries of the Pacific Ring of Fire, are set to experience a 67% increase in 100-year flood, whilst Central Europe, Northern Africa, parts of Chile and South America could see a reduction in 100-year floods ranging from > 33% to > 67%. Such reduction in 100-year flood frequency – especially in the Amazon basin – will have arguably deeper and longer impacts than the increase of flood frequency.

In Europe, flood scenarios show variations of 100-year flood return reaching a 40% increase in Western Russia, Sweden and Finland. All the river basins facing the Atlantic Ocean in Portugal, Spain, France, the United Kingdom, Belgium, Holland, Germany and Norway are predicted to experience increase in the 100-year flood (Dankers & Feyen, 2009).

The flood generation process varies from one basin to another, and it highly depends on the geographic environment. Climate change will consequently impact differently floods generated by rain on snow, from snow melt floods, from floods generated by slow soil saturation or from intense rainfall, or from lake outburst floods.

Floods in urban areas – the role of man-made structures

The urban water system is a beast on its own; it combines groundwater, surface water, storm water, drinking water and wastewater in complex systems that obey different norms depending on the countries, local rules and development. Although the third first ones are directly impacted by atmospheric events, the last two can also show breakage during earthquakes and eventual malfunction of the system, in turn contributing to flooding and disease spreading. In certain cases, where the wastewater is discharged straight back to the rivers or the ocean, floods and changes in base level (low-pressure atmospheric system) can then interact directly with the wastewater system (Zevenbergen et al., 2010).

Beside this complex network, another specificity of urban areas is the high percentage of impervious surfaces that increase direct runoff. In Europe, for instance, it has been shown that 48% of the land gained by contemporary cities has occurred over agricultural land, and 36% over pastures and mixed farmland, leading to modification of the water cycle and the speed at which water reaches waterways. To accommodate those rapid surface water, urban landscapes have designed different structures, such as levees, culverts, storm water systems and pumping stations, to limit the impacts of storms. Failures of those systems, in turn, create urban flooding, which can be as follows: (1) lack of properly designed storm drainage systems; (2) rainfalls in excess of the originally designed storm water system; (3) catchment modifications in excess of what has been designed (e.g. in cities where wild urbanisation occurs); (4) blockages of the inlets and sewers and storm water pipes by tree leaves, pipe collapses due to soil pressure or tree roots, and so forth; and (5) failure of pumping stations.

In a city like Tokyo, in Japan, which extends just above (and below) sea level, and at the junction of at least three major river systems, the city has developed a network of tunnels that can transfer water from one side of the city to the other, some 30 km away, using notably several pumps that are jet engines turned into mega-pumps. The system is completed by underground water storage systems, notably underneath sport grounds near the rivers. In such an environment, the engineering failure is a bigger risk than the natural hazards.

Coastal flooding coming from the sea

Coastal areas are the strip of land around the sea or around lakes that are influenced by those waterbodies. Very often, this strip of land will be composed of sediments that have been progressively deposited by the coastal streams and the waterbody itself. Those areas are, therefore, typically low-lying and close to the water base level. It results in a high potential for river and coastal flooding (Box 9.5). Moreover, typical engineering interventions, like dredging and waterway widening, often appear to be ineffective, as the lake or seawater comes and fill those man-made depressions.

Rivers flowing in and into coastal plains follow weak gradients, with relatively little topographic variations both longitudinally and transversally, in such a way that water piles up in the channel naturally. Therefore, when heavy rainfall clogs the river system and when low-atmospheric pressure system raises the sea level, flooding occurs. Such a recipe ended in an important flood along the floodplain of the Rokkakugawa River (Kyushu Island, Japan) on 28–29 August 2019.

Coastal floods occurring from earthquakes' secondary impacts
The majority of river floods can be described from the hydrograph, which is the variable in the equation, whilst in general, the bathymetry and the local geomorphology are given data, considered as a fix set of values that do not change drastically if not for erosion and deposition processes. In the aftermath of an earthquake, however, and especially in coastal areas, the earthquake-related geomorphological change can have a major role in generating floods. As we have seen in the chapter on endogenous processes and hazards, the effect of an earthquake can be permanent ground deformation due to tectonic subsidence, liquefaction,

Box 9.5: Hurricane Katrina and New Orleans

On 29 August 2005, Hurricane Katrina made landfall as a category 3 storm near Grand Isle in Louisiana, with winds gusting at 200 km/h, after crossing the Gulf of Mexico as a category 5 hurricane. Katrina started as a storm over the Bahamas Islands on 23 August 2005. The storm merged with a tropical wave, generating a tropical depression.

Resulting fatalities across Southern US reached 1,833 people, and ranged by state from 2 in Alabama, 2 in Georgia, 14 in Florida, 238 in Mississippi and 1,577 in Louisiana. In Louisiana, over 50% of the people who perished were over 70 years old, and 11% of the fatalities were due to heart conditions, showing other underlying issues in the American population. Besides direct casualties, the storm resulted in the displacement of more than 1 million people, with shelters housing at the peak period more than 270,000 individuals.

The federal aid reached a total of US$120.5 billion, and in the first ten years, the FEMA funded US$22 billion in public work and infrastructure reconstructions, and in recovery aid. The Katrina disaster also impacted the population of New Orleans over long term, as evacuees did not all return back home. The population that was nearing 486,700 inhabitants at the 2000 census was down by half in 2006, with a little bit less than 230,200 inhabitants. The census of 2017 did show that the population had increased back to 393,292 in 2017, but even 22 years after the disaster, the population did not reach back its original level.

The flood hazard in New Orleans is both to be imputed to the hurricane and the design of the city itself (Figure 9.9), as the water level of the Mississippi River and the Lake Pontchartrain on the ocean side are all above the city land level.

lateral spreading and riverbank collapses, and even landslides entering river channels and impounding large amount of water upstream.

As the ground subsides locally, creating close or open topographic depressions, the surface runoff of precipitated water converges to those depressions and create local flood impoundments. This phenomenon can be amplified in urban areas by the deformation produced on gravity water pipes, which also will have difficulties to evacuate the water. Even if the storm water pipes do not change slope angle with the local topographic depression, the ground surface level is closer to the pipes than at other locations and will be comparatively more prone to flooding. Ground subsidence around rivers is also problematic, as it brings river levees down, allowing smaller amplitude peak discharge to generate floods as well. This issue is particularly acute in coastal areas because the sea level becomes virtually higher compared to the ground level, and it consequently makes water evacuation through the river system less efficient.

A striking example of this issue has been put to the fore by the 2010–2011 series of earthquakes that hit the city of Christchurch in New Zealand.

9.10 Dropping the temperature

Ice storms

Ice storms are liquid precipitations at freezing temperature so that they coat the objects they fall on with a glaze of ice (icing roads, trees, power lines etc.), covering all the terrestrial objects with a hard layer, having much worse impacts than snow or sleet (Changnon, 2003). Ice storms are of common occurrence in North-east United States and in Canada. They usually form when a wedge of warm air enters between the high-altitude cold air and cold air near the ground, in such a way that the originally frozen precipitation warms up when going through the warm air wedge, and then when it reaches the cold ground, it freezes again. The main hazard related to ice storms are the loss of electric power, bringing down powerlines and rendering train lines inoperable. This then often results in secondary effects, such as carbon monoxide poisoning and fire, and further impacts on all the transportation systems (icy roads, frozen airplane parts etc.). It has been shown that ice storms have a particularly severe and long-lasting impact in rural areas because they are often not prioritised during repairs (Call, 2010), which are particularly difficult to operate in rural and forested areas over the winter. In December 2013, an ice storm hit the city of Toronto in Canada, and the brittle tree branches and the trees themselves broke under the mass of ice coating, bringing down power lines, cutting down about 500 wires and leaving more than 300,000 individuals without electricity in the midst of winter. The failure of the power lines also impacted the communication towers and the mobile network (Armenakis & Nirupama, 2014). Even if the power was restored for about 225,000 customers before 25 December, 30,000 more had to wait until the 27th, and 45,000 more only had electricity back on 31 December. Proofing lifelines and settlements against ice storms would be too costly.

Snow avalanches

Snow avalanches are fast-moving masses mostly comprising of snow and, to a lesser extent, a mixture of rock, soil and vegetation debris that the avalanche can pick up when flowing downslope. Every year, an estimated 250 people perish worldwide from snow avalanches (Schweizer et al., 2015). Humans have been dealing with their effects from very early on; their impact has been recorded as early as 218 BC, notably by the Roman historian Libius, who described Hannibal's crossing of the Alps and how 12,000 soldiers lost their lives to avalanches.

Snow avalanche happens either as wet or dry snow events, and they can vary in size from a few cubic metres to hundreds of millions of cubic metres (Table 9.3). A snow avalanche consists of a starting or triggering zone, a track, and a runout zone, where the avalanche deposits its material; that's why avalanches are also sometimes treated by scientists as mass movements, except that it is not soil that moves but snow and ice.

An avalanche can start in different manners. It can be a loose snow avalanche or as a snow slab. Also, avalanches can start from either a point or scarp line, or within the snow cover from large slabs put in motion. The body of the avalanche can then either be a slab of a given thickness within the snow or it can be the full snow cover that moves, triggering what is named a full-depth avalanche. Loose snow avalanches usually start relatively small in size and incorporate slowly loose snow, whilst slab avalanches occur more like a landslide with a zone of rupture and then sliding over a plane of weakness in the snow. The avalanche flow is made of the flowing avalanche core surmounted by a powder suspension and a front. The core entrains snow from the snow cover, whilst the powder in suspension incorporates the ambient air and moves like an elutriated particle cloud.

The destructive power of avalanches is due to their impact pressure and also to the pressure difference an avalanche does create between the inside and the outside of buildings when flowing at it, causing weak structures to implode. Depending on the vertical descents, volume and impact pressures, avalanches can size from harmless to terraforming (Table 9.3), with mass from less than 10 tons to more than 100,000 tons, with runouts ranging from several tens of metres to kilometres (Table 9.4).

In Switzerland, avalanche forecasting started as soon as 1945, using a network of stations that were all originally hand operated. This system has then been completed on the French side of the European Alps with a centralised avalanche forecasting service provided by the French National Meteorological Agency. The French *Centre d'Etudes de la Neige* (Snow Study Institute) also developed a system for each region in France to make their own forecast.

The forecasting systems all rely predominantly on the weather conditions because it is very difficult to obtain extensive snowpack information and because even with such information, forecasting the

Size	*Potential effects*	*Vertical descent (m)*	*Volume (m^3)*	*Impact pressure (Pa)*
Sluffs	Harmless	10	1–1,000	< 1000
Small	Bury, injure or kill	10–100	10–1000	1000
Medium	Destroy a wood frame house or a car	100–1,000	1000–10k	10k
Large	Could destroy a village	1,000	100k–1M	100k
Extreme	Could gouge the landscape	1,000–5,000	10M–100M, including soil and rock	1M–10M

Table 9.3
Order of magnitudes of avalanches based on the vertical descent, volume and impact pressure

Size	*Mass (t)*	*Runout distance (m)*	*Impact pressure (Kpa)*
1	< 10	10–30	1
2	~ 100	50–250	10
3	~ 1000	500–1,000	100
4	~ 10,000	1,000–2,000	500
5	~ 100,000	> 2,000	1,000

Table 9.4
Avalanche size classification
Source: modified from McClung and Schaerer (2006) and Schweizer et al. (2015)

strength of the snowpack remains difficult. Using these datasets, and as in numerous forecasting exercises, two different schools are in competition, relying either on statistical modelling or on physical modelling. At this stage, both those tools are more an aid to decision than a tool that can be used on its own (Ground Failure Hazards Mitigation Research Committee, 1990), and avalanches still remain difficult to predict.

Even if prediction is a difficult exercise, avalanche control has seen a sharp increase with, notably, the development of the winter sports during the 20th century. The control exercises are mostly based on the triggering before they start, in order to secure areas, such as ski fields and road passes. They can be initiated artificially by detonating explosives in or above the snowpack; however, the snowpack can sometimes resist the explosive trigger, even when the snowpack is unstable. Another use of the explosives can be in triggering an explosion above the unstable snowpack to load it with extra snow, in order to increase the shear stress and trigger the mass movement. The explosives are either hand delivered, delivered by cable, dropped by helicopter, or prepared as systems in advance of the snow season using pre-planted explosives.

Avalanche control is also operated by structural control, based on artificial anchoring of the snowpack, using artificial terraces, wood structures and rock-filled gabions (Box 9.6).

Box 9.6: Snow avalanche in Quebec, Canada

Snow avalanches are mixtures of snow, ice, rock and debris moving rapidly on a slope, either as a loose avalanche or a slab. The factors that contribute to the formation of avalanches are the terrain, the precipitation over existing snow, the wind, the temperature and then the different layers of snow accumulated over one another (Schweizer et al., 2003). Snow avalanches triggered naturally are often a hazard to dwellings and infrastructures, whilst human-triggered avalanches are usually a hazard to holidaymakers and other recreationists. In Canada alone, an estimated 700 fatalities have been associated with snow avalanches (a small number compared to 250 in Switzerland for the decade 1995–2014, due to the relatively sparse population). In the Quebec province, there has been a record of 43 deadly avalanches since 1825, with 73 casualties and over 50 people injured. Germain (2016) depicts the 8 and 10 March 1995 Blanc-Sablon avalanches in Quebec, which destroyed one house and two sheds, killing two in their dwelling. As the impacted dwellings are often isolated houses or groups of houses, the mitigation measures result in the relocation of dwellings. The movement of people and dwelling is often preferred because the period of return of avalanches is difficult to forecast, and any snow avalanche control system often requires the modification of mountain slopes, and the use of ropes and nets, which are not environmentally friendly. There is, therefore, more to hazard management and snow avalanche hazard management than the anchoring of the snowpack with terraces or other structures.

Glacier collapse-related surge floods

A process that is typical of polar environments and mountainous environments are the glacier collapse surge floods. Such events may be minor compared to other hazards, but the rise of glacier tourism and seasonal settlements in scenic areas have increased the related hazards. This issue is particularly acute, as individuals with virtually no to very little experience of the mountain and its hazards come in contact with glaciers (Purdie et al., 2015).

As glaciers advance in the valley, their lower extent, in the ablation zone, is prone to melting and front collapse. This constant variation in the geometry of the glacier is at the origin of two processes that can generate floods: (1) the containment of water underneath the ice or behind ice that would have collapsed and that is released suddenly by the destruction of the temporary ice dam, and such events generate a short-lived flood surge; or (2) when the front of the glacier enters in a proglacial lake, the collapse of its terminus can generate waves that propagate down the lake and the connected rivers. Both these processes are very rapid and do not last past several minutes, as the volume of water trapped in the ice or displaced by the ice collapse is relatively limited. The flood, therefore, materialises in a surge wave, which can notably bring ice chunks in the areas it inundates.

In New Zealand, Fox Glacier on the west coast of the South Island regularly produces such small collapses and related surge waves. At Tasman Glacier, on the eastern side of the divide, as the glacier enters the lake, terminal collapses also generate waves in the lake, sweeping the shores.

Rain on snow events

Rain on snow (ROS) is the precipitation of water over an existing snowpack. ROS events are an important hazard because the rainy event can remobilise the equivalent water present in the snowpack, growing a potential flood above the volume of water precipitated, making such event difficult to predict and monitor. This is particularly exacerbated because snowfall monitoring is not as precise as rainfall monitoring, and because the snowpack evolves during the winter.

Furthermore, a rain on snow event needs a level of maturity of the snowpack, and it is often the rainfall at the beginning of spring that can trigger the ROS flood event, when the snowpack is ready to melt. It has been demonstrated in the Swiss Alps that critical levels of precipitation over existing snowpack can generate converging water in the torrents and create important floods. Out of the 20 floods present in the database of the Sitter watershed (Switzerland), the 4 major events are ROS events (Beniston & Stoffel, 2016). Furthermore, as the climate is warming, ROS events have been shown to increase in the Sitter River and are projected to increase even further. In other parts of the world, like in the US, it has been shown that ROS can have more complex impacts, with the daily flows of > 1-year return expected to increase, whilst larger events may see little change (Surfleet & Tullos, 2013).

9.11 Droughts and the lack of water

All the previously presented hazards arise from too much water or its movement and related inundations, or changes in its phases and pH. If too much water is an issue, so is too little. Drought is a shortage of water over a prolonged period, exceeding the expected return period of precipitation and other wetting events. Drought is, therefore, not the absence of water but its depletion, a process that often occurs over a long period. If a flood hazard is the occurrence of an event that is – to some extent unexpected – not longed for, droughts are the non-realisation of an event occurring with an expected regular period of return and which is longed for. Smith and Petley (2009) wrote that "Drought is different from most other environmental hazards. It is called a 'creeping' hazard". This poses a real challenge to humans because genetically, we have been wired to respond to rapid and imminent threat, like when wild animals were threatening our cavemen ancestors for the possession of a cave or any advantageous natural feature. But like climate change, droughts come slowly and play a war game of attrition. Compared to other hazards where the energy developed (earthquake, storm, volcanic eruption, debris flow etc.) is the main factor of hazard, drought is more conditioned by the mean of access to water and wealth of a community (can it buy water from other places – e.g. countries of the African Sahel, with some of the world's lowest PNB, are not affected the same way as Qatar or any other Gulf states, boasting some of the richest PNB per citizen capita). The impact of drought is, therefore, not indexed on characteristics of the hazard itself but more on the social characteristics of the region it affects, eventually resulting in water and food shortages, as well as negative economic effects. For the same drought, it is mostly the state of economic development that will control the effects of the drought. Indeed, droughts do not take lives in the USA anymore, but at the turn of the 19th and 20th century, people living in the Great Plains of the USA were experiencing malnutrition and death due to the droughts.

9.12 Concluding remarks

In the 21st century, hydrometeorological and climatological hazards are in the bullseye of both scientists and mass media, as anthropogenic climate change impacts are being felt increasingly across the globe. Most unfortunately, as capitalist economies are ailing, with increasing disparities between the super-rich and working class – the latter are looking more and more like a surviving class – nationalism and weapon rattling have become an excuse for not doing anything (or just not enough). In Japan, climate change is fuelling warmer sea temperature for a longer period of the year, leading in a typhoon season that tends to last longer, with an expected increase in the frequency of super-typhoons, like the one that hit the Kansai area in 2017. Furthermore, it also modifies the distribution of heavy rainfalls, creating hazards in areas where they are less expected.

Atmospheric hazards, however, are not the product of climate change; it is mostly the distribution and frequency-magnitude that are changing. We have seen that atmospheric hazards can range from chemical hazards, with raindrop acidification, to mechanical hazards driven mostly by wind speeds (tornadoes and twisters) and in combination with water (typhoons/hurricanes/cyclones). Leaving the atmospheric domain strictly, for its interface with the Earth surface, hydrometeorological hazards take the shape of floods, with either slow or rapid onset, arising from river and channelised water discharge change or from coastal surges, or the combination of both.

Although we use the term 'natural' hazards, a number of those hazards are modified and sometimes amplified by anthropogenic activity. For instance, we have seen that urbanisation leads to an increased number of impervious surfaces, which in turn increase flood discharge peak and reduce the lag time between rainfalls and the peak flow. If future vulnerability seems partly shaped by climate change impacts and related governance, then the modalities and choices of development, as well as the increasing urbanisation, will play another important control on hazard exposure to flood and drought hazards (Güneralp et al., 2015).

In the present chapter, we only investigated processes occurring in the atmosphere and processes involving water at the Earth surface, but rainfalls and water exchange at the surface and underneath the surface of the Earth also lead to sediment-laden hazards, like landslides and debris flows, which are presented separately in Chapter 8, as exogenous hazards.

Take-away messages

1 Pollution of the atmosphere can lead to severe chemical hazards;
2 High-velocity winds born at sea or on land (hurricanes and twisters) are some of the most powerful atmospheric hazards and can blow at speeds exceeding several hundreds of kilometres per hours;
3 Land floods can come directly from rainfalls, from stream and rivers, and from waterbodies, such as lakes and the sea;
4 Complex interaction between the groundwater, sea level and climate change can exacerbate hazards;
5 Snow avalanches are equivalent to landslides and mass movements, except that the soil is replaced by ice and snow; and
6 Water is an essential resource to human life on Earth, but its presence is also at the origin of numerous hazards, fuelling a complex relationship.

To learn more about the topic discussed in this chapter, listen to the *Disasters: Deconstructed* interview with Dr Ilan Kelman (Figure 9.10).

Fig 9.10
QR code for Chapter 9.

Further suggested reading

Mukolwe, M. M. (2016). *Flood hazard mapping: Uncertainty and its value in the decision-making process*. CRC Press/Balkema, 148 p.

Proverbs, D., Lamond, J., Hammond, F., & Booth, C. (2011). *Flood hazards: Impacts and responses for the built environment*. Taylor & Francis, 373 p.

References

Adikari, Y., & Yoshitani, J. (2009). Global trends in water-related disasters: An insight for policymakers. In *Insight paper from United*

Nations World Water Assessment Program Third World Water Development. International Centre for Water Hazard and Risk Management (ICHARM) and UNESCO, 24 p.

Alfieri, L., Bisselink, B., Dottori, F., Naumann, G., de Roo, A., Salamon, P., Wyser, K., & Feyen, L. (2017). Global projections of river flood risk in a warmer world. *Earth's Future, 5*(2), 171–182. https://doi.org/10.1002/2016EF000485

Armenakis, C., & Nirupama, N. (2014). Urban impacts of ice storms: Toronto December 2013. *Natural Hazards, 74*(2), 1291–1298. https://doi.org/10.1007/s11069-014-1211-7

Arnell, N. W., & Gosling, S. N. (2016). The impacts of climate change on river flood risk at the global scale. *Climatic Change, 134*(3), 387–401. https://doi.org/10.1007/s10584-014-1084-5

Ashley, R., Garvin, S., Pasche, E., Vassilopoulos, A., & Zevenbergen, C. (2007). *Advances in urban flood management*. CRC Press.

Beniston, M., & Stoffel, M. (2016). Rain-on-snow events, floods and climate change in the Alps: Events may increase with warming up to 4°C and decrease thereafter. *Science of the Total Environment, 571*, 228–236. https://doi.org/10.1016/j.scitotenv.2016.07.146

Borga, M., Stoffel, M., Marchi, L., Marra, F., & Jakob, M. (2013). Hydrogeomorphic response to extreme rainfall in headwater systems: Flash floods and debris flows. *Journal of Hydrology, 518*, 194–205. https://doi.org/10.1016/j.jhydrol.2014.05.022

Call, D. A. (2010). Changes in ice storm impacts over time: 1886–2000. *Weather, Climate, and Society, 2*(1), 23–35. https://doi.org/10.1175/2009WCAS1013.1

Changnon, S. A. (2003). Characteristics of ice storms in the United States. *Journal of Applied Meteorology, 42*(5), 630–639. https://doi.org/10.1175/1520-0450(2003)042<0630:COISIT>2.0.CO;2

Chittibabu, P., Dube, S. K., MacNabb, J. B., Murty, T. S., Rao, A. D., Mohanti, U. C., & Sinha, P. C. (2004). Mitigation of flooding and cyclone hazards in Orissa, Idia. *Natural Hazards, 485*, 31–455.

Curry, J. A., & Webster, P. J. (1999). *Thermodynamics of atmospheres and oceans*. International Geophysics, 65. London Academic Press.

Dankers, R., & Feyen, L. (2009). Flood hazard in Europe in an ensemble of regional climate scenarios. *Journal of Geophysical Research, 114*(D16). https://doi.org/10.1029/2008JD011523

Emmanuel, K. A. (2005). *Divine wind: The history and science of hurricanes*. Oxford University Press.

Federal Emergency Management Agency. (1989). *Alluvial fans: Hazards and management*. FEMA Report 165.

Germain, D. (2016). Snow avalanche hazard assessment and risk management in northern Quebec, Eastern Canada. *Natural Hazards, 80*(2), 1303–1321. https://doi.org/10.1007/s11069-015-2024-z

Ground Failure Hazards Mitigation Research Committee. (1990). *Snow avalanche and mitigation in the United States* (pp. 47–50). National Academies Press.

Güneralp, B., Güneralp, İ., & Liu, Y. (2015). Changing global patterns of urban exposure to flood and drought hazards. *Global Environmental Change, 31*, 217–225. https://doi.org/10.1016/j.gloenvcha.2015.01.002
Katurji, M., & Zawar-Reza, P. (2011). *Meso-vorticity around the Taranaki Region—Simulations of the 5th of July 2007 tornado spawning storm*. Extreme Weather Conference, Wellington, New Zealand (PowerPoint presentation available on www.academia.edu).
Kondolf, G. M., Piégay, H., & Landon, N. (2007). Changes in the riparian zone of the lower Eygues River, France, since 1830. *Landscape Ecology, 22*(3), 367–384. https://doi.org/10.1007/s10980-006-9033-y, PubMed: 367384
McClung, D. M., & Schaerer, P. (2006). *The avalanche handbook*. The Mountaineers Books, 342.
Noji, E. K., & Lee, C. Y. (2005). Disaster preparedness. In H. Frumkin (Ed.), *Environmental health: From global to local* (1st ed., pp. 745–780). Jossey-Bass.
Purdie, H., Gomez, C., & Espiner, S. (2015). Glacier recession and the changing rockfall hazard: Implications for glacier tourism. *New Zealand Geographer, 71*(3), 189–202. https://doi.org/10.1111/nzg.12091
Ramsbottom, D., Floyd, P., & Penning-Rowsell, E. (2003). *Flood risks to people: Phase 1*. Technical Report FD2317. Department of the Environment, UK Environment Agency.
Sandra, P. (1984). Air pollution, acid rain, and the future of forests. *Worldwatch Paper, 58*, 54p.
Schumann, G. J. -P., Stampoulis, D., Smith, A. M., Sampson, C. C., Andreadis, K. M., Neal, J. C., & Bates, P. D. (2016). Rethinking flood hazard at the global scale. *Geophysical Research Letters, 43*(19), 10,249–10,256. https://doi.org/10.1002/2016GL070260
Schweizer, J., Bartelt, P., & van Hervijnen, A. (2015). Snow avalanches. In W. Haeberli & C. Whiteman (Eds.), *Snow and ice-related hazards, risks, and disasters*. Elsevier Hazards and Disasters Series (pp. 395–436). Elsevier.
Schweizer, J., Bruce Jamieson, J. B., & Schneebeli, M. (2003). Snow avalanche formation. *Reviews of Geophysics, 41*(4). https://doi.org/10.1029/2002RG000123
Smith, K., & Petley, D. N. (2009). *Environmental hazards—Assessing risk and reducing disaster* (5th ed). Routledge Publisher.
Stoffel, M., Wyżga, B., & Marston, R. A. (2016a). Floods in mountain environments: A synthesis. *Geomorphology, 272*, 1–9. https://doi.org/10.1016/j.geomorph.2016.07.008
Stoffel, M., Wyżga, B., Niedźwiedź, T., Ruiz-Villanueva, V., Ballesteros-Cánovas, J. A., & Kundzewicz, Z. W. (2016b). Floods in mountain basins. In Z. W. Kundzewicz et al. (Eds.), *GeoPlanet: Earth and planetary sciences* (pp. 23–37). Springer Publishing. https://doi.org/10.1007/978-3-319-41923-7_2
Surfleet, C. G., & Tullos, D. (2013). Variability in effect of climate change on rain-on-snow peak flow events in a temperate climate.

Journal of Hydrology, *479*, 24–34. https://doi.org/10.1016/j.jhydrol.2012.11.021

Tomkinson, F. (2009). How raindrops fall. *Nature*. https://doi.org/10.1038/news.2009.705

Zevenbergen, C., Cashman, A., Evelpidou, N., Pasche, E., Garvin, S., & Ashley, R. (2010). *Urban flood Management*. CRC Press.

Zhang, Y.-Y., & Chang, H.-R. (2012). The impact of acid rain on China's socioeconomic vulnerability. *Natural Hazards*, *64*(2), 1671–1683. https://doi.org/10.1007/s11069-012-0319-x

CHAPTER 10

Socio-natural hazards

10.1 Introduction

The impact of hazards on society has been of great concern since ancient times (Figure 10.1). Volcanic activity, thunder, earthquakes and tsunamis were amongst the phenomena that were attributed to the wrath or whims of gods and goddesses. The metamorphosis of ancient civilisations into modern societies was shaped through a series of societal processes deemed to represent progress. Such modifications involved profound changes in the relationships between human beings and nature, involving transformations that were cultural, social, economic, technological, political and institutional.

In warning that humans had introduced on the planet a "new telluric force which in power and universality may be compared to the greater forces of the earth", Stoppani (1873) (in Zalasiewicz et al., 2012) coined the term 'Anthropozoic Era' to raise concerns about the impact of societies on the environment. From a more recent perspective, although originally in the context of atmospheric impacts, an initial expression of such transformation was embedded within the term 'Anthropocene', "the current epoch in which humans and our societies have become a global geophysical force" (Steffen et al., 2007, p. 614). It could be said the Anthropocene began at the end of the 18th century, when analyses

Fig 10.1
Sistine Chapel ceiling: *The Flood* by Michelangelo (1512).

DOI: 10.4324/9781315469614-14

of the air trapped in polar ice showed the beginning of increasing global concentrations of carbon dioxide and methane (Crutzen, 2002).

Several opinions have been expressed regarding the start date for the Anthropocene as a geological epoch dating from the beginning of substantial human impact on Earth's geology and ecosystems (Crutzen & Stoermer, 2000; Steffen et al., 2007; Lewis & Maslin, 2015). Quite recently, Steffen and colleagues argued that of all the candidates, the beginning of the Great Acceleration (the acceleration in the use of global resources after World War II, beginning in the 1950s) is by far the most convincing from the perspective of Earth system science: "it is only beyond the mid-20th century that there is clear evidence for fundamental shifts in the state and functioning of the Earth System that are (1) beyond the range of variability of the Holocene, and (2) driven by human activities and not by natural variability" (Steffen et al., 2015a).

However, some have suggested that the Great Acceleration was neither spatially nor temporally homogeneous, and therefore, the attempt to explain the socio-ecological dynamics that are causing the crisis of the Earth system requires a more interdisciplinary approach; this should integrate different perspectives from natural, social and cultural sciences, and help to clarify the political relevance of this concept (Görg et al., 2020). Of similar concern is the need to further debate societal transformations, sustainable futures and mechanisms or strategies for achieving these (Bai et al., 2016).

Perhaps one of the clearest pointers to the significance of human perturbations at planetary scale is related to the planetary boundaries' framework developed by Rockström et al. (2009), which explains the environmental limits within which humankind can securely function. Seven of the nine boundaries envisaged in the framework have been quantified so far; of these, four (climate change, biosphere integrity, biogeochemical flows and land-system change) have exceeded accepted levels (Steffen et al., 2015b) (Figure 10.2).

Common to all these concepts is the view that ongoing and potential impacts of the transformations of the Earth associated with human activities at different scales reflect disequilibrium in the interplay between humankind and nature, and hence the failure to achieve sustainable development. It is imperative that impending problems be considered together with the actions needed to enhance the conservation and sustainable use of the planet; over the last five decades of the 20th century, and beyond, there has been a considerable and largely irreparable degradation of ecosystems and loss in the diversity of life on Earth that has led to the exacerbation of poverty and inequalities (Box 10.1) (Millennium Ecosystem Assessment, 2005).

Undoubtedly, hazards are inseparable from human action. Present-day UN Landmark Agreements have indeed highlighted the complexity of the interactions amongst social inequality, poverty, ill-advised development and the depletion of Earth's resources, as well as the resulting environmental and societal problems inherent in global environmental

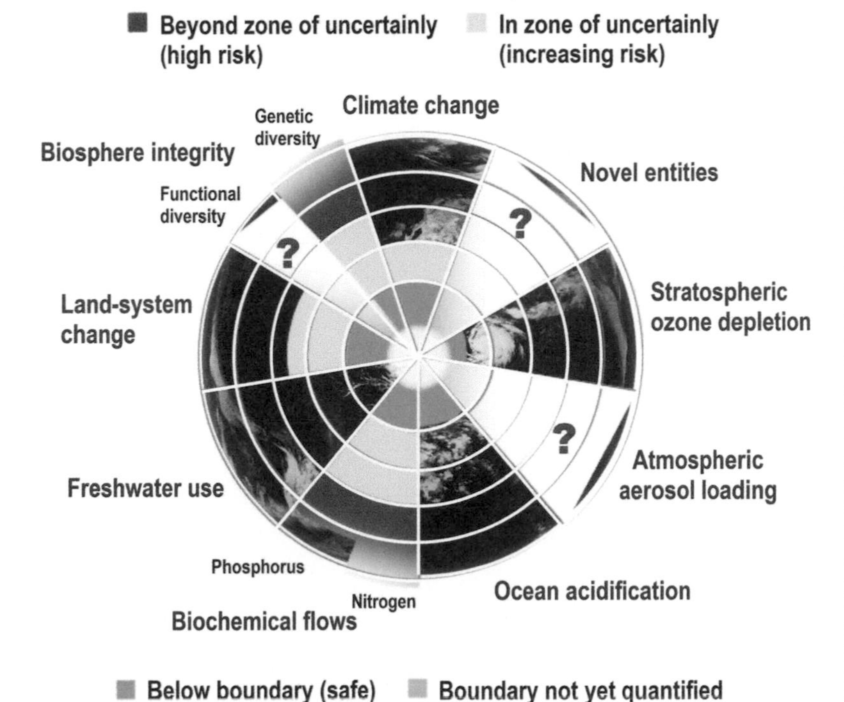

Fig 10.2
Current status of the control variables for seven of the planetary boundaries.
Note: the green zone is the safe operating space, the yellow represents the zone of uncertainty (increasing risk) and the red is a high-risk zone. The planetary boundary itself lies at the intersection of the green and yellow zones. Processes for which global-level boundaries cannot yet be quantified are represented by grey wedges; these are atmospheric aerosol loading, novel entities and the functional role of biosphere integrity.
Source: adapted from **Steffen et al. (2015b)**

Box 10.1: Four main findings of the Millennium Ecosystem Assessment

1 Over the past 50 years, humans have changed ecosystems more rapidly and extensively than in any comparable period in human history, largely to meet rapidly growing demands for food, fresh water, timber, fiber and fuel. This has resulted in a substantial and largely irreversible loss in the diversity of life on Earth;
2 The changes that have been made to ecosystems have contributed to substantial net gains in human well-being and economic development, but these gains have been achieved at growing costs in the form of the degradation of many ecosystem services, increased risks of nonlinear changes, and the exacerbation of poverty for some groups of people. These problems, unless addressed, will substantially diminish the benefits that future generations obtain from ecosystems;
3 The degradation of ecosystem services could grow significantly worse during the first half of this century and is a barrier to achieving the Millennium Development Goals; and
4 The challenge of reversing the degradation of ecosystems whilst meeting increasing demands for their services can be partially met under some

scenarios that the Millennium Assessment has considered, but these involve significant changes in policies, institutions, and practices that are not currently under way. Many options exist to conserve or enhance specific ecosystem services in ways that reduce negative trade-offs or that provide positive synergies with other ecosystem services.

Source: Millennium Ecosystem Assessment (2005)

Fig 10.3
Landslides are common hazards in areas subjected to land degradation (Zoquitlán, Guerrero, Mexico).
Source: photo courtesy of Ricardo J. Garnica

change, which are strongly intertwined with the occurrence of different types of hazards (Figure 10.3).

Whilst the Paris Agreement on Climate Change focuses on undertaking ambitious efforts to combat climate change and strengthen adaptation (UNFCCC, 2015), the New Urban Agenda acknowledges different dimensions of unsustainability in both urban and rural spheres, and their links to a series of vulnerability factors, including climate change, which drive disaster risk and determine the impact of disasters:

> We recognize that cities and human settlements face unprecedented threats from unsustainable consumption and production patterns, loss of biodiversity, pressure on ecosystems, pollution, natural and human-made disasters, and climate change and its related risks, undermining the efforts to end poverty in all its forms and dimensions and to achieve sustainable development. Given cities' demographic trends and their central role in the global economy, in the mitigation and adaptation efforts related to climate change, and in the use of resources and ecosystems, the way they are planned, financed, developed, built, governed and managed has a direct impact on sustainability and resilience well beyond urban boundaries.

> We also recognize that urban centres worldwide, especially in developing countries, often have characteristics that make them and their inhabitants especially vulnerable to the adverse impacts of climate change and other natural and human-made hazards, including earthquakes, extreme weather events, flooding, subsidence, storms, including dust and sand storms, heatwaves, water scarcity, droughts, water and air pollution, vector-borne diseases and sea level rise, which particularly affect [sic] coastal areas, delta regions and small island developing States, among others.
>
> (Habitat III, 2017)

Likewise, the 2030 Agenda for Sustainable Development mirrors efforts towards an integrated and transformative perspective for a better world, by including people, planet, prosperity, peace and partnership as the main benchmarks. Core challenges for achieving environmental, social and economic sustainability are as follows: to eradicate poverty and hunger; to preserve the planet from degradation; to guarantee inclusive well-being of societies in parallel with harmonising development and nature; to promote peaceful, just and inclusive societies; and to create global partnerships (United Nations, 2015).

In 2004, the International Strategy for Disaster Reduction provided a technical interpretation of hazards by defining the concept as "A potentially damaging physical event, phenomenon or human activity that may cause the loss of life or injury, property damage, social and economic disruption or environmental degradation. Hazards can include latent conditions that may represent future threats and can have different origins: natural (geological, hydrometeorological and biological) or induced by human processes (environmental degradation and technological hazards)" (UNISDR, 2004). This definition emphasises the multidimensional nature of hazards and was also adopted by the HFA (UNISDR, 2005), whose scope encompassed disasters caused by hazards of natural origin and related environmental and technological hazards and risks.

In a similar vein, the SFDRR (2015–2030) applied to "the risk of small-scale and large-scale, frequent and infrequent, sudden and slow-onset disasters caused by natural or man-made hazards, as well as related environmental, technological and biological hazards and risks. It aims to guide the multi-hazard management of disaster risk in development at all levels as well as within and across all sectors" (UNISDR, 2015a, p. 11).

Although none of these frameworks included specifically the socio-natural hazards concept, the UNISDR terminology incorporated this significant notion, as "the phenomenon of increased occurrence of certain geophysical and hydrometeorological hazard events, such as landslides, flooding, land subsidence and drought, that arise from the interaction of natural hazards with overexploited or degraded land and environmental resources" (UNISDR, 2017). However, Lavell (1996), who coined the term, provides a more accurate perspective.

10.2 Defining socio-natural hazards

On the basis of analysis of urbanisation in Latin America, Lavell (1996) was the first to suggest that some phenomena that appear to be typical of natural hazards have an expression or incidence that is socially induced because they are produced or exacerbated by human intervention in nature; these, he termed 'socio-natural' hazards. Within the sphere of DRM, the causality of this type of hazard may be interpreted in different ways, and consequently, the viewpoints of individuals or authorities may differ regarding responsibility for their occurrence and options for control. If they are simply interpreted as acts of nature or acts of God, there is no incentive to encourage preventive measures. Hence, failure to understand causality and accept responsibilities repeatedly leads to a focus exclusively on mitigation of effects and emergency response efforts.

Examples of socio-natural hazards include some floods, landslides, subsidence, droughts (and desertification), coastal erosion, rural fires and aquifer depletion. Explanatory variables of these phenomena may include deforestation and degradation of catchments (Figure 10.4), destabilisation of slopes by mining, dumping of industrial and domestic waste into river channels, over-exploitation of land, and destruction of mangroves (Lavell, 1996).

With regard to socio-natural hazards within the urban dimension, the most acute and growing problems emerging in the Latin American context were floods, landslides, subsidence and drought (due to depletion of aquifers, lack of economic options to exploit nearby sources and waste in pipelines) (Lavell, 1996). Nowadays, this situation is clearly recognised as a worldwide reality. Despite the effects of

Fig 10.4
Destabilisation of slopes due to deforestation and construction of roads (Sierra Norte de Puebla, Mexico).
Source: photo courtesy of Ricardo J. Garnica

deforestation, the factors with most influence in terms of urban floods were those intrinsic to urbanisation: the location of buildings and asphalt in places of natural storm infiltration, and the absence of sufficient and adequate storm drainage systems. In addition to the socio-natural hazards already known, future scenarios could involve exacerbation of existing problems and appearance of new ones caused by climate change (Lavell, 1996, pp. 11–12).

From the perspective of DRM, it was clear that recognition of the existence of socio-natural hazards involved a series of conceptual considerations, problems and reflections (Lavell, 1996):

1. Socio-natural hazards result from the impact of certain social practices often associated with profitable economic activities. They can derive from practices that constitute expressions of need or are themselves vulnerabilities. Causes may include the following: survival strategies amongst poor groups (e.g. the cutting of mangroves or felling of forest trees for firewood), fiscal crises in state or municipal governments (e.g. lack of storm drainage infrastructures, combined with densification of land use), or bad practices associated with the absence of adequate public services (e.g. dumping of rubbish in river channels, causing artificial dams or, in the streets, blocking the sewers).
2. The concept of socio-natural hazards emphasises the need to consider and assign responsibilities to certain social agents (not to God or nature); unfortunately, the social agents who have been responsible for the hazards are not always or necessarily the ones who suffer the effects.
3. Socio-natural hazards highlight the role of education and awareness as core requirements in DRM. There is an important difference between assigning responsibilities to a divine God and an unimpeachable nature, on the one hand, and assigning responsibilities to concrete social agents, on the other. Awareness leads to the difference between resignation and conscious action.
4. Even though there is often a marked correspondence between the space of causality of socio-natural hazards and the space of impact, this is not always the case. Deforestation in the upper river catchments, for example, increases runoff and can cause flooding much further down the catchment. Similarly, the depletion of aquifers in areas that surround cities may give rise to urban drought.

These four concepts emphasise two basic issues. The concepts reveal the lack of perception or knowledge of causalities that the population or the authorities possess; an understanding of the relationship between certain construction patterns in the city, and the increase in the incidence and intensity of some floods does not necessarily bring recognition of the relationship between processes that occur in distant territories and the floods that may arise. Likewise, the character of socio-natural hazards challenges the notion of a system or institutional framework for DRM that is limited, territorially, to the one city (Lavell, 1996).

10.3 Significant drivers of socio-natural hazards: key issues and information

Socio-natural hazards arising from intervention in the environment can be seen as an added dimension in the dynamic relationship between natural hazards and vulnerability (Lavell, 1996) (Figure 10.5 and Box 10.2).

Generated by social practices and socially induced, socio-natural hazards are linked to a series of disaster risk drivers related to the vulnerability domain. Hence, although establishing a typology of socio-natural hazards would be a pointless and never-ending exercise, a series of significant drivers can be recognised, with population growth, urbanisation, ecosystem deterioration and climate change being the most important (Table 10.1). The present section builds on earlier studies and considers fundamental issues and information related to the dynamics of those drivers.

Assessments made by the United Nations indicate that from an estimated 7.7 billion people worldwide in 2019, the global population is expected to reach 8.5 billion in 2030, 9.7 billion in 2050 and 10.9 billion in 2100 (UN-DESA, 2019). This population growth is expected to be assimilated in urban areas. Whereas in 1950, 30% of the global population was urban (751 million inhabitants); by 2018, it had increased to 55% (4.2 billion); and by 2050, it is expected to reach 68% (United Nations, 2018).

Fig 10.5
Vulnerable groups living in abandoned houses on slopes affected by landslides (Tuxtla Gutiérrez, Chiapas, Mexico).
Source: photo courtesy of Irasema Alcántara-Ayala

Box 10.2: Terminology associated with ecosystem degradation

Land degradation is defined as the many human-caused processes that drive the decline or loss in biodiversity, ecosystem functions or ecosystem services in any terrestrial and associated aquatic ecosystems;
Deforestation is the direct human-induced conversion of forested land to non-forested land;
Forest degradation is a reduction in the biomass, productivity or benefits from the forest;
Land abandonment can be caused by changes in economic conditions, policies or political circumstances, or by changes in the soil, making it unsuitable for cropping; and
Soil degradation includes loss of soil through erosion at a rate faster than it is formed; nutrient removal in harvest greater than it is replaced; depletion of soil organic matter, surface sealing, compaction, increasing salinity, acidity, metal or organic toxicity to the point where it cannot support former uses.

Source: IPBES (2018)

Table 10.1 Significant drivers of socio-natural hazards

Drivers	*Socio-natural hazards*
Population growth Urbanisation Forest clearing (cutting and burning) Forest degradation Deforestation Land-use change Land-cover conversion Land degradation Desertification Land abandonment Mining Damming of rivers Dams and impoundments Groundwater removal Lack of capacity and/or maintenance of sewers, drainage and pipe systems Climate change	Flash floods, exacerbated floods and landslides, sewer flooding, droughts, wildfires (i.e. forest and rural fires), land subsidence, coastal erosion, and aquifer depletion

Urban areas are systems that mirror the complex and dynamic interactions amongst socio-economic, geopolitical, and environmental processes that occur at local, regional and global scales; urbanisation "is one of the most powerful, irreversible, and visible anthropogenic forces on Earth" (Sánchez-Rodríguez et al., 2005). In 2018, 529 million people lived in 33 megacities concentrated in only 20 countries; they represented 13% of the world's total population, whilst large cities with 5 to 10 million inhabitants accounted for 4%, and medium-sized cities with 1 to 5 million had 12% of the total population of the world. It is estimated that, by 2030, there will be 43 megacities in the world, the majority of them in developing regions. Current estimates suggest that in low-income countries, the number of cities with 300,000 to 500,000 inhabitants will increase from 33 in 2018 to 61 in 2030 (United Nations, 2018).

Owing to the higher pressure on natural resources, urban areas are a major source of environmental problems that exacerbate the impact of hazards on society; these include pollution, congestion, urban heat effect, crime, informal (unauthorised) settlements, lack of affordability and accumulation of waste. Some 40% of worldwide energy use is associated with the construction and maintenance of buildings, and buildings are also responsible for 33% of greenhouse gas emissions globally (Berardi, 2012; Berardi et al., 2014).

Urban areas of high relevance to socio-natural hazards include cities that lack planning and those unregulated and unplanned urban sprawls characterised by inefficient use of or absence of resources (Figure 10.6). For instance, notwithstanding the decrease in slum dwellers in developing countries from 39% of the urban population in 2000 to 30% in 2014, absolute numbers continue to mount (UN-Habitat, n.d.).

An increase in the global demand for food is a central consequence of population growth. This involves a major challenge especially in terms of the use of productive land in developing countries where more than 800 million poor and hungry people are concentrated (FAO, 2018).

It has been argued that land degradation not only possesses social consequences, but that it has a social origin (Figure 10.7). Nonetheless,

Fig 10.6
Urbanisation is one of the main drivers of socio-natural landslides in Teziutlán, Puebla.
Source: photo courtesy of Ricardo J. Garnica

Fig 10.7
Land degradation in Zoquitlán, Guerrero, Mexico. Note soil degradation near the river and landslides in the mountains.
Source: photo courtesy of Ricardo J. Garnica

although land degradation is often caused or exacerbated by population pressure, it can also arise when there is a significant decrease in population density; in this case, poorly managed land favours human-induced degradation. When land is degraded, its intrinsic qualities are lost or a decline in capability takes place; degradation can, thus, be defined as a reduction in the capability of land to satisfy a particular use. For this reason, land degradation is commonly evaluated with reference to the changing benefits and costs for the people at present and in the future (Blaikie & Brookfield, 1987).

The significance of the global decrease of forest in recent decades has been emphasised in a series of assessments. The world's forest area decreased from 31.6% of global land area to 30.6% between 1990 and 2015 (FAO, 2018). During the period 1990–2016, 1.3 million km^2 of forests were lost (World Bank, 2017). Further estimations that focused on the period between 2001 and 2015 suggested that 27% of global forest loss was due to permanent land-use change for commodity production, 26% to forestry operations, 24% to shifting agriculture and 23% to wildfire (Curtis et al., 2018).

Other assessments indicated that 2.3 million km^2 of forest were lost during 2000 and 2012, whilst only 0.8 million km^2 of new forest were established; 32% of the net loss in global forest cover occurred in the tropical rainforest, with nearly half of this being in South America (Hansen et al., 2013). More than 15 billion trees are cut down annually, and the global number of trees is estimated to have decreased by approximately 46% since the start of human civilisation (Crowther et al., 2015).

The Intergovernmental Science-Policy Platform on Biodiversity and Ecosystem Services (IPBES) has estimated that, by 2050, less than 10% of the Earth's land surface will remain substantially free of direct human impact (IPBES, 2018) (Box 10.3). Deforestation associated with

Box 10.3: Land degradation

> Land degradation is a pervasive, systemic phenomenon: it occurs in all parts of the terrestrial world and can take many forms. Unless urgent and concerted action is taken, land degradation will continue to accelerate in the face of continued population growth, unprecedented consumption, an increasingly globalized economy and climate change. The implementation of known, proven actions to combat land degradation and thereby transform the lives of millions of people across the planet will become more difficult and costly over time.
>
> (IPBES, 2018, pp. XXVIII, XXXVIII, XLII)

> Land degradation is an increasing threat for human well-being and ecosystems, especially for those in rural areas who are most dependent on land for their productivity. Land degradation hotspots cover approximately 29 per cent of land globally, where some 3.2 billion people reside.
>
> (UN Environment, 2019)

ecosystem degradation is one of the major drivers of socio-natural hazards. It results from the conversion of forest land to accommodate crops and livestock, and it has adverse consequences for the livelihoods of foresters, forest communities and indigenous peoples, and for the variety of life on our planet (IPBES, 2018).

The pervasive character of the global combination of changes in land use and land cover influences key aspects of Earth system functioning. These changes are not simply related to population growth, poverty and infrastructure, but related to a series of pathways generated by markets and policies, increasingly influenced by global factors, including "weak state economies in forest frontiers; institutions in transition or absent in developing regions; induced innovation and intensification, especially in peri-urban and market accessible areas of developing regions; urbanised aspirations and income with differential rural impacts; new economic opportunities linked to new market outlets, changes in economic policies or capital investments; and inappropriate intervention giving rise to rapid modifications of landscapes and ecosystems" (Lambin et al., 2001, p. 267).

A considerable amount of care needs to be taken to recognise the multiple and interacting drivers influencing the creation of socio-natural hazards. Major concern and transformative action should be directed towards ecosystem deterioration or environmental degradation and unsustainable use of natural resources (e.g. forest clearing, forest degradation, deforestation, land-use change, land-cover conversion, land degradation, desertification, land abandonment), population growth, population density, urban processes, unplanned urbanisation, and climate change.

10.4 Socio-natural hazards: some examples

Floods

Box 10.4: Leonardo da Vinci's perception of the nature of water

Amid all the causes of the destruction of human property, it seems to me that rivers on account of their excessive and violent inundations hold the foremost place. And if as against the fury of impetuous rivers any one should wish to uphold fire, such a one would seem to me to be lacking in judgment, for fire remains spent and dead when fuel fails it, but against the irreparable inundation caused by swollen and proud rivers no resource of human foresight can avail; for in a succession of raging and seething [waves], gnawing and tearing away the high banks, growing turbid with the earth from the ploughed fields, destroying the houses therein and uprooting the tall trees, it carries these as its prey down to the sea which is its lair, bearing along with it men, trees, animals, houses and lands, sweeping away every dike and every kind of barrier, bearing with it the light things, and devastating and destroying those of weight, creating big landslips out of small fissures, filling up with its floods the low valleys, and rushing headlong with insistent and inexorable mass of waters.

What a need there is of flight for whoso is near! O how many cities, how many lands, castles, villas and houses has it consumed!

How many of the labours of wretched husbandmen have been rendered idle and profitless! How many families has it brought to naught, and overwhelmed! What shall I say of the herds of cattle which have been drowned and lost!

And often issuing forth from its ancient rocky beds it washes over the tilled [lands].

Source: notebook c.a. 361 r.b. in MacCurdy (1955)

Approximately 90% of all disasters are water related. During the period 1995–2015, floods were associated with 43% of all documented disasters. Consequences included 2.3 billion affected people (of whom 95% lived in Asia), 157,000 fatalities and economic damage in the order of US$662 billion. Although a high proportion of floods occur in Asia and Africa, consequences are also considerable in other places. For example, each year, between 1995 and 2004, 560,000 people on average were affected by floods in South America. A decade later, in the period 2005–2014, some 2.2 million per year were adversely affected, approaching a four-fold increase (UNISDR, 2015b).

The European Environment Agency (EEA, 2016) indicated that approximately 1,500 floods had occurred in Europe since 1980, more than 50% of them since 2000. It is forecast that by 2050, some 2 billion people will be at risk of floods owing to population increase in areas susceptible to floods, climate change, deforestation, loss of wetlands and rising sea levels (WWAP, 2012).

Floodwaters can originate from the sea, from glacial melt, from snowmelt or rainfall, from ground infiltration, and as a consequence of failure of watercourses or dams, reservoirs and pumping systems (Box 10.5). This must be borne in mind in the evaluation of dams, which, despite their significant contribution to human development, can quite often have adverse consequences particularly within the social and environmental spheres, including involuntary displacement of people, and considerable impact on communities downstream and on the natural environment (World Commission on Dams, 2000).

Box 10.5: The first comprehensive flood control project of Mesoamerica

"*Tlaloc*, the main water god, was the god of rain and lightning as well as of floods and droughts. This god was probably the ancient and most famous water god of Mesoamerica, because the people of Teotihuacan (200 BC–AD 600) venerated this god, and they were the first settlers in the Valley of Mexico . . .

. . . *Chalchihuitlicue* was the flowing waters goddess and was represented by a jade skirt; she was the sister and probably the consort of Tlaloc . . .

. . . Flood control works were started early in the history of the Aztec Empire, given the need that they had to protect lives and goods against excess of water. In spite of the original purpose that they served, roads played an important role as structural measures for flood control . . .

Everything seemed to be running smoothly in the Aztec Empire up to AD 1449, when a very wet rainy season produced a large flood in the capital city of the empire. Water rose to a depth of about 2 m, flooding all the places where the Aztec people could normally walk. Moctezuma Ilhuicamina, the ruler at that time, sought the help of his cousin the king of Texcoco, Netzahualcoyotl, who immediately began hydrological works that would protect Mexico-Tenochtitlán from the floods coming from the surrounding lakes. He proposed to Moctezuma the construction of the most important flood control work of the Aztec culture: the Netzahualcoyotl's dike or dike of the indians. The dike was 16 km long and 20 m wide. The dike was completed with the construction of the Cuitlahuac road, which divided the waters of the lakes of Chalco and Xochimilco, and with the construction of the Mexicaltzingo road, which divided the waters of the lakes of Mexico and Xochimilco.

Those hydrological works had gates to control the flow of water from the lakes and to allow canoes to pass. During the ruling of Moctezuma, the Tepeyac road was also built: it had the double purpose of containing the waters coming from the northern lakes and to open the access to the lands in the north of the valley. This set of hydrological works was the first comprehensive flood control project of Mesoamerica".

Source: Raynal-Villaseñor (1987, pp. 3, 6–8)

Before the Anthropocene, humankind was not so socio-ecologically rapacious. Amongst ancient civilisations, for example, the Aztecs recognised the vital necessity to live in harmony with the environment; for example, their pantheon was rich in water gods. It was not until the Spanish conquest that the balance with the generous land was broken. The introduction of systemic power and exploitation was accompanied by desiccation of the lakes of the old Tenochtitlán (Figure 10.8), deforestation, and the socio-cultural and environmental transformation of the territory; disruption of the communion between society and nature gave rise to a series of hazards, such as flooding (Boyer, 1975; Alcántara-Ayala, 2019).

In view of the need to understand and address the consequences of interventions in the environment, particularly those derived from urbanisation, attention needs to be paid to urban hydrology. Creation of impermeable areas and simplification of drainage systems in urban areas lead to increased runoff rates and volumes, and to decreased infiltration (Fletcher et al., 2013); changes in land use in areas surrounding urban development similarly influence soil properties and runoff processes. Storage of water and runoff streams are also modified (Wheater & Evans, 2009).

An increase in green cover reduces runoff volumes, whereas a decrease in green cover increases runoff volumes. For example, it has been estimated (Gill et al., 2007) that under a 2050s high-emissions scenario with 50% probability level in Northern England, a 10% decrease in green cover would lead to an increase of 13.3% in runoff in the Lower Irwell Valley catchment (a mixed-use housing site) and an increase of 29.0% in the Mersey Multi Modal Gateway catchment (an economic development site).

Effects of urbanisation have been studied in the White Oak Bayou catchment, to the northwest of Houston, Texas, using data from 1949 to 2000 (Olivera & DeFee, 2007). Since the early 1970s, when 10% of

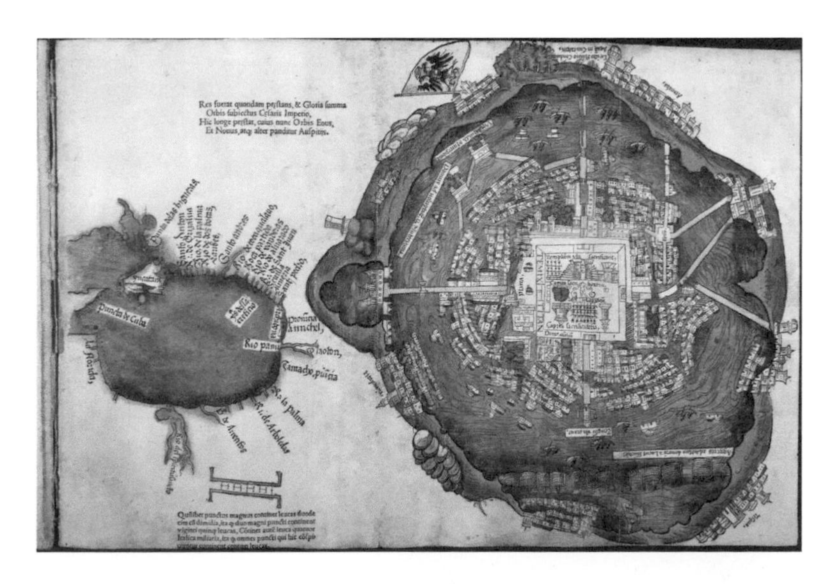

Fig 10.8
Map of Tenochtitlán (1524), the capital city of the Aztec Empire (now Mexico City), built on a series of islets in Lake Texcoco (printed in 1524, in Nuremberg, Germany; colourised woodcut; on the left, the Gulf of Mexico (south is at the top); on the right, Tenochtitlán with west at the top).

the terrain had become impervious, annual runoff depths had increased by 146%, with 77% being attributed to urbanisation and 39% to changes in precipitation; peak flows had increased by 159%, with 32% being attributed to urbanisation and 96% to changes in precipitation (Olivera & DeFee, 2007).

Increased runoff, decreased infiltration of water into the soil and a lack of vegetation can all produce a warming of the surface of the land and a progressive change in its microclimatic conditions (Kravčik et al., 2007).

The influence of soil infiltration on urban flooding and waterlogging has been demonstrated in a model for Wuhan City, China (Ren et al., 2020); this suggested that under low rainfall intensity and urban impermeability, there is a critical value of the stable infiltration rate, and that an increase in this or in the overall infiltration rate in urban green space can greatly reduce runoff coefficient and runoff peak, and thereby mitigate urban flooding.

The urban hydrology balance is also affected by ageing infrastructure, establishment of rural and urban settlements on floodplains, and interactions with the sanitary sewer systems (Box 10.6). Quite commonly, sewer and drainage systems are overwhelmed by rainfall so that urban floodwaters are frequently contaminated with sewage. This can bring risks to health and higher costs of repair to properties (Wheater & Evans, 2009).

Box 10.6: The historical development of sewers worldwide

Although there is evidence of surface-based storm drainage systems in early Babylonian and Mesopotamian Empires in Iraq (ca. 4000–2500 BC), it is not until after ca. 3000 BC that we find evidence of the well organized and operated sewer and drainage systems of the Minoans and Harappans in Crete and the Indus valley, respectively. The Minoans and Indus valley civilizations originally, and the Hellenes and Romans thereafter, are considered pioneers in developing basic sewerage and drainage technologies, with emphasis on sanitation in the urban environment. The Hellenes and Romans further developed these techniques and greatly increased the scale of these systems. Although other ancient civilizations also contributed, notably some of the Chinese dynasties, very little progress was made during the Dark ages from ca. 300 AD through to the middle of the 18th century. It was only from 1850 onwards that modern sewerage was "reborn", but many of the principles grasped by the ancients are still in use today. . . .

The Romans are usually cited for the magnificence of their aqueducts in terms of hydraulic infrastructure, but they also gave an extraordinary contribution to the development of sewerage systems. The name of the main collector of Rome's sewerage system is known all over the world: the Cloaca Maxima ("Cloacina" was the goddess of sewers).

Source: De Feo et al. (2014, pp. 3936–3937, 3951)

The future is likely to bring higher rates of urbanisation, with demands on an outstripped drainage capacity being exacerbated by intense storms induced by climate change; this could overwhelm the DRM in cities and lead to frequent and more severe flooding of sewers. This scenario would not be exclusive to developing nations; it would also take place in the developed world, particularly in megacities characterised by higher levels of inequality and vulnerability of people, and lacking sustainability.

Timely preparation and DRR has, therefore, become an urgent requirement, especially in urban areas. For instance, in the United Kingdom, the need for long-term policies has prompted models of major flood and coastal defence scenarios between 2030 and 2100, through a series of analyses within the Foresight Future Flooding (FFF) project. From this perspective, flooding is perceived as two sub-systems: in the catchment and coastal flooding system, water enters urban areas from outside; in the intra-urban system, flooding arises from events within urban areas (Wheater & Evans, 2009).

A series of drivers influencing future flood risk were considered for the FFF scenarios. The major groups of drivers were taken to be climate change, socio-economic factors linked to vulnerability of people and value of assets at risk, and governance aspects (Table 10.2). Resulting estimations suggested that numbers of people at high risk in the United Kingdom from river and coastal flooding could increase from the current 1.6 million to between 2.3 and 3.6 million by the 2080s. As a consequence of intra-urban flooding, triggered by short-duration events, the number of people at risk could rise from 200,000 today to between 700,000 and 900,000 in the 2080s. Of particular significance

Table 10.2
Combined list of fluvial/coastal and intra-urban drivers
Source: Wheater and Evans (2009)

Driver group	*Driver*	*Explanation*
Climate change	Precipitation	Changes in all aspects of precipitation (amount, intensity, duration, location, seasonality and clustering).
	Temperature	Influence of temperature on soil moisture and, hence, runoff.
	Relative sea level rise	Rising relative sea level is due to climate change–induced melting of ice caps and thermal expansion in conjunction with land subsidence or uplift. Rising relative sea level makes coastal flooding more frequent and allows more energy to reach the shoreline. Long-term effects include morphological change as the coastline adjusts.

(*Continued*)

Table 10.2
(Continued)

Driver group	*Driver*	*Explanation*
	Waves	Offshore waves are generated by winds and increase in height with storminess and fetch length. Increases in wave height and changes in wave direction due to climate change may affect transmission of wave energy to the shoreline. Impacts will be influenced by increases in near-shore depth caused by changes in next two drivers.
	Surges	Increases in surge levels are expected due to climate change–induced increases in storminess. Stronger surges mean that higher extreme water levels with more energy reach the shoreline, increasing the risks of breaching or overtopping of coastal defences.
Catchment runoff	Urbanisation	Changes in the catchment that increase the area of impermeable surfaces and extent of storm water drainage systems to increase surface runoff.
	Rural land management	Effects of land management practices on agricultural and other managed rural land, including conservation and recreational areas and wetlands that affect runoff generation.
Groundwater systems and processes	Groundwater flooding	Groundwater flooding occurs when the water table reaches the elevation of the land surface (waterlogging) or by the emergence of water originating from subsurface permeable strata.
Fluvial systems and processes	Environmental regulation	Future legislation intended to increase biodiversity and habitat protection may influence policy on flood management, with implications for river and floodplain morphology, vegetation, conveyance and flood storage.

Driver group	*Driver*	*Explanation*
	River morphology and sediment supply	Changes in river channel morphology (size and shape) and sediment supply that alter attributes of the river channel and floodplain to influence flood conveyance, routing and storage.
	River vegetation and conveyance	Vegetation and micro-morphology influence velocity distributions and turbulence levels in flows significantly. Hence, changes may affect flood conveyance.
	Urbanisation and intra-urban runoff	A change in land management with green field and previous surfaces covered by less-pervious materials (buildings and infrastructure) and associated new conveyance systems.
Urban systems and processes	Sewer conveyance, blockage and sedimentation	Processes associated with processes that occur in below-ground drainage systems, including performance, maintenance and operation.
	Impact of external flooding on intra-urban drainage systems	Loss of conveyance and serviceability in below-ground drainage systems due to flooding from external sources.
	Intra-urban asset deterioration	Changes in the performance, condition and serviceability of urban drainage assets (ageing, performance wear and tear, and rehabilitation management).
Coastal processes	Coastal morphology and sediment supply	Changes in the near-shore seabed, shoreline, and adjacent coastal land, coastal inlets, and estuaries will, in the short term, affect the wave and surge energies that affect the shoreline. In the long term, the coastline adjusts to changes in coastal processes.

(*Continued*)

Table 10.2
(Continued)

Driver group	*Driver*	*Explanation*
Human behaviour	Stakeholder behaviour	Stakeholders may influence flood risk in many ways, ranging from preflood preparedness to self-help after an event. Corporate and government stakeholders influence availability of insurance, agricultural practices, food production and pursuance of ecological (or other) aims. Future changes in stakeholder behaviour will be strongly linked to societal values and goals.

is urbanisation in areas susceptible to floods, where increased rainwater runoff could increase flooding risk by up to three-fold. Future urban developments could similarly increase risk if planning control is weak or absent (Evans et al., 2004).

Flash floods

The term 'flash flood' has been defined in a variety of ways (Table 10.3). Flash floods are often triggered by short, high-intensity rainfalls, mostly of convectional type; they have a considerable impact on catchments less than 1,000 km^2, in which their effects may evolve rapidly within a few hours or less. Such response depends on the size of catchment and on the surface runoff that results from the intensity of rainfall, soil moisture and soil hydraulic properties, land-use changes, and modifications induced by urbanisation and fires (Marchi et al., 2010).

In a comprehensive review of an international disaster database for the period 1975–2002, Jonkman (2005) noted that although flash floods generally affect a limited number of people, average mortality is high, at 3.6% per event, because these are swiftly developing events whose effects, although over a relatively small area, can be intense. During the period, average mortality was 5.6% in Europe, 4.2% in Africa, 3.2% in Asia and 2.7% in the Americas.

Analysis of 21,549 flash flood events that occurred in the US between 2006 and 2012 (Špitalar et al., 2014) suggested that the factors with the most influence on the effects of flash floods are short flood durations, small catchment sizes in rural areas, shortage of vehicles and occurrence at night with low visibility.

Between 1950 and 2006, 40% of the flood-related human fatalities in Europe were associated with flash floods (Barredo, 2007). In Hungary, for example, some 700,000 inhabitants of more than 400 settlements in the hilly region of Southern Transdanubia live in areas potentially affected by flash floods (Lóczy et al., 2012).

Table 10.3 Approaches to defining flash floods
Source: Sene (2013)

Definition	*Reference*
A flash flood can be defined as a flood that threatens damage at a critical location in the catchment, where the time for the development of the flood from the upstream catchment is less than the time needed to activate warning, flood defence or mitigation measures downstream of the critical location. Thus, with current technology, even when the event is forecast, the achievable lead time is not sufficient to implement preventive measures (e.g. evacuation, erecting of flood barriers).	ACTIF (2004)
Flash floods occur as a result of the rapid accumulation and release of runoff waters from upstream mountainous areas, which can be caused by heavy rainfall, cloud bursts, landslides, the sudden break-up of an ice jam or failure of flood control works. They are characterised by a sharp rise followed by relatively rapid recession, causing high flow velocities. Discharges quickly reach a maximum and diminish almost as rapidly. Flash floods are particularly common in mountainous areas and desert regions but are a potential threat in any area where the terrain is steep, surface runoff rates are high, streams flow in narrow canyons and severe thunderstorms prevail. They are more destructive than other types of flooding because of their unpredictable nature and unusually strong currents carrying large concentrations of sediment and debris, giving little or no time for communities living in its path to prepare for it and causing major destruction to infrastructures, humans and animals, rice and crop fields, and whatever stands in their way.	APFM (2006)
A rapid and extreme flow of high water into a normally dry area, or a rapid water level rise in a stream or creek above a predetermined flood level, beginning within six hours of the causative event (e.g. intense rainfall, dam failure, ice jam). However, the actual time threshold may vary in different parts of the country. Ongoing flooding can intensify to flash flooding in cases where intense rainfall results in a rapid surge of rising floodwaters (US National Weather Service).	NOAA (2010)

(*Continued*)

Table 10.3 (Continued)

Definition	*Reference*
Flash floods are rapidly rising floodwaters that are the result of excessive rainfall or dam break events. Rain-induced flash floods are excessive water flow events that develop within a few hours – typically less than 6 h – of the causative rainfall event, usually in mountainous areas or in areas with extensive impervious surfaces, such as urban areas. Although most of the flash floods observed are rain induced, breaks of natural or human-made dams can also cause the release of excessive volumes of stored water in a short period, with catastrophic consequences downstream. Examples are the break of ice jams or temporary debris dams.	World Meteorological Organisation (2009)

Analysis of a series of flash-flood events that occurred during 2005–2014 in the Attica catchment, an area of Greece with high population density and levels of urbanisation, indicated that depth of rainfall accumulated over a short time interval was a good indicator of the impact of a flash flood at local scale (Papagiannaki et al., 2015); quantitative thresholds depended on local conditions.

In Metro Manila, flash floods result from infiltration loss associated with urban concrete, a century-old drainage system and clogged streams. High-resolution topography derived from Light Detection and Ranging (LiDAR) and a modelling approach (Lagmay et al., 2017) showed that most flood-prone areas were along the intersection of creeks and streets situated in topographic lows.

The critical role played by anthropogenic activities in the occurrence of flash floods has been demonstrated. Predisposing factors include the following (Sene, 2013):

1. Increased settlement and recreational uses in mountain areas;
2. Encroachment of housing and infrastructure onto existing floodplains in lower-lying areas;
3. Development in urban areas affecting drainage paths and increasing the proportion of paved and other relatively impervious areas, exacerbated by a lack of maintenance in some cases;
4. Widening ownership of cars and other vehicles with extensions of road networks into mountainous areas and coastal zones; and
5. Catchment degradation, leading to increased sedimentation and reduced river channel carrying capacities, and increased risk of debris flows.

Because of the combined scenarios of climate change and increasing urbanisation throughout the world, flash floods are of major concern

(Gruntfest & Handmer, 2001). In the course of an urban flash flood, streets can easily be transformed into rapidly moving watercourses affecting infrastructure, at the same time that trunk roads, railway lines, large buildings, concrete walls and pavements can act as temporary embankments (Hapuarachchi et al., 2011); hence, the number of people affected is expected to rise. As these socio-natural hazards can develop rapidly, warning and preparation are essential but not easy to accomplish (Montz & Gruntfest, 2002).

In September 1995, 65 people died in a disaster triggered by a rainfall-induced flash flood on an alluvial fan in Himachal Pradesh, India. Owing to the confinement of the Beas River, coupled with urbanisation and mounting encroachment of development in the constrained valley floor, the destruction caused by the flash flood in the Kulu Valley involved extensive damage to private and government properties, as massive buildings were washed away (Sah & Mazari, 1998).

In December 1999, rainfall-induced landslides, debris flows and flash floods in the state of Vargas in Venezuela caused major damage, including more than 8,500 buildings destroyed or damaged, in addition to disruption of roads, telephone lines, electricity, water and sewage systems, with economic losses in the order of US$1.79 billion (Salcedo, 2000). The number of fatalities was initially estimated between 25,000 and 50,000; this would have been one of the worst disasters in recent decades. However, further investigations have suggested that such calculations lacked evidence and that the number of fatalities did not exceed 800 persons; nevertheless, the societal impact on the community was indeed severe (Altez, 2010).

Landslides

Links between anthropogenic activity and landslides include land-use changes (Glade, 2003; Van Beek & Van Asch, 2004; Alcántara-Ayala et al., 2012); land management (Bruschi et al., 2013); poor terracing practices and absence of terrace maintenance (Giordan et al., 2017); the impact of roads on slope instability (Borga et al., 2004); mining and quarrying (Xu et al., 2019); fire damage to hillslopes (Cannon et al., 2001); and climate change (Geertsema et al., 2007; Shan et al., 2015).

The complex human-induced environmental modifications on hillslopes have been classified (Jaboyedoff et al., 2016) into seven categories (Table 10.4).

Awareness of the human influence on landslides is leading to growing concern. It has been noted, for instance, that in the Himalayas, landsliding has inadvertently been initiated and accelerated by undercutting and removal of the toe of slopes for the cutting of roads and paths; in practice, these works eliminate a key ingredient of slope cohesion and strength, and contribute to slope failure (Barnard et al., 2001). Quarrying has been a major factor in landslide occurrence on the eastern Tibetan Plateau, where, in 2019, a landslide produced a 10 m high dam and consequently formed a lake (Xu et al., 2019).

Table 10.4
Seven human-induced changes or actions potentially affecting slope stability
Source: Jaboyedoff et al. (2016)

1 Slope re-profiling;
   a Embankments;
   b Fill slopes;
   c Cut slopes;
   d Construction work;
   e Excavation work; and
   f Tailing hills.
2 Groundwater flow perturbation and fast pore pressure changes;
   a Pipe leaks;
   b Dam reservoirs;
   c Leaks in old canalisation networks; and
   d Pipe bursts.
3 Surface water overland flow modifications;
   a Deficient drainage system; and
   b Diversion of river.
4 Land-use changes and land degradation;
   a Deforestation;
   b Forest fire; and
   c Urbanisation.
5 Inappropriate artificial structures;
   a Inappropriate retaining wall; and
   b Infrastructure break.
6 Vibration and explosive;
   a Blasting; and
   b Heavy traffic.
7 Ageing and degradation of infrastructure;
   a Weakening of terraced wall; and
   b Filling of torrential check dams.

In an analysis of recently burned hillslopes of Colorado (Cannon et al., 2001), the dynamics of rainfall-induced debris flows were more strongly influenced by runoff-dominated than by infiltration-dominated processes; these fire-related debris flows were initiated by significant sheetwash, rill and rainsplash erosion, and by transport of burned mineral soil and dry-ravel material from the hillslopes high within the contributing areas.

By examining the interplay between landslides and urban development in two densely populated settings in the Campania region, Italy, Di Martire et al. (2012) concluded that regardless of the long history of hillslope instability in the area, urban sprawl, often illegal, has involved the occupation of unsafe areas; this has shaped to a great extent the occurrence and impact of landslides.

In May 2007, 250 rainfall-induced landslides in the Rio Taraza catchment, Colombia, triggered a disaster that involved the loss of 13 lives, the evacuation of 67 people and the temporary relocation of 600 families. Infrastructure, lowland crops and private buildings were also affected. Deforestation was the major factor influencing the occurrence of the landslides; the 3,000 ha affected were slash-and-burn areas that had been used for illegal crops (López-Rodríguez & Blanco-Libreros, 2008).

Deforestation can lead to landslides even in areas created to protect the natural environment (Alcántara-Ayala et al., 2012). In February 2010, a series of rainfall-induced landslides occurred in deforested areas in the Monarch Butterfly Biosphere Reserve, Michoacán, Mexico. The ensuing disaster involved the loss of 19 lives and substantial damage to roads, electricity and the water supply system, with indirect repercussions on crop production, cattle farming and tourism. During the same extreme event, 54–57 ha of forest were either degraded or destroyed because of landslides and winds (López-García & Alcántara-Ayala, 2012) (Figure 10.9).

A diachronic analysis of hillslope instability distribution in the Sierra Norte, Puebla, Mexico (Alcántara-Ayala et al., 2006) recorded a higher landslide concentration on bare surfaces that had low vegetation density (0–50%), and suggested that land-use change and land degradation are precursors to landsliding, and have marked consequences for regional population distributions and economic viability.

An assessment of the role of land-use changes in shallow landsliding in the northern Apennines (Persichillo et al., 2017) noted that the areas most susceptible to landslides were those abandoned cultivated lands that had been progressively recolonised by natural grasses, shrubs and trees.

The links between landsliding and climate change are widely documented. However, the response varies as a function of the landslide type and processes involved; for example, in northern British Columbia, Canada, the incidence of landslides has increased as a result

Fig 10.9
Damage caused by the main debris flows along the San Luis River in Angangueo, Michoacán, Mexico.
Source: photo courtesy of José López

of the debuttressing of valley walls due to thinning and retreat of glaciers, thawing of permafrost under a warming climate and increased precipitation (Geertsema et al., 2007).

Analysis of the significance of human activities in the occurrence of landslides in the central Swiss Alps (Meusburger & Alewell, 2008) found that the area affected by landslides increased by 92% between 1959 and 2004; the increase was attributed to both climate change and land-use change.

The effects of climate change have been examined in high-latitude permafrost regions of China, where increases in temperature and human construction activity are accelerating the thawing of permafrost; this in turn causes landslides. Permafrost degradation will lead to further temperature rise in this region, and the southern boundary of the permafrost region will recede northwards even more rapidly (Shan et al., 2015).

Droughts

'Drought' is defined by the Intergovernmental Panel on Climate Change (IPCC) as "a period of abnormally dry weather long enough to cause a serious hydrological imbalance", and the concept is further clarified as follows: "Drought is a relative term, therefore any discussion in terms of precipitation deficit must refer to the particular precipitation-related activity that is under discussion. For example, shortage of precipitation during the growing season impinges on crop production or ecosystem function in general (due to soil moisture drought, also termed agricultural drought), and during the runoff and percolation season primarily affects water supplies (hydrological drought). Storage changes in soil moisture and groundwater are also affected by increases in actual evapotranspiration in addition to reductions in precipitation [Figure 10.10]. A period with an abnormal precipitation deficit is defined as a meteorological drought. A megadrought is a very lengthy and

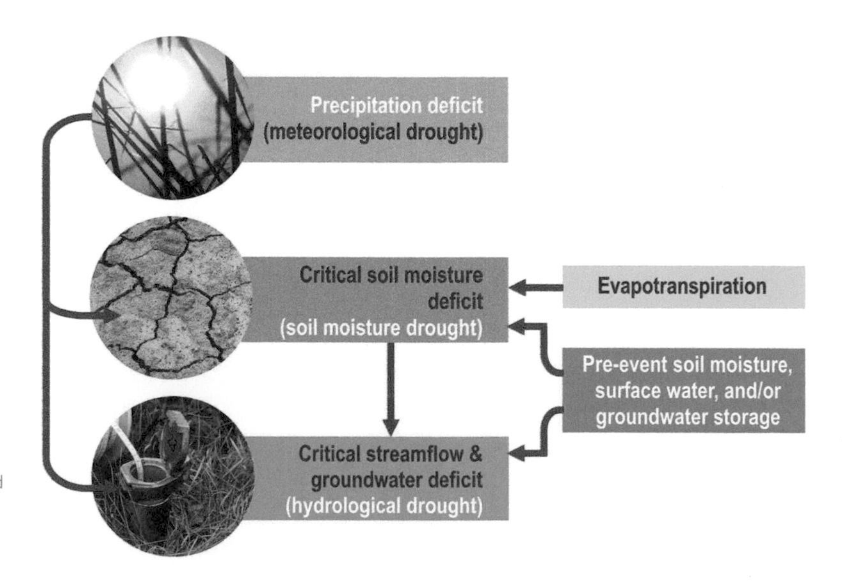

Fig 10.10
Simplified sketch of processes and drivers relevant for meteorological, soil moisture (agricultural) and hydrological droughts. Source: adapted from Seneviratne et al. (2012)

pervasive drought, lasting much longer than normal, usually a decade or more" (Seneviratne et al., 2012, p. 167).

Even though droughts accounted for only 5% of the total number of disasters that occurred during the period 1994–2013, more than 1.1 billion people were affected by them. This represented 25% of the total number of people affected by disasters, placing droughts as the second most significant type after floods. Approximately 41% of drought disasters were in Africa; of the 131 droughts registered, 75 occurred in East Africa. In 2002 alone, some 300 million people in India were affected by drought (UNISDR, 2015b).

Droughts are of the uttermost significance given their high toll with regard to hunger, poverty and the perpetuation of underdevelopment. Droughts are closely associated with widespread agricultural failures, loss of livestock, water shortages and outbreaks of epidemic diseases. When they persist for several years, they can cause extensive and long-term economic disruption, and displacement of people (UNISDR, 2015b).

Drought hazard is spreading partly as a result of changes in the global climate; El Niño, the warm phase of the oscillation of southern oceanic waters, has increased in frequency and strength in recent decades (Trenberth & Hurrell, 1999), leading to more extensive drought conditions overall and increased frequency of severe drought in the tropics (Lyon, 2004).

Climatic effects can be exacerbated by human actions, sometimes with disastrous ramifications: human responses to the climate in Syria resulted in a more severe and persistent drought, which, during 2007–2010, contributed to the conflict in that country (Kelley et al., 2015).

Owing to the significant socio-economic costs, the impact of global warming on the occurrence of droughts has been the subject of a series of model scenarios developed by the scientific community. For example, Lehner et al. (2017) analysed changes in aridity and the risk of consecutive drought years. They demonstrated that, for the Southwest and Central Plains of the USA, warming by 2 degrees C would have no major effect on drought risk. In contrast, in the Mediterranean and Central Europe, even 1.5 degrees C warming would considerably intensify drought risk, and an additional 0.5 degrees C to 2 degrees C would further increase the risk significantly.

Current and potential future effects of droughts arise from the interlinkages between the hazard (the drought itself), vulnerability and exposure conditions. This involves different sectors, including the following: environment (e.g. forests, wildfires, wetlands, biodiversity); agriculture (including crop and livestock production) and forestry; public water supply (water supply reduction and greater demand for various uses); hydro, thermal and nuclear power generation (hydroelectricity production is decreased by drought); buildings and infrastructure (damage to buildings derived from the swelling and shrinkage of soils); tourism and recreation; commercial shipping (ships may struggle when navigating rivers during low-flow conditions); industry; and social conditions (people's health and safety etc.) (United Nations Office for Disaster Risk Reduction, 2021).

10.5 Concluding remarks

Despite the understanding of disaster risk and of disasters as social constructs, there remains a reluctance to acknowledge the creation of socio-natural hazards. Instead of recognising the occurrence of hazards of particular character as a consequence of human interventions in the environment that induce new or exacerbated dynamics and consequences, it has been easier to point the finger at the occurrence of extreme natural events.

The impact of hazards is closely related to the level of vulnerability and exposure of communities, whose dynamics vary across temporal and spatial scales as a function of factors that may be economic, social, geographic, demographic, cultural, institutional, governmental or environmental (Cardona et al., 2012).

The concept of socio-natural hazard coined by Lavell (1996) provides important insights into the role of human actions in the environment, and in shaping vulnerability and exposure.

The socio-economic effects of flooding have increased since the mid-20th century, owing mainly to greater exposure and vulnerability (Jiménez Cisneros et al., 2014), and climate change will cause more frequent and severe droughts and extreme flash floods. The consequent increase in adverse effects on societies (IPCC, 2014) should lead to intensified effort to identify and address the drivers of socio-natural hazards, and thereby, to avoid future risk and incidence of disasters.

DRM could be strengthened by knowledge gained from a greater focus on the human-induced dynamics of socio-natural hazards, such as flash floods, exacerbated floods and landslides, sewer flooding, droughts, fires, land subsidence, coastal erosion, and aquifer depletion. It is imperative that this be achieved.

Take-away messages

1. Ongoing and potential impacts of the transformations of the Earth associated with human activities at different scales reflect disequilibrium in the interplay between humankind and nature;
2. Socio-natural hazards are phenomena that appear to be typical of natural hazards but have an expression or incidence that is socially induced because they are produced or exacerbated by human intervention in nature, often associated with profitable economic activities;
3. The most common socio-natural hazards include flash floods, exacerbated floods and landslides, sewer flooding, droughts, wildfires (i.e. forest and rural fires), land subsidence, coastal erosion, and aquifer depletion;
4. Since socio-natural hazards are generated by social practices and are socially induced, they are linked to a series of disaster risk drivers related to the vulnerability domain;

5 There is an urgent need for politicians at all levels to recognise the need to assess disaster risk before grandiose schemes are executed; hence, their urgent need for more informed advice from specialists;
6 More specialists are needed to participate in interdisciplinary and transdisciplinary research; and
7 DRR requires science-based policymaking and practice.

To learn more about the topic discussed in this chapter, listen to the *Disasters: Deconstructed* interview with Dr Susanna Hoffman (Figure 10.11).

Fig 10.11
QR code for Chapter 10.

Further suggested reading

Blaikie, P., & Brookfield, H. (1987). *Land degradation and society*. Methuen.

Cardona, O., Van Aalst, M., Birkmann, J., Fordham, M., McGregor, G., Perez, R., Pulwarty, R. S., Schipper, E. L. F., Tan Sinh, B., Décamps, H., Keim, M., Davis, I., Ebi, K. L., Lavell, A., Mechler, R., Murray, V., Pelling, M., Pohl, J., Oliver-Smith, A., & Thomalla, F. (2012). Determinants of risk: Exposure and vulnerability. In C. Field, V. Barros, T. Stocker, & Q. Dahe (Eds.), *Managing the risks of extreme events and disasters to advance climate change adaptation: Special report of the intergovernmental panel on climate change* (pp. 65–108). Cambridge University Press.

Crutzen, P. J. (2002). Geology of mankind-the. *Nature*, *415*(6867), 23. https://doi.org/10.1038/415023a

Lavell, A. (1996). Degradación ambiental, riesgo y desastre urbano. Problemas y conceptos: Hacia la definición de una agenda de investigación M. A. Fernández (Comp.). In *Ciudades en riesgo. Degradación ambiental, riesgos urbanos y desastre*. LA RED/USAID.

Montz, B. E., & Gruntfest, E. (2002). Flash flood mitigation: Recommendations for research and applications. *Environmental Hazards*, *4*(1), 15–22. https://doi.org/10.3763/ehaz.2002.0402

Ostrom, E. (1999). Coping with tragedies of the commons. *Annual Review of Political Science*, *2*(1), 493–535. https://doi.org/10.1146/annurev.polisci.2.1.493

Raynal-Villaseñor, J. A. (1987, April). The remarkable hydrological works of the Aztec civilization. In *Water for the future: Hydrology in perspective*. IAHS Publication, 164 (Proceedings of the Rome Symposium).

References

ACTIF. (2004). *Some research needs for river flood forecasting in FP6.* Achieving Technological Innovation in Flood Forecasting. European Commission Project EVK1-CT-2002-80014.

Alcántara-Ayala, I. (2019). Desastres en México: Mapas y apuntes sobre una historia inconclusa [Disasters in Mexico: Maps and notes about an unfinished story]. *Investigaciones Geográficas, Boletín del Instituto de Geografía UNAM*, *100*, 1–17.

Alcántara-Ayala, I., Esteban-Chávez, O., & Parrot, J. F. (2006). Landsliding related to land-cover change: A diachronic analysis of hillslope instability distribution in the sierra Norte, Puebla, Mexico. *Catena*, *65*(2), 152–165. https://doi.org/10.1016/j.catena.2005.11.006

Alcántara-Ayala, I., López-García, J., & Garnica, R. J. (2012). On the landslide event in 2010 in the monarch butterfly biosphere reserve, Angangueo, Michoacán, Mexico. *Landslides*, *9*(2), 263–273.

APFM. (2006). *Social aspects and stakeholder involvement in integrated flood management.* WMO/GWP Associated Programme on Flood Management, Technical Document No. 4, Flood Management Policy Series, WMO-No. 1008.

Altez, R. (2010). Lo que puede aprenderse de un desastre de muertes masivas: La experiencia de Vargas [What can be learned from a mass death disaster: The Vargas experience]. In J. L. López (Ed.), *Lecciones Aprendidas del Desastre de Vargas* (pp. 127–144). Edición Fundación Polar-UCV.

Bai, X., van der Leeuw, S., O'Brien, K., Berkhout, F., Biermann, F., Brondizio, E. S., Cudennec, C., Dearing, J., Duraiappah, A., Glaser, M., Revkin, A., Steffen, W., & Syvitski, J. (2016). Plausible and desirable futures in the Anthropocene: A new research agenda. *Global Environmental Change*, *39*, 351–362. https://doi.org/10.1016/j.gloenvcha.2015.09.017

Barnard, P. L., Owen, L. A., Sharma, M. C., & Finkel, R. C. (2001). Natural and human-induced landsliding in the Garhwal Himalaya of northern India. *Geomorphology*, *40*(1–2), 21–35. https://doi.org/10.1016/S0169-555X(01)00035-6

Barredo, J. I. (2007). Major flood disasters in Europe: 1950–2005. *Natural Hazards*, *42*(1), 125–148. https://doi.org/10.1007/s11069-006-9065-2

Berardi, U. (2012). Sustainability assessment in the construction sector: Rating systems and rated buildings. *Sustainable Development*, *20*(6), 411–424. https://doi.org/10.1002/sd.532

Berardi, U., GhaffarianHoseini, A., & GhaffarianHoseini, A. (2014). State-of-the-art analysis of the environmental benefits of green roofs. *Applied Energy*, *115*, 411–428. https://doi.org/10.1016/j.apenergy.2013.10.047

Blaikie, P., & Brookfield, H. (1987). *Land degradation and society.* Methuen.

Borga, M., Tonelli, F., & Selleroni, J. (2004). A physically based model of the effects of forest roads on slope stability. *Water Resources Research*, *40*(12), 1–9. https://doi.org/10.1029/2004WR003238

Boyer, R. E. (1975). *The great flood: Life and society in Mexico (1629–1638) (La gran inundación: Vida y Sociedad en México (1629–1638))*. SEP (In Spanish).

Bruschi, V. M., Bonachea, J., Remondo, J., Gómez-Arozamena, J., Rivas, V., Barbieri, M., Capocchi, S., Soldati, M., & Cendrero, A. (2013). Land management versus natural factors in land instability: Some examples in Northern Spain. *Environmental Management*, *52*(2), 398–416. https://doi.org/10.1007/s00267-013-0108-7

Cannon, S. H., Kirkham, R. M., & Parise, M. (2001). Wildfire-related debris-flow initiation processes, Storm King Mountain, Colorado. *Geomorphology*, *39*(3–4), 171–188. https://doi.org/10.1016/S0169-555X(00)00108-2

Cardona, O., Van Aalst, M., Birkmann, J., Fordham, M., McGregor, G., Perez, R., Pulwarty, R. S., Schipper, E. L. F., Tan Sinh, B., Décamps, H., Keim, M., Davis, I., Ebi, K. L., Lavell, A., Mechler, R., Murray, V., Pelling, M., Pohl, J., Oliver-Smith, A., & Thomalla, F. (2012). Determinants of risk: Exposure and vulnerability. In C. Field, V. Barros, T. Stocker, & Q. Dahe (Eds.), *Managing the risks of extreme events and disasters to advance climate change adaptation: Special report of the intergovernmental panel on climate change* (pp. 65–108). Cambridge University Press.

Crowther, T. W., Glick, H. B., Covey, K. R., Bettigole, C., Maynard, D. S., Thomas, S. M., Smith, J. R., Hintler, G., Duguid, M. C., Amatulli, G., Tuanmu, M. N., Jetz, W., Salas, C., Stam, C., Piotto, D., Tavani, R., Green, S., Bruce, G., Williams, S. J.,. . . Bradford, M. A. (2015). Mapping tree density at a global scale. *Nature*, *525*(7568), 201–205. https://doi.org/10.1038/nature14967

Crutzen, P. J. (2002). Geology of mankind-the. *Nature*, *415*(6867), 23. https://doi.org/10.1038/415023a

Crutzen, P. J., & Stoermer, E. F. (2000). The "anthropocene". *IGBP Newsletter*, *41*, 12.

Curtis, P. G., Slay, C. M., Harris, N. L., Tyukavina, A., & Hansen, M. C. (2018). Classifying drivers of global forest loss. *Science*, *361*(6407), 1108–1111. https://doi.org/10.1126/science.aau3445

De Feo, G., Antoniou, G., Fardin, H., El-Gohary, F., Zheng, X., Reklaityte, I., Butler, D., Yannopoulos, S., & Angelakis, A. N. (2014). The historical development of sewers worldwide. *Sustainability*, *6*(6), 3936–3974. https://doi.org/10.3390/su6063936

Di Martire, D., Rosa, M., Pesce, V., Santangelo, M., & Calcaterra, D. (2012). Landslide hazard and land management in high-density urban areas of Campania region, Italy. *Natural Hazards and Earth System Sciences*, *12*, 905–926. https://doi.org/10.5194/nhess-12-905-2012

European Environment Agency. (2016). *River floods* [WWW Document]. https://www.eea.europa.eu/data-and-maps/indicators/river-floods-2/assessment

Evans, E. P., Ashley, R., Hall, J. W., Penning-Rowsell, E. C., Saul, A., Sayers, P. B., Thorne, C. R., & Watkinson, A. R. (2004). *Foresight future flooding, scientific summary* (p. 1). Future Risks and their Drivers Office of Science and Technology. http://www.foresight.gov.uk/OurWork/CompletedProjects/Flood/index.asp

Food and Agriculture Organization. (2018). *The state of the world's forests 2018*. Forest Pathways to Sustainable Development, FAO.

Fletcher, T. D., Andrieu, H., & Hamel, P. (2013). Understanding, management and modelling of urban hydrology and its consequences for receiving waters: A state of the art. *Advances in Water Resources*, *51*, 261–279. https://doi.org/10.1016/j.advwatres.2012.09.001

Geertsema, M., Egginton, V. N., Schwab, J. W., & Clague, J. J. (2007). Landslides and historic climate in Northern British Columbia. In R. McInnes, J. Jakeways, M. Fairbank, & R. Methie (Eds.), *Landslides and climate change: Challenges and solutions* (pp. 9–16). Taylor & Francis.

Gill, S. E., Handley, J. F., Ennos, A. R., & Pauleit, S. (2007). Adapting cities for climate change: The role of the green infrastructure. *Built Environment*, *33*(1), 115–133. https://doi.org/10.2148/benv.33.1.115

Giordan, D., Cignetti, M., Baldo, M., & Godone, D. (2017). Relationship between man-made environment and slope stability: The case of 2014 rainfall events in the terraced landscape of the Liguria region (northwestern Italy). *Geomatics, Natural Hazards and Risk*, *8*(2), 1833–1852. https://doi.org/10.1080/19475705.2017.1391129

Glade, T. (2003). Landslide occurrence as a response to land use change: A review of evidence from New Zealand. *Catena*, *51*(3–4), 297–314. https://doi.org/10.1016/S0341-8162(02)00170-4

Görg, C., Plank, C., Wiedenhofer, D., Mayer, A., Pichler, M., Schaffartzik, A., & Krausmann, F. (2020). Scrutinizing the great acceleration: The anthropocene and its analytic challenges for social-ecological transformations. *Anthropocene Review*, *7*(1), 42–61. https://doi.org/10.1177/2053019619895034

Gruntfest, E., & Handmer, J. (2001). Dealing with flash floods: Contemporary issues and future possibilities. In E. Gruntfest, & J. Handmer (Eds.), *Coping with flash floods* (pp. 3–10). Kluwer Academic Publishers.

Habitat III. (2017). *The new urban agenda*. Retrieved January 12, 2020, from http://habitat3.org/the-new-urban-agenda

Hansen, M. C., Potapov, P. V., Moore, R., Hancher, M., Turubanova, S. A., Tyukavina, A., Thau, D., Stehman, S. V., Goetz, S. J., Loveland, T. R., Kommareddy, A., Egorov, A., Chini, L., Justice, C. O., & Townshend, J. R. G. (2013). High-resolution global maps of 21st-century forest cover change. *Science*, *342*(6160), 850–853. https://doi.org/10.1126/science.1244693

Hapuarachchi, H. A. P., Wang, Q. J., & Pagano, T. C. (2011). A review of advances in flash flood forecasting. *Hydrological Processes*, *25*(18), 2771–2784. https://doi.org/10.1002/hyp.8040

IPBES. (2018). *Summary for policymakers of the assessment report on land degradation and restoration of the intergovernmental science policy platform on biodiversity and ecosystem services* (R. Scholes, L. Montanarella, A. Brainich, N. Barger, B. ten Brink, M. Cantele, B. Erasmus, J. Fisher, T. Gardner, T. G. Holland, F. Kohler, J. S. Kotiaho, G. Von Maltitz, G. Nangendo, R. Pandit, J. Parrotta, M. D. Potts, S. Prince, M. Sankaran, & L. Willemen, Eds., p. 44). IPBES Secretariat.

Intergovernmental Panel on Climate Change] (2014). *Impacts, adaptation and vulnerability: Working group II contribution to the fourth assessment report of the IPCC*. Cambridge University Press.

Jaboyedoff, M., Michoud, C., Derron, M., Voumard, J., Leibundgut, G., Sudmeier-Rieux, K., Nadim, F., & Leroi, E. (2016, June 12–19). Human-Induced landslides: Toward the analysis of anthropogenic changes of the slope environment. In S. Aversa, L. Cascini, L. Picarelli, & C. Scavia (Eds.), *Proceedings of the landslides and engineered slopes: Experience, theory and practice of the 12th international symposium on landslides* (p. 217). CRC Press.

Jiménez Cisneros, B. E., Oki, T., Arnell, N. W., Benito, G., Cogley, J. G., Döll, P., Jiang, T., Mwakalila, S. S., Fischer, T., Gerten, D., Hock, R., Kanae, S., Lu, X., Mata, L. J., Pahl-Wostl, C., Strzepek, K. M., Su, B., & van den Hurk, B. (2014). *Freshwater resources. Climate change 2014: Impacts, adaptation, and vulnerability. Part A: Global and sectoral aspects* (pp. 229–269). Contribution of Working Group II to the Fifth Assessment Report of the Intergovernmental Panel on Climate Change.

Jonkman, S. N. (2005). Global perspectives on loss of human life caused by floods. *Natural Hazards*, *34*(2), 151–175. https://doi.org/10.1007/s11069-004-8891-3

Kelley, C. P., Mohtadi, S., Cane, M. A., Seager, R., & Kushnir, Y. (2015). Climate change in the fertile crescent and implications of the recent Syrian drought. *Proceedings of the National Academy of Sciences of the United States of America*, *112*(11), 3241–3246. https://doi.org/10.1073/pnas.1421533112

Kravčik, M., Pokorný, J., Kohutiar, J., Kovác, M., & Tóth, E. (2007). *The new water paradigm—water for the recovery of the climate*. Krupa Print.

Lagmay, A. M., Mendoza, J., Cipriano, F., Delmendo, P. A., Lacsamana, M. N., Moises, M. A., Pellejera, N., Punay, K. N., Sabio, G., Santos, L., Serrano, J., Taniza, H. J., & Tingin, N. E. (2017). Street floods in Metro Manila and possible solutions. *Journal of Environmental Sciences*, *59*, 39–47. https://doi.org/10.1016/j.jes.2017.03.004

Lambin, E. F., Turner, B. L., Geist, H. J., Agbola, S. B., Angelsen, A., Bruce, J. W., Coomes, O. T., Dirzo, R., Fischer, G., Folke, C., George, P. S., Homewood, K., Imbernon, J., Leemans, R., Li, X., Moran, E. F., Mortimore, M., Ramakrishnan, P. S., Richards, J. F.,. . . Xu, J. (2001). The causes of land-use and land-cover change: Moving beyond the myths. *Global Environmental Change*, *11*(4), 261–269. https://doi.org/10.1016/S0959-3780(01)00007-3

Lavell, A. (1996). Degradación ambiental, riesgo y desastre urbano. Problemas y conceptos: Hacia la definición de una agenda de investigación M. A. Fernández (Comp.). In *Ciudades en riesgo: Degradación ambiental, riesgos urbanos y desastre*. LA RED/USAID.

Lehner, F., Coats, S., Stocker, T. F., Pendergrass, A. G., Sanderson, B. M., Raible, C. C., & Smerdon, J. E. (2017). Projected drought risk in 1.5°C and 2°C warmer climates. *Geophysical Research Letters*, *44*(14), 7419–7428. https://doi.org/10.1002/2017GL074117

Lewis, S. L., & Maslin, M. A. (2015). Defining the anthropocene. *Nature*, *519*(7542), 171–180. https://doi.org/10.1038/nature14258

Lóczy, D., Czigány, S., & Pirkhoffer, E. (2012). Flash flood hazards. In M. Kumarasamy (Ed.), *Studies on water management issues*. In Tech. https://doi.org/10.5772/28775.

López-García, J., & Alcántara-Ayala, I. (2012). Land-use change and hillslope instability in the monarch butterfly biosphere reserve, Central Mexico. *Land Degradation and Development*, *23*(4), 384–397. https://doi.org/10.1002/ldr.2159

López-Rodríguez, S. R., & Blanco-Libreros, J. F. (2008). Illicit crops in tropical America: Deforestation, landslides, and the terrestrial carbon stocks. *Ambio*, *37*(2), 141–143. https://doi.org/10.1579/0044-7447(2008)37[141:icitad]2.0.co;2

Lyon, B. (2004). The strength of El Niño and the spatial extent of tropical drought. *Geophysical Research Letters*, *31*(21). https://doi.org/10.1029/2004GL020901

MacCurdy, E. (Ed.). (1955). *Notebooks of Leonardo da Vinci*. George Braziller.

Marchi, L., Borga, M., Preciso, E., & Gaume, E. (2010). Characterisation of selected extreme flash floods in Europe and implications for flood risk management. *Journal of Hydrology*, *394*(1–2), 118–133. https://doi.org/10.1016/j.jhydrol.2010.07.017

Meusburger, K., & Alewell, C. (2008). Impacts of anthropogenic and environmental factors on the occurrence of shallow landslides in an alpine catchment (Urseren Valley, Switzerland). *Natural Hazards and Earth System Sciences*, *8*(3), 509–520. https://doi.org/10.5194/nhess-8-509-2008

Millennium Ecosystem Assessment. (2005). *Ecosystems and human well-being: Biodiversity synthesis*. World Resources Institute.

Montz, B. E., & Gruntfest, E. (2002). Flash flood mitigation: Recommendations for research and applications. *Environmental Hazards*, *4*(1), 15–22. https://doi.org/10.3763/ehaz.2002.0402

National Oceanic and Atmospheric Administration. (2010). *Flash flood early warning system reference guide*. University Corporation for Atmospheric Research. http://www.meted.ucar.edu

Olivera, F., & DeFee, B. B. (2007). Urbanization and its effect on runoff in the Whiteoak Bayou watershed, Texas. *American Water Resources Association*, *43*, 170–182.

Papagiannaki, K., Lagouvardos, K., Kotroni, V., & Bezes, A. (2015). Flash flood occurrence and relation to the rainfall hazard in a highly

urbanized area. *Natural Hazards and Earth System Sciences*, *15*(8), 1859–1871. https://doi.org/10.5194/nhess-15-1859-2015
Persichillo, M. G., Bordoni, M., & Meisina, C. (2017). The role of land use changes in the distribution of shallow landslides. *Science of the Total Environment*, *574*, 924–937. https://doi.org/10.1016/j.scitotenv.2016.09.125
Raynal-Villaseñor, J. A. (1987, April). The remarkable hydrological works of the Aztec civilization. In *Water for the future: Hydrology in perspective*. IAHS Publication, 164 (Proceedings of the Rome Symposium.
Ren, X., Hong, N., Li, L., Kang, J., & Li, J. (2020). Effect of infiltration rate changes in urban soils on stormwater runoff process. *Geoderma*, *363*. https://doi.org/10.1016/j.geoderma.2019.114158, PubMed: 114158
Rockström, J., Steffen, W., Noone, K., Persson, Å., Chapin, F. S. I., III, Lambin, E., Lenton, T. M., Scheffer, M., Folke, C., Schellnhuber, H. J., Nykvist, B., de Wit, C. A., Hughes, T., van der Leeuw, S., Rodhe, H., Sörlin, S., Snyder, P. K., Costanza, R., Svedin, U.,. . . Foley, J. (2009). Planetary boundaries: Exploring the safe operating space for humanity. *Ecology and Society*, *14*(2), 32. https://doi.org/10.5751/ES-03180-140232
Sah, M. P., & Mazari, R. K. (1998). Anthropogenically accelerated mass movement, Kulu Valley, Himachal Pradesh, India. *Geomorphology*, *26*(1–3), 123–138. https://doi.org/10.1016/S0169-555X(98)00054-3
Salcedo, D. A. (2000). Los flujos torrenciales catastróficos de Diciembre de 1999. En *el estado Vargas y en Caracas: Características y lecciones aprendidas: Memorias XVI seminario venezolano de geotecnica* (pp. 128–175). Caracas.
Sánchez-Rodríguez, R., Seto, K., Simon, D., Solecki, W., Krass, F., & Laumann, G. (2005). *Science plan*. Urbanization and Global Environmental Change, IHDP Report No. 15.
Sene, K. (2013). *Flash floods: Forecasting and warning*. Springer. https://doi.org/10.1007/978-94-007-5164-4
Seneviratne, S. I., Nicholls, N., Easterling, D., Goodess, C. M., Kanae, S., Kossin, J., Luo, Y., Marengo, J., McInnes, K., Rahimi, M., Reichstein, M., Sorteberg, A., Vera, C., & Zhang, X. (2012). Changes in climate extremes and their impacts on the natural physical environment. In C. B. Field et al. (Eds.), *Managing the risks of extreme events and disasters to advance climate change adaptation* (pp. 109–230). A Special Report of Working Groups I and II of the Intergovernmental Panel on Climate Change (IPCC). Cambridge University Press.
Shan, W., Hu, Z., Guo, Y., Zhang, C., Wang, C., Jiang, H., Liu, Y., & Xiao, J. (2015). The impact of climate change on landslides in Southeastern of high-latitude permafrost regions of China. *Frontiers in Earth Science*, *3*(7). https://doi.org/10.3389/feart.2015.00007
Špitalar, M., Gourley, J. J., Lutoff, C., Kirstetter, P., Brilly, M., & Carr, N. (2014). Analysis of flash flood parameters and human impacts in the US from 2006 to 2012. *Journal of Hydrology*, *519*(A), 863–870. https://doi.org/10.1016/j.jhydrol.2014.07.004

Steffen, W., Broadgate, W., Deutsch, L., Gaffney, O., & Ludwig, C. (2015a). The trajectory of the anthropocene: The great acceleration. *Anthropocene Review*, *2*(1), 81–98. https://doi.org/10.1177/2053019614564785

Steffen, W., Crutzen, J., & McNeill, J. R. (2007). The Anthropocene: Are humans now overwhelming the great forces of nature? *Ambio*, *36*(8), 614–621. https://doi.org/10.1579/0044-7447(2007)36[614:taahno]2.0.co;2

Steffen, W., Richardson, K., Rockström, J., Cornell, S. E., Fetzer, I., Bennett, E. M., Biggs, R., Carpenter, S. R., de Vries, W., de Wit, C. A., Folke, C., Gerten, D., Heinke, J., Mace, G. M., Persson, L. M., Ramanathan, V., Reyers, B., & Sörlin, S. (2015b). Sustainability planetary boundaries: Guiding human development on a changing planet. *Science*, *347*(6223). https://doi.org/10.1126/science.1259855

Stoppani, A. (1873). *Corsa di geologia*. Bernardoni and Brigola.

Trenberth, K. E., & Hurrell, J. E. (1999). commentary and analysis. *Bulletin of the American Meteorological Society*, *80*(12), 2721–2728. https://doi.org/10.1175/1520-0477-80.12.2721

UN DESA. (2019). *World population prospects 2019: Highlights* (ST/ESA/SER.A/423). United Nations Department of Economic and Social Affairs Population Division.

UN Environment. (2019). *Global environment outlook—GEO-6: Healthy planet, healthy people*. UNE. https://doi.org/10.1017/9781108627146

UNFCCC. (2015). *Adoption of the Paris Agreement, COP21*. https://unfccc.int/resource/docs/2015/cop21/eng/l09r01.pdf

UN-Habitat. (n.d.). *Slum Almanac 2015/2016 – tracking the lives of slum dwellers*. UN-Habitat. Retrieved March 31, 2020, from https://unhabitat.org/wp-content/uploads/2016/02-old/Slum%20Almanac%202015-2016_EN.pdf

UNISDR. (United Nations International Strategy for Disaster Reduction). (2004). *Living with risk: A global review of disaster reduction initiatives*. UNISDR.

UNISDR (United Nations International Strategy for Disaster Reduction). (2005). Hyogo framework for action 2005–2015: Building the resilience of nations and communities to disasters. In *Extract from the final report of the world conference on disaster reduction* (A/CONF. 206/6, Vol. 380). The United Nations International Strategy for Disaster Reduction.

UNISDR (United Nations International Strategy for Disaster Reduction). (2015a). *Sendai framework for disaster risk reduction*. Proceedings of the 3rd United Nations World Conference on DRR, Sendai, pp. 2015–2030.

UNISDR (United Nations International Strategy for Disaster Reduction). (2015b). *Human cost of weather-related disasters 1995–2015*. UNISDR.

UNISDR. (United Nations International Strategy for Disaster Reduction). (2017). *Report of the open-ended intergovernmental expert working group on indicators and terminology relating to disaster risk reduction*. UNISDR.

United Nations. (2015). *Transforming our world: The 2030 agenda for sustainable development*, A/RES/70/1. United Nations.

United Nations. (2018). *World urbanization prospects: The 2018 revision: Key facts*,. United Nations, Department of Economic and Social Affairs, Population Division. https://population.un.org/wup/Publications/Files/WUP2018-KeyFacts.pdf

United Nations Office for Disaster Risk Reduction. (2021). *GAR special report on drought 2021*. UN.

Van Beek, L. P. H., & Van Asch, T. W. J. (2004). Regional assessment of the effects of land-use change on landslide Hazard by means of physically based modelling. *Natural Hazards*, *31*(1), 289–304. https://doi.org/10.1023/B:NHAZ.0000020267.39691.39

Wheater, H., & Evans, E. (2009). Land use, water management and future flood risk. *Land Use Policy*, *26*, S251–S264. https://doi.org/10.1016/j.landusepol.2009.08.019

World Bank. (2017). *World development indicators 2017*. The World Bank Group.

World Commission on Dams. (2000). *Dams and development: A new framework for decision-making*. Earthscan Publications.

World Meteorological Organization. (2009). *Guide to hydrological practices Volume II management of water resources and application of hydrological practices* (6th ed.). WMO No. 168.

World Water Assessment Programme. (2012). *The United Nations world water development report 4: Managing water under uncertainty and risk*. UNESCO.

Xu, C., Cui, Y., Xu, X., Bao, P., Fu, G., & Jiang, W. (2019). An anthropogenic landslide dammed the Songmai River, a tributary of the Jinsha River in Southwestern China. *Natural Hazards*, *99*(1), 599–608. https://doi.org/10.1007/s11069-019-03740-y

Zalasiewicz, J., Crutzen, P., & Steffen, W. (2012). The anthropocene. In F. M. Gradstein et al. (Eds.), *A geological time scale 2012* (pp. 1033–1040). Elsevier.

PART IV

People's response to and resilience during and after disasters

CHAPTER 11

People's behaviour in times of disaster

This chapter will be about people's behaviour in times of disasters (Figure 11.1). The previous chapters explained how disasters happen and identified the factors that determine their consequences, that is, types and gravity of hazards, exposure of people to hazards, and their vulnerabilities and capacities. This chapter will focus on people's behaviour and how they influence the response of people in times of disasters, that is, activities and actions that are taken by individual, groups and organisations upon perceiving the threat of a hazard, or during and immediately after occurrence of a hazard or a disaster event. Discussions in the existing literature on the common myths and misconceptions about people's behaviour in times of disasters being anti-social as a means to overcome or escape a crisis situation, or recover from the negative impacts of disasters, will be reiterated. And instead, this chapter will reassert more common observations supported by scientific research pointing to people's behaviour in times of disasters as more prosocial, altruistic and based on common good.

Fig 11.1
Evacuees assisted by local volunteers after the eruption of Bulusan Volcano in Sorsogon, Philippines, on 21 February 2011.
Source: photo courtesy of Jake Rom Cadag

DOI: 10.4324/9781315469614-16

11.1 Questioning the mainstream narrative

During and immediately after the devastation of Typhoon Haiyan (local name Typhoon *Yolanda*) in the Philippines, both local and international news media organisations persistently reported on the situation in the affected areas particularly in the provinces of Samar and Leyte, Central Philippines. One of the highlights of many reports both in live news and newspapers were looting incidences and aggravating problems on peace and order. In one report, a popular local news anchor reported what appears to be a picture of total chaos in the typhoon-affected area:

> At this moment, our fellow citizens are appealing for help particularly for water and food after the devastation of Typhoon Yolanda. People in Tacloban City (capital city of the Province of Leyte, Central Philippines) are now helpless and they need food and water, and this situation explains why looting in many establishments could not be avoided. We talked to the Mayor of Tacloban City, Mayor Martin Romualdez, and he said that law and order must be restored, and that they are asking the National Government to send military personnel to contain the situation.
>
> (GMA News TV, 2013)

During the live news report on television, additional text is displayed saying that "several establishments are being looted because of hunger". The live news report was then uploaded online with a title "Hungry, desperate victims of Typhon Yolanda resort to looting in Leyte". Other incidents, such as shooting, rape and child abuse, amongst other crimes, were associated with the disaster event, and reported live on television and circulated in online news and social media platforms. These law-breaking conducts by survivors and affected people in the aftermath of Typhoon Haiyan had also been highlighted by media organisations and government authorities during and after Hurricane Katrina devastated the South-eastern United States, particularly in the densely populated city of New Orleans (Figure 11.2). Very recently, looting, particularly mass looting, has become the focus of reports in Palu City, Indonesia, in 2018, after the impact of a major tsunami. This aspect of a disaster situation characterised by chaos, looting, panic behaviour and some anti-social conducts resulting to crime is a common picture of people's behaviour highlighted by news media during and after disasters, especially major disasters.

But are these observations about people's behaviour in times of disasters factually accurate? Do people's behaviour, from being socially conformed and law-abiding in normal days or everyday life change to anti-social and law-breaking in times of disasters? If yes, in terms of occurrences, are they widespread or just isolated cases? And if no, what are then the predominant people's behaviour in times of disasters? A scientific or academic enquiry to these questions is necessary to scrutinise our understanding of people's behaviour in times of disasters.

Fig 11.2
Storefront in New Orleans, Louisiana, after Hurricane Katrina in 2005.
Source: Sheridan (2005)

More importantly, the answers to these questions are vital in the practice of DRR, particularly in disaster preparedness and crisis management. There are, therefore, intellectual and practical reasons for enquiring on people's behaviour in times of disasters. The intellectual reasons refer to the opportunity to study people's behaviour from different angles, perspectives and frameworks. The practical benefit points to the opportunity to understand how people behave and how such behaviour manifests in times of disasters. This knowledge is useful to reduce disaster risk, particularly in planning disaster preparedness.

Many scientific studies have produced findings and conclusions about people's behaviour in times of disasters that can be used in reducing disaster risk. For example, Frey et al. (2011), in their analysis of behaviour and chance of survival of both survivors and victims in the famous Titanic disaster, concluded that "under extreme situations, the behaviour of human beings is not random or inexplicable, but can be explored and, at least in part, explained by economic analysis". And indeed, Quarantelli (1999, p. 2) noted that "typical patterns of behaviour at the individual, organizational, community and mass media levels" can be observed and studied in times of disaster. This knowledge could save people's lives if properly used in planning and implementing actions for disaster preparedness and crisis management. In the case of Typhoon Haiyan, for instance, prior to the impact of the storm surge brought by the typhoon to identified coastal communities, foreseeing people's behaviour before, during and after the disaster, authorities and disaster managers could have provided hazard and risk information that have positive effects on people's preparedness and response (i.e. readiness for evacuation, preparation of emergency kit (or go bag), prepositioning of relief goods and other possible actions for disaster preparedness).

It can be argued that the lack of understanding of people's behaviour has aggravated negative impacts of many disasters that had happened in the past. At the very least, the prevalence of misconceptions and myths, and lack of consideration for people's behaviour in times of disasters, did not help in DRR. Yet there is so much to learn from already existing wide and rich literature on the topic (e.g. a hundred-year research on the topic since the study by Samuel Prince of the Halifax disaster in 1917; see further discussion about this in the succeeding sections of this chapter). Thus, this chapter provides a review of literature and reasserts scientific findings about people's behaviour. It brings into the discussion the recent and relevant research findings on the topic in light of major disasters that happened in the last decade. Recommendations for DRR that consider people's behaviour in times of disaster are also provided.

11.2 What is people's behaviour?

Referring to a dictionary definition, human 'behaviour' is "the way in which one acts or conducts oneself, especially towards others" or a "particular situation or stimulus" (Merriam-Webster, 2019). Such actions and conducts are considered by psychologists as manifestations of the mind, be it the conscious or unconscious mind, based on an individual personality and social situation (Rhodewalt, 2008). Therefore, an enquiry on human behaviour is an enquiry into the human mind. Psychologists refer to that method as 'introspection', where the human mind is considered as a stage or arena that could be examined for the purpose of understanding how mental processes lead one to act or behave in ways that may be similar or different to others (Baum, 2017, p. 6). In the landmark publication of the book *Psychology as the behaviourist views* by John B. Watson (1879–1958), in 1913, introspective psychology was criticised for being too subjective and unreliable as a scientific method. Methodological approaches, such as objective and comparative psychology, have then been developed to further study human behaviour following a more objective and scientific approach (i.e. laboratory experiments and statistical analysis). All these efforts in the last century to understand human behaviour suggests, as Baum (2017) emphasised, that a science of behaviourism is "possible".

One of the debates on behaviourism in the social science, particularly in psychology, is the explanation of human behaviour, whether they are based on free will or determined by factors that are external to the individual. Deterministic approach suggests that human behaviour is shaped and eventually caused by environmental factors (Ogletree & Oberle, 2008). In this argument, behaviour can be explained by known or unknown variables that lie in the environment or by the genetic ancestry of that person. Therefore, regardless of the consequences of one's actions being good or bad to other people or the society, a person's behaviour is not entirely his or her own will but is motivated by many factors in the environment (i.e. socio-economic condition, cultural background, consequences of other's actions etc.) or by a person's biology, hence the term 'environmental and biological determinism' (Baer et al., 2008). Contrary to deterministic approach to human behaviour, humanists believe that a person has a human agency or the capacity to choose and make actions based on free will. In this view, people are free agents and can be held accountable for their actions because "they are a partial first cause of their own actions" (Myers, 2008, p. 32). Behaviourists have proposed middle-ground perspectives to compromise determinism and free will. One proposition is soft determinism, where people can invoke their own "conscious choices between different courses of actions" as a manifestation of free will but only within the bounds of their past and present environmental situation and experience, and genetic inheritance (Sappington, 1990, p. 20).

This debate on human behaviour is also at the core of discussions in many social science disciplines. In geography, particularly in geographical philosophy, environmental determinism is a deep-rooted theory that is heavily influenced, and still continues to influence, many geographical studies. The main argument of this view is that human cultures and organisations, especially those activities that engage with the natural environment (e.g. settlement pattern, agricultural practices, and language similarities and differences) are determined by environmental conditions (Ernste & Philo, 2009). In the past decades, however, environmental determinism has been resolutely rejected in geography and in many disciplines because of its utility in justifying racism, colonialism and imperialism (Gilmartin, 2009). Alternative frameworks that arguably remain fundamentally based on environmental determinism, such as possibilism and probabilism, were proposed and developed to partially accommodate or emphasise the influence of human agency in the human-environment relationship (Ernste & Philo, 2009). In contradiction to arguments of environmental determinism and in agreement to that of free will, behavioural and humanistic geographers argued that people are agents capable of making their own decisions with or without the influence of the environment and social structures (Gregory, 1981).

For the clarity of the use of concept in this paper, 'people's behaviour' is used as a synonymous term to 'human behaviour' but pertaining not only to a single person but a group of persons or people. People's behaviour captures both an individual's and a group's behaviour in times of disasters. The concept also pertains to as a singular noun, hence the used of 'behaviour' instead of 'behaviours', to avoid association with issues related to multiple behaviour disorder and other associated terms. The concept of behaviour, as defined and explained from different perspectives in social science disciplines, is fundamental to the enquiry on people's behaviour in times of disasters. In relation to the discussion on the nature of people's behaviour are decisions, actions and responses of people that are environmentally determined or based purely on an individual choice (i.e. free will and human agency), or a combination of both, as proposed in the idea of soft determinism, possibilism and probabilism.

Understanding people's behaviour in a disaster situation necessitates an understanding of everyday life in specific spatio-temporal context. Everyday life is characterised by the ordinariness of human situation or routine activities of people. Everyday life is defined in the context of everyday social interactions that are characterised by mundane, unexciting and uninteresting happenings in people's lives (Orleans, 2000). The definition of 'everyday life' can then be elusive and multiple, depending on one's perspective or experience of 'ordinary', 'routine activity' and 'mundane happenings', amongst other terms that define everyday life. But perhaps everybody agrees that going to work or schools during weekdays, eating three times (or more) a day, doing household chores and doing many other activities are examples of activities in the everyday life. These everyday activities may be conducted in different or similar manners across cultures and contexts, a topic of great interest in cultural and behavioural geography (refer to Holloway & Hubbard, 2013). And it is in the conduct of these everyday life activities that people's behaviour in everyday life can be observed. A deeper discussion on 'everyday life' and associated terms that define it is beyond the scope of this paper, and readers are referred to relevant references about the topic (i.e. Eyles (1989) and *Journal of Mundane Behaviour* (2000–2004)). But it is important to get a sense of people's behaviour in everyday life, however contentious academically the task is, to get a sense of people's behaviour in a disaster situation.

It may help better appreciate everyday life when contrasted with situations that appear to be extraordinary or do not represent the everyday life. At the scale of the community or country, these extraordinary situations can refer to war or disaster events, and similar situations that do not happen regularly or could have too many impacts, both positive and negative, in people's lives. During and after major earthquake or flooding events, for instance, everyday activities, such as going to work and school, and performing household chores, are disturbed or totally disrupted. In a disaster situation, therefore, the order of activities that is usually experienced or conducted by people in everyday life are not the same or no longer possible to implement because of the circumstances where people's lives are threatened and physical disorganisation affects many aspects of their lives (e.g. school classes are suspended, roads and houses are partially or totally destructed, and agriculture and people's livelihoods are disrupted).

Therefore, during or after a disaster or experience of a hazard, people cope with the situation or make decisions and actions to go back to normal or move forward to a better condition (e.g. recover or find a more secure shelter and livelihood). Observations from previous disasters suggest that there is a social tendency for individuals, groups or societies affected by disasters to return to normalcy or to the scenario of everyday life where lives of people are protected, and their goals and aspirations, or of the society as a whole, are better pursued. Theoretically, this tendency to protect oneself and ensure common or collective survival by an individual or group under threat or at risk agrees with the principles of self-preservation (see Karni & Schmeidler, 1986). In practice, returning to normal or moving forward in a better condition during or after disaster is captured in the principles of 'build back better' which is one of the four main priorities for action of the SFDRR (UNDRR, 2015).

People's behaviour in times of disasters may then refer to decisions, actions or conducts of individuals or groups in response to the threat of hazards or impacts of disasters in a specific place and period. It should be emphasised that people's behaviour in times of disasters are geared towards protection of individual and group, or of a society, in general (i.e. self-preservation, collective survival and build back better principles). The analysis of people's behaviour in times of disasters, however, is not straightforward and often require some theoretical framing. For one, are people's behaviours exhibited in disaster situations fundamentally different or similar to those exhibited in everyday life? Who are the individuals and groups that are being referred to and what is their common or distinct behaviour in times of disasters? These questions are tackled in the succeeding sections where people's behaviour, as observed in previous disasters and interpreted from different perspectives, are discussed.

11.3 People's behaviour in times of disasters

The analysis of people's behaviour in facing hazards and disasters is not limited at the time and place of a particular disaster event. The scale of analysis can be temporally and geographically wider (i.e. analysis beyond the actual disaster occurrence in the impact area and at pre- and post-disaster context). For instance, the loss of lives and significant damages to properties in Florida (USA) in 2004, after the impacts of hurricanes, are partly explained by the lack of culture of disaster preparedness (Kapucu, 2008). Many households and individuals were found complacent in response to hazards and disasters because of their level of awareness of hazards and experiences of previous disasters (Kapucu, 2008, p. 526). In this study, the analysis of people's behaviour does not focus on the reaction of people as disaster happens but on what people do in preparation for a disaster that is likely to happen in the future or in response to a recent or past experience of disasters.

People's behaviour in facing hazards and disasters can also be analysed by looking at some or many aspects of culture and human activities (Bankoff, 2017). The analysis of people's behaviour can extend to the analysis of human adjustments to hazards, which can be observed in terms of established settlement patterns, choice of location of human activities, type of architecture and construction materials of structures, and characteristics of livelihood. Some or many observable characteristics of society may be associated to people's awareness and experiences of hazards and disasters in the past or even speculation of their occurrences in the present time or in the future. In earthquake-prone regions, such as Gujarat and Kashmir, Jigyasu (2008, p. 77) found out that vernacular structures make use of building techniques (i.e. such as *Taq* (timber-laced masonry bearing wall) and *Dhajji Dewari* (timber frame with masonry infill)), making them more resilient to earthquakes compared to modern structures. The traditional architecture of the structures also appears to be more responsive to the space requirements of people and more adapted to the local climate. The analysis of people's behaviour in times of disasters can then be situated in specific spatio-temporal and cultural contexts.

One of these contexts is people's behaviour before, during and immediately after the actual hazard or disaster event, where reactions and responses of an individual, organisations and the community as they try to cope and survive are observed. Quarantelli (1999) has provided a summary of observation of this behaviour in times of disasters based on what he claimed to be 50 years of social science research on the topic at the time. He organised his observations into four "social times" or phases of a disaster (i.e. mitigation, preparedness, response and recovery), arguing that people's behaviour is fundamentally different but related in each phase. He also deconstructed the generic term 'people' or 'human' behaviour into four social levels; that is, individual, organisational, community and mass media (as surrogate to society) behaviour (Quarantelli, 1999, p. 2) (Table 11.1). Quarantelli (1999) noted, however, that these observations are generalisations and are intended only to provide an overall picture of inherently complex people's behaviour in times of disasters.

Since the 1950s, the analysis of people's behaviour in times of disasters expanded to investigation of many related concepts (e.g. panic). The behaviour of different people involved (i.e. stakeholders, such as individuals, groups, communities, and government and non-governmental organisations, including humanitarian organisations) is scrutinised in many studies. These studies have been done at different spatio-temporal scales where people's behaviour is examined before, during and after disasters at different geographical scales (i.e. international and national level, down to local communities and actual impact areas). Two key institutions on disaster studies in the United States have conducted extensive research on the topic, that is, the Committee on Disaster Studies of the National Academy of Sciences – National Research Council (Washington, DC) and Disaster Research Center (DRC) of the University of Delaware (Newark, Delaware). The discussion on specific topics on disaster-related materials focused on social and behavioural science aspects of disasters are beyond the scope of this paper. Readers are referred to consult the research by E.L. Quarantelli and his colleagues at Disaster Research Center.

For the purpose of this paper, it is important to elaborate on the influence of the nature of hazard to people's behaviour, particularly on the emergent behaviour. The nature of hazard and magnitude of disaster involved appear to have a great influence on people's behaviour,

Table 11.1
Summary of people's behaviour in times of disasters based on 50 years of social science research
Source: derived from **Quarantelli (1999)**

	Mitigation (before disaster, normal days)	*Preparedness (before disaster, probability of hazard occurrence or disaster is perceived)*	*Response (during disaster)*	*Recovery (after disaster)*
Individual	Typically not much interested or concerned about disasters	React rationally and in a socially oriented manner Take warnings seriously and make actions (i.e. evacuation)	React very well Rarely engage in panic flight and looting	Rarely experience lasting behavioural consequences, including post-traumatic stress disorder
Organisational	Seldom engage in activities towards mitigation	Focused on written plans, which is less important compared to actual planning process, with the different stakeholders involved	Difficulty in coping with their own crisis situation Encounter issues on inter-organisational coordination, prohibiting efficient response when providing assistance	Engage in proactive planning to improve crisis management Dependent on the leadership in realising proactive plans
Community	Give low priority to community-wide mitigation activities Exemption may be observed in communities that are usually affected by disasters or at risk or are supported by government programmes on DRR	Encounter coordination and planning issues due to pre-disaster relationship problems and differences between individuals and organisations	Emergent behaviour towards self-protection and collective survival amongst individuals and organisations are observed depending on the extent or magnitude of disasters	New and old problems are uncovered, making disaster recovery a challenging task
Mass media	Perceive disasters as low probability event Give little attention to mitigation activities	Play a role of both insider and outsider when providing report and warning about the threats or hazards	Provide useful information but often an incomplete and unbalanced picture of the crisis event	Tend to focus on conflict and out-of-the-ordinary situations

particularly during and immediately after the crisis. Natural hazards can be loosely categorised in terms of onset of occurrence as rapid and slow. Rapid-onset hazards include earthquake and volcanic eruption hazards (i.e. tsunami, groundshaking, pyroclastic flow etc.) and hydrometeorological hazards (typhoon, floods, landslides, storm surges and tornadoes). On the one hand, earthquakes and tornadoes are some of the rapid-onset hazards that have no forewarnings or provide limited windows for planning at the individual or organisational level. Often, there is very little lead time to acquire sufficient information about the intensity, magnitude and location or place of impact of the hazard. On the other hand, typhoons and volcanic eruptions (depending on the type and nature of eruption), can be detected earlier, and sufficient information may be gathered for early warning purposes. Still, these hazards cannot be predicted with complete certainty, and there is often improbability about their potential impacts and consequences. And this is why most early warning systems are designed to detect potential occurrence of these rapid-onset hazards in a specific place and period (Villagrán de León, 2012). Slow-onset hazards, such drought, sea level rise and other climate change–related stimuli (precipitation and temperature changes), take years or decades to occur, and impacts are not felt immediately. It should be noted, however, that climate changes are also associated with the occurrences of extreme events, such as stronger typhoons and more frequent droughts (see the Fifth Assessment Report of the International Panel for Climate Change accessible at https://www.ipcc.ch/assessment-report/ar5/).

The timing and speed of occurrence of the hazard involved has implication on the nature of resulting crisis and, eventually, on people's behaviour. A crisis is a situation where there is a threat to human lives (and their properties) that needs to be urgently addressed using available information (Boin et al., 2018). In the event of a rapid-onset hazard, such as earthquake characterised by ground shaking and other related hazards (i.e. ground rupture, landslide and tsunami), crisis is expected to rapidly escalate in the affected area. In this scenario, people's behaviour, or the decisions and actions of people in response to the crisis, is constrained in terms of time and space. Depending on the availability and reliability of information, such as emergency warning and evacuation procedure, decisions must be made and actions implemented within a certain amount of time and specific places by concerned people and organisations (i.e. evacuate to safer place, secure family members and friends, implement contingency plans, etc.). It is in this crisis situation, particularly, when sudden hazard is involved that decisions and actions of concerned people, such as being in temporary shock, running away from the perceived impact area and shouting to seek help, amongst other reactions to a hazard, that people's behaviour is conceived as being in a state of panic. Immediately after the crisis or impact of the hazard, people's response, such as securing food and water from any available sources, and scavenging for any useful resources, are also portrayed usually in news reports (Box 11.1 and Figure 11.3) as anti-social and law-breaking.

Box 11.1: Disaster reporting

Fig 11.3
QR code for Box 11.1. In this episode of *Disasters: Deconstructed* podcast, Dr Samantha Montano, Assistant Professor of Emergency Management and Disaster Science at University of Nebraska Omaha, talked about disaster reporting and how it shapes people's perception of risk and disaster. She also discusses the importance of building relationship between scientists and journalists, and why it is necessary in disaster reporting.
Available at https://disastersdecon.podbean.com/e/s3e2-disaster-reporting/

11.4 Scientific rebuttals to myths about people's behaviour in times of disasters

Depictions of people's behaviour in times of disasters being anti-social are considered in the disaster studies literature as myths and misconceptions. They form part of the disaster mythologies, where realities about people's behaviour in times of disasters are distorted by perception based on interpretation of selected facts and unrealistic events (Fischer, 2001). These myths are portrayed and perpetuated by news reports, films and other multimedia platforms. Some historical accounts of disasters mostly based on eyewitness accounts are also supportive of these myths. For instance, in the plague that affected Athens in 430 BCE and the eruption of Vesuvius in 79 CE, Hughes (2013) described responses of people as being in a state of "confusion" and "uncertainty about what to do to survive". Hughes (2013, p. 134) added that people's responses in those two events are a picture of social disorganisation, where "most individuals tried to save themselves and perhaps also those closest to them without concern for the larger community and also without knowing what was the best course of action". Indeed, these earlier records of disasters during the Greek and Roman periods left us stories of great tragedies inspiring many movie productions, drama series and novels.

Films, particularly Hollywood films, about disasters of different kinds usually revolve on a common story. Disasters triggered by natural or human-induced hazards kill people and destroy physical structures and the natural environment. Hazards are usually overwhelming in terms of magnitude and scope, and beyond the capacity of the characters of the film to avoid. To name a few, remarkable movies that successfully depicted this storyline are *Dante's Peak* in 1997 (volcanic eruption; https://www.imdb.com/title/tt0118928/), *Into the Storm* in 2014 (storms and tornadoes; https://www.imdb.com/title/tt2106361/) and *San Andreas* in 2015 (earthquake; https://www.imdb.com/title/tt2126355/). In these films, and in many disaster-related films, the aftermath of disasters results to the breakdown of social order where self-interest over common good is the rule in order to survive, and peace and order are disrupted or no longer exist. These movies tend to perpetuate the notion that the general public or the people (i.e. characters in the movie) are powerless (i.e. shocked or traumatised), and the heroism of the main character and performance of government response are the only solutions to such a hopeless scenario (Mitchell et al., 2000).

To a large extent, these images about people's behaviour are reinforced by news media reports about disasters disseminated in different platforms, such as televisions, newspaper and social media. But this time, images and information are based on real disaster events affecting communities and places that are familiar or known to viewers and listeners. Reports, such as people being in a state of panic, prevalence of looting and problems of peace and order increasing in the affected area, concretise such image of people's behaviour becoming anti-social in times of disasters. The facts about those news reports are difficult to challenge because many of them are factual information documented on videos or photos taken before, during and after a disaster event. However, these images about people's behaviour can also be selected facts or situations, or selected facts of a particular situation. They can even be an isolated but exaggerated fraction of a disaster event. Thus, our understanding of people's behaviour in times of disasters based only on the angle emphasised in news media reports are likely to be incomplete, exaggerated and based on vested interest. Alexander (1980), in his study of the press coverage of the Italian floods of 1966, concluded this:

> The first lesson to be drawn from press coverage of the Italian floods of 1966 is that the preoccupations and preconceptions of writers will often tend to divert them from straightforward accounts of events – perhaps toward a lengthy and tenuous comparison with events elsewhere. Second, there is frequently a strong tendency for reports to decline into cozy accounts of exaggerated human drama – "how the ordinary people are coping" – which further distort the picture. It is clear that journalists will write about the kind of events that they imagine will interest their readers.
>
> (Alexander, 1980, p. 33)

There is a tendency for news media reports to highlight, intentionally or unintentionally, a particular fact or situation to catch the attention of the public (e.g. portrayal of a chaotic situation of a disaster event). Thus, people's behaviour in times of disasters as seen on news reports may not represent the entirety of a disaster event, including those images about people's behaviour. But because they are recorded and shown with the intention to be appealing to the public viewers, those images can become the truth and, later, the basis of understanding of people's behaviour in the face of disasters (Box 11.2 and Figure 11.4). In the last decade, dissemination of online information has become faster and has become

Box 11.2: Fight or flight: coastal community adaptation

In this episode of *The Climate Ready Podcast*, Elizabeth Rush, Assistant Professor of the Practice, Nonfiction Writing Program at Brown University, Providence, Rhode Island, and author of *Rising: Dispatches from the New American Shore*, talked about the experiences of people living in the coastal communities of sea level rise and other extreme weather events, and how they adapt and perceive the risk. This episode highlights how the changing environment due to climate change makes or forces people to make hard decisions that have great impacts on their lives (e.g. fight or flight, or retreat or perish in place).

Fig 11.4
QR code for Box 11.2.
Available at https://www.podbean.com/ew/dir-6bmqd-4e02742

more accessible, particularly to netizens (or citizens of the internet) through social media platforms (i.e. Facebook, YouTube and Twitter, amongst other platforms). And like many other consumable multimedia materials, online news about people's response and behaviour reported by media organisations and even by netizens, usually accompanied with photos and videos, are distributed and consumed by the public regardless of the reliability of the source of information.

Many of these accounts of disasters provide us description and, to some extent, analysis of people's behaviour in times of disasters. Disaster events in different historical periods (i.e. ancient, pre-modern and modern history) documented in a variety of ways give us a picture of how people, organisations and communities, and governments and non-governmental organisations, behaved and responded to disasters, particularly in the affected area. Narratives of these disasters are open to interpretations and assumptions of scholars who later examined them mainly for academic reasons, and individuals and organisations who reported them. Many of them had emphasised the negative behaviour of people in response to the threats and dangers, which is then exaggerated and misconceived as the totality of people's behaviour in times of disasters. Consequently, disaster myths previously identified and studied by Quarantelli and his colleagues at Disaster Research Center in the last five decades remain a dominant narrative of many contemporary disasters (Box 11.3 and Figure 11.5). Quarantelli and Dynes (1972) identified six common myths about people's behaviour in times of disaster. They include the following:

- People panic in times of disasters or when facing a threat or hazard;
- Disasters make a "large numbers of persons dazed, shocked and unable to cope with the new realities of the situation";

Box 11.3: E.L. Quarantelli Resource Collection

Enrico L. Quarantelli and his colleagues at Disaster Research deserve a commendation for their significant intellectual contribution on understanding people's behaviour in times of disasters and crisis situations. A more intensive and complete literature on the subject can be found in the E.L. Quarantelli Resource Collection (https://www.drc.udel.edu/elq-collection/about) at DISCAT, the catalog database for the E.L. Quarantelli Resource Collection, available at https://infolab.ece.udel.edu/solr/new_drc/browse.

Fig 11.5
QR code for Box 11.3.

- Organisations become "ineffective" in times of disasters because they have to deal with their own crisis as well as the "irrationality" of other people;
- Disasters results to disorganisation of the community wherein anti-social behaviour is prevalent;
- In a post-disaster situation, recovery is difficult because people in the community exemplify irrational behaviour, are disorganised, are helpless and are in immobilised conditions; and
- The community is likely to indulge in a "total personal and social chaos" because of lack of resources to recover from the impacts.

These disaster myths have consistently been perpetuated in news media, films and social media platform. And amongst these disaster myths, the notion of panic during crisis situation and looting, including other anti-social conducts in post-disaster period remain persistent. These myths persist because of a lack of understanding of those concepts. For instance, panic is commonly but incorrectly used to pertain to actions of people during crisis situations. Reactions and responses, such as fear and running away from the impact area, are regarded as panic behaviour, when in fact, they are "often the most rational course of action" (Auf der Heide, 2004, p. 342). Further, looting (including rioting) is often regarded as the tendency that affected people would engage in, in the aftermath of a disaster. This presumption is often based on the fear that disaster causes people to change behaviour from socially conforming individuals and groups to selfish and anti-social, capable of committing crimes. López-Carresi (2014) refutes such perceived anti-social behaviour of people (e.g. prevalence of looting after disasters). He argues that what really happens in times of disasters are mostly appropriation and salvaging of useful resources and materials for the purpose of sustenance, and actual looting events are rare and isolated cases, and happen even in a pre-disaster context, and do not represent the totality of behaviour of all affected people. The perception of mass looting is, thus, often an exaggeration of what really happens in times of disasters perpetuated by the media and used by the authority to justify policing or military control (López-Carresi, 2014).

Many studies confirmed that panic behaviour, looting and other myths are mostly misconceptions, and are rare occurrences or isolated events (Jacob et al., 2008; Mawson, 2005; Auf der Heide, 2004). Evidence and findings in disaster literature are consistent that people's behaviour in times of disasters is not at all characterised by panic or decisions and actions that are anti-social behaviour (see studies by Fritz & Marks, 1954; Quarantelli & Dynes, 1972; Nogami, 2018). These studies, mostly based on empirical research, have relegated panic, looting, increasing disaster-related crimes and several disaster-related mental health issues (e.g. PTSD) as myths and misconceptions (see recent studies by Fischer (2002) on the 11 September 2001 terrorist attacks (United States); Jacob et al. (2008) on Hurricane Katrina in 2004 (United States); Gaillard et al. (2013) on Typhoon Haiyan in 2013 (Philippines); Nogami (2018) on Great East Japan disaster in 2011 (Japan)).

11.5 Reasserting the prosocial people's behaviour in times of disasters

In the disaster events that occurred in the modern era, there are more narratives of disasters gathered through eyewitness accounts and other methods, where a picture of chaos (i.e. disaster myths about people's behaviour in times of disasters) can also be presumed. But some scholars have observed and emphasised the positive aspects of people's behaviour in times of disasters. In the great flood in November 1333, that killed about 300 people and completely devastated the city of Florence (Italy), several accounts and stories of people drowning and fleeing the flooded area were documented. Schenk (2007) also pointed out in his examination of the event how the government responded and provided assistance to the affected people. In the aftermath of the event, charitable works by private citizens were organised to help the poor and pilgrims. The causes of the disaster were also investigated by the government and later explained as partially a result of natural occurrence, a divine punishment and a failure of government to protect its citizens (Schenk, 2007). These reactions and responses at the scale of the individual and organisation during and after a disaster point to altruistic and rational behaviour, where people try to survive and cope together, and understand both the physical and social causation underlying a disaster event.

The characterisation of people's behaviour in the face of disasters based on secondary archival research may be refutable because findings and conclusions are derived from the subjectivities of the scholars who wrote about them. In the 20th century, the basis of many scientific investigations of disaster events are first-hand data gathered by the researchers themselves immediately after the disasters. A wealth of scientific methods, both qualitative and quantitative, have become available, allowing researchers to gather verifiable data. Study of people's behaviour in a disaster situation, like many other fields in the physical and social sciences, had a more rigorous scientific basis. Samuel H. Prince, in his doctoral dissertation about the Halifax disaster in 1917, conducted what he claimed to be the first attempt to "present a purely scientific and sociological treatment of any great disaster", which capitalised mostly on qualitative methods available at the time (Prince, 1920, p. 7). Prince (1920) elaborated on the social changes succeeding the explosion. The initial reaction of people, both victims and survivors, to the explosion was described as a shock, characterised by several physiological responses (i.e. blindness, paralysis, helplessness and fear) followed by the instinct to protect loved ones and of flight for survival (Prince, 1920, p. 36) (Figure 11.6). But such a state of human behaviour was temporary and had been immediately replaced by a collective response for survival in the forms of mutual aid, acts of heroism, sympathy and "springs of generosity" to others (see Prince, 1920). The Halifax disaster then presents a succession or a shift of human behaviour from what appeared to be individual disorder during the actual explosion followed by social reorganisation towards self-protection and collective survival.

The study by Prince (1920) presents us a characterisation of people's behaviour at different spatio-temporal context and scales of analysis. Human behaviour appears to be non-linear and complex, and changes based on one's situation (e.g. a shock reaction and fight or flight instincts replaced by heroism and sympathy for others). Further research on human behaviour in a disaster situation had presented results that are consistent to Prince's study. In the 1950s, many of these researches had been pioneered by known scholars mostly in the United States. One of the pioneering works is by Dr Anthony F.C. Wallace of the Behavioural Research Council (University of Pennsylvania) who conducted the first extensive review of the literature on people's behaviour in extreme situations (e.g. disasters, wars and famine) in 1956. This literature review consisted of about 13,000 bibliographic entries and became the basis of discussion of the nature of people's behaviour at the time. Wallace (1956) confirmed the complexity of human behaviour in extreme situations, which is consistent with the earlier observation by Prince (1920) and many other authors who wrote on the topic. Wallace (1956, p. 19) had pointed out that people's behaviour in extreme situations is better understood when "plotted in both time and space". For example, disaster syndrome, a behaviour associated with apathy or shock, is usually apparent in the impact area of a disaster during or immediately after the event, as observed in a tornado that affected Worcester, Massachusetts, in 1953. But immediately after the event, individuals and groups responded within their capacity, without "panic", as noted by Wallace (1956), to help themselves and others.

In most cases, people's behaviour in times of disasters is manifested through decisions and actions that are already part of the existing roles and practices embedded in the daily routine of the people or

The Boston Daily Globe

BOSTON, SATURDAY MORNING, DECEMBER 8, 1917—FOURTEEN PAGES

BLIZZARD CUTS OFF HALIFAX
20,000 SURVIVORS DESTITUTE

Trains Carrying Injured to Nearby Places For Treatment and Rushing Aid to Homeless Thousands Stalled in Deep Snow Drifts—
List of Dead May Reach 2800 With 3000 or More Hurt

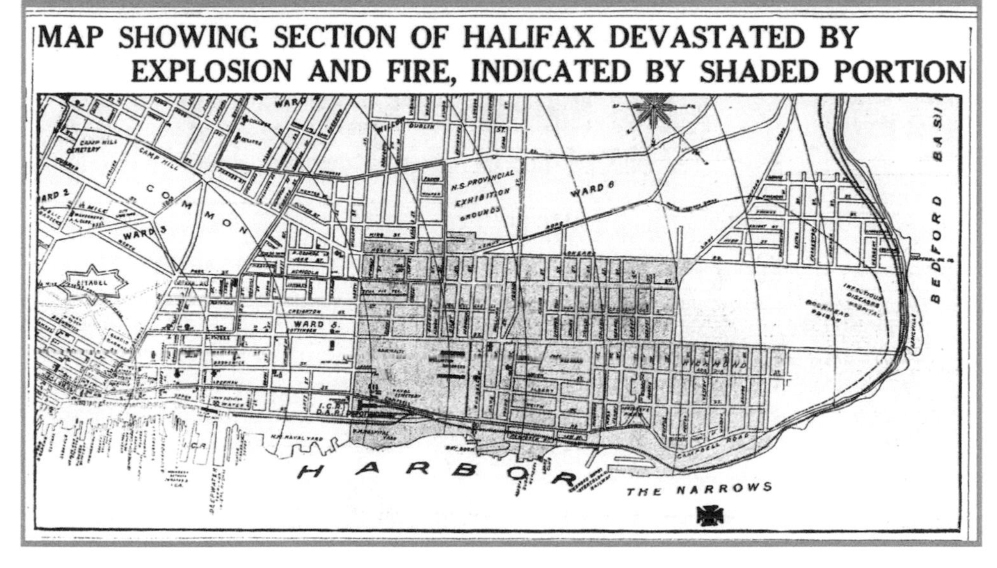

Fig 11.6
Newspaper headlines and map reporting the aftermath of the Halifax Explosion of 6 December 1917.
Source: *The Boston Daily Globe* (1917)

organisations in a particular society. In the typology of responses proposed by Quarantelli (1983) and his colleagues in the DRC, responses can be categorised based on the tasks that are performed and the structures where they are conducted (Table 11.2). Notably, most of the responses are already considered organised responses by existing groups. Emergent behaviour of people in times of disasters are also observed mostly through the formation of informal small organisation of which functions are unique to respond to the needs and circumstances of a particular disaster (Provitolo et al., 2011, p. 52). Emergent behaviour

Table 11.2
Typology of organised responses
Source: adapted from **Quarantelli (1983**, p. 4); **Provitolo et al. (2011**, p. 52)

	Old tasks	*New tasks*
Old structures	(1) Established groups, regular tasks and old structures; This group exists before an event, and many of its actions are planned (hospitals, emergency medical services etc.).	(2) Expanding groups, regular tasks and new structures; This group exists before an event, but many of its actions are not predetermined (the government agencies that aid in managing the removal of debris and help in rebuilding operations, as in Haiti, for example).
New structures	(3) Established groups, regular tasks and old structures; Many of the actions are planned, but the basic structure of the organisation shifts from a small group of professionals to a larger group of volunteers (e.g. Red Cross and National Guard).	(4) Emergent groups, new tasks and new structures. The emergent organised response (people sometimes speak of 'emergent groups', too) is related to the idea of non-traditional and new behaviour. Its existence and activities are ad hoc and, therefore, unique to the event.

at different social levels (i.e. individual and organisational) should not be misconceived and relegated as an expression of panic and anti-social behaviour because they are often necessary to cope with the crisis or recover from the impacts of disasters. (Quarantelli, 1983).

Since the pioneering work by Prince (1920) on the Halifax disaster, more studies have confirmed disaster myths about human behaviour in times of disasters. The same studies have also observed that people's behaviour in times of disasters tends to be more prosocial rather than anti-social (Box 11.4). In this perspective, it can be argued that if people change in times of disasters in terms of behaviour, they change for the better. In a recent study, López-Carresi (2014) confirmed the disaster myths and misconceptions identified in the previous studies and highlighted prosocial behaviour of people in times of disasters (Table 11.3). The prosocial behaviour of people in times of disasters are supported by the actual experiences of government and non-governmental organisations (NGOs) (including humanitarian and grassroots organisations) who worked with communities on the ground (Cretney, 2016; Institute of Philippine Culture, 2012; Shepherd et al., 2010; Rodríguez, 2006) (Figure 11.7).

Box 11.4: Blinded by the ashes, bonded by the ashes

On 21 February 2011, at 9:12 a.m., Bulusan Volcano, one of the most active volcanoes in the Philippines, located in the Bicol Region (Central Philippines), suddenly erupted and ejected volcanic ashes. Government advisories indicated that ash columns had reached 3 kilometres above the summit and drifted towards southwest. Within 15 minutes, the nearest village (Village of Cogon) on the foot of the volcano about 5 kilometres southwest of the summit was engulfed by ash. At least 200 families evacuated in the primary school buildings located in the municipal centre situated 5 kilometres away from village. Other villages within the 10 kilometre radius southwest of the volcano was covered by ash within 30 minutes after the eruption.

The volcanic eruption was photo and video documented by the author of this chapter, who was at the time facilitating a community-based disaster risk reduction activity in one of the villages in the area. During the eruption, which lasted for at least 20 minutes, the author had observed the reactions of people and activities that occurred in response to the event.

During the eruption:

- Based on the interviews, most people in the Village of Cogon hid in their houses whilst some immediately evacuated to the municipal centre on foot and in vehicles;
- At least two vehicular accidents had been recorded in the village near the volcano during the evacuation; and
- In other villages in the municipal centre about 8 kilometres from the summit, many people were observing the eruption, and most of them appeared to be amazed by the picturesque view of the volcanic ash columns.

15–30 minutes after the initial eruption:

- In the villages and city centre within the 10 kilometre radius, people were continuously moving from one place to another on foot or in vehicles. Many of them are farmers, students and workers who were trying to get home.
- In schools, children in primary grades were kept inside the classrooms. Some of them were crying and wanted to go home.
- Older people, teachers and village officials had initiated safety measures by advising people to stay inside the house and wait for advice.

Three hours after the eruption in the evacuation centre:

- With the help from municipal rescue units and other volunteers, residents in the village affected by ashfall were evacuated to schools.
- Some people, including people with disability, refused to evacuate or were not evacuated.

- The author had observed several problems in the evacuation centres, such as a lack of rooms and toilets, delays in food distribution, and waste management and other sanitation problems.
- There was an apparent lack of coordination and communication amongst the affected populations, and the authorities were quite evident.
- In the late afternoon of the same day, village officials implemented a contingency plan. Firstly, coordination was established between local officials, school coordinators and municipal officials. Secondly, rooms were arranged and organised to fit the needs of the evacuees. Thirdly, village police and health workers (assigned to particular hamlets) monitored peace and order, and sanitation issues.
- Many of the evacuees had also assisted in the evacuation management. For example, women helped in food preparations and made food distribution "easier, faster and more efficient".
- People with special needs were given attention (e.g. pregnant and nursing women), and older and sick people were allocated rooms.
- Local officials "gave regular updates to the evacuees on the situation in the evacuated village, particularly on the damages incurred".
- Non-governmental organisations and concerned citizens provided relief goods to affected people in the evacuation centres.

This first-hand account of the author of a crisis situation provides evidence that people affected by disasters do not become anti-social or misbehave in times of disaster. The situations observed in this disaster event, where people were seen running or trying to get home, and problems and challenges in the evacuation centres, are reactions that are to be expected, especially when protocol are not in place or disaster preparedness activities are not practiced. Instead, as observed by the author, people, especially those affected by the disaster, tend to help each other and exemplify solidarity towards a common goal of overcoming a disaster situation. This account is also discussed in Cadag et al. (2017).

Table 11.3 Disaster realities and prosocial behaviour of people in times of disasters as opposed to disaster myths
Note: some sentences in the original source are not included in the table.
Source: adapted from **López-Carresi (2014, p. 144)**

Disaster myths	*Disaster realities/prosocial behaviour of people in times of disasters*
When disaster strikes, panic is a common reaction.	Most people behave rationally in a disaster.
People will flee in large numbers from a disaster area.	Usually, there is a convergence reaction, and the area fills up with people.
After disaster has struck, survivors tend to be dazed and apathetic.	Survivors rapidly start reconstruction. Even in the worst scenarios, only 15–0% of survivors show passive or dazed reaction.

Disaster myths	*Disaster realities/prosocial behaviour of people in times of disasters*
Looting is common and a serious problem after disasters.	Looting is rare and limited in scope.
Disasters cause a great deal of chaos and cannot possibly be managed systematically.	There are excellent theoretical models of how disasters function and how to manage them.
Disasters usually give rise to widespread, spontaneous manifestation of anti-social behaviour.	Generally, they are characterised by great social solidarity, generosity and self-sacrifice, perhaps even heroism.

Fig 11.7
A local organisation (i.e. *Buklod Tao*, which literally means "United People") in Rizal, Philippines, that provides assistance before, during and after disasters in the communities.
Source: photo courtesy of Manuel Abinales (2020)

11.6 Explanations to people's behaviour in times of disasters

People's behaviour in times of disasters observed at different social levels and specific spatio-temporal and cultural contexts can be explained from different angles. In a more theoretical level, continuous debates in the social science (particularly in psychology and sociology) provides light to the nature of people's behaviour in times of disasters, whether they are determined by social and environmental factors or based on free will, or a combination of both. In the disaster studies literature, explanations to people's behaviour in times of disasters have been explored and examined within the frame of two research paradigms – (1) behavioural and hazard-centric, and (2) structuralist and vulnerability-centric paradigms.

The human adjustment and response to a disaster based on risk perception and hazard awareness are central to the hazard and behavioural approach in disaster studies (Box 11.5). In this approach, the understanding of the characteristics of hazards, and the level of awareness and experience of people of their potential impacts form risk perception. In turn, risk perception largely determines people's adjustment and response to hazards (i.e. people's behaviour). This view is rooted to Gilbert White's prominent ideas in his foundational thesis "Human Adjustments to Flood", where people's adjustment to flood rely on their perception of risk (White, 1945). In this view, disaster occurs because people are not able to adjust or respond properly because of lack of awareness and experience of hazards and, thus, low perception of risk. This is why behavioural and hazard-centric approaches are also called dominant-adjustment paradigm or perception-adjustment paradigm (Gaillard & Texier, 2008). Central to the argument of this paradigm is the assumption that human adjustment to hazards is, at best, always based on "bounded rationality" and "satisficing" (Kates, 1971, p. 446). Burton and Kates (1964) indicated this:

> To know and to fully understand these natural phenomena is to give to man the opportunity of avoiding or circumventing the hazard. To know fully, in this sense, is to be able to predict the location in time and space and the size or duration of the natural phenomenon potentially harmful to man. Despite the sophistication of modern science or our ability to state the requirements for such a knowledge system, there seems little hope that basic geophysical phenomena will ever be fully predictable.
>
> (Burton & Kates, 1964, p. 424)

Therefore, the optimal adjustment and response to hazards cannot be achieved because of the imperfect knowledge of many factors owing to the vagaries of both natural hazards and human behaviour. In a more recent publication, Kates (2012, p. 52) reinforce the same ideas in the context of climate change, suggesting that extreme weather and climatic events resulted to increased exposure and sensitivities of people and places to hazards, thereby complicating human adjustment and response in the form of "collective, anticipatory, and transformational adaptations", which are required to survive and prosper in a given territory. It is in this context that Kates (2012) has also explained how "adjustments", originally proposed by Gilbert White (1945), relates to what is currently known as "adaptations" (e.g. climate change adaptation (CCA)). In the last decades, the two concepts are mingled in the literature, particularly in the climate change literature, with the former being a keyword in the definition of the latter; that is, 'adaptation' is "the process of adjustment to actual or expected climate and its effects" (IPCC, 2014, p. 117). The adjustment model has, therefore, developed into transformational adaptation and forms part of

Box 11.5: Human adjustment to natural hazards

Kates (1971) has proposed a general systems model that elaborates on the human adjustment to hazards (refer to human adjustment to natural hazards: a general systems model by Kates (1971, p. 444). According to Kates (1971), adjustment to hazards is a complex process of decision-making revolving around human-environment interactions. In the event of hazard occurrence, an individual's or a group's awareness and experiences of those hazards determine the perception of risk and, eventually, the adjustments to conduct (Figure 11.8). An individual's or a group's attributes, such as personality, roles and responsibilities, access to communication, and some socio-economic factors (Kates, 1971) are factors affecting identification, evaluation and choice of adjustment. People's behaviour is manifested through a variety of adjustments and responses before, during and immediately after disasters.

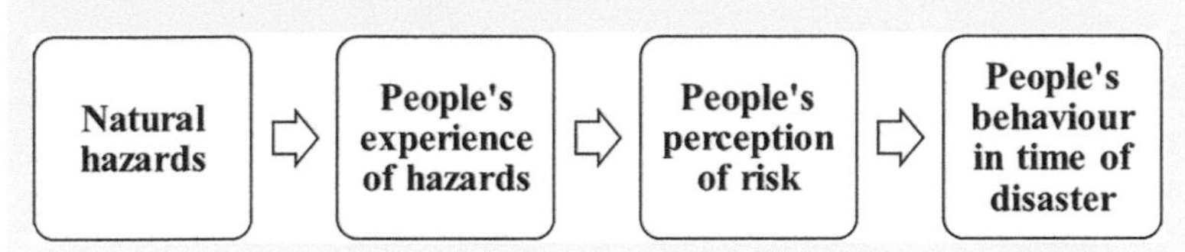

Fig 11.8 Simplified behaviourist model of human adjustment to natural hazards.
Source: modified from the human adjustment to natural hazards: a general systems model by Kates (1971, p. 444)

the broader resilience theory in disaster and climate change literature. For over four decades, both in the context of adjustment and adaptation model, Kates (2012) maintained his earlier insight that "people encounter hazard . . . in the search for the useful". Therefore, human adjustment (or adaptation) and response to hazards are determined not only by the perception of risk but also by the benefits that one obtains from occupying and exploiting the resources in a certain place and time. In fact, in a person's lifetime, many people repeatedly experience the same hazards and disasters in certain places. And they tend to develop a disaster subculture, a memory of past disaster experiences, which influence risk perception and, if properly used in DRR, proved important and useful as basis of planning and actual response in times of disasters (Granot, 1996).

The hazard and behavioural approach provides a framework to understanding the causes of a disaster and how people adjust and respond to them. This approach, however, is criticised for being too hazard-centric, with strong emphasis on the rare and extreme dimensions of hazards. Hilhorst and Heijmans (2012) indicated that this hazard-centric explanation of people's behaviour is aimed at preventing and mitigating hazards, and improving people's risk perception and behavioural response to hazards. This approach is consistent with the ideology of liberal social theories, which consider members of the society as rational individuals (Wisner, 2009). Given sufficient information on the nature of hazards that may affect them, people are able to decide rationally and avoid hazards and, thus, eliminate or reduce impacts of disasters. For the purpose of reducing risk, it follows that hazard prediction and forecasting should be improved, and people need to be taught of the risk they face for them to adjust and respond appropriately.

Since the 1970s, critics of the hazard and behavioural approach have been proposing an alternative explanation to the causes of disasters as well as to the behaviour and response of people to them. They argue hazards per se as the explanation to people's behaviour in times of disasters that undermine the social structures of the society that make people vulnerable (Wisner et al., 2004). This structuralist and vulnerability-centred approach (or vulnerability paradigm) puts unfair power relation, marginalisation and access to resources at the centre of discussion aimed at understanding the underlying causes of disasters. 'Vulnerability', as defined by Wisner et al. (2012, p. 22), refers to the "degree to which one's social status (e.g. culturally and socially constructed in terms of roles, responsibilities, rights, duties and expectations concerning behaviour) influences differential impact by natural hazards and the social processes which led there and maintain that status". The term 'vulnerability' has multiple definitions and should be analysed in conjunction to other related concepts, such as capacity, risk and resilience (refer to Kelman, 2018; Gaillard et al., 2019).

The state of being vulnerable is characterised by fragile livelihoods and occupation of unsafe locations of certain individuals and groups (refer to Wisner et al., 2012). The vulnerable sectors of the society usually consist of the poor, women, children, older people, gender minorities, persons with disability, ethnic minorities, amongst other groups (Wisner, 2009). In most societies, these groups are those who lack access to resources because of their marginal positions in the existing social structures (i.e. being economically poor, belonging to a cultural minority group, being politically neglected or segregated, etc.). Marginalisation, thus, may take different forms, and all lead to conditions of vulnerability. People are "marginalized geographically because they live in hazardous places, socially because they are members of minority groups, economically because they are poor, and marginalized politically because their voice is disregarded by those with political power" (Gaillard & Cadag, 2009, p. 197).

Both the state of being vulnerable and the processes that led there are keys to understanding people's behaviour (i.e. short-term response and coping strategies as well as long-term adjustment and adaptive strategies). In pre-disaster context, when people decide to occupy unsafe places and dangerous zones in exchange of livelihood opportunities, the decisions and patterns of behaviour they exhibit are influenced by the limited choices they have. For instance, individuals and families who died and survived after the tragic Payatas trash slide in the Philippines in 2000 had no choice but to settle on the foot of the dumpsite in exchange for easier access to trash (Gaillard & Cadag, 2009). In many cases, despite warnings from the authorities and apparent threats of hazards, many people are hesitant to leave their homes because of concerns related to livelihood, discrimination and fear of prosecution in the evacuation area, and many other cultural factors. As an example, National Dalit Watch (NDW) has reported instances of caste-based discrimination during and after Cyclone Fani in April 2019, where Dalit members of the community, considered lower-caste ethnic group in India, are denied entries in shelter homes by higher-caste groups and were least prioritised during recovery efforts (National Dalit Watch, 2019). In one way or another, gender-based, age-based and ethnicity-based discrimination, amongst other kinds of discrimination, intersect with each other, and make some people and groups more or less vulnerable than others.

People's behaviour based on human adjustment and response to hazards, and the factors that led to them, can be analysed historically using the lens of vulnerability paradigm. Oliver-Smith (1995), in his landmark publication "Peru's five hundred year earthquake", provided a detailed account of the local adaptation developed by the Andes people, which proved to be effective against earthquakes and other natural hazards. The author elaborated on the subversion of local cultures by the forces of colonialism and historical processes of marginalisation, which led to the vulnerable state of the Peruvian society. The colonial history of Peru that lasted 500 years under Spanish rule is characterised by changes in settlement pattern (i.e. policy of reduction), use of weaker construction materials and unsafe livelihoods for the local people. All these factors contributed to the gravity of impacts of the 7.7 magnitude earthquake on 31 May 1970, where at least 70,000 people died and 140,000 people injured. In a more recent study on the Haiti earthquake in 2010, Oliver-Smith (2010) reiterated the same arguments about historical construction of vulnerability and further emphasised how neoliberal policies have resulted to "individualist gain-seeking rather than collective responses to crisis". These situations of vulnerability of people that are socio-economic and cultural in nature, and usually independent of natural hazards, contribute to shaping a person's or group's response and adjustment (i.e. people's behaviour) in times of disasters.

11.7 Concluding remarks

This chapter provides a non-exhaustive review of disaster studies literature about people's behaviour in times of disasters. Many disciplines in the social sciences, particularly psychology, sociology and geography, have conducted their own research focusing on some aspects of the topic based on their disciplinary expertise. It has become a challenge how to make sense of the wealth of information and research findings about people's behaviour in times of disasters without a framework or, as Wallace (1956) pointed out, without plotting them in time and space in which they can be organised as a knowledge. Further, the debates surrounding people's behaviour in times of disasters seems to be endless in light of the convergence and divergence of theoretical models and perspectives (i.e. environmental determinism, free will and voluntarism) and research paradigms (i.e. behavioural and hazard-centric, and structuralist and vulnerability-centric paradigms) that try to explain it. But such seemingly tedious task should be seen as necessary to encourage further research and eventually understanding on the subject.

This chapter emphasises that the enquiry on people's behaviour in times of disasters is necessary for intellectual and developmental reasons. The lack of academic understanding on the topic is a gap in disaster studies. Images that depict people's behaviour in times of disasters as anti-social, law-breaking and chaotic, amongst other mythologies and misconceptions about people's behaviour, remain the dominant narratives of news reports, films and other multimedia platforms (e.g. social media). Without scientific scrutiny, such images and narratives become facts and truth about people's behaviour in times of disasters. In the worst-case scenario, they become the basis of actions of people when put in a similar situation. This presumption is the central argument of Kenneth Boulding, a known social scientist and interdisciplinary philosopher, in the monograph *The Image*, where it is argued that images or impressions of the world gathered through experience become the basis of one's behaviour (Boulding, 1956).

Persistence of disaster mythologies and misconceptions on people's behaviour in times of disasters have negative consequences to the society, particularly to communities affected by disasters. Disaster myths do not only impede implementation of effective DRR but also mislead its practice in many ways. Disaster managers, such as local government authorities, community leaders and disaster responders, are confronted with real-life problems, such as difficulties of evacuating people, inadequacy of facilities and relief needs, lack of human resources, failure of communication, and many other management challenges and problems encountered in a disaster situation. Coupled with the uncertainty of the occurrence of a disaster and the many probable scenarios that may arise from it, this complicates disaster management, particularly disaster response and crisis management. The belief that the vagaries of both nature and human behaviour would result to an unpredictable outcome predominate understanding of disasters as well as DRR.

Disaster managers then tend to be in a position where they have to exert authority and take control of a disaster situation in order to perform their duties and roles. And such a scenario is not hypothetical but a common observation in many disaster events. For one, the misconception that looting is prevalent after disasters, particularly those described as catastrophic disasters lead to plans and actions that focus on military interventions aimed to control the public (Sun, 2011). The literature on disaster studies, however, is consistent on the findings that looting, panic and related anti-social behaviour are isolated events, and happen only in specific places, and thus, not prevalent or widespread as often portrayed by the mass media. Therefore, a military intervention in disaster situations should be considered as an exemption rather than as a rule.

Taken together, the key points from the literature and arguments presented in this paper have academic and practical implications and recommendations. Firstly, any attempts to propagate disaster myths and misconceptions about people's behaviour in times of disasters should be disqualified and demystified, underlining the findings from the solid disaster studies literature on the topic. Secondly, further enquiry on the topic in light of the recent and future disasters should seek to be interdisciplinary, holistic and analytical, and aimed at understanding the causation of event and factors that shape people's behaviour. Thirdly, the practice of DRR should capitalise on collaborative works between stakeholders where people's capacities in the communities (inside stakeholders) are integrated in the protocols, and standard operating procedures of government and non-governmental organisations (outside stakeholders). This collaboration is proven to have positive effects on people's behaviour in times of disasters. Finally, stakeholders of DRR, especially disaster managers and news media organisations, who engage in disaster response and monitoring must be aware of the scientific rebuttals to disaster myths and misconceptions. Hopefully, it would lead to decisions and actions for DRR in times of disasters that are based on the conviction that people's behaviour in times of disasters is not anti-social and selfish, but prosocial and altruistic.

Take-away messages

1 A scientific or academic enquiry to understand people's behaviour in times of disasters is important for intellectual and practical reasons. The former refers to the opportunity to study people's behaviour from different angles, perspectives and frameworks. And a better understanding of how people behave and how such behaviour manifests in times of disasters is very useful in formulating and implementing plans for DRR.
2 People's behaviour in times of disasters refer to decisions, actions or conducts of individuals or groups in response to the threat of hazards or impacts of disasters in a specific place and time.
3 There are four social times or phases of disaster (i.e. mitigation, preparedness, response and recovery) and four social levels (i.e. individual, organisational, community and mass media) in which people's behaviour can be studied and explained.
4 Depictions of people's behaviour in times of disasters being anti-social are considered in the disaster studies literature as myths

and misconceptions. These disaster myths have consistently been perpetuated in news media, films and social media platform.

5 There is a solid scientific literature pointing to people's behaviour in times of disasters as more prosocial, altruistic and based on common good.

To learn more about the topic discussed in this chapter, listen to the *Disasters: Deconstructed* interview with Dr Samantha Montano (Figure 11.9).

Fig 11.9
QR code for Chapter 11.

Further suggested reading

Cadag, J. R. D., Driedger, C., Garcia, C., Duncan, M., Gaillard, J. C., Lindsay, J., & Haynes, K. (2017). Fostering participation of local actors in volcanic disaster risk reduction. In C. Fearnley, D. Bird, G. Jolly, K. Haynes, & B. K. McGuire (Eds.), *Advances in volcanology* (pp. 1–17). Springer.

Gaillard, J. C. (2015). *People's response to disasters in the Philippines: Vulnerability, capacities and resilience* (p. 216). Palgrave Macmillan.

Kates, R. W. (1971). Natural hazard in human ecological perspective: Hypotheses and models. *Economic Geography*, *47*(3), 438–451. https://doi.org/10.2307/142820

Mawson, A. R. (2005). Understanding mass panic and other collective responses to threat and disaster. *Psychiatry*, *68*(2), 95–113. https://doi.org/10.1521/psyc.2005.68.2.95

Nogami, T. (2018). What behaviour we think we do when a disaster strikes: Misconceptions and realities of human disaster behaviour. In P. Samui, D. Kim, & C. Ghosh (Eds.), *Integrating disaster science and management global case studies in mitigation and recovery* (pp. 343–362). Elsevier. https://doi.org/10.1016/B978-0-12-812056-9.00020-8

Prince, S. H. (1920). *Catastrophe and social change: Based upon a sociological study of the Halifax disaster*. Columbia University.

Quarantelli, E. L. (1999). *Disaster related social behaviour: Summary of 50 years of research findings* (Preliminary Paper 280). University of Delaware, Disaster Research Center.

Wallace, A. F. C. (1956). *Human behaviour in extreme situations: A study of the literature and suggestions for further Research*. National Academy of Sciences, National Research Council.

White, G. F. (1945). *Human adjustment to floods: A geographical approach to the flood problem in the United States* (Research Paper no. 29). Department of Geography University of Chicago.

References

Alexander, D. (1980). The Florence floods—what the papers said. *Environmental Management*, *4*(1), 27–34. https://doi.org/10.1007/BF01866218

Auf der Heide, E. (2004). Common misconceptions about disasters: Panic, the disaster syndrome, and looting. In M. O'Leary (Ed.), *The first 72 hours: A community approach to disaster preparedness* (pp. 340–381). Universe.

Baer, J., Kaufman, J. C., & Baumeister, R. F. (2008). *Are we free?: Psychology and free will*. Oxford University Press.

Bankoff, G. (2017). Living with hazard: Disaster subcultures, disaster cultures and risk-mitigating strategies. In G. Schenk (Ed.), *Historical disaster experiences* (pp. 45–59). Springer. http://doi:10.1007/978-3-319-49163-9_2

Baum, W. M. (2017). *Understanding behaviourism: Behaviour, culture, and evolution* (3rd ed.). Wiley-Blackwell.

Boin, R. A., Hart, P., & Kuipers, S. L. (2018). The crisis approach. In H. Rodriguez, W. Donner, & J. Trainor (Eds.), *The handbook of disaster research* (pp. 23–38). Springer.

The Boston Daily Globe. (1917). *Newspaper headlines and map (The Boston Daily Globe, December 8, 1917) reporting aftermath of the Halifax Explosion of December 6, 1917*. Retrieved December 8, 1917, from https://commons.wikimedia.org/wiki/File:19171208_Halifax_explosion_with_map_-_The_Boston_Daily_Globe.jpg

Boulding, K. E. (1956). *The image: Knowledge in life and society*. University of Michigan Press.

Burton, I., & Kates, R. (1964). The perception of natural hazards in resources management. *Natural Resources Journal*, *3*, 412–441.

Cadag, J. R. D., Driedger, C., Garcia, C., Duncan, M., Gaillard, J. C., Lindsay, J., & Haynes, K. (2017). Fostering participation of local actors in volcanic disaster risk reduction. In C. Fearnley, D. Bird, G. Jolly, K. Haynes, & B. K. McGuire (Eds.), *Advances in volcanology* (pp. 1–17). Springer.

Cretney, R. M. (2016). Local responses to disaster. *Disaster Prevention and Management*, *25*(1), 27–40. https://doi.org/10.1108/DPM-02-2015-0043

Ernste, H., & Philo, C. (2009). Determinism/environmental determinism. In N. J. Thrift & R. Kitchin (Eds.), *International encyclopedia of human geography*. Elsevier.

Eyles, J. (1989). The geography of everyday life. In D. Gregory & R. Walford (Eds.), *Horizons in human geography horizons in geography*. Palgrave.

Fischer, H. W. (2002). Terrorism and 11 September 2001: Does the "behavioral response to disaster" model fit? [International journal]. *Disaster Prevention and Management*, *11*(2), 123–127. https://doi.org/10.1108/09653560210426803

Frey, B., Torgler, B., & Savage, D. (2011). Behavior under extreme conditions: The Titanic disaster. *Journal of Economic Perspectives*, *25*, 209–222. http://doi.org/10.2307/23049445

Fritz, C. E., & Marks, E. S. (1954). The NORC studies of human behavior in disaster. *Journal of Social Issues*, *10*(3), 26–41. http://doi.org/10.1111/j.1540-4560.1954.tb01996.x

Gaillard, J. C., & Cadag, J. R. D. (2009). From marginality to further marginalization: Experiences from the victims of the July 2000 Payatas trashslide in the Philippines. *Jàmbá: Journal of Disaster Risk Studies*, *2*(3), 195–213. https://doi.org/10.4102/jamba.v2i3.27

Gaillard, J. C., Cadag, J. R. C., Abinales, N., Anacta, V. J. A., Batario, R., Binoya, C. S., Felizar-Cagay, F., Dalisay, S. N., Delica-Willison, Z., Luna, E. M., Occeña-Gutierrez, D., Sanz, K., Soriaga, R., Viado, R. P., & Victoria, L. P. (2013). Coping, not looting. *Inquirer*. https://opinion.inquirer.net/65551/coping-not-looting

Gaillard, J. C., Cadag, J. R. D., & Rampengan, M. M. F. (2019). People's capacities in facing hazards and disasters: An overview. *Natural Hazards*, *95*(3), 863–876. https://doi.org/10.1007/s11069-018-3519-1

Gaillard, J. C., & Texier, P. (2008). Natural hazards and disasters in Southeast Asia: Guest editorial. *Disaster Prevention and Management*, *17*(3), 3.

Gilmartin, M. (2009). Colonialism/imperialism. In C. Gallaher, C. T. Dahlman, & M. Gilmartin (Eds.), *Key concepts in political geography* (pp. 115–123). Sage Publications Ltd. https://www.doi.org/10.4135/9781446279496.n13

GMA News TV. (2013). Hungry, desperate victims of Typhon Yolanda resort to looting in Leyte (Balitanghali). *GMA News TV*. https://www.youtube.com/watch?v=8RMEXcvul-0

Granot, H. (1996). Disaster subcultures. *Disaster Prevention and Management*, *5*(4), 36–40. https://doi.org/10.1108/09653569610127433

Gregory, D. (1981). Human agency and human geography. *Transactions of the Institute of British Geographers*, *6*(1), 1–18. https://doi.org/10.2307/621969

Hilhorst, D., & Heijmans, A. (2012). University research's role in reducing disaster risk. In B. Wisner, J. C. Gaillard, & I. Kelman (Eds.), *Handbook of hazards and disaster risk reduction* (pp. 739–749). Routledge.

Holloway, L., & Hubbard, P. (2013). *People and place: The extraordinary geographies of everyday life*. Routledge.

Hughes, J. D. (2013). Responses to natural disasters in the Greek and Roman world. In K. Pfeifer & N. Pfeifer (Eds.), *Forces of nature and cultural responses*. Springer.

Institute of Philippine culture. (2012). *The social impacts of tropical storm Ondoy and Typhoon Pepeng: The recovery of communities in Metro manila and Luzon*. Ateneo de Manila University. http://siteresources.worldbank.org/INTPHILIPPINES/Resources/TheSocialImpactsofTropicalStormOndoyandTyphoonPepengFINAL.pdf

IPCC. (2014). Annex II: Glossary. In K. J. Mach, S. Planton, & C. von Stechow (Eds.), *Climate change: Synthesis report* (pp. 117–130). Contribution of Working Groups I, II and III to the Fifth Assessment Report of the Intergovernmental Panel on Climate Change [Core Writing Team]. IPCC.

Jacob, B., Mawson, A. R., Payton, M., & Guignard, J. C. (2008). Disaster mythology and fact: Hurricane Katrina and social attachment. *Public Health Reports*, *123*(5), 555–566. https://doi.org/10.1177/003335490812300505

Jigyasu, R. (2008). Structural adaptation in South Asia: Learning lessons from tradition. In L. Bosher (Ed.), *Hazards and the built environment* (pp. 74–95). Routledge.

Kapucu, N. (2008). Culture of preparedness: Household disaster preparedness. *Disaster Prevention and Management*, *17*(4), 526–535. https://doi.org/10.1108/09653560810901773

Karni, E., & Schmeidler, D. (1986). Self-preservation as a foundation of rational behavior under risk. *Journal of Economic Behavior and Organization*, *7*(1), 71–81. https://doi.org/10.1016/0167-2681(86)90022-3

Kates, R. W. (1971). Natural Hazard In Human Ecological Perspective: Hypotheses and models. *Economic Geography*, *47*(3), 438–451. https://doi.org/10.2307/142820

Kates, R. W. (2012). Natural hazards, climate change, and adaptation: Persistent questions and answers. *South Australian Geographical Journal*, *111*, 43–55. https://doi.org/10.21307/sagj-2012-002

Kelman, I. (2018). Lost for words amongst disaster risk science vocabulary? *International Journal of Disaster Risk Science*, *9*(3), 281–291. https://doi.org/10.1007/s13753-018-0188-3

López-Carresi, A. (2014). Common myths and misconceptions in disaster management. In A. López-Carresi, M. Fordham, B. Wisner, I. Kelman, & J. C. Gaillard (Eds.), *Disaster management: International lessons in risk reduction, response and recovery* (pp. 142–159). Routledge.

Mawson, A. R. (2005). Understanding mass panic and other collective responses to threat and disaster. *Psychiatry*, *68*(2), 95–113. https://doi.org/10.1521/psyc.2005.68.2.95

Merriam-Webster.com (2019). Retrieved June 11, 2019, https://www.merriam-webster.com/dictionary/behavior

Mitchell, J. T., Thomas, D. S. K., Hill, A. A., & Cutter, S. L. (2000). Catastrophe in reel life versus real life: Perpetuating disaster myths through Hollywood films. *International Journal of Mass Emergencies and Disasters*, *18*, 383–402.

Myers, D. G. (2008). Determined and free. In J. Baer, J. C. Kaufman, & R. F. Baumeister (Eds.), *Are we free?: Psychology and free will* (pp. 32–43). Oxford University Press.

National Dalit Watch. (2019). *Cyclone Fani: Tracking inclusion of Dalits, Adivasis, minorities and other marginalized communities in the disaster response*. National Campaign on Dalit Human Rights. http://ncdhr.org.in/wp-content/uploads/2019/08/Fani-IMfor-Web_public.pdf

Nogami, T. (2018). What behaviour we think we do when a disaster strikes: Misconceptions and realities of human disaster behaviour. In P. Samui, D. Kim, & C. Ghosh (Eds.), *Integrating disaster science and management global case studies in mitigation and recovery* (pp. 343–362). Elsevier. https://doi.org/10.1016/B978-0-12-812056-9.00020-8

Ogletree, S. M., & Oberle, C. D. (2008). The nature, common usage, and implications of free will and determinism. *Behavior and Philosophy*, *36*, 97–111.

Oliver-Smith, A. (1995). Peru's five hundred year earthquake: Vulnerability to hazard in historical context. In A. Varley (Ed.), *Disasters, development and environment* (pp. 31–48). John Wiley and Sons.

Oliver-Smith, A. (2010). Haiti and the historical construction of disasters. *NACLA Report on the Americas*, *43*(4), 32–36. https://doi.org/10.1080/10714839.2010.11725505

Orleans, M. (2000, February). Why the mundane? Or, my 'unassailable advantage': Reflections on Wisemnan's Belfast, Maine. *Journal of Mundane Behaviour*, 7–13.

Prince, S. H. (1920). *Catastrophe and social change: Based upon a sociological study of the Halifax disaster*. Columbia University.

Provitolo, D., Dubos-Paillard, E., & Muller, J. P. (2011, September 15). *Emergent human behaviour during a disaster: Thematic versus complex systems approaches*. A paper presented in the ECCS'11 Emergent Properties in Natural and Artificial Complex Systems, pp. 47–57. https://agritrop.cirad.fr/561127/1/document_561127.pdf

Quarantelli, E. L. (1983). *Emergent behaviour at the emergency time periods of disasters* (Final Project Report 31). University of Delaware, Disaster Research Center.

Quarantelli, E. L. (1999). *Disaster related social behaviour: Summary of 50 years of research findings* (Preliminary Paper 280). University of Delaware, Disaster Research Center.

Quarantelli, E. L., & Dynes, R. R. (1972). *Images of disaster behaviour: Myths and consequences* (Preliminary Paper 5). University of Delaware, Disaster Research Center.

Rhodewalt, F. (2008). Personality and social behavior: An overview. In F. Rhodewalt (Ed.), *Personality and social behavior* (pp. 1–8). Taylor & Francis Group.

Rodríguez, H., Trainor, J., & Quarantelli, E. L. (2006). Rising to the challenges of a catastrophe: The emergent and prosocial behavior following Hurricane Katrina. *Annals of the American Academy of Political and Social Science*, *604*(1), 82–101. https://doi.org/10.1177/0002716205284677

Sappington, A. A. (1990). Recent psychological approaches to the free will versus determinism issue. *Psychological Bulletin*, *108*(1), 19–29. https://doi.org/10.1037/0033-2909.108.1.19

Schenk, G. J. (2007). . . . prima ci fu la cagione de la mala provedenza de' Fiorentini . . .' Disaster and 'life world' – reactions in the Commune of Florence to the flood of November 1333. *Medieval History Journal*, *10*(1–2), 355–386.
Shepherd, J., Leslie, T., & Watters, A. (2010). *From tragedy to recovery: Samoa tsunami response 2009–2010*. Oxfam New Zealand. https://www.oxfam.org.nz/report/from-tragedy-to-recovery-samoa-tsunami-response-2009-2010
Sheridan, R. (2005, June 20). *Door: Storefront after Katrina*. Flicker.
Sun, L. G. (2011). Disaster mythology and the law. *Cornell Law Review*, *96*(5), 1131–1207. http://scholarship.law.cornell.edu/clr/vol96/iss5/8
United Nations Office for Disaster Risk Reduction (UNDRR) (2015). *Sendai framework for disaster risk reduction 2015–2030* (p. 37). United Nations International Strategies for DRR. https://www.unisdr.org/files/43291_sendaiframeworkfordrren.pdf
Villagrán de León, J. C. (2012). Early warning principles and systems. In B. Wisner, J. C. Gaillard, & I. Kelman (Eds.), *The Routledge handbook of hazards and disaster risk reduction* (pp. 481–492). Routledge.
Wallace, A. F. C. (1956). *Human behaviour in extreme situations: A study of the literature and suggestions for further research*. National Academy of Sciences, National Research Council.
White, G. F. (1945). *Human adjustment to floods: A geographical approach to the flood problem in the United States (Research Paper no. 29)*. Department of Geography University of Chicago.
Wisner, B. (2009). Vulnerability. In K. R. Thrift (Ed.), *International encyclopaedia of human geography* (pp. 176–182). Elsevier.
Wisner, B., Blaikie, P., Cannon, T., & Davis, I. (Eds.). (2004). *At risk: Natural hazards, People's vulnerability and disasters*. Routledge.
Wisner, B., Gaillard, J. C., & Kelman, I. (2012). Framing disaster: Theories and stories seeking to understand hazards, vulnerability and risk. In B. Wisner, J. C. Gaillard, & I. Kelman (Eds.), *Handbook of hazards and disaster risk reduction* (pp. 18–33). Routledge.

CHAPTER 12

People's resilience

Fig 12.1
Enlightened.
Source: EAGER Project, North-West University

This theoretical chapter introduces the concepts of resilience and complexity. The dimensions of resilience (i.e. personal, community and physical) enjoys attention. The difference between resilience and vulnerability is discussed. Subsequently, the linkages between complexity and resilience are made, and the chapter focuses on how resilience features within complex systems. The characteristics of resilience systems are then highlighted within the context of complex adaptive systems. Furthermore, an understanding of resilience pivots if it allows connecting theory with practice. To tie the theoretical and practical

DOI: 10.4324/9781315469614-17

application of resilience and resilience systems, this chapter proposes a conceptual model for resilience in complex systems in which all the preceding elements of the chapter are drawn together. In conclusion, the chapter argues for the use of resilience not as a way of return to normality but rather as a way of enhancing development outcomes for the better.

12.1 Introduction

The world has changed, and the impact that humans have on planet Earth is immense. Humans have the power and knowledge not only to sustainably transform our natural and built environments but also to cause harm to the resources and systems on which our livelihoods depend. Inequality and unjust global economic systems force millions of people into poverty, trapping them for generations. This heightened vulnerability is furthermore exploited by hazards – natural or otherwise – greatly induced by human behaviour and exacerbated by unsustainable development tracks. Globally, we have seen the rise in disasters of all kinds and the emergence of new, unknown ones. There is now global agreement that humans are pushing the boundaries of planetary resources, and if we maintain the status quo, many necessary systems will fail. Certain scholars believe we have entered a new geological epoch, called the Anthropocene (the age of man). In this period, it is believed that human activity has been the dominant influence on the environment and the climate. Now, more than ever, humans need to seriously consider our anticipated future, our desired development and the costs involved. It would be naïve to assume that the human collective will unanimously, and within a short period, make the needed changes to radically mitigate our impact on Earth. Rather, with political will and leadership, this will be a gradual process that requires humans, and the systems on which we depend, to become more resilient.

This chapter will provide insight into the concept of resilience, focusing on people and systems. Research over the past 15 years into resilience has shown that it is extremely useful to examine resilience through a systems perspective lens. The concept of complexity and complex adaptive systems will be used as a foundation from which resilience will be explained. Personal, community and physical resilience as the broader dimensions of resilience will be investigated. One of the common errors made is treating vulnerability and resilience as opposites. This phenomenon enjoys attention, and the chapter argues why these two terms are not mutually exclusive. However, the multi-, inter- and transdisciplinary nature of resilience have identified several characteristics of resilience systems. Although not present in all resilience systems, these characteristics provide a unit of analysis. The notion of resilience pivots is also introduced. In this chapter, a conceptual model is presented, which aim to provide an understanding of the intricacies of resilience within a systems perspective. An explanation of the resilience/DRR link is provided, and in conclusion, the chapter investigates how disasters are opportunities for communities to bounce forward.

12.2 Resilience and complexity

Over the past few decades, resilience thinkers have been studying resilience within complex systems and, in particular, complex adaptive systems (Boulding, 1956; Ashby, 1960; Buckley, 1968; Becker, 2009; Skyttner, 2005). A complex system is not necessarily complicated. Complicated problems can be hard to solve, but they mostly follow some logical rules. For instance, understanding how a computer works can be very complicated if you are not an engineer or IT hardware specialist. There are many parts, each fulfilling a specific function. Some of these parts are much more critical than others, whilst most function interdependently. The absence of one component does not always mean the system ceases to function but can mean the system functions sub-optimally. For example, the hard drive of a computer is an important component that must be present for the computer to function. On the other hand, a computer can function just fine without a USB port, although it might mean a sub-optimal functioning. As long as we understand how a computer works as a complicated system, we can troubleshoot and, hopefully, in time, identify the problem to be fixed. Complex systems, however, do not follow linear logic. In such a system, there are many unknowns and factors that directly impacts on the perceived rules and processes of the system. Such systems consist of many known and unknown parts, some of which might seem insignificant but could be the anchor point that holds the system in some state of function. For instance, understanding contemporary global economic inequality is a very complex problem. There are many different systems that are interconnected and dependent on each other, various historical events that build the system and there are different sets of rules that govern portions of this big system. Philip Newton (2015) aptly explains that complex is the opposite of simple, and complicated is more the opposite of easy. Therefore, understanding resilience is complex, but it is not hard. However, it is also not simple (Figures 12.2a and 12.2b).

Complexity theory lends itself well towards understanding resilience. This theory allows us to better understand systems, which are only partially understood by traditional reductionist scientific methods (i.e. reducing a system to its individual parts in order to understand the system, much like taking a car apart). Complexity theory, as with resilience thinking, spans across a variety of disciplines and provides alternative insights into real-world problems, their behaviours and patterns (Holland, 1992). Within complexity theory, the study of complex adaptive systems (CAS) is, moreover, enjoying attention. CAS, as the name implies, are systems with the ability to adapt to stimuli

Figs 12.2a and 12.2b
Understanding how all the parts of a car fit together is complicated, and understanding a forest habitat is complex.
Source: photo courtesy of Paul Povoroznuk and Casey Horner, Unsplash

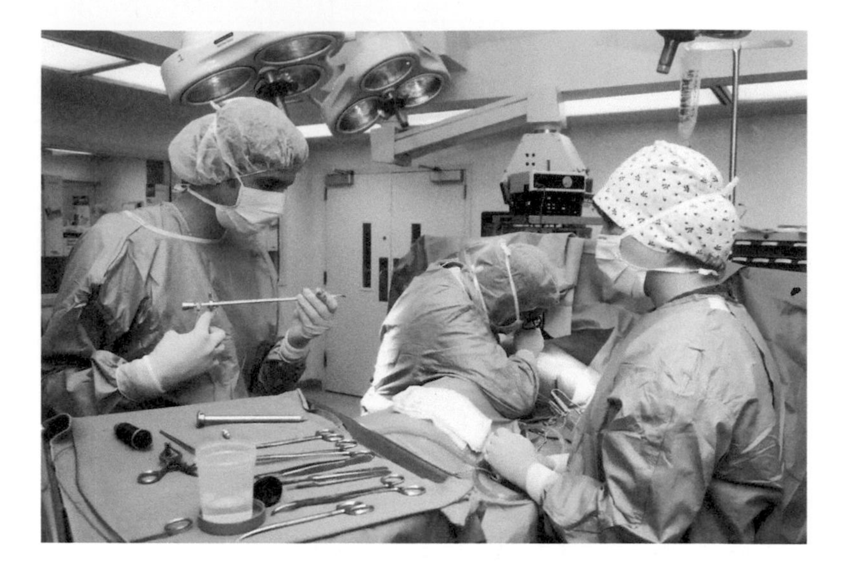

Fig 12.3
Interdisciplinary healthcare teams function as complex adaptive systems.
Source: Pype et al. (2018); photo credit: National Cancer Institute, USA

in their environment. The human body is a good example of a CAS. When one gets cold, you start to shiver, your heart rate lowers and blood is supplied to the areas that most need it. Similarly, humans in groups (communities) can react to stimuli. For example, communities can organise themselves around a common cause (protests against an increase of rates and taxes, or to protect a certain habitat, or maybe to help the elderly), or develop in a certain way, depending on their access to resources (like multidisciplinary healthcare workers, all with special skills, yet working together towards a common cause). Understanding CAS helps us understand what makes them resilience, or not. Within CAS, diversity is a common characteristic. As the system evolves, it adapts to its environment. CAS also exhibits decentralised (rather than centralised) behaviour and does not follow cause and effect assumptions. In this way, a CAS can adapt and react much quicker to stimuli. Railsback (2001) believes that at the core of CAS analysis is an attempt to show how the simple interaction between individual elements at a micro-level leads to very complex behaviours at a macro level (for example, the influence of social media posts in shaping national political thought and actions). Therefore, using a complexity approach helps to understand resilience within communities and how this contributes to DRR (Figure 12.3).

12.3 Coming to terms: defining resilience

Literature indicates that the term 'resilience' has far over 100 different definitions (Manyena et al., 2011) and means different things to different people (Manyena, 2006). The history of resilience dates back as far as AD 35, first being defined in Latin as *resilio* (Alexander, 2013) or *resalire* (Gunderson, 2010), meaning to bounce back or to walk/leap back. Resilience has been used in several disciplines, which range from engineering, urban planning, ecology, political science, psychology, sociology, anthropology, disaster risk sciences, geography and geology (Alexander, 2013). 'Resilience' has almost become a go-to term, which, like 'sustainable development', runs the risk of losing its meaning altogether. Contemporary understandings of resilience are starting to recognise that resilience and the building thereof is a much more complicated process than previously anticipated. Resilience, of the social, ecological and built systems, is a fundamental measure of sustainable development. One can almost argue that evidence of resilience is a barometer for successful development. However, some of the complications lie in the fact that resilient systems are the epitome of anticipation, change, transformation, adaption and so forth. This makes resilience not a static goal to achieve but, instead, a moving target that changes constantly as the variables that make up the social, built and ecological systems are changed during periods of stress or prosperity.

The UNDRR defines 'resilience' as the "ability of a system, community or society exposed to hazards to resist, absorb, accommodate, adapt to, transform and recover from the effects of a hazard in a timely and efficient manner, including through the preservation and restoration of its essential basic structures and functions through risk management" (United Nations, 2016). Understandably, the UNDRR's definition has a very specific hazard, prevention and response focus. Folke et al. (2010) define resilience as the ability of a system to absorb disruption and to change the way it is organised so that it can still have the necessary function, structure, feedbacks and identity. This means that a system can experience change yet still retain the same identity. Cote and Nightingale (2012) agree with the earlier point and describe resilience as the ability of a system to absorb distressing events without going into another state or stage, thus, in other words, remaining as is. Moberg and Simonsen (2011) take a bit of a different stance and believe resilience to be the capacity of a wide variety of systems to continue developing whilst experiencing change, thus having the ability to use distressing events (e.g. hazards and disasters) to bring about innovative thinking as well as renewal. Béné et al. (2012) combine the previously mentioned definitions to present a more holistic view by defining resilience as the ability of a system to accommodate positively adverse events by either buffering an event, adapting to it or being transformed by it. It is evident that all the authors view resilience in relation to a system or systems; however, their view of the way that the system should function after an event, differ somewhat.

According to Folke et al. (2010) resilience thinking is central to three aspects: persistence, adaptability and transformation. They highlight that resilience, in the context of socio-ecological systems, continuously adapt to changes to remain within critical thresholds, whereas Moberg and Simonsen (2011) refer to resilience thinking as generating knowledge on how people can strengthen their capacity to cope with environmental

Box 12.1: Defining resilience

Fig 12.4
QR code for Box 12.1. Available at https://www.youtube.com/watch?v=k_KQCqcb7EQ&t=21s

Steve Lade from the Stockholm Resilience Centre explain in this video their understanding of resilience within socio-ecological systems.

stressors. Resilience may, thus, be understood as incorporating both nature and society. These definitions shift our understanding of resilience from an individual concept to a more system-based approach. That does not only focus on the individual but includes society and the environment. In this way, resilience can be seen as a social construct within complex socially and environmentally linked systems. Resilience, therefore, emphasises complex dynamics, such as the interactions between ecological, political, economic and social dimensions, and the importance of non-linear dynamics, thresholds, pivots, tipping points and poverty traps (Béné et al., 2012; Rotarangi & Stephenson, 2014). In most of the earlier definitions, it is clear that the presence of a shock or stress within a system is seen as the defining factor to determine resilience. However, resilience is more than just the ability to withstand perturbations (i.e. a reactive focus on how resilience is determined). A system, moreover, can apply foresight that will lead to adaptation and/or transformation. Thus, the manifestation of a hazard, shock and/or stress should not be seen as the endgame to determine resilience (Box 12.1 and Figure 12.4).

For resilience to be understood in complex linked systems, the definition of the term must make provision for a much more inclusive understanding, application and implementation across various sectors, and at multiple scales. Resilience is about people, and the systems on which they depend, having the capacity to continue to learn, self-organise, exert agency (human agency in this context means the capacity of individuals to act independently and to make their own free choices, based on their will) (Béné et al., 2012) and "develop in dynamic environments faced with true uncertainty and the unexpected, like steering a vessel in turbulent waters" (Folke et al., 2016). This definition makes provision for two very important aspects of resilience. The system's ability to adapt and change before being exposed to shocks and stressors due to foresight and anticipative ability (Poli, 2010, 2014), and also recognise that people must be the centre of resilience-building, recognising their agency. The notion of power and agency is, thus, a much needed and inclusive part of how resilience needs to be understood (Béné et al., 2012). Thus, true resilient systems and people, like a vessel on the water, can navigate turbulent conditions that involve more than hazards and shocks.

12.4 Dimensions of resilience

The multidisciplinary, multilayered and multiconnected nature of resilience makes it problematic to easily identify its various dimensions. However, literature over the past 20 years has gravitated towards three broader perspectives, that of personal resilience (mental, emotional), community resilience and physical resilience (Shimizu, 2012).

Personal resilience

Personal resilience is mostly defined as a person's ability to cope with adverse circumstance invoked by tragedy, threats, shocks, stressors and significant external agencies, such as the impacts of natural hazards, economic downturn or even crime. Such resilience is related to a person's ability to cope and adapt to such events (Major et al., 1998). Maddi (2004) refers to this ability as "hardiness" – a set of attitudes or beliefs about oneself that provides courage and motivation to turn stressful circumstances into opportunities. Ali et al. (2019), picking up from the work of Maddi, identifies commitment, challenges and control as hardy attitudes and components of existential courage and motivation. *Commitment* relates to the ability to be connected, present and involved with other people no matter the stressful circumstances. "*Challenge* is the desire to learn continuously from experiences (positive or negative)" (Ali et al., 2019). Lastly, *control* is the ability to influence the outcomes of stressful situations in the face of possible powerlessness and insecurity. How someone copes and/or adapts is directly related to a person's ability to control oneself, respond to events and control responses to people. It is, thus, adapting well in the face of adversity. Personal resilience can, thus, be psychological, emotional and physical. Personal resilience is an innate human capacity that can be developed in all persons (Ahmed et al., 2004). It, therefore, stands to reason that within a community context, the more individual resilience presents, the more resilience a community will be, irrespective of their socio-economic status.

Community resilience

It is logical to assume that personal and community resilience are closely related. Magis (2010) says that community resilience is "the existence, development, and engagement of community resources by community members to thrive in an environment characterized by change, uncertainty, unpredictability, and surprise". Communities' resilience is dynamic and changes with external and internal factors, linked to resilience-building. Social cohesion; sharing of resources, information and expertise; ability to organise, make connections and network, trust and adapt, are all traits of such resilience. The need for community resources is stressed by Magis (2010). Therefore, although all the personal attributes, as highlighted in the earlier section, might be present, community resilience still requires some *common community resources*. One could, thus, argue that community resilience will be less in communities that are socially and economically less well off. As will be argued later on in this chapter, it is the capital domains (e.g. social, human, economic, physical and natural) that form the pivots around which resilience can be built.

Physical resilience

The best-known dimension of resilience is physical resilience. Within science, one of the earliest measurements of resilience came from the field of metallurgy. In this case, resilience was seen as a metal's ability to withstand deformation when subjected to external forces (Hickford et al., 2018). The notion of bouncing back (elasticity) was important. This early thinking assisted greatly in understanding and forming more transdisciplinary understandings of resilience. However, physical resilience has moved beyond a pure focus on metals and now includes a broader systems-based approach. Physical resilience, therefore, not only includes the elasticity of metals but all physical forms, organic and not. All these physical systems (e.g. human body, infrastructure, agriculture, urban centres, wildlife and biodiversity) must be understood for their interlinkages, dependency, complexity and complementarity. The resilience in physical systems is a cornerstone to sustainable development and, in the end, contributes to DRR. Thus, just like humans (personal and community resilience) can develop resilience, so, too, should we design our physical environment around resilience principles if we want to safeguard our development and limit vulnerabilities (Figure 12.5).

Fig 12.5
Zimbabweans have shown remarkable resilience over decades within an unjust political environment.
Source: photo courtesy of Peter Kvetny, Unsplash

12.5 Resilience and vulnerability

One of the biggest errors in the resilience-vulnerability debate is treating these two concepts as opposites. It is a common cause to think that if people or systems are not resilient then they must be vulnerable. Or vulnerability erodes resilience. Such arguments fail to appreciate the complex nature of resilience and vulnerability. For example, communities living in poverty-stricken informal settlements throughout the globe are quite susceptible to almost every form of natural hazard impact (Zweig, 2015; Flanagan et al., 2011). Yet research has shown that these communities are surprisingly adaptable to changes and can reorganise quite quickly (Zweig, 2017; Usamah et al., 2014) – a characteristic of a resilient system. Similarly, bushfires over the past few decades in Australia had a significant impact due to the lack of social cohesion and connectedness of well-off communities (Gibbs et al., 2013). Vulnerabilities are generally the attributes and capabilities of individuals and households, yet resilience is coming together and using the capacities of individuals and households to effect sustainable change within linked systems. The level of vulnerability of a system can, thus, be mitigated to enable resilience, and similarly, building the resilience of a system might lead to less vulnerability to shocks and stressors. However, lessening the vulnerability in a system can bring about unintended consequences that lead to less resilience. For instance, for years the people of Zimbabwe have demonstrated their resilience within a complex and unjust political system. The political system in itself proved to be extremely resilient, although the majority of the population suffered under the latter rule of Robert Mugabe. With his handover of power, the economy did not recover as hoped, leaving many Zimbabweans in a worse economic state, and thus more vulnerable, than before. In both cases, vulnerability and resilience are very context-specific. One, therefore, needs to understand the complexities of (and in) a system and relate these complexities to the characteristics of a resilient system. In doing so, a better understanding of the dynamics driving resilience and vulnerability can be gained.

12.6 Characteristics of resilient systems

Although one finds several resilience measurement instruments, the complex nature of resilience does not allow for a uniform application of these. One can also debate the logic and value of measuring resilience because it is not an input to a system. However, a much more valuable way of thinking about resilient systems is to understand the characteristics that make systems resilient. By focusing on the characteristics, one is in a much better position to determine the variables that constitute the characteristics. This allows for practical action in enhancing or creating these characteristics, which in turn will lead to a resilient system.

Anticipative

An anticipative system is a system that has (or can envisage) a future time-bound blueprint model of itself. Such a system can take action to achieve the envisaged future state (Van Niekerk & Terblanché-Greeff, 2017). When a resilient system is anticipatory, it can expect both known and unprecedented shocks with the ultimate goal of minimising adverse effects (Kerner & Thomas, 2014). In being anticipatory, a system mitigates against the shortcomings of not being able to predict all shocks by being prepared for any eventuality, with all necessary strategies put in place. Thus, anticipatory resilient systems avoid reliance on reactive strategies that often prove costly. For example, a community would be able to anticipate a possible future state that requires them to resettle due to their exposure to a natural hazard, which threatens their property and livelihood (e.g. sinkholes). It would be easy for such a community to envisage what a future settlement would look like, how to make it safe, what amenities they require – infrastructure, green areas and economic opportunity. The presence of a blueprint that reduces their risk makes such anticipation possible. Should we require the same community to anticipate resettlement on Mars, then the anticipatory ability is severely curtailed due to too many unknowns in the system and the lack of any blueprint to guide anticipatory behaviour. Anticipation in a system is built through the many feedback loops that are created in time. Ultimately, the anticipatory nature of resilience ensures that not only will lives be saved, but livelihoods can be protected resulting in more gains for resilience-building efforts. Additionally, in being anticipatory, systems can also be able to plan for long-term prevention of shocks and devastation, well before calamity unfolds.

Adaptive

Adaptive capacity rests on people's agency (i.e. their ability to make informed choices and to develop and successfully execute their plans) (Levine et al., 2011). Adaptive capacity is the property of a system in which structures are modified to prevent future disasters (Lorenz, 2013; Norris et al., 2007). Resources with dynamic attributes, institutional memory, innovative learning and connectedness determine the foundation of adaptive capacity at the community level (Longstaff et al., 2010; Norris et al., 2007). Therefore, a system's adaptive capacity influences the ultimate potential for implementing sustainable alternatives (Engle, 2011). Moreover, adaptive capacity is a property that can facilitate transitions to a new system state when the current state is untenable (Folke, 2006). It should, however, be noted that human agency, particularly power structures, can limit both transformation and adaptation. This is so because changes often challenge the status quo, threatening those who benefit from current systems and structures (Béné et al., 2012). When an individual or system can reduce its exposure, or respond to a shock by making various adjustments to its functions or identity, without loss of its distinctive functions or identity,

Fig 12.6
Quiver trees (*Aloe dichotoma*) in Namibia exhibit adaptive characteristics. The quiver tree is dying over large areas of its range because of climate change. However, there is a gradual poleward and upslope range shift that is now happening.
Source: Hannah (2010); photo courtesy of Bjørn Christian Tørrissen (CC BY-SA 4.0)

then that system or individual is said to be adaptable (Headey et al., 2014; Fafchamps & Lund, 2003). Usually, the adaptive capacity of the individual or system arises when its absorptive capacity (see later) if a shock is exceeded, requires instead that incremental adjustments be made (Cutter et al., 2008). Being adaptable applies experimental or experiential learning to make the necessary adjustments to exploit opportunities or respond to shocks without loss of core functions (Berkes et al., 2003). It is important to note that adaptable resilience (Figure 12.6) can be continuous and multiscalar, ranging from smaller intra-household to larger community-wide adaptive capacity given the idiosyncratic and covariate shocks that affect societies.

Transformative

Transformability is the "capacity to create a fundamentally new system when ecological, economic or social structures make the existing system untenable" (Walker et al., 2004, p. 5). CAS might require structural changes, reorganisation or reinvention that reconfigure old traditional system elements into newer approaches that may be different altogether. Hence, resilient systems need to be transformative and, in some instances, transcend multiple sectors within different societies, creating new resilient developmental pathways. Transformation often results when the intensity of the shock is so high that it exceeds the system or individuals' absorptive or adaptive capacities (Béné et al., 2016). As a result, drastic or permanent alterations to the functioning of the system may be instituted depending on both the nature of the shock in a proactive manner (Kroeze et al., 2017). In being transformative, resilient systems challenge the status quo, offering an opportunity to address vulnerability, social injustices and inequalities that compromise the capacity to contend with shocks (Béné et al., 2012). In that regard, when a resilient system is transformative, it can tackle social dynamics, such as agency and power disparities specific to that system. As a result, transformation cannot be crafted by external actors who could fail to understand the intricate and sensitive justice and inequality issues characteristic of a particular system. It is the local stakeholders with the best knowledge of the region who should drive transformation (Kerner & Thomas, 2014). Taken together, to be transformative, resilience requires political will and leadership from various disciplines, entrenching home-grown, contextualised transformative pathways. Once again, the COVID-19 pandemic can be used as an example. In several developing states, governments seized the opportunity created by the crisis to institute structural changes to the economy and healthcare systems to address inequalities.

Absorptive

The absorptive characteristic means the capacity of a system to absorb the impacts of negative events to preserve and restore its structure and basic functions. It involves intentional protective action against shocks (Jeans et al., 2017). The absorptive characteristic ensures stability in a system and prevents the system from reaching its system thresholds (see the section later). This characteristic has been associated with medium- to long-term actions and putting in place of necessary mitigation measures against known shocks and stressors. In essence, it means minimising exposure on the fly. For example, a government might consider implementing farmers' drought assistance programmes as a preventive measure against an intense El Niño season to ensure that losses are minimised. Or in the case of COVID-19, many governments implemented lockdowns to absorb possible future risks.

Resistant

A system that exhibits a resistant characteristic is a system able to withstand all stressors, shocks or impacts without suffering any loss (Lake, 2012), thus remaining unchanged. Resistance also related to the ease of which a system can be changed (or not). This concept aligns closely with the characteristic of robustness. Resistance in CAS can take on many forms. Resistance to change is often addressed in terms of recovery (Folke, 2006). Thus, the focus is on the time it takes a system to return to a previous state following a disturbance. However, focusing on one singular event is deceptive. The frequency and extent of disturbances need to be understood in the context of system resistance. A CAS might well exhibit resistant characteristics for frequent idiosyncratic risks and shocks, yet be less well adapted to resist longer-terms stresses. Although this characteristic can have major positive benefits for a system, it can also be undesirable. This is particularly true if a system needs to be changed to allow for resilience to occur. Certain systems are so ingrained in our development that changing them to unleash the resilient potential for other systems might not only be problematic but almost impossible. In such a case, additional adaptation is needed to circumvent the negative resistance of such systems.

Reflective

Resilient systems need to learn from their past to inform the future (Béné et al., 2012). It could have been a bad past, such as a devastating drought, for which institutions and communities were ill-prepared for, from which lessons can be drawn. Past experiences also create opportunities for improvements to be made when individual states or regions at large are presented with similar shocks in future. Whilst not limited to drawing lessons from bad cases, in being reflective, resilient systems can use good practices as well. Governments, institutions and communities can adopt best practices from the past that can be used to formulate actions to tackle future calamities. The merits of reflective resilient systems are that it creates an opportunity for active learning (Kerner & Thomas, 2014) and room for the review of the effectiveness and efficiency of existing processes in the face of new shocks.

Resourceful

In an environment where there are limited resources in terms of time, human, financial, technological and natural resources, resilient systems need to be resourceful. Such a situation requires that the little available resources are allocated and used effectively, and where possible, best alternative pathways for resource use are identified and pursued in resilience-building (Kerner & Thomas, 2014). Also, in being resourceful, systems should be able to make provision for the required materials without any devastating disruptions in service provision or resource access in the event of a shock. In that regard, it is important to note the importance of time as a resource to be used efficiently. Inefficient quick fixes to shocks achieved over a short space of time often prove costly in the long-term, and wastage of resources may be recorded. Therefore, a resourceful resilient system ensures there is a sustainable use of resources (Figure 12.7).

Inclusive

In being inclusive, resilient systems realise that whilst approaches and strategies may be developed and disseminated in a top-down route, there needs to be provision for bottom-up consultative processes that feed into the strategies, planning and decision-making for resilience-building. Inclusiveness also requires that at all rungs of society acknowledge the diversity of social groups and their specific needs (Béné et al., 2012). Resilience-building should endeavour to include contributions from

Fig 12.7
Subsistence farmers in the Shire River basin in Malawi show immense resourcefulness and learning from the past by using portable water pumps to irrigate their fields, thus not losing in-built pumps once a flood occurs.
Source: photo courtesy of Dewald van Niekerk

key population groups, such as women, youth, people living with disability and children. Mainstreaming and participatory processes can be applied to ensure that contributions are drawn from all sectors of society, regardless of gender, wealth status, education or any other form of stratification in society. When resilience-building consults widely, it has the benefit that it builds a sense of ownership where individuals and institutions all realise they have the mandate to fulfil and are contributing to a shared common vision. Furthermore, inclusive resilient systems can be achieved through the creation of equal opportunities, promotion of participatory governance and social justice.

Innovative

Reinmoeller and van Baardwijk (2005) emphasise four innovation strategies leading to more resilient systems, that is, knowledge management, exploration, cooperation and entrepreneurship. Kroeze et al. (2017) say that a resilient system that is innovative is one that can combine exploration and exploitation. Exploration related to how well a system can identify developments to which it must adapt and transform, and in doing so, in an unconventional manner. Exploitation is "about fine-tuning the functionality of the system under current conditions" to ensure gains in effectiveness and efficiency (Kroeze et al., 2017). Innovation means the introduction of something new and novel. Being innovative means a CAS can draw on "un-like-minded" individuals and structures to solve issues. However, there is also a risk involved, and innovation can also mean unanticipated outcomes (positive and negative). Innovation in CAS require constant iteration. Resilience strategies must, therefore, be aligned with non-conventional partners to make a difference and impact.

Integrated

Due to their complexities and intricacies, shocks and stressors may reinforce or confound each other's effects. This means that for systems to be resilient, diverse actors from across geographic space and disciplines need to work together because integration allows for solving multiple problems by many actors. Thus, an integrated resilient system holistically brings together institutions, stakeholders and different actors across the diversity of their disciplines in polycentric management and implementation approach (Béné et al., 2016). Additionally, integration contributes to efficient resource use as it taps into the resource-pool of the various stakeholders working together to achieve desired resilience goals, which in some cases could be too difficult to be achieved by a single actor (Kerner & Thomas, 2014).

Robust

If a system is robust, it means that its design is well-conceived, constructed and managed, and includes making provision to ensure failure is predictable, safe and not disproportionate to the cause. Thus, its assets and systems are designed to withstand shocks and hazards (Walker & Salt, 2006). This means that resources need to be invested in design processes, construction and management of physical structures. Robustness means systems, technological and asset development incorporate monitoring and surveillance to be able to predict potential failures and address them in advance. For example, protective infrastructure that is robust will not fail catastrophically when design thresholds are exceeded. It can, thus, be seen as a fail-safe mechanism. Safe failure means that the system can absorb sudden shocks (including those that exceed design thresholds) or the cumulative effects of slow-onset stress in ways that avoid catastrophic failure. Safe failure also refers to the interdependence of various systems, which support each other; failures in one structure or linkage being unlikely to result in cascading impacts across other systems. Ahern (2011) adds dimension to the notion of robustness. He argues that resilient CAS must move beyond the notion of fail-safe but become safe-to-fail. In doing so, CAS, especially in the urban context, must become multifunctional and have built-in redundancy and modularisation, diversity (bio and social), multiscale networks and connectivity, and adaptive planning and design. Ahern (2011) argues that if these aspects are built into a CAS, then the system will become safe-to-fail as well. Thus, the thinking shifts from a system well-conceived for its purpose to one that can fail but not have catastrophic consequences as a result.

Flexible

Flexibility refers to the willingness and ability to adopt alternative strategies in response to changing circumstances or crises. With flexibility, resilience systems can make appropriate adjustments relating to changes in the social, economic and environmental contexts. This allows resilient systems to accommodate newly developed knowledge and technological innovations (Béné et al., 2012). Furthermore, flexibility means having the ability to address change and stagnation to the benefit of the system. It also means the will to transition towards new, different alternative approaches and strategies. Flexibility is also referred to as the ability to perform essential tasks under a wide range of conditions and to convert assets or modify structures to introduce new ways of doing so. A resilient system has key assets and functions physically distributed so that they are not all affected by a given event at any one time (spatial diversity) and has multiple ways of meeting a given need (functional diversity). Ultimately, flexibility allows for context-specific adjustments. For example, many public health systems, as well as the manufacturing sector, had to quickly make adjustments to mitigate the COVID-19 pandemic (Figure 12.8). Some health systems showed immense flexibility towards quick adaptation, and factories converted their manufacturing lines to supply needed medical goods in minimal time.

Fig 12.8
The COVID-19 pandemic required systems to exhibit flexibility. Many institutions adopted Fourth Industrial Revolution technologies to ensure their operations can continue within this "new normal".
Source: photo courtesy of Chris Montgomery, Unsplash

Redundancy
Kerner and Thomas (2014) argue that redundancy means putting in place alternative options to deal with potential disruptions that may arise due to a shock. It entails having options that may be undisturbed in the event of shocks or stressors. Redundancy in resilient systems is, thus, the ability to offer numerous options to achieve desired goals or functions. The presence of diverse options or actors with overlapping functions ensures that when one area or component fails, there is no detrimental system collapse, as other components may be able to compensate for the failure or loss (Béné et al., 2012). Redundancy adds to the robustness of the system. Redundancy may be multiscalar, starting at the very localised household scale, where it may mean households explore various livelihood options relevant to their contexts, such as switching to climate-smart agriculture (Khoza et al., 2020), or scaling down and moving into cheaper urban accommodation. At another scale, and as the COVID-19 crisis has shown, industries could consider repurposing their manufacturing to address a pressing issue (like clothing factories making face masks) or diversifying their manufacturing lines. At a national level, economic diversification options could be considered to ensure resilience to shocks. For example, a country that was traditionally reliant on rain-fed agricultural production may move towards improving its tourism industry. Thus, for such a country, in the event of an extended drought, income could still be earned from tourism.

Self-organised
Self-organisation refers to a system's ability to make its structure more complex given its system's rules. Positive self-organisation allows for the creation of heterogeneity. When a CAS self-organises, it comes up with new structures and ways of achieving its objectives. Thus, a measure of freedom and experimentation is needed. Disorder and bounded chaos must be permissible, and this relates to the concept of systems functioning at the edge of chaos (Coetzee et al., 2018). Urban growth in developing countries can be used as an example. Urban planning cannot keep up with the rate of expansion, which leads to urban fringe communities being forced to find new ways of settling, building, developing and connecting. Self-organising can also have negative consequences in the sense that if a CAS becomes overly organised, it starts to erode its robustness, and becomes less flexible to change and adaptation. This phenomenon is called self-organised criticality. This means that a system becomes so organised that a small change can have massive implications (for example, the global financial crisis of 2007–2008).

Connected
Connectedness describes the quantity and quality of relationships between system elements (O'Sullivan et al., 2013). It also relates to the paths of interaction between system elements and other systems and

their elements. Connectedness can be economic, social, psychological or physical. Connectedness in systems allows for the flow of information and knowledge, goods and services. It contributes to the linkages between systems and, in many instances, enhances innovation and inclusiveness. As a resilient characteristic, connectedness allows for expansion of heterogeneity and diversity. As with almost all other characteristics, too much connectedness can have negative consequences. Overconnectedness within and between systems hinders adaptation, transformation and flexibility. High and weak connectedness imparts diversity and flexibility to the system, whereas low and strong impart dependency and rigidity.

Besides the various and interdependent resilience characteristic as discussed, research by Rotarangi and Stephenson (2014) show that, in many resilience systems, there are also other elements present that allow the system to retain its identity. These are called resilience pivots.

12.7 Resilience pivots

Resilience pivots are those elements of a resilient system that remain stable despite adaptation or even transformation of other elements of that system and, in doing so, support the maintenance of the system's distinctive identity (Rotarangi & Stephenson, 2014). It is thus "a person, thing, or factor having a major or central role, function, or effect" on the functioning of a system. Pivots can, thus, be an indispensable part of a system key to its integrity, and/or it can bring needed stability in the system for its continued existence. Rotarangi and Stephenson (2014) refer to this phenomenon as the "stable core of a system" (see Figures 12.9 and 12.10). In some instances, one can argue that a fundamental law or rule of a system is to protect and ensure the continued existence of its pivots. The loss of such pivots might mean the integration of the system into an undesirable state. The complexity of resilience allows for the creation of an understanding of pivots. Understanding the local context in terms of its geography (physical, social, economic, environmental, spatial and political characteristics) helps to reveal the relations between the different elements of a system, thereby identifying factors that contribute to emergent patterns of vulnerability and resilience (Moench, 2014). As an example of the pivots for understanding community resilience, the five capital domains (human, social, economic, physical and natural), as the stability core pivots, can be used. The term 'capital' implies a usable productive resource that can be harnessed for human development (Šlaus, & Jacobs, 2011), which also gives meaning to a person's world (Bebbington, 1999). All the five capitals are important, although the extent of their importance in any system will change over time. Most importantly, these capitals interact across space and time, and may reduce or increase some at the expense of others. As such, systems may sacrifice some capital for others if it deems it more appropriate for survival, and that switching may reverse at another time (Bebbington, 1999). Weighing the trade-offs between these five types of capital is an ongoing process for CAS. However, a system is assumed to need a balance of these five capitals to maintain adaptive capacity and well-being.

Human capital

Human capital is the aggregate of innate abilities, an individual's intrinsic potential to acquire skills. It relates to the knowledge and the skills that individuals acquire and develop throughout their lifetime, and this includes physical, intellectual and psychological capacities that individuals possess (Laroche et al., 1999). Similarly, Šlaus and Jacobs (2011) define human capital to include the knowledge, skills, attitudes and capacities of individuals, as well as the social and cultural endowments of the collective, including the capacity for discovery, invention, innovation and resourcefulness. It is important to note and acknowledge that the concept of human capital is complex and multifaceted (Laroche et al., 1999) and that is interwoven with the sustainability of all other forms of capital (Šlaus & Jacobs, 2011). Human capital is less tangible, being embodied in the skills and knowledge acquired by an individual. It is probably one of the most important determinants of resilience amongst other forms of capital (Mayunga, 2007).

Social capital

Social capital is the aggregate of the actual or potential resources that are linked to possession of durable networks of more or less institutionalised relationships of mutual acquaintance and recognition, which provides each of its members with the backing of the collectively owned capital. According to Twigg (2001), social capital refers to the social resources upon which people draw in pursuit of livelihood objectives, such as networks and connections, membership of groups, relationships of trust, reciprocity and exchanges. For Adger (2003), social capital captures the nature of social relations and uses it to explain outcomes in society. In the context of community resilience, social capital reflects the quantity and quality of social cooperation (Mayunga, 2007). Whereas social capital has been defined in a variety of ways, there is a common emphasis on the aspect of social structure, trust, norms and social networks that facilitate collective action (Mayunga, 2007), and contributes to resilience-building. At its core, the social capital theory provides an explanation for how individuals use their relationships with other actors in systems and societies for their own and the collective society good (Adger, 2003). This is important because a persons' actions are shaped, redirected and constrained by the social context. Norms, interpersonal trust, social networks and social organisation are important in the functioning not only of the society but also of the economy. As Hanifan (1916) indicated, when an individual comes into contact with a neighbour, and them with other neighbours, it becomes an accumulation of social capital that will satisfy an individual need and ultimately improve the living conditions of the whole. The whole community benefits by the cooperation of all its parts. Thus, common to all theories of social interaction is the recognition that collective action requires networks and flows of information between individuals, groups and systems. These sets of networks are usefully described as an asset of an individual or a society and are increasingly termed social capital (Adger, 2003). What is important, though, is to note that social capital is productive but not fungible (maybe specific to certain actions in a certain time-space), making possible the achievement of certain ends that in its absence would not be possible. As such, a given form of social capital that is valuable in facilitating certain actions may be useless or even harmful to others.

Economic capital

Economic capital denotes financial resources, including savings, income, investments and credit, that people use to achieve their livelihoods (Mayunga, 2007). According to Twigg (2001), economic or financial capital includes savings and credit, and other inflows of money other than earned income, such as pensions and remittances. Economic capital is immediately and directly convertible into money and may be institutionalised in the form of property rights. Amongst other factors, economic capital can, thus, be measured through household income, property value, employment and investments (Mayunga, 2007). Economic capital and the use of such capital constitutes how one can attain many human and social goals. A level of economic prosperity is needed that is sufficient to provide how everyone can be fed, clothed and housed. It is also the provision of clean water and proper sanitation, ensuring of universal education, and provision of health and social services accessible to all. Economic capital can and should create healthy jobs, and its equitable distribution ensures that people's basic needs are met. At the same time, the means of increasing economic capital must not threaten either human capital or the environmental and social capital, which influences health and well-being.

Physical capital

Physical capital is the basic infrastructure, which includes affordable transport, secure shelter, adequate water supplies and sanitation, access to information and producer goods needed to support livelihoods, such as the tools and equipment that people use to function more productively (Twigg, 2001). Pandey et al. (2017) share similar views and indicate that physical capital comprises the essential infrastructure, skilled personnel and goods required for sustainability. According to Mayunga (2007), physical capital also refers to the built environment, which comprises of residential housing, public buildings, business/industry, dams and levees, shelters, and lifelines, such as electricity, water, telephone, and critical infrastructures, such as hospitals, schools, fire and police stations, and nursing homes. As evidenced from these definitions, physical capital is tangible; being embodied in observable material form (Laroche et al., 1999). The list of these tangible things can also include machinery, factories, plants, patented processes, raw materials, inventories held by producers or traders, and means of transportation and communication.

Natural capital

Natural capital is essential in sustaining all forms of life (Mayunga, 2007). It includes natural resources on which users depend, and these cover a wide range of tangible and intangible goods and services (Pandey et al., 2017). Twigg (2001) refers to natural capital as the natural resource stocks from which resource flows and services are derived, such as land, forests, marine/wild resources, water, protection from storms and erosion. Similarly, Mayunga (2007) refers to natural capital as the natural resources, such as water, minerals and oil, land which provides space on which to live and work, and the ecosystems that maintain clean water, air and a stable climate. Natural capital can comprise goods and services, such as the soil for growing crops and trees; water for drinking, washing and cooking; uncultivated plants for food and medicine; wild animals for food. Examples of ecosystem services include the temporary storage of floodwater by wetlands, long-term storage of climate-altering greenhouse gases in forests and dilution and assimilation of wastes by rivers. The attainment of these benefits is as a result of the conversion of natural capital into a wide range of permanent, more or less human-dominated systems, such as agriculture, aquaculture and plantation forestry. This conversion of natural ecosystems into cultivated and human-made capital has greatly reduced the quantity (spatial coverage) and quality (or integrity) of the natural capital on Earth. It is important to note that ecosystem degradation and the loss of biodiversity undermine ecosystem functioning and resilience, and thus, threaten the ability of ecosystems to continuously supply the flow of ecosystem services for present and future generations. It is, therefore, important to preserve the stocks of natural capitals to assure the continued provisioning of ecosystem services to society.

Looking at all the earlier characteristics and pivots can be daunting. It leaves one wondering how all this fit together. The following section provides a conceptual model of a complex adaptive resilient system, and in particular, the focus will be on a community level.

12.8 Resilience in complex systems: a conceptual model

Providing a visual representation of resilience in systems is not only extremely complex, but it is also virtually impossible. The following model is, thus, an elementary attempt at explaining these integrated and dependent processes (linked to a more in-depth explanation of the components that follows this section). To form a visual picture of a resilient system, one must imagine standing in three-dimensional space

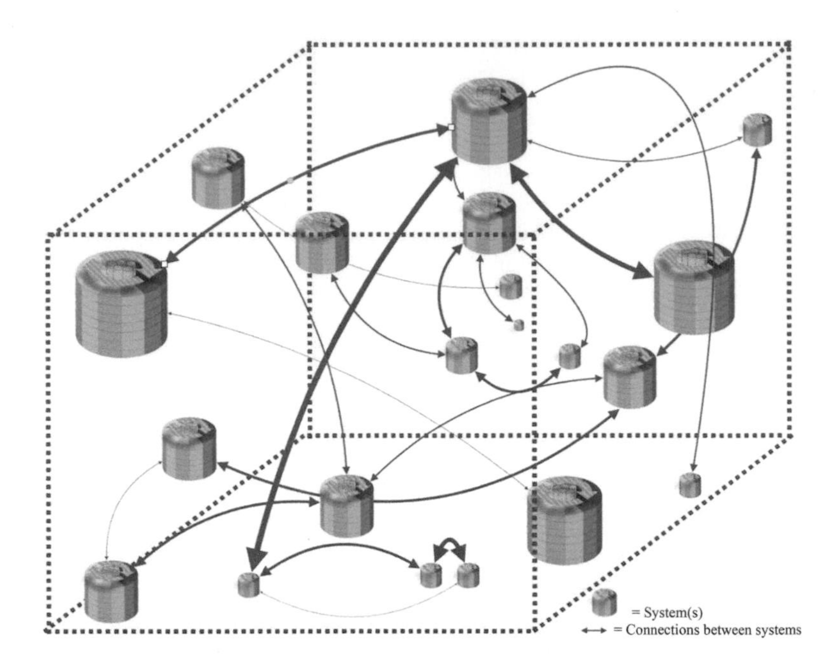

Fig 12.9
Conceptual model of linked systems.
Source: Dewald van Niekerk

(red cube in Figure 12.9). This space represents the supra-system or bigger system (of which the system being examined could be but one part). The aim of this supra-system can be various, from providing safe and sustainable urban centres, to reducing disaster risks, to developing a certain community, and so forth. Within this space, several systems exist (cylinders), which are connected in some way (arrow lines).

Within this space, various systems are needed for social, ecological, economic, political and physical functioning, and to achieve certain objectives (e.g. the Sustainable Development Goals). One can typically argue that everything in this room constitutes the resources, institutions (public and private), governments, policies, projects, communities, individuals and so forth needed within the broader objectives. These systems function at various levels and scales. These systems are linked, (inter)dependent, nested and/or integrated. The dependencies between these systems are determined by a complex web of rules, much of which is informal, non-written and not always well understood.

The main feature of the model is the representation of a system as a hollowed pie (Figure 12.10), divided into many wedges. Each of these wedges represents variables, processes, sectors or components (and even other systems), which in essence constitutes a system (or system of systems). Figure 12.10 represents a resilient community.

However, for the system to exhibit resilient behaviour, certain characteristics must be present, as discussed in the sections earlier. In the model, these characteristics are represented by the vertical layers in each wedge (Figure 12.11) (making the hollowed pie a hollowed cylinder).

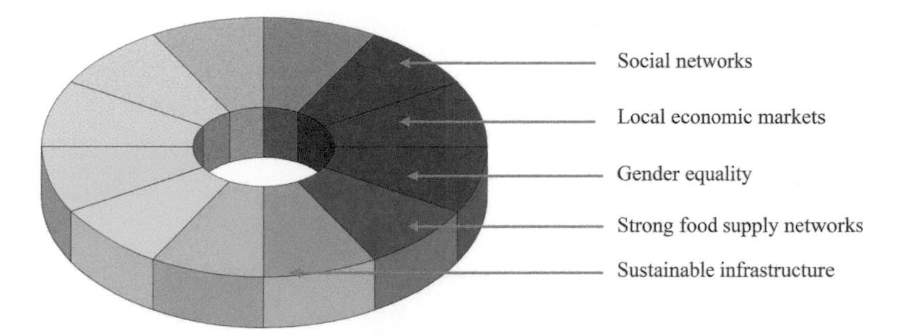

Fig 12.10
A community system comprising of various elements/components. Source: Dewald van Niekerk

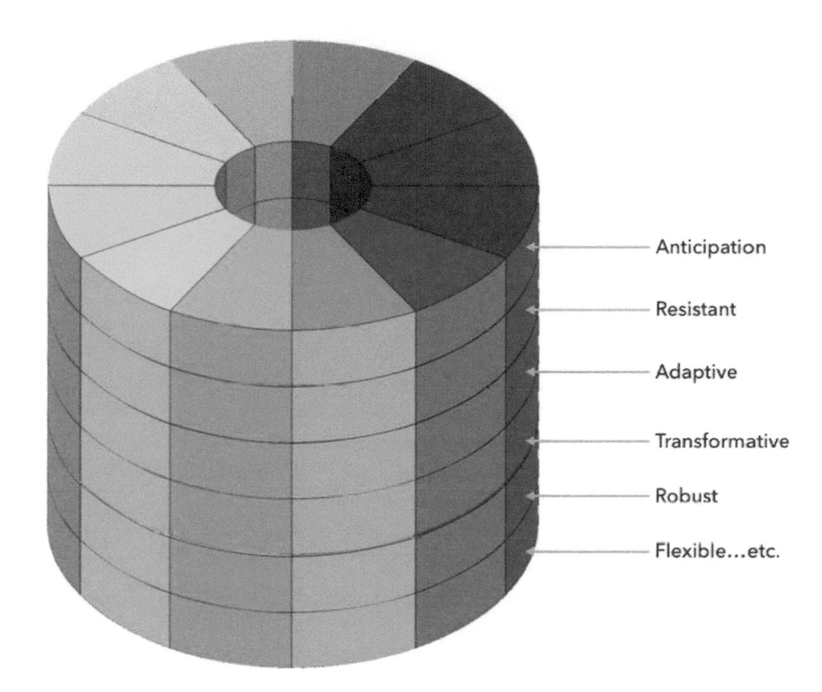

Fig 12.11
Resilient characteristics of a system. Source: Dewald van Niekerk

Literature indicates that the more of these characterise are present, the more resilient a system will be.

Therefore, the various resilient characteristics of each component (wedge) of a system can be determined, which in term will tell one a lot about how resilient this specific element/component of the system is (or even the system as a whole). Therefore, the assumption in our ideal model is that each of these elements is totally resilience (all the same size).

However, this is in the ideal world, and not all characters will be equally present or functional. Therefore, our component might exhibit various resilience characteristics but not all to the same degree (see the thickness of the wedges in Figure 12.13).

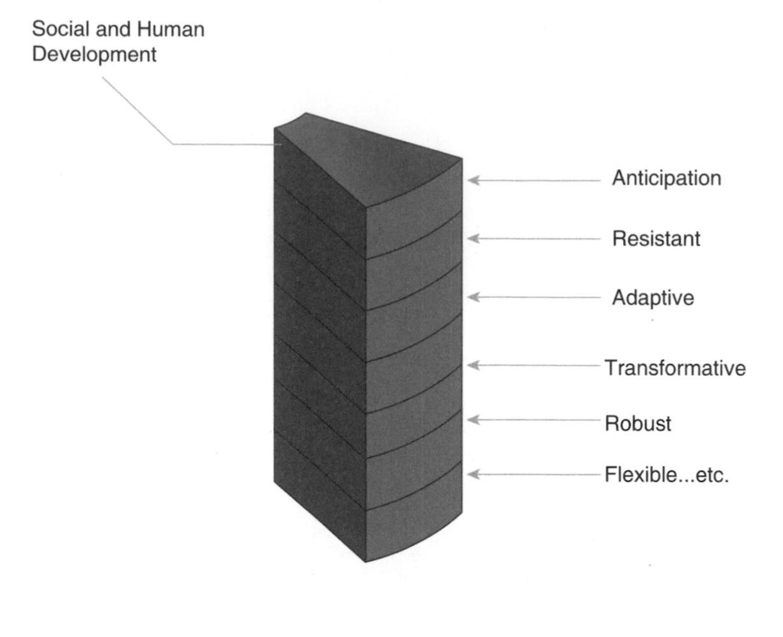

Fig 12.12
A social and human development system with all resilient characteristics present.
Source: Dewald van Niekerk

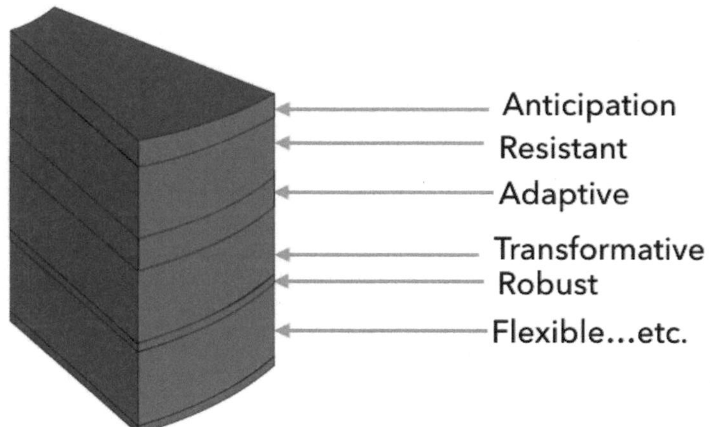

Fig 12.13
Varied levels of characteristics within a system/component.
Source: Dewald van Niekerk

The foundation of any system is related to the resilience pivots, which it needs for functioning. In the case of this explanation, the five domains of capital (social, natural, physical, economic and human) form the pivots of our community systems (Figure 12.14).

The various characteristics of a resilient system should also be present in each of these capital domains because they function as systems on their own and provide inputs (and/or outputs and feedbacks) to other systems (Figure 12.15).

One of the characteristics of a resilience system is the various connections between components, processes and other systems (see the varied lines connecting systems in Figure 12.9; Some of these connections are stronger than others). These connections can be input, output or

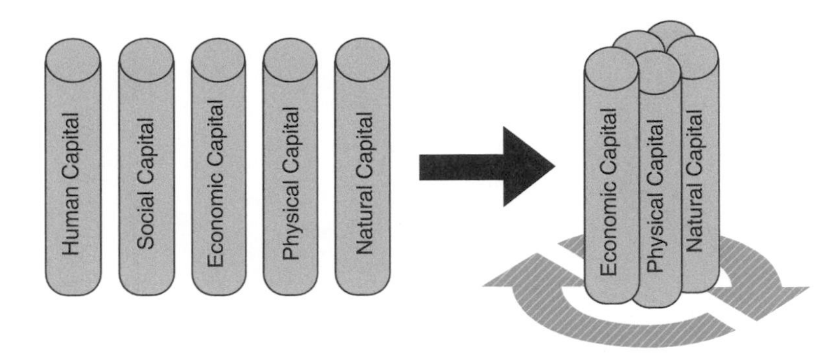

Fig 12.14
Capital domains as the pivots of a resilient community system.
Source: adapted from Rotarangi and Stephenson (2014)

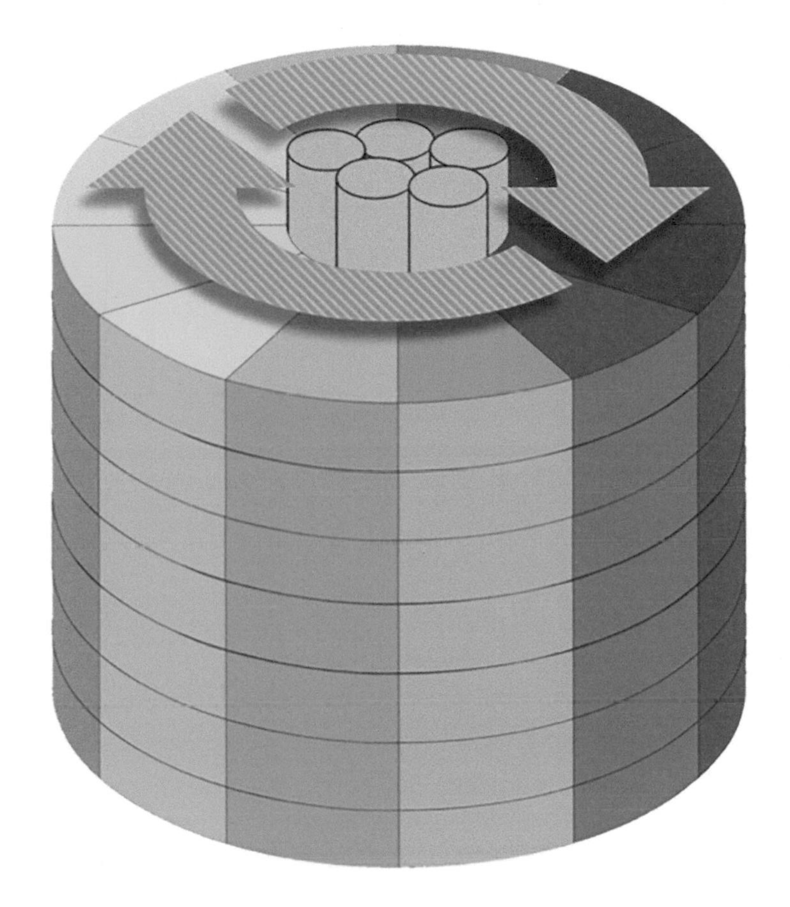

Fig 12.15
Pivots form an integrated part of any system.
Source: Dewald van Niekerk

feedback loops. The stronger and/or more rigid these connections, the less resilient the system might be.

Resilience thinking also alludes to the fact that each system finds itself within some form of an adaptive cycle (Figure 12.16). The adaptive cycle is characterised by four components of exploitation (e.g. the start of a new system or implementation of a project), conservation (e.g. implementing or working towards sustainability of a system), release

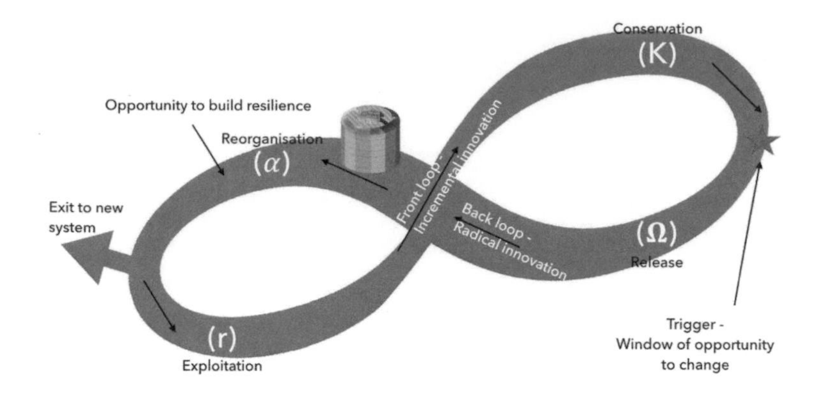

Fig 12.16
The adaptive cycle.
Source: adapted from Scheffer et al. (2002)

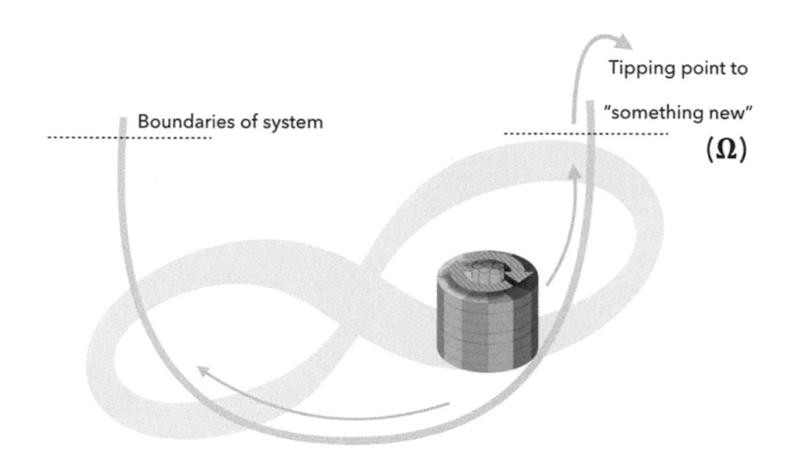

Fig 12.17
Resilience bowl.
Source: adapted from Holling (1973)

(e.g. approaching the end of a system) and reorganisation (e.g. finding a new iteration due to learning and feedback loops).

The adaptive cycle shows that all systems follow a similar path that might lead to sustainability through reorganisation or even a total change of a system (which will bring about a new system). However, there is a window of opportunity that might occur once a system finds itself stressed or under pressure to change. Such a window might also relate to a trigger, such as a natural hazard, which forces the system into reorganisation or a new system state. However, one can also imagine that the system, on its adaptive cycle path, also functions within a resilience bowl (Figure 12.17) or different development pathways (Figure 12.18). This bowl is an indication of the depth of resilience and the boundaries or thresholds of the system. So, too, the width of the tunnel in Figure 12.18 represents the tolerance of the system to external disturbances (Elmqvist et al., 2019). If a system progresses beyond the boundaries of the system, it runs the risk of tipping into something else – a new system with new characteristics.

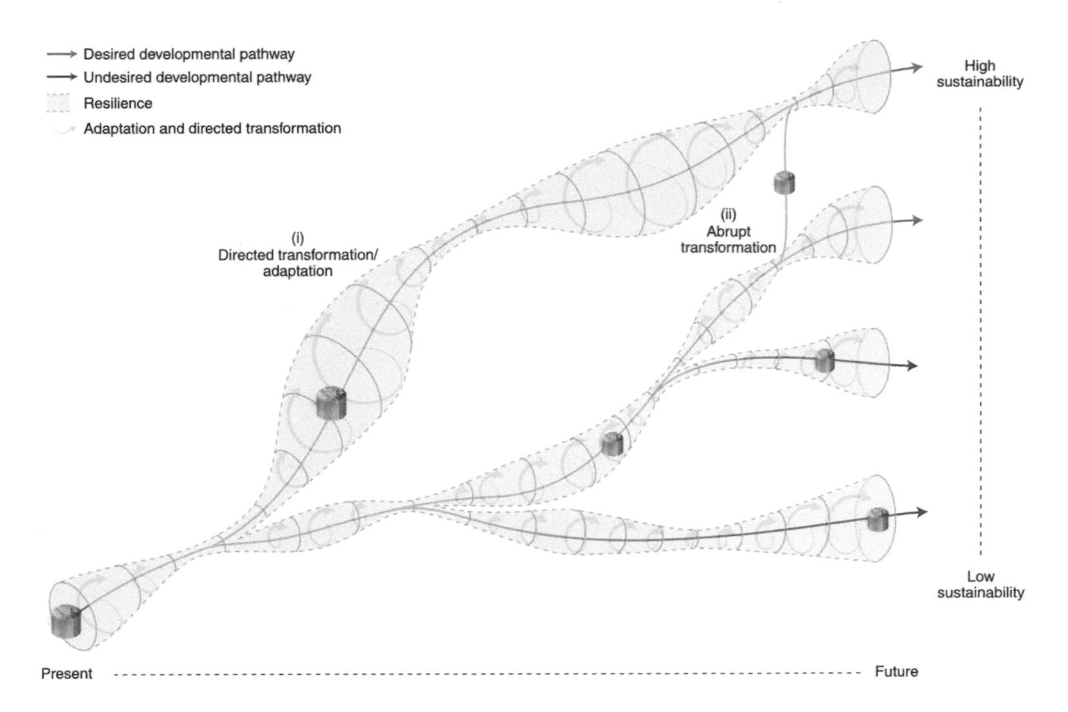

Fig 12.18
Interlinkages between sustainability, resilience and transformations.
Source: adapted from Elmqvist et al. (2019)

Figure 12.18 shows how a resilient system can follow various developmental pathways. Resilience is, therefore, the capacity to "adhere to, or simply strengthen" various development pathways (Elmqvist et al., 2019). These pathways may all be quite resilient; however, some might lead to less desirable development outcomes. The presence of the discussed resilience characteristics will allow a system to transform, adapt and change according to the feedback loops and signals in its external and internal environments. Various systems also exhibit dependencies on other systems or components of other systems. System dependencies can have a profound impact on the adaptiveness of a system (within its adaptive cycle/development pathway).

Resilience in CAS is, therefore, a key feature of which offsets disaster risks. The section that follows will delve more into why resilience for DRR is needed.

12.9 Resilience for disaster risk reduction

Resilience thinking is shaping current DRR discourse, as seen firstly in the HFA (2005–2015) and its successor, the SFDRR (2015–2030). Resilience is also understood to be the common thread across the major post-2015 global development frameworks that also focus on CCA and sustainable development through the Paris Agreement on

Climate Change and the 2030 Agenda for Sustainable Development, respectively. Globally, the concept of risk-informed sustainable development is increasingly gaining traction in policy, research and practice, and its bedrock is to build resilience. The orientation of development towards resilience is compelled by the imminent need to curtail human and financial losses associated with disasters. More so, the projected negative impacts of climate change are likely to further debilitate development efforts across the globe, with most developing regions likely to be disproportionately affected. The complex interactions between various forms of risk mandate an amalgamated approach, which can be adequately driven by a resilience perspective. When viewed within the context of DRR, resilience helps shape long-term sustainable development. Resilience allows the anticipation of disaster impacts, which requires the application of DRM. As such, steps are taken to understand and address anticipated disaster risks, and interactions thereof, to make risk-informed decisions. One key tenet of resilience that comes in relevant to this aspect is the issue of social learning and the ability to reorganise based on feedback loops (Mayunga, 2007). Whilst generally within DRR, emphasis is placed on experiential learning, the systems perspective of resilience illuminates the importance of social learning, where risk information and knowledge may be passed across heterogeneous social groups at multiple scales. This means that in any system, experience does not necessarily have to be the best teacher, but rather the experience and knowledge of others can also teach and inform decisions on how disaster risks may be prevented or reduced or managed. On this aspect, it is also important to pay attention to issues of context-specificity (Boal & Schultz, 2007; Zhou et al., 2010), which should further guide ways in which social learnings may be contextualised and applied to different risk scenarios. This illustrates one way of taking a proactive anticipatory strategy for resilience in DRR, and ultimately, when adequate good resilience measures are put in place, it helps minimise the recovery time should disaster strike. Hence, it is always said that a resilient system or community builds back or bounces forward faster (Mayunga, 2007), getting back on its outlined or desired development trajectory.

The intention to bounce forward or build back better needs to also pave the way for embracing transformation, without simply being limited to persistence and adaptability (Béné et al., 2016). In this way, communities and systems should be aiming to surpass the previously existing normalcy before a disaster event occurred (Manyena et al., 2011; Mayunga, 2007). When viewed in the context of DRR where vulnerability reduction is the core, this means that resilience efforts need to also confront and tackle any inequalities and injustices driving vulnerability and the resultant diverse forms of risk interactions existing in a system. In this context then, bouncing forward means that steps are taken to address inequality and injustice so that those on its receiving end, in terms of risk and vulnerability, emerge in a better place after a disaster. For instance, concerning access to, control and ownership of resilience capitals, the transformation may mean policy reforms that create equal and equitable opportunities for disadvantaged groups. Increasingly, discourse on equitable resilience to reduce disaster risk for diverse disadvantaged groups is gaining traction (Khoza et al., 2020; Matin et al., 2018). This is also valuable in the context of long-term risk-informed sustainable development, where the aim is to leave no one behind. To ensure that no one is left behind, resilience efforts may need to disrupt the status quo in certain contexts.

The relationship between resilience and DRR is also in terms of preparedness for disaster that may arise from residual risk. This means that any intention to build back better or move on after a disaster event needs to be anchored by adequate preparedness of what may unfold when a hazard strikes where residual risk exists. Failure to prepare most likely means that the resilience process is likely to ultimately fail. Therefore, for the entire process of resilience-building to succeed, it needs to be informed by various preparedness activities that ensure there are generation, management and appropriate flow of risk information throughout the system (Coetzee et al., 2016; Opitz-Stapleton, 2019). This can be achieved through preparedness activities, such as functional early warning systems, drills, simulations, preparedness plans, amongst others, which, when appropriately undertaken, help build efficiency and effectiveness into the response and recovery phase. In so doing, allowing for learning and improvements to be made in different components of the system is needed.

It is worth noting that resilience is often cited as a moving target and, hence, cannot be taunted as any specific static outcome but rather a process involving diverse components and actors within a system that is affected by multiple hazards (Norris et al., 2008). In light of this, the impetus is to appreciate that resilience can be construed differently across heterogeneous groups of people and system components, and therefore, resilience needs may vary even within the same community based on context. As such, it is important to involve the communities targeted for resilience-building to decipher what their resilience needs and resilience capital gaps are at any given time. This brings in the important role that can be played by DRR concepts, such as community-based DRR (CBDRR), which facilitate participatory community engagement. CBDRR is an appropriate vehicle within which the traditional unidirectional top-to-bottom resilience approach may be opened up for better community participation. Aspects of DRR, such as ecosystems-based DRR (eco-DRR), are also critical to understanding resilience from a systems perspective, as they acknowledge the interdependence of humans and the natural resource endowments and ecosystem services that should be exploited sustainably.

After a disaster has occurred, the recovery phase also needs to be informed by prospective risk management to ensure that new risks are not created, or that previous risks are not exacerbated. However, uncertainty, especially in the face of changing global climatic conditions, means that no single system may possess foolproof resilience. Therefore, as disasters occur, systems are likely to persist, adapt, fail and change, and possible instances for new opportunities may even be present (Coetzee et al., 2016). The pursuit of new opportunity trajectories (Figure 12.18) still requires that a resilience lens be applied to ensure that even when future hazards strike, communities may have the ability and capacity to bounce forward. Generally, most of the natural hazards tend to be geographically specific, and we have developed various ways to build resilience to these; for example, to cyclones and hurricanes in coastal zones. The events of the 2020 COVID-19 global pandemic, which brought almost the entire world to a standstill, decimated economies in developed and developing countries alike. The 2020 global pandemic panned out differently in each of the affected countries, but one thing made apparently was that a system may be resilient to one hazard and still be vulnerable to another, especially when resilience is pursued in a hazard-specific approach, as has tended to be the case in many situations. For instance, in parts of Africa, where countries are dependent on rain-fed agriculture, resilience has been singularly focused on building the resilience of the agriculture sector to climate risks and has remained a preserve of the agriculture discipline. This is insufficient, as it maintains the traditional silo-based approach that maintains risk reduction as discipline-specific. A resilience lens helps bring together various players, as a system is recognised to be open, with dynamic interactions occurring and influencing each other at various scales. The 2020 global pandemic gives credence to the advocates for a multistakeholder approach to resilience-building so that disaster risks of any form may be tackled and development gains protected. That said, it is important to assert that dominant dependence on hazard stimulus to initiate resilience-building may be myopic on its concern to sustainable development. Resilience must be incorporated across all sectors, with recognition of their interlinkages, which must be facilitated by enabling disaster risk governance conditions (Opitz-Stapleton, 2019).

12.10 Resilience following disasters: bouncing forward, or is it?

> Horrible in itself, disaster is sometimes a door back to paradise, the paradise in which at least we are who we hope to be, do the work we desire and each our sister's and brother's keeper.
>
> (Solnit, 2009, p. 3)

The elasticity of an elastic band, on which the notion of bouncing back in material science and engineering resilience, is what makes it return to its original position (equilibrium) after being stretched. Elastic can be stretched (not necessarily in a disaster situation) and can return to its normal position without change (Manyena et al., 2011). The focus of this form or resilience is on the virtues of stability, the level of resistance to external disturbances and the speed of return to the equilibrium. This makes the resilience of social systems a far cry from the resilience defined in material science and engineering. Instead, the resilience of social systems is related in some (still undefined) way to the resilience of the ecological systems on which social systems depend (Adger, 2000). However, just simply applying resilience concepts from the ecological sciences to social systems may fail to recognise the essential differences that exist in behaviour, processes and structures between social systems and ecological systems (Adger, 2000). This is so because social dimensions of resilience are often reflected mainly in material terms: resources, governance structures, physical infrastructure and institutions (rules, networks) (Armitage et al., 2012). Moreover, resilience is not only about being persistent or robust to disturbance, but it is also about the opportunities that disturbance opens up in terms of recombination of evolved structures and processes, renewal of the system, and emergence of new trajectories (Folke, 2006).

Unlike elastic bands or ecological systems, social systems do not have a single equilibrium or zone of stability; they are characterised by rational agents. For Cannon and Muller-Mahn (2010), the origin of the concept in physics or engineering and ecosystem makes the concept inadequate to be applied to social phenomena as is because human systems embody power relations. In this regard, Folke (2006) advocates for the integration of social dimension within resilience, to help bridge social-ecological systems thinking. This is so because humans hold the unique capacity for insight, imagination and creativity, which aid humanity in pre-empting and preparing for a change in ways that ecological communities are unable to do (Cretney, 2014). For Lister (2004), individuals are autonomous, purposive and creative actors, capable of making choices. Moreover, social systems are aware of being within an environment with a given history and with certain expectations of the future and can learn and act forward-looking in anticipation of future states (Young et al., 2006). These characteristics of social systems make changes that take place after a disaster not to be by chance; they are a result of rational choices made by the affected communities and are transformative (Manyena et al., 2011). This gives social systems an advantage to function effectively and adapt successfully in the aftermath of disasters (Norris et al., 2008). It is unimaginable for social systems to return to the pre-existing state after a disaster.

Opportunities for change

That disasters are accompanied by change (Manyena et al., 2011) or are catalysts for change (Paton, 2006) is undeniable. Indeed disasters are extraordinary events, as they force people to relinquish their usual behavioural patterns and instead follow different patterns (Kotani & Yokomatsu, 2016). Disasters also bring the opportunity for new insight and a chance to rethink basic principles (Hayward, 2013). In a resilient social-ecological system, disturbance (in this case, disaster) has the potential to create an opportunity for doing new things, innovating and developing (Folke, 2006). As such, the bounce back notion does not seem to acknowledge that disasters are accompanied by change and neither captures the changed reality nor encapsulates the new possibilities opened by the changes brought by a disaster (Manyena et al., 2011; Paton, 2006). Even if people wanted to return to a previous state, changes to the physical, social and psychological reality of social life emanating from the disaster can make this untenable (Paton, 2006). The bounce back notion does not reflect whether resilience is an active or passive action and does not denominate whether resilience happens before or after the impact (Bogardi & Fekete, 2018). Besides, the post-disaster reality, irrespective of whether it reflects the direct consequences of disaster or recovery and rebuilding activities, will present community members with a new reality that may differ in several fundamental ways from that prevailing pre-disaster. Bouncing back to a status quo that degrades the environment, increases greenhouse gases and widens inequality makes social systems more vulnerable in the long term. Simply put, bouncing back might mean a return to vulnerability and a bounce back to the conditions that caused the disaster in the first place (Manyena et al., 2011).

In her book *In a paradise built in hell: The extraordinary communities that arise in disaster*, Solnit (2009) notes that "in the moment of disaster, the old order no longer exists and people improvise rescues, shelters and communities. Thereafter a struggle takes place over whether the old order with all its shortcomings will be reimposed (bouncing back) or a new one perhaps more oppressive (worse) or perhaps more just and free (better) (bouncing forward) like a disaster utopia, will arise" (Solnit, 2009, p. 6). This assertion is in line with the third conceptualisation of resilience as presented by Vallance (2012) and Elmqvist et al. (2019), which recognises that a system may have multiple ideal states. The ability to bounce back to normal might be less important – or even less desirable – than the ability to adapt and thrive in new conditions (bouncing forward) (Vallance, 2012; Lorenz, 2013; Sudmeier-Rieux, 2014). Whether it is bouncing forward, moving on or thriving, some within the social system, when affected by disasters, will become worse than they were before the disaster, whilst others will flourish.

Social change

Participative capacity is a measure of the system's ability to change its structures about interventions by other systems decreasing the system's resilience (Lorenz, 2013). Such social resilience demands the actual community participation in response and recovery dynamics. Participative capacity speaks to the flow and distribution of power and resources in a social system (Vallance, 2012). In this regard, Stark and Taylor (2014) argue that policy designed to promote community resilience requires a redefined principle of subsidiarity so that government resources can drop down and out beyond the current borders of the state. Social change is not itself tenuous for a system but can be essential for the future persistence of the system (Lorenz, 2013). Thus, the ability to bounce forward depends as much on the collective capacity for abstract thought as it does on the technical ability and/ or financial resources (Vallance, 2012). Political, economic and natural forces operating at larger ecological levels undoubtedly influence these capacities that operate at the community level (Norris et al., 2008). In most instances, the thriving or moving on after a disaster is shaped by more traditional, top-down responses and by the dominance of managerial or technical solutions (Shaw & Maythorne, 2013), with little or no involvement of those that are affected. As Walker et al. (2004) posit, it is important to nurture and preserve the elements that enable the system (social system) to renew and reorganise itself following a massive change, and to manage resilience to prevent the system from moving to unintended system configurations in the face of external stresses and disturbance. No matter the magnitude, disasters are extraordinary events that overwhelm the vulnerable. As a process, the resilience of the social system must be built before the disaster event (preparedness), during the disaster (response) and after the disaster (recovery). Moreover, it is important to harness the changes brought by disasters during response and recovery to direct the trajectory to a positive direction so that those who are affected can become well off and flourish.

12.11 Concluding remarks

Building people's resilience is not only one way to ensure sustainable development; it contributes to reducing vulnerability. This chapter aimed to explain the concept of resilience though multiple dimensions and explore the various characteristics as identified through years of research on the topic, based on several disciplines. A conceptual model of resilience within a complex adaptive system was provided, which contextualised the interrelationship of resilience pivots and resilience characteristics. It was argued that community resilience relates to people's ability to reflect on their past, foresee alternative futures and are empowered to take steps to realise such ideals. This also means that people and the systems on which they depend must be able to adapt and even transform. However, a resilience system is also one that can resist and absorb shocks and stressors without collapsing into a new state or change its identity. This assumes a level of resourcefulness, innovation and inclusivity of all stakeholders. Resilience systems are robust systems with built-in redundancy. Therefore, they are well conceived, and they are safe-to-fail. Most importantly, resilience systems are connected but not overly connected and can self-organise as the need arise. The function of resilience in complex adaptive systems on various development paths also enjoyed attention. It was argued that using a complex adaptive systems' lens to resilience-building has multiple gains and can spearhead social and political change. Similarly, disasters provide periods of opportunity to address undesirable systems and to change systems towards better thoughtful development. In conclusion, resilience is not an end goal. Rather, it is refinement in systems that will lead to better development outcomes and alternatives.

Take-away messages

1. Resilience does not only relate to our ability to bounce back from shocks and stressors but also how we are able to anticipate and foresee such events and take corrective actions.
2. To better understand resilience, we need to understand the complexity in and between systems.
3. Complex does not always mean complicated.
4. Resilience can be understood within the dimensions of personal, community and physical resilience.
5. Resilience is not the opposite of vulnerability.
6. A resilience system has certain characteristics that could be more, or less, developed.
7. All resilience systems have certain pivots that is necessary to keep the system in its current state. Within socio-ecologically linked systems, the various capitals (human, social, economic, physical and natural) can be used to explain this phenomenon.
8. Resilience systems need to be able to adapt and change to continue functioning but should not change into another state in doing so.
9. A resilient system can follow various developmental pathways. Resilience is, therefore, the capacity to "adhere to, or simply strengthen" various development pathways.

To learn more about the topic discussed in this chapter, listen to the *Disasters: Deconstructed* interview with Dr Zenaida Delica-Willison (Figure 12.19).

Fig 12.19
QR code for Chapter 12.

Further suggested reading

Ahern, J. (2011). From fail-safe to safe-to-fail: Sustainability and resilience in the new urban world. *Landscape and Urban Planning*, *100*(4), 341–343. https://doi.org/10.1016/j.landurbplan.2011.02.021

Alexander, D. E. (2013). Resilience and disaster risk reduction: An etymological journey. *Natural Hazards and Earth System Sciences*, *13*(11), 2707–2716. https://doi.org/10.5194/nhess-13-2707-2013

Béné, C., Wood, R. G., Newsham, A., & Davies, M. (2012). *Resilience: New utopia or new tyranny? Reflection about the potentials and limits of the concept of resilience in relation to vulnerability reduction programmes* (pp. 1–61). IDS Working Papers, 2012(405). https://doi.org/10.1111/j.2040-0209.2012.00405.x
Cannon, T., & Müller-Mahn, D. (2010). Vulnerability, resilience and development discourses in context of climate change. *Natural Hazards*, *55*(3), 621–635. https://doi.org/10.1007/s11069-010-9499-4
Coetzee, C., van Niekerk, D., & Raju, E. (2018). Should all disaster risks be reduced? A Perspective from the systems concept of the edge of chaos. *Environmental Hazards*, *17*(5), 470–481. https://doi.org/10.1080/17477891.2018.1463912
Elmqvist, T., Andersson, E., Frantzeskaki, N., McPhearson, T., Olsson, P., Gaffney, O., Takeuchi, K., & Folke, C. (2019). Sustainability and resilience for transformation in the urban century. *Nature Sustainability*, *2*(4), 267–273. https://doi.org/10.1038/s41893-019-0250-1

References

Adger, W. N. (2000). Social and ecological resilience: Are they related? *Progress in Human Geography*, *24*(3), 347–364. https://doi.org/10.1191/030913200701540465
Adger, W. N. (2003). Social capital, collective action, and adaptation to climate change. *Economic Geography*, *79*(4), 387–404. https://www.jstor.org/stable/30032945. https://doi.org/10.1111/j.1944-8287.2003.tb00220.x
Ahern, J. (2011). From fail-safe to safe-to-fail: Sustainability and resilience in the new urban world. *Landscape and Urban Planning*, *100*(4), 341–343. https://doi.org/10.1016/j.landurbplan.2011.02.021
Ahmed, R., Seedat, M., van Niekerk, A., & Bulbulia, S. (2004). Discerning community resilience in disadvantaged communities in the context of violence and injury prevention. *South African Journal of Psychology*, *34*(3), 386–408. https://doi.org/10.1177/008124630403400304
Alexander, D. E. (2013). Resilience and disaster risk reduction: An etymological journey. *Natural Hazards and Earth System Sciences*, *13*(11), 2707–2716. https://doi.org/10.5194/nhess-13-2707-2013
Ali, A. D., Harder, A., Lindsey, A., Lundy, L., & Roberts, T. G. (2020). Personal resilience and coping abilities of extension agents post-hurricane. *Journal of Agricultural Education and Extension*, *26*(1), 97–112. https://doi.org/10.1080/1389224X.2019.1690013
Armitage, D., Béné, C., Charles, A. T., Johnson, D., & Allison, E. H. (2012). The interplay of well-being and resilience in applying a social-ecological perspective. *Ecology and Society*, *17*(4). https://doi.org/10.5751/ES-04940-170415
Ashby, W. R. (1960). *Design for a brain: The origin of adaptive behaviour*. Wiley.

Bebbington, A. (1999). Capitals and capabilities: A framework for analyzing peasant viability, rural livelihoods and poverty. *World Development*, *27*(12), 2021–2044. https://doi.org/10.1016/S0305-750X(99)00104-7

Becker, P. (2009). Grasping the Hydra: The need for a holistic and systematic approach to disaster risk reduction. *Jàmbá: Journal of Disaster Risk Studies*, *2*(1), 1–13. https://doi.org/10.4102/jamba.v2i1.12

Béné, C., Headey, D., Haddad, L., & von Grebmer, K. (2016). Is resilience a useful concept in the context of food security and nutrition programmes? some conceptual and practical considerations. *Food Security*, *8*(1), 123–138. https://doi.org/10.1007/s12571-015-0526-x

Béné, C., Wood, R. G., Newsham, A., & Davies, M. (2012). *Resilience: New utopia or new tyranny? Reflection about the potentials and limits of the concept of resilience in relation to vulnerability reduction programmes* (pp. 1–61). IDS Working Papers, 2012(405). https://doi.org/10.1111/j.2040-0209.2012.00405.x

Berkes, F., Colding, J., & Folke, C. (2003). *Navigating social-ecological systems: Building resilience for complexity and change*. Cambridge University Press.

Boal, K. B., & Schultz, P. L. (2007). Storytelling, time, and evolution: The role of strategic leadership in complex adaptive systems. *Leadership Quarterly*, *18*(4), 411–428. https://doi.org/10.1016/j.leaqua.2007.04.008

Bogardi, J. J., & Fekete, A. (2018). Disaster-related resilience as ability and process: A concept guiding the analysis of response behaviour before, during and after extreme events. *American Journal of Climate Change*, *7*(1), 54–78. https://m.scirp.org/papers/83376. https://doi.org/10.4236/ajcc.2018.71006

Boulding, K. E. (1956). General systems theory—the skeleton of science. *Management Science*, *2*(3), 197–208. https://www.jstor.org/stable/2627132. https://doi.org/10.1287/mnsc.2.3.197

Buckley, W. F. (1968). Society as a complex adaptive system. In W. F. Buckley (Ed.), *Modern systems research for the behavioral scientist: A source book*. Aldline Publishing.

Cannon, T., & Müller-Mahn, D. (2010). Vulnerability, resilience and development discourses in context of climate change. *Natural Hazards*, *55*(3), 621–635. https://doi.org/10.1007/s11069-010-9499-4

Coetzee, C., Van Niekerk, D., & Raju, E. (2016). Disaster resilience and complex adaptive systems theory: Finding common grounds for risk reduction. *Disaster Prevention and Management*, *25*(2), 196–211. https://doi.org/10.1108/DPM-07-2015-0153

Coetzee, C., Van Niekerk, D., & Raju, E. (2018). Should all disaster risks be reduced? A Perspective from the systems concept of the edge of chaos. *Environmental Hazards*, *17*(5), 470–481. https://doi.org/10.1080/17477891.2018.1463912

Cote, M., & Nightingale, A. J. (2012). Resilience thinking meets social theory: Situating social change in socio-ecological systems (SES)

research. *Progress in Human Geography*, *36*(4), 475–489. https://doi.org/10.1177/0309132511425708
Cretney, R. (2014). Resilience for whom? Emerging critical geographies of socio-ecological resilience. *Geography Compass*, *8*(9), 627–640. http://doi.org/10.1111/gec3.12154
Cutter, S. L., Barnes, L., Berry, M., Burton, C., Evans, E., Tate, E., & Webb, J. (2008). A place-based model for understanding community resilience to natural disasters. *Global Environmental Change*, *18*(4), 598–606. https://doi.org/10.1016/j.gloenvcha.2008.07.013
Elmqvist, T., Andersson, E., Frantzeskaki, N., McPhearson, T., Olsson, P., Gaffney, O., Takeuchi, K., & Folke, C. (2019). Sustainability and resilience for transformation in the urban century. *Nature Sustainability*, *2*(4), 267–273. https://doi.org/10.1038/s41893-019-0250-1
Engle, N. L. (2011). Adaptive capacity and its assessment. *Global Environmental Change*, *21*(2), 647–656. https://doi.org/10.1016/j.gloenvcha.2011.01.019
Fafchamps, M., & Lund, S. (2003). Risk-sharing networks in rural Philippines. *Journal of Development Economics*, *71*(2), 261–287. https://doi.org/10.1016/S0304-3878(03)00029-4
Flanagan, B. E., Gregory, E. W., Hallisey, E. J., Heitgerd, J. L., & Lewis, B. (2011). A social vulnerability index for disaster management. *Journal of Homeland Security and Emergency Management*, *8*(1), 1–22. https://doi.org/10.2202/1547-7355.1792
Folke, C. (2006). Resilience: The emergence of a perspective for social–ecological systems analyses. *Global Environmental Change*, *16*(3), 253–267. https://doi.org/10.1016/j.gloenvcha.2006.04.002
Folke, C., Biggs, R., Norström, A. V., Reyers, B., & Rockström, J. (2016). Social-ecological resilience and biosphere-based sustainability science. *Ecology and Society*, *21*(3), Art. 41. https://doi.org/10.5751/ES-08748-210341
Folke, C., Carpenter, S. R., Walker, B., Scheffer, M., Chapin, T., & Rockström, J. (2010). Resilience thinking: Integrating resilience, adaptability and transformability. *Ecology and Society*, *15*(4). http://www.ecologyandsociety.org/vol15/iss4/art20/; https://doi.org/10.5751/ES-03610-150420
Gibbs, L., Waters, E., Bryant, R. A., Pattison, P., Lusher, D., Harms, L., Richardson, J., MacDougall, C., Block, K., Snowdon, E., Gallagher, H. C., Sinnott, V., Ireton, G., & Forbes, D. (2013). Beyond bushfires: Community, resilience and recovery—a longitudinal mixed method study of the medium to long term impacts of bushfires on mental health and social connectedness. *BMC Public Health*, *13*(1), 1036. https://doi.org/10.1186/1471-2458-13-1036
Gunderson, L. (2010). Ecological and human community resilience in response to natural disasters. *Ecology and Society*, *15*(2), 18. https://doi.org/10.5751/ES-03381-150218
Hannah, L. (2010). A global conservation system for climate-change adaptation. *Conservation Biology*, *24*(1), 70–77. https://doi.org/10.1111/j.1523-1739.2009.01405.x

Hanifan, L. J. (1916). The rural school community center. *Annals of the American Academy of Political and Social Science*, *67*(1), 130–138. https://www.jstor.org/stable/1013498; https://doi.org/10.1177/000271621606700118

Hayward, B. M. (2013). Rethinking resilience: Reflections on the earthquakes in Christchurch, New Zealand, 2010 and 2011. *Ecology and Society*, *18*(4). https://doi.org/10.5751/ES-05947-180437

Headey, D., Taffesse, A. S., & You, L. (2014). Diversification and development in pastoralist Ethiopia. *World Development*, *56*, 200–213. https://doi.org/10.1016/j.worlddev.2013.10.015

Hickford, A. J., Blainey, S. P., Ortega Hortelano, A., & Pant, R. (2018). Resilience engineering: Theory and practice in interdependent infrastructure systems. *Environment Systems and Decisions*, *38*(3), 278–291. https://doi.org/10.1007/s10669-018-9707-4

Holland, J. (1992). Complex adaptive system theory. *Dædalus*, *121*(1), 17–30. https://www.jstor.org/stable/20025416

Holling, C. S. (1973). Resilience and stability of ecological systems. *Annual Review of Ecology and Systematics*, *4*(1), 1–23. https://doi.org/10.1146/annurev.es.04.110173.000245

Jeans, H., Castillo, G. E., & Thomas, S. (2017). *Absorb, adapt, transform: Resilience capacities*. Oxfam GB. https://policy-practice.oxfam.org.uk/publications/absorb-adapt-transform-resilience-capacities-620178

Kerner, D. A., & Thomas, J. S. (2014). Resilience attributes of social-ecological systems: Framing metrics for management. *Resources*, *3*(4), 672–702. https://doi.org/10.3390/resources3040672

Khoza, S., Van Niekerk, D., & Nemakonde, L. D. (2020). Rethinking climate-smart agriculture adoption for resilience-building among smallholder-farmers: Gender-sensitive adoption framework. In W. Leal (Ed.), *African handbook on climate change adaptation*. Springer Nature.

Kotani, H., & Yokomatsu, M. (2016). Natural disasters and dynamics of "a paradise built in hell": A social network approach. *Natural Hazards*, *84*(1), 309–333. https://doi.org/10.1007/s11069-016-2432-8

Kroeze, C., Caniëls, M. C., Huitema, D., & Vranken, H. (2017). Editorial overview: Learning and innovation in resilient systems. *Current Opinion in Environmental Sustainability*, *28*, iv–vi. https://doi.org/10.1016/j.cosust.2017.11.004

Lake, P. S. (2012). Resistance, resilience and restoration. *Ecological Management and Restoration*, *14*(1), 20–24. https://doi.org/10.1111/emr.12016

Laroche, M., Mérette, M., Ruggeri, G. C., & Merette, M. (1999). On the concept and dimensions of human capital in a knowledge-based economy context. *Canadian Public Policy/Analyse de Politiques*, *25*(1), 87–100. https://doi.org/10.2307/3551403

Levine, S., Ludi, E., & Jones, L. (2011). *Rethinking support for adaptive capacity to climate change—the role of development interventions*.

A report for the Africa Climate Change Resilience Alliance. Overseas Development Institute.
Lister, R. (2004). *Poverty*. Polity Press.
Longstaff, P. H., Armstrong, N. J., Perrin, K., Parker, W. M., & Hidek, M. A. (2010). Building resilient communities: A preliminary framework for assessment. *Homeland Security Affairs*, *6*(3), 1–23. https://www.hsaj.org/articles/81
Lorenz, D. F. (2013). The diversity of resilience: Contributions from a social science perspective. *Natural Hazards*, *67*, 7–24. https://doi.org/10.1007/s11069-010-9654-y
Maddi, S. R. (2004). Hardiness: An operationalization of existential courage. *Journal of Humanistic Psychology*, *44*(3), 279–298. https://doi.org/10.1177/0022167804266101
Magis, K. (2010). Community resilience: An indicator of social sustainability. *Society and Natural Resources*, *23*(5), 401–416. https://doi.org/10.1080/08941920903305674
Major, B., Richards, C., Cooper, M. L., Cozzarelli, C., & Zubek, J. (1998). Personal resilience, cognitive appraisals, and coping: An integrative model of adjustment to abortion. *Journal of Personality and Social Psychology*, *74*(3), 735–752. https://doi.org/10.1037//0022-3514.74.3.735
Manyena, S. B. (2006). The concept of resilience revisited. *Disasters*, *30*(4), 433–450. https://doi.org/10.1111/j.0361-3666.2006.00331.x
Manyena, S. B., O'Brien, G., O'Keefe, P., & Rose, J. (2011). Disaster resilience: A bounce back or bounce forward ability? *Local Environment*, *16*(5), 417–424. https://doi.org/10.1080/13549839.2011.583049
Matin, N., Forrester, J., & Ensor, J. (2018). What is equitable resilience? *World Development*, *109*(9), 197–205. https://doi.org/10.1016/j.worlddev.2018.04.020
Mayunga, J. S. (2007). *Understanding and applying the concept of community disaster resilience: A capital-based approach*. Paper presented at the Summer academy for Social Vulnerability and Resilience Building.
Moberg, F., & Simonsen, S. H. (2011). *What is resilience?* Stockholm Resilience Centre. http://www.stockholmresilience.org/download/18.10119fc11455d3c557d6d21/1459560242299/SU_SRC_whatisresilience_sidaApril2014.pdf
Moench, M. (2014). Experiences applying the climate resilience framework: Linking theory with practice. *Development in Practice*, *24*(4), 447–464. https://doi.org/10.1080/09614524.2014.909385
Newton, P. (2015). What's the difference between complex and complicated? *Quora*. https://www.quora.com/Whats-the-difference-between-complex-and-complicated-1
Norris, F. H., Stevens, S. P., Pfefferbaum, B., Wyche, K. F., & Pfefferbaum, R. L. (2008). Community resilience as a metaphor, theory, set of capacities, and strategy for disaster readiness. *American*

Journal of Community Psychology, *41*(1–2), 127–150. https://doi.org/10.1007/s10464-007-9156-6

Opitz-Stapleton, S. N. R., Kellett, J., Calderone, M., Quevedo, A., Peters, K., & Mayhew, L. (2019). *Risk-informed development: From crisis to resilience*. https://www.odi.org/sites/odi.org.uk/files/resource-documents/12711.pdf

O'Sullivan, T. L., Kuziemsky, C. E., Toal-Sullivan, D., & Corneil, W. (2013). Unraveling the complexities of disaster management: A framework for critical social infrastructure to promote population health and resilience. *Social Science and Medicine*, *93*, 238–246. https://doi.org/10.1016/j.socscimed.2012.07.040

Pandey, R., Jha, S. K., Alatalo, J. M., Archie, K. M., & Gupta, A. K. (2017). Sustainable livelihood framework-based indicators for assessing climate change vulnerability and adaptation for Himalayan communities. *Ecological Indicators*, *79*, 338–346. https://doi.org/10.1016/j.ecolind.2017.03.047

Paton, D. (2006). Disaster resilience: Building capacity to co-exist with natural hazards and their consequences. *Disaster Resilience: An Integrated Approach*, 3–10.

Poli, R. (2010). An introduction to the ontology of anticipation. *Futures*, *42*(7), 769–776. https://doi.org/10.1016/j.futures.2010.04.028

Poli, R. (2014). Anticipation: A new thread for the human and social sciences? *Cadmus*, *2*(1), 23–36. https://cadmusjournal.org/node/419

Pype, P., Mertens, F., Helewaut, F., & Krystallidou, D. (2018). Healthcare teams as complex adaptive systems: Understanding team behaviour through team members' perception of interpersonal interaction. *BMC Health Services Research*, *18*(1), 570. https://doi.org/10.1186/s12913-018-3392-3

Railsback, S. F. (2001). Concepts from complex adaptive systems as a framework for individual-based modelling. *Ecological Modelling*, *139*(1), 47–62. https://doi.org/10.1016/S0304-3800(01)00228-9

Reinmoeller, P., & van Baardwijk, N. (2005, Summer). The link between diversity and resilience. *MIT Sloan Management Review*. https://sloanreview.mit.edu/article/the-link-between-diversity-and-resilience/.

Rotarangi, S. J., and J. Stephenson. (2014). Resilience pivots: stability and identity in a social-ecological-cultural system. *Ecology and Society*, *19*(1): 28. http://dx.doi.org/10.5751/ES-06262-190128

Scheffer, M., Westley, F., Brock, W. A., & Holmgren, M. (2002). Dynamic interaction of societies and ecosystems—linking theories from ecology, economy and sociology. In L. H. Gunderson & C. S. Holling (Eds.), *Panarchy: Understanding transformations in human and natural systems*. Island Press.

Shaw, K., & Maythorne, L. (2013). Managing for local resilience: Towards a strategic approach. *Public Policy and Administration*, *28*(1), 43–65. https://doi.org/10.1177/0952076711432578

Shimizu, M. (2012). Resilience in disaster management and public policy: A case study of the Tohoku disaster. *Risk, Hazards and Crisis in Public Policy*, *3*(4), 40–59. https://doi.org/10.1002/rhc3.17

Skyttner, L. (2005). *General system theory: Problems, perspective, practice*. World Scientific Publishing.

Šlaus, I., & Jacobs, G. (2011). Human capital and sustainability. *Sustainability*, *3*(1), 97–154. https://doi.org/10.3390/su3010097

Solnit, R. (2009). *A paradise built in hell: The extraordinary communities that arise in disaster*. Penguin Books.

Stark, A., & Taylor, M. (2014). Citizen participation, community resilience and crisis-management policy. *Australian Journal of Political Science*, *49*(2), 300–315. https://doi.org/10.1080/10361146.2014.899966

Sudmeier-Rieux, K. I. (2014). Resilience—an emerging paradigm of danger or of hope? *Disaster Prevention and Management*, *23*(1), 67–80. https://doi.org/10.1108/DPM-12-2012-0143

Twigg, J. (2001). *Sustainable livelihoods and vulnerability to disasters*. Benfield Greig Hazard Centre, Disaster Management Working Paper 2/2001. http://lib.riskreductionafrica.org/bitstream/handle/123456789/1120/sustainable%20livelihood%09s%20and%20vulnerability%20to%20disasters.pdf?sequence=1

United Nations. (2016). *General assembly: Report of the open-ended intergovernmental expert working group on indicators and terminology relating to disaster risk reduction*. https://www.preventionweb.net/files/50683_oiewgreportenglish.pdf

Usamah, M., Handmer, J., Mitchell, D., & Ahmed, I. (2014). Can the vulnerable be resilient? Co-existence of vulnerability and disaster resilience: Informal settlements in the Philippines. *International Journal of Disaster Risk Reduction*, *10*, 178–189. https://doi.org/10.1016/j.ijdrr.2014.08.007

Vallance, S. A. (2012, April 16–18). *Urban resilience: Bouncing back, coping, thriving*. Paper presented at Earth: FIRE and Rain, Australian & New Zealand Disaster and Emergency Management Conference.

Van Niekerk, D., & Terblanché-Greeff, A. (2017). Anticipatory risk reduction. In R. Poli (Ed.), *Handbook of anticipation*. Springer.

Walker, B., Holling, C. S., Carpenter, S. R., & Kinzig, A. P. (2004). Resilience, adaptability and transformability in social-ecological systems. *Ecology and Society*, *9*(2). http://www.ecologyandsociety.org/vol9/iss2/art5/; https://doi.org/10.5751/ES-00650-090205

Walker, B., & Salt, D. (2006). *Resilience thinking: Sustaining ecosystems and people in a changing world*. Island Press.

Young, O. R., Berkhout, F., Gallopin, G. C., Janssen, M. A., Ostrom, E., & van der Leeuw, S. (2006). The globalization of socio-ecological systems: An agenda for scientific research. *Global Environmental Change*, *16*(3), 304–316. https://doi.org/10.1016/j.gloenvcha.2006.03.004

Zhou, H., Wang, J., Wan, J., & Jia, H. (2010). Resilience to natural hazards: A geographic perspective. *Natural Hazards*, *53*(1), 21–41. https://doi.org/10.1007/s11069-009-9407-y

Zweig, P. J. (2015). Everyday hazards and vulnerabilities amongst backyard dwellers: A case study of Vredendal North, Matzikama Municipality, South Africa. *Jàmbá: Journal of Disaster Risk Studies*, 7(1). https://doi.org/10.4102/jamba.v7i1.210

Zweig, P. J. (2017). Collaborative risk governance in informal urban areas: The case of Wallacedene temporary relocation area. *Jamba*, 9(1), 386. https://doi.org/10.4102/jamba.v9i1.386

CHAPTER 13

Moving towards disaster recovery

Fig 13.1
Nepal reconstruction after 2015 earthquake.
Source: photo courtesy of Emmanuel Raju

13.1 Introduction

Recovery is described as a complex process (Rubin et al., 1985; Llyod-Jones, 2006; Berke et al., 1993; Tierney & Oliver-Smith, 2012). The various "built, natural, and social environments, as well as institutions and economies" are "interrelated in complex ways" (Tierney & Oliver-Smith, 2012, pp. 123–124). It is synthesised by three consistent themes in the literature, namely its "non-linearity, holistic and dynamic nature" (Johnson & Hayashi, 2012, p. 228). Therefore, we know that it is not linear but must encompass the myriad complexities of various temporal and spatial scales (Tierney & Oliver-Smith, 2012; Johnson & Hayashi, 2012). Also, "recovery is not a final, identifiable state, but evolves from decisions made over time and is achieved most readily when local organisations are free to respond to their specific circumstances" (Olshansky, 2006, p. 148). It is a field where there is little consensus between authors. Bates and Peacock (1989) argue that one of the reasons for this is the interdisciplinary nature of disaster research. Many scholars

DOI: 10.4324/9781315469614-18

highlight that although physical reconstruction (Figure 13.1) after a disaster is an important component of the process, it is not the only one (Smith & Wenger, 2007), and Nigg (1995) frames disaster recovery as a social process.

Recovery is defined in many different ways. Recovery, for example, is seen as "a differential process of restoring, rebuilding and reshaping the physical, social, economic and natural environment through pre-event planning and post event actions" (Smith & Wenger, 2007, p. 237). This definition emphasises that recovery is a process shaped by several conditions occurring both before and after the disaster. The UNDRR defines recovery as "The restoring or improving of livelihoods and health, as well as economic, physical, social, cultural and environmental assets, systems and activities, of a disaster affected community or society, aligning with the principles of sustainable development and 'build back better', to avoid or reduce future disaster risk". (UNDRR, 2016). Root causes of disasters must be addressed during recovery.

Disaster recovery is one of the least well-understood aspects of DRM (Smith & Wenger, 2007; Berke et al., 1993). This may be attributed to the huge focus on post-disaster relief, which leaves little room for attention to long-term recovery or, at best, a fragmented approach (Llyod-Jones, 2006). However, at present, "notions of recovery have evolved in ways that recognise the non-linear and often iterative character of recovery" (Tierney and Oliver-Smith, 2012, p. 126). As there are no distinct boundaries between response and recovery, the Gujarat earthquake of 2001 showed that it is important to "improve coordination between a wide range of local, regional, national and international partners" (UNDP, 2001, p. 5) during the transition. Llyod-Jones (2006) highlights that despite huge improvements in the emergency response to disasters, permanent reconstruction is often inefficiently managed, uncoordinated and slow to get off the ground. Coordinating disaster recovery is one such process that requires communication and participation from many governmental departments, which is discussed in Chapter 16 in detail. Coordination between entities corresponds to activities that cannot be undertaken in isolation, and "the multiple actors involved have at least partially differing values; usually no single individual or organisation can control the process" (Robinson et al., 2000). Authors have highlighted that the failure to involve a wide range of stakeholders and poor decision-making during disaster recovery lead to disasters that are even more destructive in the future (Smith & Wenger, 2007). Further, many scholars, such as James Lewis and David Alexander, have highlighted the need to discuss corruption in disasters. The L'Aquila reconstruction, for example, was undermined with corruption and poor governance. The legal proceedings in the aftermath of the earthquake in Italy had attention, cases were resolved but families affected were left with recovery dilemmas (Alexander, 2019).

Berke and Campanella (2006, p. 194) outline the importance of disaster recovery planning in terms of providing a vision for the future; it frames future goals, builds long-term resilience and can "represent a big picture of the community that is related to broader regional, state, and national disaster response and reconstruction policies". Moreover, consistent with Smith and Birkland (2012, p. 150), effective recovery goes along with planning and coordination, and "cooperative partnerships among the actors involved in the disaster recovery network". Decisions taken during disaster recovery have long-term implications (Olshansky, 2006). Furthermore, planners have an "obligation" to play a crucial role in recovery, as affected communities "will reconstruct their lives whether or not planners participate" (ibid., p. 147). Recovery is more than just reconstruction, and "it is not neatly separable from either the response or the mitigation processes of disasters" (Dynes & Quarantelli, 1989, p. 2). Recovery is bound to conflict and bureaucracy, and studies have highlighted the need for research into the effects of institutional arrangements that may prove to be incentives or barriers to recovery (Berke et al., 1993). This is discussed in Chapter 16, on fostering disaster recovery.

Rubin (2009) noted the disappointing fact that recovery had lacked attention from researchers for over twenty years; and as stated earlier, a similar observation was made by Smith and Wenger (2007). Rubin (1985, 2009) identified leadership, the ability to act and knowledge as the three key elements of the recovery process. Moreover, Rubin (ibid.) highlighted intergovernmental relationships as crucial for effective recovery. In a compilation of case histories of recovery from disasters across the world, Johnson and Olshansky (2013) highlighted the key lessons learned by recovery organisations. The first of these is to ensure sufficient funding and its management. The second is to increase the flow of information between the various actors to ensure effective decision-making. The third is to enable collaboration and coordination between different levels of government. And the last is to handle time constraints by prioritising both immediate and long-term recovery needs. It is also a time to include planning for the future (e.g. Alexander, 2002).

A key disaster recovery principle involves taking a comprehensive integrated approach and giving importance to stakeholder participation in the process (Duxbury & Dickinson, 2007). After Hurricane Mitch struck Nicaragua in 1998, one of the key issues was a failure of the international community to understand local institutional frameworks. One amongst the many factors that contribute to successful disaster recovery depends on "how effectively many different sets of organizational relationships are able to be coordinated and managed" (IRP, 2007, p. 34). Furthermore, research has highlighted that "multi-agency collaboration is crucial to effective decision making in all aspects of disaster risk management" (Gopalakrishnan & Okada, 2007, p. 366). It is already known that confusion during reconstruction may be due to the failure of government agencies to coordinate their efforts. Similarly, Duyne Barenstein (2010, p. 150) argues that "local stakeholders, including state governments, civil society organisations and local communities, have more influence on reconstruction approaches and outcomes than international actors do". Further, recent research shows how post-Irma in Antigua and Barbuda, one must not forget historical standpoints of colonial legacies with regard to land tenure, for example, and not look at the disaster as the starting point to address recovery (Look et al., 2019). Rather, root causes and concerns of why the hazard became a disaster must be addressed. This chapter will focus on the need for vision and leadership in recovery; recovery as a social process; and how recovery must not rebuild the status quo in society.

13.2 Vision and leadership

Sustainable disaster recovery is not possible to achieve without vision and good leadership (Davis & Alexander, 2016). In disaster and humanitarian interventions, there continues to be a Band-Aid approach – needless to say, the immense focus on disaster response and relief, and less on risk reduction and avoidance of risk creation. Particularly, if disaster recovery is seen as a potential time to invest in stronger DRR measures, it is a missed opportunity if that does not materialise. This missed opportunity could potentially be due to the lack of a clear vision for disaster recovery or the combination of lack of vision and strong leadership driving the disaster recovery process in the direction of DRR.

Governments potentially have the ability to synchronise all the efforts going on the make of disaster recovery. From my own experience, after the Indian Ocean tsunami, whilst there was strong leadership within the state bureaucracy of Tamil Nadu in India, there was a lack of leadership to see disaster recovery as one single process. What happened in many instances was duplication of efforts due to almost two parallel processes running at the same time – one by the state government and other by the NGOs. This lack of synchronisation could be influenced by a lack of emphasis within institutional frameworks or legal mechanisms, or even the mere lack of an institutional setup to address recovery (Garnett & Moore, 2010). In a very interesting article Boin et al. (2013) highlight that leaders can be declared winners and losers based on their performance in crises and disaster situations. Whilst there is increasing work on accountability of states in the context of risk creation and risk reduction, there are very few examples of political leadership taking complete responsibility for failures during disaster recovery. There are some examples from the United States where governors did not run for elections based on their performance after the Hurricane Katrina (ibid.). This drives home the fact that political power can no longer be viable after an unsuccessful disaster recovery process. However, this is only visible during mega disasters, such as hurricanes and tsunamis. What we need to be speaking of more and more is about the small-scale floods that get ignored from media or find a small corner on an irrelevant page of the newspaper. One such example are the Assam floods of India. In 2019 July, the floods displaced nearly 400,000 people, and this continues to be an event every year (*The Hindu*, 2019). The larger vision for disaster recovery should be to avoid this kind of massive displacement and build stronger risk reduction initiatives by focusing on developmental failures and vulnerability of communities being affected.

"The involvement of nongovernmental actors builds the capacity of communities to deal with future disasters. The disaster experience can speed recovery and make communities more resilient when disaster strikes again" (Waugh & Streib, 2006). Whilst this statement could be true for many organisations, it is hard to believe that this is the bare truth in the context of recovery and resilience-building. For example, there were more than 700 organisations that landed in Indonesia in the aftermath of the Indian Ocean tsunami (Comfort & Kapucu, 2006). Whilst many organisations had no prior experience of the cultural or national context, many organisations were there for a short-term project. However, there are positive examples of organisations that stayed on for many years after their arrival and continue to work in the area post-tsunami. One such example that comes to mind is from India. In Nagapattinam (the worst affected district of the Indian mainland), the NGO Coordination and Resource Centre (NCRC), which was formed to coordinate activities of different NGOs in the region in disaster relief and recovery, transformed itself into an organisation called Building and Enabling Disaster Resilience of Coastal Communities (BEDROC). There seems to be a clear message of learning and vision in the way the coordination centre (which was primarily a network) was transformed and set up into a long-term organisation (Nagarajan, 2016).

Crisis management (Boin et al., 2013, p. 86) encompasses "the capacity to improvise, discover, and experiment". Leadership needs to take this as a positive space for creating new ways of addressing risk, taking stock of previous measures that have not worked in light of development and vulnerability reduction. Clear vision and leadership are keys to successful recovery in all sectors – social, economic, physical and environmental. "The failure to establish clear recovery goals and an effective implementation strategy can lead to shoddy reconstruction, a loss of jobs, a reduction in affordable housing stock, missed opportunities to incorporate mitigation into the rebuilding process, and an inability to assist the neediest recover" (Smith & Wenger, 2007, p. 239). Goals are crucial for disaster recovery, and this is discussed in Chapter 16.

13.3 Resources and organisation for recovery

Between 2008 and 2012, during the author's fieldwork in Tamil Nadu, lines between DRR and development were very blurry. In interactions with different NGOs during the tsunami recovery, many respondents from different organisations highlighted that it was extremely difficult to convince donors about the need for funding to extend recovery into development and risk reduction. This could have very well changed now with more clear visions on DRR and legislations in different countries. However, what continues to be the challenge is to draw more attention to DRR than disaster response. There are many positive initiatives in legal terms and institutionally that guide organisations and countries to focus on DRR.

Economic resources play a crucial role during recovery. In the aftermath of the Nepal earthquake of 2015, many families were struggling to find resources to rebuild their houses. The government announced a scheme where financial reimbursements for earthquake resistant/disaster-resilient housing would be done in instalments. Just outside the Kathmandu Valley, many families were struggling to build their houses. For example, in Namobuddha Municipality, families that survived the earthquake mentioned (during a fieldwork in February 2018) that since it was hard to rebuild the entire house at the same time, many of them continued to use their old/damaged house as a residence. Whilst the owner-driven approach may have given choice and power, the fact that economic resources were unavailable made it difficult for housing recovery to be a reality for some affected families. However, a great resource that cannot be measured is social in nature. A study after the Canterbury earthquakes in New Zealand reveals that pre-existing ties play a great role in responding to disasters. This goes to show the importance of building ties during recovery that benefit communities in the long run (Thornley et al., 2015).

Nepal is known to benefit from remittances of many men working in different countries in the Middle East. After the Nepal earthquake of 2015, many men returned to assist their families in the aftermath of the disaster (He, 2019). However, in the long run, the lack of remittances from abroad affected the overall economy of the household. "What is the role of economic and political power in shaping people's inability to cope with natural hazards and limiting initiatives to modify or remove root causes of vulnerability? A clue to the answer is the fact that even less discussed as 'disasters' are the day-to-day outcomes of political, social and economic policies in the disaster literature" (Mascarenhas & Wisner, 2012). Disasters, therefore, arise from failed development and must consider the so-called small-scale events, such as annual rains and flooding, in recovery planning. "Resilient disaster management seeks to embed positive economic growth as a positive adaptation from exposure to disorderly crises or disaster events while protecting human life" (Rogers, 2015, p. 62).

13.4 Recovery is beyond physical reconstruction

Post-disaster assessments have largely focused on tangible damages. This completely ignores intangible losses and most assessments. Research by Sou and Webber (2019) shows how the majority of research about home has been on displacement and very less on what it means to rebuild or remain in the same place. Further, the research shows "in the aftermath of disasters, the performance and behaviour of external actors such as the State and NGOs, significantly shape how intangible resources are transformed, as well as households' capacity to maintain emotional geographies of home" (ibid., p. 192). It is important for post-disaster needs assessments and discussions within the loss and damage networks in the context of climate change to focus on non-tangible losses as well.

Many times, relocation is seen as a strategy during disaster recovery. Questions are raised about why people live here. Why do people choose to live in hazardous locations? The initial chapters of the book clearly establish that vulnerability plays a key factor in disasters. People do not choose to live in hazardous locations, but the lack of choice and access to resources push them to such a brink. Relocation solves the immediate concerns of housing. However, what does it mean for the long term in terms of children's education, livelihood options/ alternate livelihoods, access to healthcare and similar development concerns? Further, much less spoken about is either the creation of new vulnerabilities and disaster risk, or even reproduction or exacerbation of existing risk.

"Temporary or permanent displacement is one of the most damaging consequences of disaster, and the exacerbating influence of climate change means that displacement is an increasingly significant concern" (Lewis et al., 2016, p. 327). Research after the Nepal earthquake of 2015 shows that even temporary relocation can contribute to exacerbating existing vulnerabilities (He et al., 2018). This research concludes "post-disaster reconstruction should be a process aimed at building resilience at affected communities, which must necessarily include consideration of the importance of dwellings as well as issues such as access to infrastructure, community cohesion, and rebuilding livelihoods" (He et al., 2018, p. 74).

In recovery, relocation studies mostly cover long-term displacement/ recovery after disasters. Many relocation strategies have failed due to non-participation or approval from local communities (Oliver-Smith, 1991). However, very little focus has been paid to temporary displacement after disasters. This is not discussed much because affected community members move back into their homes after a few weeks or months. However, these displacements, on a regular basis (sometimes yearly in many cases; for example, in flood-prone areas), can never reach the stage of recovery and DRR because disaster relief is the constant. This process of normalisation of risk is seen in many disaster-prone communities. In Bangladesh, whilst the author was leading a team conducting vulnerability and capacity assessments in communities that were heavily affected by Cyclone Sidr of 2008, even the worst affected communities ranked the traditional natural hazards (floods, cyclones) quite low as a problem. To our surprise, these communities, which were in the Sunderban region (largest mangrove area in the world), ranked tigers as the biggest threat/hazard. Traditional forms of VCAs, primarily approaching discussions with the community with a hazard lens, ignores risk perception from a broader community perspective. VCAs, as long as they are conducted through a hazard lens and not a vulnerability or development lens, will continue to ignore such details and plan a hazard-centric approach to addressing disasters (Figure 13.2). In the long run, this approach does not address structural issues of why communities are impacted in a certain manner during disasters. Why do communities have to relocate or get displaced every year during the monsoons? Which communities are worst affected by these displacements and forced relocations in the aftermath of disasters? These difficult questions do not have answers in a hazard paradigm of studying and addressing disasters.

In the wake of the 2004 tsunami recovery, housing reconstruction was seen in many places as asset replacement. This would mean that owing a house/land would be a prerequisite for a house, or most housing recovery was based on being affected by the disaster. In Nagapttinam, an interesting case of hope and strong leadership was shown. Whilst most of the houses destroyed by the tsunami itself were on the coast of Nagapttinam, many other disasters (mostly floods and droughts) affect the region on a regular basis. Many villages that were not affected by the tsunami have other hazards to think about. Housing recovery

Fig 13.2
A Vulnerability and Capacity Assessment conducted outside Dhulikhel in Nepal, in 2019, shows animal attacks as everyday disasters and realities.
Source: photo courtesy of Emmanuel Raju

Box 13.1: Dedicated to those fighting for their land

The seashore, its water and the sand,
It belonged to my community . . .
Ages that my forefathers have lived here,
Now it all may be washed away . . .
Our hope in the sea never dies,
For she is our mother, our bread . . .
The waves came to take away our homes,
Another wave to take away our people . . .
Here comes a greater wave,
That they call as 'recovery' . . .
In time, our homes lost,
Soon our hope will be taken away . . .
Can our cries be heard?
In the crests of the second waves . . .
In hope we live,
Our prayer will be,
Let our sea be ours . . . Our hope, our life!

Emmanuel Raju

Source: written by Emmanuel Raju during fieldwork in 2009

and reconstruction did not necessarily consider those outside of the beneficiary list created based on the tsunami-affected families. As indicated earlier, the region is wrecked by many other disasters, and many of those affected by floods and droughts on a regular basis were not necessarily affected by the tsunami. In one instance, there were a small number of families living by the river in a village in Nagapattinam. The tsunami housing discussions, as one of the stakeholder recalls, explained that this was seen as an opportunity to include these families (not on the tsunami beneficiary list for housing). The local NGO leadership took this matter to the local government officials. This was seen in a positive light, and discussions started about including the small number of families (not affected by the disaster) into the tsunami recovery process (Raju, 2013a). Much of the recovery process (in theory) had to do with DRR and strengthening the community to be better prepared for future disasters. In a hope context (Box 13.1), this sheds light on possible alternatives to the traditional beneficiary list recovery and how a multi-hazard approach to recovery is crucial.

In a lesser-known disaster, Typhoon Morakot, which affected Taiwan in 2009, killing more than 600 people, relocation strategies undertaken by the government during recovery has proven unsuccessful. Along with relocation came many measures of changing cropping patterns that did not yield results. Further, research highlights that community members were unhappy with these top-down decisions and found it hard "to adapt their lifestyle and customs to a new environment, causing severe emotional distress, and nearly spiralling into social disintegration" (Taiban et al., 2020, p. 13). Despite the fact that many previous disasters have shown that relocation is not the most viable first option, governments continue to look at relocation as an option, taking primarily only physical safety into account and, thus, ignoring the whole of society or a holistic approach to disaster recovery.

Whilst recovery has been the most understudied aspect in disaster studies, recovery has also been a neglected area operationally. For example, in the United States, it was only after Hurricane Katrina that a National Recovery Framework came into discussion (Garnett & Moore, 2010). During the author's fieldwork in India after the Indian Ocean tsunami between 2008 and 2012, many actors who worked in the recovery highlighted the attempt of the United Nations (UN) – one UN approach that was seen in the form of the United Nations Tsunami Recovery Support (UNTRS). However, the idea of a one-recovery approach was far from true in reality. This was seen in detail in the author's PhD thesis (Raju, 2013a). The idea was for different agencies to work together in a collaborative mode. However, this setup did not go beyond the basic form of coordination. In the long run, this created a fragmented approach to recovery, which affected the long-term strategic vision of DRR and, thereby, missing the focus on building resilience towards future disasters.

Discussions with communities in Nagapattinam in 2018 (one of the worst affected districts of the Indian Ocean tsunami and continues to see many disasters that wreck along the coast), whilst conducting vulnerability and capacity assessments in two different villages, highlighted that the housing approach was not seen as a process. The project-oriented approach to housing had clearly ignored a futuristic perspective. For example, one of the main concerns raised by community members was about maintenance and retrofitting of post-tsunami housing. When the houses were constructed in the aftermath of the tsunami over a decade ago, there were many consultations about design, culture, land size and other crucial issues. There were many communities where consultation has been very minimal (Chandrasekhar, 2010). Overall, however, there were no discussions about how to maintain the new houses and the cost involved in such an exercise. There were many houses by the coast (in Nagapattinam) that had developed many cracks in the wall. Housing was not seen as an ongoing process of development that needed consistent maintenance and retrofitting some years later. How does this affect long-term recovery? Housing is a matter that usually takes maximum prominence during disasters, given the tangible nature of the sector. So was the case after the tsunami as well. Walking through many of these housing projects from 2008 to 2012, at least two years after the houses were built, many of them had concerns on location, size and design, and more importantly, that these concerns were a result of a top-down decision-making in disaster recovery. In the long run, challenges created for communities during recovery become their problem, as many organisations leave as soon as the projects have been completed.

The Nepal earthquake of 2015 shows an interesting story of how post-earthquake reconstruction has taken a single-hazard approach. This is detrimental to DRR in the long run. A study shows that many informal construction workers had high awareness of hazards but mostly related to the built environment after the earthquake (Chmutina & Rose, 2018). The study highlights that informal construction cannot alone be held responsible for damages post-disasters with regard to the built environment but also the urban planning processes that create risk in cities. Looking at the traditional vulnerability and risk assessments, we need to develop newer tools to understand urban risk and how recovery could be used as a window to address some these risks. Further, recovery is also an opportunity to align different global frameworks to the local context, such as the climate goals, DRR (SFDRR) and sustainable development goals. "The advantages of Disaster Risk Management and Climate Change Adaptation should be presented as contributions that go beyond risk reduction and include potential benefits such as wealth creation, reduction of inequalities, job creation, improved access to basic services as well as other aspects of DRR's triple dividend" (Lavell, 2017, p. 14).

13.5 Self-recovery

Self-recovery is an even more understudied sub-topic of disaster recovery. Whilst there is always a focus on the government and the NGOs working post-disasters, there is a mass of recovery efforts that run without any form of support from any local or national government or any external organisation. "In lower-income countries, and indeed in many middle-income countries, most disaster-affected families rebuild their homes relying on their own resources, with little or no external assistance. This is commonly referred to as 'self-recovery' " (Twigg et al., 2017, p. 9). For example, in Dhaka, whilst conducting risk assessments in urban settlements in 2011, community members expressed that after the monsoon season and seasonal flooding (given the location of the community), self-recovery is the way communities get back to normalcy. This can be seen increasingly in many urban settings (for example, in Mumbai and Jakarta). The concept of build back better in the majority of cases in disasters is a parlance seen only in mega disasters. Recovery must become an integral part of DRM mandates within governmental institutions. Recovery can be used as the window of opportunity described by many researchers in the past. The research on self-recovery highlights the changing needs and priorities of the affected community through the process of recovery (ibid.). This was seen in the case of the

Fig 13.3
Everyday disasters: street markets in Hanoi, where street vendors face flooding annually, causing disruption to livelihoods. Source: photo courtesy of Emmanuel Raju

relocation study post-tsunami in India (Raju, 2013b). Moving through different junctures of disaster recovery, communities do highlight varying priorities of change, reflection, hope or discontent. In a mega disaster, recovery spans over a number of years planning housing and livelihoods, and even more particularly, if communities are being relocated. However, in a seasonal flood (which is almost never declared as a disaster), there is no clear indication of what recovery is or what the different junctures of recovery are (Figure 13.3). In any case, the lines between the phases in a traditional disaster management cycle are blurred. With the adoption of the International Cluster System, the concept of early recovery was introduced to ensure the transition between response and recovery. However, the cluster system has been heavily critiqued for not living up to the needs for what is was created (Stoddard et al., 2007).

On housing and self-recovery, it is noted that "if disaster recovery is to remain a normative and sustainable agenda, it is important to recognize and begin conversations about how to operationalize the myriad needs and concerns of households into recovery policies" (Sou, 2019, p. 152). Self-recovery is an economically taxing exercise overall. For instance, if we take the case of urban Mumbai, every year, if communities have to invest in repairing damages caused by the flood (basically without an insurance plan/assistance), that is the normal, and recovery in a true sense will be hard to achieve. Mumbai, a city that has seen displacement of slum dwellers very often since Indian independence. To add to the misery, slum development does not feature in the city's development agenda/strategy (Bhide, 2009). This affects not only housing but livelihoods, health and well-being in general. Mumbai's slums stand as a testimony to the basic premise that these inequalities in society come striking when a disaster occurs. In the Philippines, for example, a study about micro-, small- and medium-scale enterprises reported that in the aftermath of disasters, "they recovered from disasters by working longer and harder, and often using informal loans for recovery capital". (UNDRR, 2019). This form of a loan culture, in some instances, help to recover from disasters only until the next flood could potentially put communities in debt traps. One of the worst factors of keeping communities in poverty are the debt traps created, which puts land ownership and livelihoods at a huge risk. These "debt burdens can exacerbate disasters and cripple recovery efforts, further hampering development" (Abramovitz, 2001, p. 131). The majority of housing needs post-disaster are met by the affected community themselves, which begs the question of how we achieve resilient housing, as most of the affected populations would use their minimal resources to rebuild within a short duration (Parrack et al., 2014). Further, a key question of self-recovery that needs to be examined is "Can families really be said to be recovering if using salvaged and damaged materials for reconstruction leaves them more vulnerable than they were prior to the disaster?" (Humanitarian Policy Network, 2017).

Housing is a symbol of social status in many societies. Recent research highlights household recovery needs are also associated with the aspects of beauty of a house; status in society and competition in the neighbourhood (Sou, 2019). The research by Sou (2019) also highlights that we need a more anthropocentric understanding that does not aid an approach where housing is reduced to only brick and mortar (i.e. recovery and housing is beyond the physical).

13.6 Disaster recovery as a social process

Visible and invisible forms of recreating existing inequalities or creating new ones happen during the process of disaster recovery. Social inequalities are either re-created or exacerbated, or new ones are created during disaster recovery when the norm is to return to normal functioning of a society by re-establishing existing social and institutional arrangements (Gupta, Pouw and Ros-Tonen, 2015). An example of this is seen when post-disaster scenarios present an opportunity for land-grabbing through preventing the return of people back to the places affected by a disaster, as they do not have legal rights to be there, or through relocation – thus, taking away their established livelihoods (Raju, 2013b; Uson, 2017). Social pressures and neoliberal development policies (Gupta et al., 2015) exacerbate unequal access to welfare and critical facilities, such as healthcare, denying vulnerable sections of society from political, social and economic participation in disaster recovery processes. In Germany, for example, post-disaster housing reconstruction has created inequalities in post-disaster rentals for tenants, on the one hand, and influenced inequality where the application process for government funding for recovery increased marginalisation of the elderly and migrants (Kammerbauer & Wamsler, 2017).

One of the core features of disaster recovery governance must be social inclusion. Inequalities are never unidimensional. "Unresolved developmental practices" (Lavell & Maskrey, 2014, p. 271) manifest themselves in an array of economic, social and political inequalities in a myriad ways, such as access to basic services, and this phenomenon has to be explored from different perspectives. Disasters are socio-political, and economic processes are often driven by development processes and failed development, as rapid reconstruction often implies a failure to consult adequately with the affected. Little attention is paid to the disaggregation of these failures, and to pre-existing and newly created inequalities. Instead, disasters are often politicised or de-politicised (Cretney, 2017) for the wrong reasons, moving the emphasis from the root causes to climate change or other natural causes. Poverty does play a key role in the exacerbation of disaster impacts. Poverty is not only economic but has many different dimensions, including the lack of agency and voice (Krishna et al., 2018). Unless the politicisation of disasters is about shifting or negotiating power, ensuring justice and human rights, advocating for change in structural problems and dynamic pressures (not forgetting the root causes) (Wisner et al., 2004), disaster recovery will continue to be an exercise of recreating vulnerability (along with new forms of vulnerability), which in itself is the recipe for the next disaster. Recovery can very well reinforce differences and disagreements between and within communities. As mentioned about two decades ago, "the life-history of a disaster begins prior to the appearance of a specific event-focused agent" (Oliver-Smith, 1999, pp. 29–30), and in the aftermath of disasters, "vicious cycle of property loss, borrowed money, and improvisation creates a downward spiral of socioeconomic status" (Choudhury & Haque, 2016, p. 154).

13.7 Disaster recovery and the environment

Ryder argues that environmental justice and disaster vulnerability "both attend primarily to societal inequalities which manifest as environmental inequalities, and the way these environmental inequalities are reproduced across time and space. There are obvious detriments to the advancement of the literature if we continue to pursue these as separate social justice issues, but there are also practical problems when we do not pre-emptively conceptualize and address the roots of environmental oppression at their core" (Ryder, 2017, p. 97). These arguments very well extend to looking at disaster recovery from a justice perspective (Box 13.2). Recovery cannot be seen as a returning to normal only activity. Recovery is also about addressing inequalities (social, environmental and other forms of inequalities).

Box 13.2: Disaster law and policy: DRR and the environment (EIA: Environmental Impact Assessment)

In most cases, the implementation of environmental laws is the responsibility of environment ministries; hence, they are administered separately from much of the building and spatial planning regulation, and also from DRM laws. Indeed, few links are made between these sectors, even though each has a role in DRR. Environmental management is potentially a key element of DRR, especially regarding emerging risks related to climate change. Mechanisms for cross-sectoral coordination with DRM systems and mainstreaming of DRR principles into environmental laws and institutions could greatly enhance the DRR potential of such laws. A particular aspect of environmental regulation deserving of more study is the potential use of EIA as a DRR tool in the approval process for new developments. EIA of planned developments are provided for in the laws of the vast majority of the sample countries, though only Ukraine featured criteria specific to natural hazards. EIA may provide a vehicle for communities to influence planned developments and, thus, prevent or reduce the creation of risks, especially when these laws provide for community objections, consultation or participation in EIA. However, since such broad requirements do not provide specific mandates on DRR, the inclusion of DRR criteria in EIA is left to the discretion of those implementing the laws. It would, therefore, be preferable if DRR criteria were explicitly incorporated into principles for environmental management and EIA.

Source: IFRC and UNDP (2014, p. 63)

Disaster recovery must not be seen as a linear process. One of the reasons for recovery to fail is not acknowledging the complexity of interdependencies in the process (Blackman et al., 2017; Raju et al., 2018). The authors argue that disaster relief is organised with a main aim of managing the disaster. However, recovery goals need to be different (if development is the trajectory chosen). This is echoed by others to highlight that recovery needs a developmental approach, and the different set of actors need to have clear visions of long-term recovery (Raju & Becker, 2013). One such interdependency is of the linkages between the physical, social and natural environment. In Mumbai, after the 2005 floods, the Chittale committee was formed to assess the cause of the floods. This committee was aimed at understanding why the impact was so high and what measures need to be taken. The majority of the actions taken based on the recommendations of the committee were responsive in nature (how to respond better) with the exception of an action of decongesting Mumbai (Gupta, 2007). That requires heavy planning in terms of infrastructure but also developing inclusive programmes to address vulnerability. As Revi (2008) highlights, in the case of Indian cities, the majority of the problem lies with vulnerability

Box 13.3: Five reasons why ecosystems are central to disaster risk reduction

1 Human well-being depends on ecosystems that provide multiple livelihood benefits. They also increase the resilience of vulnerable people to withstand, cope with and recover from disasters resulting from hazard events, such as droughts, hurricanes, earthquakes and others.
2 Ecosystems, such as wetlands, forests and coastal systems, can provide cost-effective natural buffers against natural events and the impacts of climate change.
3 Healthy and diverse ecosystems are more resilient to extreme weather events.
4 Ecosystem degradation, especially when related to forests and peatlands, reduces the ability of natural ecosystems to sequester carbon, increasing the incidence and impact of climate change and climate-related disasters.
5 Human conflicts can cause devastation to communities similar to the effects of natural hazards and are often caused by competition over scarce natural resources. These conflicts cause further environmental degradation. Environmental management is, therefore, essential to both decrease risk of conflict and allow post-conflict recovery.

Source: International Union for Conservation of Nature and Natural Resources (IUCN) (n.d.)

compared to exposure to hazards. The opportunity of flood resilience by addressing the reasons of the flood (hazard to disaster) is lost by focusing on response-oriented measures. This case is not an exception but yet another window of opportunity lost. From a study, Platt et al. (2016, p. 459) argue that post-recovery resilience is "a trade-off between risk and economic prosperity and decisions about resilience need integrating into wider questions of economic and social futures".

In the case of Japan in the aftermath of one of the worst tsunamis of 2011, Takeuchi et al., (2014, p. 513) argue that decision makers at all levels "need to take a holistic approach based on sustainability science to understand the inter-relationships between these landscapes and ecosystems to develop a robust rebuilding plan for the affected communities". During fieldwork in 2011, in Bangladesh, in some of the worst affected cyclone areas, many people from various disaster-affected villages highlighted the need for more focus on the ecosystems in disaster recovery (Box 13.3). The local and indigenous communities in Sunderbans (in India and Bangladesh) experience multidimensional vulnerabilities in their day-to-day life. Environmental degradation, risk and hazards (particularly cyclone, tidal inundation and saline water intrusion) further expose them to disasters, which compound their socio-cultural, political and economic vulnerabilities. Sunderbans is the world's largest mangrove forest, with rich biodiversity, located at the northern tip of the Bay of Bengal, falls in both Bangladesh and

India. It is a confluence of several major river systems from both the countries. More than 100 species of mangroves, and a number of medicinal plants exist in this ecologically sensitive area in about 102 islands formed by different rivers before meeting the Bay of Bengal. The major part, about two-thirds of Sunderbans, falls under the geographical territory of Bangladesh, whilst the remaining one-third falls under India. The Sunderbans is known for the Bengal tigers. As highlighted in previous sections, during our fieldwork, whilst conducting vulnerability assessments, to our surprise, it was tigers that was ranked very high as a potential risk by the people. The main reason identified for this was the depletion of the ecosystems in the area, putting fishermen at risk on a regular basis. This brings home the need for eco-systems-based DRR to be taken into account during disaster recovery. This example also highlights the nexus between ecological and human security, which is a much-neglected subject in disaster recovery and risk reduction.

13.8 Disaster recovery and the role of law

Can law play a role in affecting disaster recovery and avoiding disaster risk creation in the aftermath of disasters? Can actors be held responsible for the negligence of not addressing vulnerability to disasters? In the aftermath of the Japanese Fukushima double disaster, a group called Complainants for the Criminal Prosecution of the Fukushima Nuclear Disaster was formed. The main aim of the joint effort was to prosecute the negligence of the government officials and the executives of the Tokyo Electric Power Company (TEPCO). The complainant's arguments were about criminal negligence of TEPCO for not taking efforts to do anything about reports that warned about tsunamis that could cause damage (Herber, 2016). Such initiatives prove to the world that legal courses of action post-disasters can help with reducing disaster risk creation (Box 13.4).

Many countries have adopted disaster law programmes across the world. These legal frameworks provide the basis not only for DRR but also to ensure and minimise/avoid disaster risk creation. Whilst implementation challenges emerge and continue to dominate, a legal framework is remarkable. For example, in Peru, the Law No. 29664 passed in 2011 created the National Disaster Risk Management System. Along with it came the strategy lines, which states, "Promote the assessment, prevention, and reduction of disaster risk, as well as emergency preparedness through financial mechanisms within the results-based budget framework and incorporate disaster risk management into public investment" (World Bank, 2016, p. 8). Similarly, in September of 2018, Mexico also passed a second version of the 2012 Law of Civil Protection, which similarly forbids the construction of risk.

Land and legal ownership of land are always issues in the making of disaster risk. Research shows that people who do not own land and mainly squatters take longer to recover from disasters (Walch, 2018). The study by Walch (ibid.) also highlights that there has been violence in the

Box 13.4: Four potential implications for future research in law and disasters

First and foremost, modern disaster research is interdisciplinary research or, at least, research requiring an interdisciplinary attitude. Disasters are, by definition, interdisciplinary objects, and studying their legal implications requires fundamental knowledge of the affected societies, technologies and natures. This is already clear from the present body of scholarship, but in line with the overall trajectories of the global research agenda, it seems reasonable to assume that we will see even more interdisciplinary research projects with implications for, and hopefully involvement of, law and disasters.

Second, as the body of scholarship on law and disasters reaches critical mass, a more comprehensive scientific landscape will emerge; a landscape where different schools or approaches to the field become clear. This will make it easier for the new reader to navigate in the scholarship and open the possibility of further exploring a number of understudied aspects of disaster regulation.

Third, the vertical (conflicts within jurisdictions) as well as horizontal (conflicts between jurisdictions) dimensions of jurisdictional issues in disaster response will increasingly put pressure on legal scholarship to form bridges between different (national, regional, global) jurisdictions. This need is already clear from the ongoing discussion on international disaster response law; however, the other disaster management phases will also increasingly be subject to discussion; for instance, as the post-2015 framework for DRR gets traction.

Final, a number of new actors and technologies have entered the disaster management arena, some of them challenging traditional types and modes of regulation. Not least, increasing reliance on technology calls for innovative legal ideas – and for a critical investigation of the relationships between law, science and technology. This might be relevant for the use of drones/robots, satellite technology and social media strategies in the response phase but also more generally for technology in protection of critical infrastructure.

Source: Lauta (2016, pp. 104–105)

context of land tenure (for example, in the Philippines and Sri Lanka) perpetrated by the state and non-state actors (also see Klein, 2008). It is extremely important to address structural issues through law/legal mechanisms. Walch (2018) notes that resilience programmes initiated by the government have done very little to address social vulnerability, and resilience is used as a political rhetoric, where the responsibility largely is shifted to vulnerable populations to draft their own path for development and to get out of vulnerability (which is largely a structural problem) (also see Béné et al., 2018). Due to the physical nature of the housing sector, 'recovery' many a time is used as a synonym for 'reconstruction'. However, there is a clear linkage between housing and livelihoods, for example, that completely gets ignored (Pomeroy et al., 2006; Raju, 2013b).

In China, for example, Xu and Shao (2020, p. 541) write that "the state also reinvents its structures and practices, such as simplifying planning procedures, breaking administrative boundaries to relocate factories from high-risk areas while securing profits for the original areas, and sending cadres to work in disaster-devastated regions to foster local planning capacity". During disaster recovery, it was found that the state became an unparalleled agent. However, it also raises crucial questions of leaving no one behind and the role of voice in the recovery process. Whilst vulnerability is an often-discussed issue in disasters (in general), recovery very often becomes about rebuilding or building back the status quo that is considered normal.

13.9 Voice in disaster recovery

Disaster recovery literature not only highlights the weakness in the way recovery is approached but also the opportunities it provides. Can geographies of hope provide a platform for disaster recovery to be used as a window of opportunity? Disaster studies is critiqued for the lack of theoretical developments in many decades. This allows us to use/borrow frames of analysis from different disciplines given the interdisciplinary nature of disaster studies. Digging deeper into the material on geographies of hope (Anderson & Fenton, 2008; Cretney, 2017) brings to mind a case after the Indian Ocean tsunami of 2004. Geographies of hope are about daily experiences at possibly a very local level, and disaster recovery could be a point of hope for transformation, contestation and engagement in a political debate of structure and power.

Research suggests that people's voice plays a role in giving power if voice is recognised and valued (Madianou et al., 2015). The old expression of "being the voice of the powerless" has to shift to ensuring people have their own voice and affected populations are able to make their position for needs in disaster recovery clear. Disaster recovery is a clear matter of human rights, and therefore, law can be a powerful tool in ensuring a space to dissent and creating an environment where people affected by the disaster have a say in the way their lives are shaped in the recovery process. The notions of build back better on paper is very hard to achieve in reality on the ground. The slogan that started as a campaign during tsunami recovery left people wondering, is this not a given, and why the hue and cry around a new slogan? It certainly had potential to talk about the structural changes needed to build societies better to avoid disasters from happening. However, the major focus was on building built environment better. Existing socio-cultural norms can be a major hindrance to the ideal recovery process. For example, in the Philippines, a study shows that unmarried single mothers were kept out of the beneficiary lists (Madianou et al., 2015). Similarly, in India, studies have shown caste plays a major role in inequality before and after disasters (Bosher et al., 2007; Jha, 2015). Whilst, as suggested earlier, geographies of hope literature presents optimism for change, recovery must be dealt with more seriously to address issues of inequality stemming from structural notions of patriarchy, caste and so forth. "We cannot understand voice without understanding voicelessness" (Madianou et al., 2015, p. 3031). This voicelessness arises from generations of structural problems of society. The silence of many affected communities post-disasters is not only because of the lack of space for voice but also because of the dominance of powerful voices. During a proposed post-tsunami relocation in Chennai in India, when the local fishing communities resisted, during the author's fieldwork, a government official remarked that the tsunami has changed people's lives because it brings a concrete roof for many of them. This remark explains the response to the voices of resistance in post-disaster recovery processes.

Participation in disaster recovery can be in its complete sense of the word, that is, in being part of framing problems and finding durable solutions to long-term recovery versus a token approach to participation (Vallance, 2015). 'Participation' is one of the most abused words by different actors in DRM. This interesting paper uses "local government elections as an indicator of residents' and communities' satisfaction with recovery outcomes and processes" (ibid., p. 1299). There are different views on whether elections and voting behaviour are influenced by disasters. A recent study in Japan showed that voters prefer local leaders in power when a disaster strikes (Nakajo, 2017). "In Christchurch, it is possible to see how the engagement of community organisations has repoliticised the recovery to challenge the priorities of government led recovery and to provide alternatives not only to participation but also to the shape and form of post-disaster recovery and reconstruction" (Cretney, 2018, p. 127). In a study in Timor Leste, it is highlighted that "devastation resulting from loss of land and livelihoods, for some people, was compounded by a lack of voice or representation in the political processes that shaped the regional context" (Bovensiepen & Meitzner Yoder, 2018, p. 387). One could argue that disaster relief (that most governments turn immediate attention to) carried more weight than disaster recovery. Whilst memories play a great role in disasters and in elections, disaster relief continues to dominate the scene. Token participation in disasters has gone on for a very long time. This is seen in cases where participation is merely a discussion, and not a collaborative effort of rebuilding livelihoods and the well-being of society from a holistic perspective. Taking a normative stand, participation is not an exercise to tick boxes but rather to drive home that community ownership is a key to building resilience in the wake of disaster recovery.

13.10 Disaster memory and recovery

"Given that commemoration is not static, particularly in the digital age, which quickens reinterpretation, we question how digital reinterpretations will remember and frame recovery" (Zavar &

Fig 13.4
A tsunami memorial in Karaikal, Tamil Nadu, India.
Source: photo courtesy of Emmanuel Raju

Schumann, 2019, p. 176). Interestingly, all words associated with recovery – reconstruction, restoration, rebuilding, rehabilitation (all *r* words) – remembering seems to be missing (ibid.) (Figure 13.4 and Box 13.5). Remembering is also closely associated with learning. Remembering past disasters in a way that facilitates learning is crucial. Commemoration is important and integral to recovery (Haas et al., 1977). In Gall, in Sri Lanka, a beautiful clock memorial stands with the time that the tsunami struck on 26 December 2004. One of the key

Box 13.5: Social vulnerability, memory and disasters

Using the case of Argentina, Ullberg writes this:

> Normalisation also refers to the process that turns a particular extraordinary condition into a regular accepted fact of the ordinary state of things. Normalisation is here inherent in the concept of adaptability or adaptation to hazard, related both to the objective and to the subjective side of things.
>
> (p. 252)

While social vulnerability is recognised as constitutive of disaster risk, responsibilities are not located with and actions are not aimed at those economic and political processes that can be said to produce vulnerability to disasters in the first place. Economic and social inequalities are rarely addressed, nor are the ecological processes that exacerbate the crises. Rather, policies have shifted to focus on the community in general and on the vulnerable subjects themselves in particular, presumably a more doable project for governments.

> Discourses of resilient communities are more often than not framed as a form of local empowerment, yet in practice they often serve the purposes of maintaining the established social order at the local level through normalization.
>
> (p. 255)
>
> Memory within the public administration was shaped by practice of exchanging administrative staff within the public agencies following the electoral time cycle, not because individual experiences are forgotten when people are exchanged, but because there are incentives to not remember past decisions and arguments. This pattern of forced exclusion also involved materialised memory of the bureaucracy, that is, documents such as plans and maps in public archives, in processes of selective remembering and forgetting. Public works of infrastructure also operated according to this logic, constituting the material traces of past disastrous events at the same time as they evoked such events. In addition, works of infrastructure are techniques to control hazardous forces such as flooding and forged the memory of past floods as well as oblivion of future risks. Technocratic narratives and calculations of risk framed how the flooding past was addressed. Future oriented contingency plans omitted addressing root causes of social vulnerability to flooding and instead framed the problem as one of human obstruction to the course of nature. The responsibility for this infliction on the environment was placed with particular people, practices and places, most notably at the urban outskirts.
>
> (pp. 183–184)

Source: Ullberg (2013)

questions in this regard is whether such a memorial impedes (as a sign of what happened) or strengthens (moving on) recovery.

Very often, we see celebrations marking anniversaries and success stories of recovery, which in many ways marks the end of recovery of a certain disaster. This is primarily in the eyes of the governments or international actors. However, this leaves the question unanswered: whose recovery, and have affected populations truly and fully recovered?

13.11 Recovery and local capacities

The late former Indian President APJ Abdul Kalam said, "Building capacity dissolves differences. It irons out inequalities". Recovery processes have the potential, if utilised well, to invest in local capacities and capacity development. Capacity development refers to "the process through which individuals, organisations and societies obtain, strengthen and maintain the capabilities to set and achieve their own development objectives over time" (UNDP, 2008, p. 4). During recovery, introduction of new technologies brings in resources from outside but ignores local capacities (Fayazi et al., 2018). In the context of capacity development, ownership of local communities is key to the process (Hagelsteen & Becker, 2013). However, in many cases of recovery where international agencies and many times state actors have been involved, recovery is one-sided and driven with a top-down attitude. One of the most abused words in this regard is 'ownership'. What does 'ownership' mean? As compiled by Hagelsteen and Becker (2013, p. 5), inspired by SIDA

Box 13.6: I woke up to the waves!!

I woke up to the waves,
Not in my backyard but in my home.
I woke up to the waves,
Not as the evening walk on the beach
But to see my own depart.
They brought bread, they brought linen,
They brought water in its cleanest form I have even seen,
They brought linen I would never wear, bread my children don't eat,
They left and came again,
When I woke up to the waves.
The waves took my home,
But why did I lay my foundation here,
Nobody asked me why, but they brought bread,
Nobody asked me why only me?
Nobody asked me why my neighbor?
But they brought bread
When I woke up to the waves.
Democracy needed another term,
They brought bread but for a favour,
This time for a ballot, even more bread and more promises,
Promises I have heard, promises I have read,
Promises that didn't answer my woes,
They came again when I woke up to the waves.
Here they come again after the waves calmed,
This time with news they are elated about.
You never had a 'real home',
Here is a gift, move to the new house,
Remoteness matters? New jobs, new lives I am told,
Fears of my land like never before,
They brought papers to sign,
They came again before the waves with a new wave,
That they call recovery, to me a new wave.

Emmanuel Raju

Source: written by Emmanuel Raju in 2020

(Schulz et al., 2005), 'ownership' means "creating and owning ideas and strategies, development processes, resources and the result of the development process". From the authors' experiences of walking and talking through villages that were undergoing post-tsunami recovery in India from 2008 to 2012, the strong presence of the state and the massive influx of humanitarian organisations undermined both capacities of local communities and that of local organisations. Whilst the introduction of a discussion on safe housing was brought in, the lack of ownership contributes to a mismatch in knowledge of maintenance, which causes depletion of the asset/resource (Fayazi et al., 2018) (Box 13.6).

13.12 Concluding remarks

More research is required to understand and find out what tools best support recovery governance. It is extremely important to note that early recovery and response interventions have long-term repercussions. Recovery is "profoundly developmental in nature, risky because it deals with transformations and equality, and multiple because it takes place across different domains" (Boano & Hunter, 2010, p. 1). This developmental character in disaster recovery is a unique opportunity to align thinking, working paradigms and approaches between the disaster-development actors. These efforts must primarily be to address the development (or re-development) vision of the disaster-affected areas and populations. It is extremely important to remember that disasters contribute to reshaping the landscape of society environmentally, geographically and socially. These changes do not (most of the time) close the chasm that exist in the development patterns in society, which are the very reason disasters happen. Recovery must, therefore, fundamentally challenge the "status quo" (Von Meding et al., 2020). Recovery, therefore, should not be seen as a physical process but a social process that takes time, must ensure human rights and dignity are protected, and must utilise every opportunity to address vulnerability and reduce risk.

Take-away messages

1. Disaster recovery is beyond the physical reconstruction and must be seen as a socio-political process;
2. Disaster recovery must ensure a truly inclusive and participatory process of the affected populations;
3. Local capacities must be recognised and strengthened during recovery processes; and
4. Recovery must be seen as a long-term process, and avoid creating any new and/or rebuilding old risks and vulnerabilities.

To learn more about the topic discussed in this chapter, listen to the *Disasters: Deconstructed* interview with Dr Danielle Rivera and Dr Emmanuel Raju (Figure 13.5).

Fig 13.5
QR code for Chapter 13.

Further suggested reading

Davis, I., & Alexander, D. (2016). *Recovery from disaster*. Routledge.

Hilhorst, D., Boersma, K., & Raju, E. (2020). Research on politics of disaster risk governance: Where are we headed? *Politics and Governance*, *8*(4), 214–219. https://doi.org/10.17645/pag.v8i4.3843

Jha, M. K. (2015). Liquid disaster and frigid response: Disaster and social exclusion. *International Social Work*, *58*(5), 704–716. https://doi.org/10.1177/0020872815589388

Kelman, I. (2017). Linking disaster risk reduction, climate change, and the sustainable development goals. *Disaster Prevention and Management*, *26*(3), 254–258. https://doi.org/10.1108/DPM-02-2017-0043

Oliver-Smith, A. (1991). Successes and failures in post-disaster resettlement. *Disasters*, *15*(1), 12–23. https://doi.org/10.1111/j.1467-7717.1991.tb00423.x

Raju, E., & Schmid, B. (2021). COVID-19 and urban informal settlements – time to rethink vulnerability! In N. Palacios (Ed.), *Pandemic resilient cities* (1st ed.). Emergency Architecture and Human Rights. https://secureservercdn.net/160.153.137.99/pho.521.myftpupload.com/wp-content/uploads/2021/04/Pandemic-Resilient-Cities.pdf

Ryder, S. (2017). A bridge to challenging environmental inequality: Intersectionality, environmental justice and disaster vulnerability. *Social Thought and Research Journal*, *34*.

Shaw, R. (2013). *Disaster recovery: Used or misused development opportunity (Disaster Risk Reduction)*. Springer.

Smith, G., & Wenger, D. (2007). Sustainable disaster recovery: Operationalizing an existing agenda. In H. Rodriguez, E. L. Quarantelli, & R. R. Dynes (Eds.), *Handbook of disaster research* (pp. 234–257). Springer.

Sou, G. (2019). Sustainable resilience? Disaster recovery and the marginalization of sociocultural needs and concerns. *Progress in Development Studies*, *19*(2), 144–159. https://doi.org/10.1177/1464993418824192

Tierney, K., & Oliver-Smith, A. (2012). Social dimensions of disaster recovery. *International Journal of Mass Emergencies and Disasters*, *30*(2), 123–146.

References

Abramovitz, J. (2001). *Unnatural disasters*. Worldwatch Paper 158.

Alexander, D. E. (2002). *Principles of emergency planning and management*. Oxford University Press.

Alexander, D. E. (2019). L'aquila, central Italy, and the "disaster cycle", 2009–2017 [International journal]. *Disaster Prevention and Management*, *28*(4), 419–433. https://doi.org/10.1108/DPM-01-2018-0022

Anderson, B., & Fenton, J. (2008). Editorial introduction: Spaces of hope. *Space and Culture*, *11*(2), 76–80. https://doi.org/10.1177/1206331208316649

Bates, F. L., & Peacock, W. G. (1989). Long-term recovery. *International Journal of Emergency Management*, *7*(3), 349–365.

Béné, C., Mehta, L., McGranahan, G., Cannon, T., Gupte, J., & Tanner, T. (2018). Resilience as a policy narrative: Potentials and limits in the context of urban planning. *Climate and Development*. Taylor & Francis, *10*(2), 116–133. https://doi.org/10.1080/17565529.2017.1301868

Berke, P. R., & Campanella, T. J. (2006). Planning for postdisaster resiliency. *Annals of the American Academy of Political and Social Science*, *604*(1), 192–207. https://doi.org/10.1177/0002716205285533

Berke, P. R., Kartez, J., & Wenger, D. (1993). Recovery after disaster: Achieving sustainable development, mitigation and equity. *Disasters*, *17*(2), 93–109. https://doi.org/10.1111/j.1467-7717.1993.tb01137.x

Bhide, A. (2009). Shifting terrains of communities and community organization: Reflections on organizing for housing rights in Mumbai. *Community Development Journal*, *44*(3), 367–381. https://doi.org/10.1093/cdj/bsp026

Blackman, D., Nakanishi, H., & Benson, A. M. (2017, March). Disaster resilience as a complex problem: Why linearity is not applicable for long-term recovery. *Technological Forecasting and Social Change. Elsevier Inc.*, *121*(2011), 89–98. https://doi.org/10.1016/j.techfore.2016.09.018

Boano, C., & Hunter, W. (2010). Risks in post disaster housing: Architecture and the production of space. *ABACUS International Journal on Architecture, Conservation and Urban Studies*, *5*(2), 23–31.

Boin, A., Kuipers, S., & Overdijk, W. (2013). Leadership in times of crisis: A framework for assessment. *International Review of Public Administration*, *18*(1), 79–91. https://doi.org/10.1080/12294659.2013.10805241

Bosher, L., Penning-Rowsell, E., & Tapsell, S. (2007). Resources accessibility and vulnerability in Andhra Pradesh. *Development and Change*, *38*(4), 615–640.

Bovensiepen, J., & Meitzner Yoder, L. S. (2018). Introduction: The political dynamics and social effects of megaproject development. *Asia*

Pacific Journal of Anthropology, *19*(5), 381–394. https://doi.org/10.1080/14442213.2018.1513553

Chandrasekhar, D. (2010). Setting the stage. *How Policy Institutions Frame Participation in Post-Disaster Recovery*, *5*(2).

Chmutina, K., & Rose, J. (2018, March). Building resilience: Knowledge, experience and perceptions among informal construction stakeholders. *International Journal of Disaster Risk Reduction. Elsevier Ltd.*, *28*, 158–164. https://doi.org/10.1016/j.ijdrr.2018.02.039

Choudhury, M. U. I., & Haque, C. E. (2016). "We are more scared of the power elites than the floods": Adaptive capacity and resilience of wetland community to flash flood disasters in Bangladesh. *International Journal of Disaster Risk Reduction*. Elsevier, *19*, 145–158. https://doi.org/10.1016/j.ijdrr.2016.08.004

Comfort, L. K., & Kapucu, N. (2006). Inter-organizational coordination in extreme events: The world trade center attacks, September 11, 2001. *Natural Hazards*, *39*(2), 309–327. https://doi.org/10.1007/s11069-006-0030-x

Cretney, R. M. (2017). Towards a critical geography of disaster recovery politics: Perspectives on crisis and hope. *Geography Compass*, *11*(1). https://doi.org/10.1111/gec3.12302

Cretney, R. M. (2018). Beyond public meetings: Diverse forms of community led recovery following disaster. *International Journal of Disaster Risk Reduction. Elsevier Ltd.*, *28*, 122–130. https://doi.org/10.1016/j.ijdrr.2018.02.035

Davis, I., & Alexander, D. (2016). *Recovery from disaster*. Routledge.

Duxbury, J., & Dickinson, S. (2007). Principles for sustainable governance of the coastal zone: In the context of coastal disasters. *Ecological Economics*, *63*(2–3), 319–330. https://doi.org/10.1016/j.ecolecon.2007.01.016

Duyne Barenstein, J. (2010). Who governs reconstruction? Changes and continuity in policies, practices and outcomes. In C. Lizarralde, G. Johnson, & C. Davidson (Eds.), *Rebuilding after disasters from emergency to sustainability* (pp. 149–176). Taylor & Francis.

Dynes, R. R., & Quarantelli, E. L. (1989). *Reconstruction in the context of recovery: Thoughts on the Alaskan earthquake*. Preliminary Paper 141. Disaster Research Centre, University of Delaware http://udspace.udel.edu/handle/19716/513

Fayazi, M. et al. (2018). *Meta-patterns in post-disaster housing reconstruction and recovery*. Routledge.

Garnett, J. D., & Moore, M. (2010). Enhancing disaster recovery: Lessons from exemplary international disaster management practices. *Journal of Homeland Security and Emergency Management*, *7*(1).

Gopalakrishnan, C., & Okada, N. (2007). Designing new institutions for implementing integrated disaster risk management: Key elements and future directions. *Disasters*, *31*(4), 353–372. https://doi.org/10.1111/j.1467-7717.2007.01013.x

Gupta, J., Pouw, N. R. M., & Ros-Tonen, M. A. F. (2015). Towards an elaborated theory of inclusive development. *European Journal of Development Research. Nature Publishing Group*, *27*(4), 541–559. https://doi.org/10.1057/ejdr.2015.30

Gupta, K. (2007). Urban flood resilience planning and management and lessons for the future: A case study of Mumbai, India. *Urban Water Journal*, *4*(3), 183–194. https://doi.org/10.1080/157306207 01464141

Haas, J. E., Kates, R. W., & Bowden, M. J. (1977). *Reconstruction following disaster*. MIT Press.

Hagelsteen, M., & Becker, P. (2013). Challenging disparities in capacity development for disaster risk reduction. *International Journal of Disaster Risk Reduction. Elsevier*, *3*, 4–13. https://doi.org/10.1016/j.ijdrr.2012.11.001

He, L. (2019). Identifying local needs for post-disaster recovery in Nepal. *World Development. Elsevier Ltd.*, *118*, 52–62. https://doi.org/10.1016/j.worlddev.2019.02.005

He, L., Aitchison, J. C., Hussey, K., Wei, Y., & Lo, A. (2018, April). Accumulation of vulnerabilities in the aftermath of the 2015 Nepal earthquake: Household displacement, livelihood changes and recovery challenges. *International Journal of Disaster Risk Reduction. Elsevier Ltd.*, *31*, 68–75. https://doi.org/10.1016/j.ijdrr.2018.04.017

Herber, E. (2016). The 2011 Fukushima nuclear disaster Japanese citizens' role in the pursuit of criminal responsibility. *Journal of Japanese Law*, *32*(2011), 87–109.

Humanitarian Policy Network. (2017). Humanitarian exchange – June 2017. *Humanitarian Exchange*, *69*.

IFRC & UNDP. (2014). *Effective law and regulation for disaster risk reduction: A multi country report*. Available at https://www.undp.org/publications/effective-law-regulation-disaster-risk-reduction

International Recovery Platform (IRP). (2007). *Learning from disaster recovery: Guidance for decision makers*. http://www.unisdr.org/eng/about_isdr/isdr-publications/irp/Learning-From-Disaster-Recovery.pdf

International Union for Conservation of Nature and Natural Resources. (n.d.). *Five reasons why ecosystems are central to disaster risk reduction*. Retrieved March 30, 2020, from https://www.iucn.org/theme/ecosystem-management/our-work/environment-and-disasters/about-ecosystem-based-disaster-risk-reduction-eco-drr/five-reasons-why-ecosystems-are-central-disaster-risk-reduction

Jha, M. K. (2015). Liquid disaster and frigid response: Disaster and social exclusion. *International Social Work*, *58*(5), 704–716. https://doi.org/10.1177/0020872815589388

Johnson, L. A., & Hayashi, H. (2012). Synthesis efforts in disaster recovery research. *International Journal of Mass Emergencies and Disasters*, *30*(2), 212–239.

Johnson, L. A., & Olshansky, R. B. (2013). The road to recovery: Governing post-disaster reconstruction. *Land Lines*, *25*(3), 14–21.

Kammerbauer, M., & Wamsler, C. (2017, June). Social inequality and marginalization in post-disaster recovery: Challenging the consensus? *International Journal of Disaster Risk Reduction. Elsevier Ltd., 24*, 411–418. https://doi.org/10.1016/j.ijdrr.2017.06.019

Klein, N. (2008). *The shock doctrine: The rise of disaster capitalism*. Knopf.

Krishna, R. N., Majeed, S., Ronan, K., & Alisic, E. (2018). Coping with disasters while living in poverty: A systematic review. *Journal of Loss and Trauma, 23*(5), 419–438. https://doi.org/10.1080/15325024.2017.1415724

Lavell, A. (2017). Preface. In V. Marchezini et al. (Eds.), *Reduction of vulnerability to disasters: From knowledge to action* (pp. 9–14). Sao Carlos, RiMa Editora. https://www.preventionweb.net/publications/view/56269

Lavell, A., & Maskrey, A. (2014). The future of disaster risk management. *Environmental Hazards, 13*(4), 267–280. https://doi.org/10.1080/17477891.2014.935282

Lauta, K. C. (2016). Legal scholarship and disasters. In R. Dahlberg, O. Rubin, & M. Vendelø (Eds.), *Disaster research: Multidisciplinary and international perspectives*. Humanitarian Studies Series (pp. 97–109). Routledge. https://doi.org/10.4324/9781315724584

Lewis, B., & Maguire, R. (2016). A human rights-based approach to disaster displacement in the Asia-Pacific. *Asian Journal of International Law, 6*(2), 326–352. https://doi.org/10.1017/S2044251315000168

Look, C., Friedman, E., & Godbout, G. (2019). The resilience of land tenure regimes during Hurricane Irma: How colonial legacies impact disaster response and recovery in Antigua and Barbuda. *Journal of Extreme Events, 06*(1), 1940004. https://doi.org/10.1142/S2345737619400049

Llyod-Jones, T. (2006). *Mind the gap! Post-disaster reconstruction and the transition from humanitarian relief*. https://www.preventionweb.net/files/9080_MindtheGapFullreport1.pdf

Madianou, M., Longboan, L., & Ong, J. C. (2015). Finding a voice through humanitarian technologies? Communication technologies and participation in disaster recovery. *International Journal of Communications, 9*(1), 3020–3038.

Mascarenhas, A., &Wisner, B. (2012). Politics: Power and disasters. In B. Wisner, J. Gaillard, & I. Kelman (Eds.), *The Routledge handbook of hazard and disaster risk reduction* (pp. 42–60). Routledge.

Nagarajan, S. (2016). *Building livelihoods*. https://www.thehindu.com/features/metroplus/building-livelihoods/article3408475.ece

Nakajo, M. (2017). *Do voters prefer local leaders from the ruling parties after natural disasters?* (Vol. 16, pp. 1–12). https://ijrdp.org/paper/mpsa2017.pdf

Nigg, J. M. (1995). *Disaster recovery as a social process*. Disaster Research Centre, University of Delaware. http://udspace.udel.edu/handle/19716/625

Olshansky, R. B. (2006). Planning after Hurricane Katrina. *Journal of the American Planning Association, 72*(2), 147–153. https://doi.org/10.1080/01944360608976735

Oliver-Smith, A. (1991) Successes and failures in post-disaster resettlement. *Disasters*, *15*(1), 12–23. https://doi.org/10.1111/j.1467-7717.1991.tb00423.x

Oliver-Smith, A. (1999). "What is a disaster?": Anthropological perspectives on a persistent question. *The Angry Earth: Disaster in Anthropological Perspective*, 18–34.

Parrack, C., Flinn, B., & Passey, M. (2014). Getting the message across for safer self-recovery in post-disaster shelter. *Open House International*, *39*(3), 47–58. https://doi.org/10.1108/OHI-03-2014-B0006

Platt, S., Brown, D., & Hughes, M. (2016, May). Measuring resilience and recovery. *International Journal of Disaster Risk Reduction*, *19*, 447–460. https://doi.org/10.1016/j.ijdrr.2016.05.006

Pomeroy, R. S., Ratner, B. D., Hall, S. J., Pimoljinda, J., & Vivekanandan, V. (2006). Coping with disaster: Rehabilitating coastal livelihoods and communities. *Marine Policy*, *30*(6), 786–793. https://doi.org/10.1016/j.marpol.2006.02.003

Raju, E. (2013a). *Exploring disaster recovery coordination*. Lund University. https://lup.lub.lu.se/search/ws/files/4305738/4180232.pdf

Raju, E. (2013b, April). Housing reconstruction in disaster recovery: A study of fishing communities post-tsunami in Chennai, India. *PLoS Currents*, *5*, 2004–2007. https://doi.org/10.1371/currents.dis.a4f34a96cb91aaffacd36f5ce7476a36

Raju, E., & Becker, P. (2013). Multi-organisational coordination for disaster recovery: The story of post-tsunami Tamil Nadu, India. *International Journal of Disaster Risk Reduction*. Elsevier, *4*, 82–91. https://doi.org/10.1016/j.ijdrr.2013.02.004

Raju, E., Becker, P., & Tehler, H. (2018). Exploring interdependencies and common goals in disaster recovery coordination. *Procedia Engineering*, *212*, 1002–1009. https://doi.org/10.1016/j.proeng.2018.01.129

Revi, A. (2008). Climate change risk: An adaptation and mitigation agenda for Indian cities. *Environment and Urbanization*, *20*(1), 207–229. https://doi.org/10.1177/0956247808089157

Robinson, D., Hewitt, T., & Harris, J. (2000). *Managing development: Understanding inter-organizational relationships*. Open University.

Rogers, P. (2015). Researching resilience: An agenda for change. *Resilience*, *3*(1), 55–71. https://doi.org/10.1080/21693293.2014.988914

Rubin, C. B. (2009). The neglected component of emergency management long-term recovery from disasters—The neglected component of emergency management. *Journal of Homeland Security and Emergency Management*, *6*(1).

Rubin, C. B., Saperstein, M., & Barbee, D. (1985). *Community recovery from a major natural disaster* (pp. 61–63). FMHI Publication. http://scholarcommons.usf.edu/cgi/viewcontent.cgi?article=1086&context=fmhi_pub

Ryder, S. (2017). A bridge to challenging environmental inequality: Intersectionality, environmental justice and disaster vulnerability. *Social Thought and Research Journal*, *34*.

Schulz, K., Gustafsson, I., & Illes, E. (2005). *Manual for capacity development*. Sida.

Smith, G., & Birkland, T. (2012). Building a theory of recovery: Institutional dimensions. *International Journal of Mass Emergencies and Disasters*, *30*(3), 147–170.

Smith, G., & Wenger, D. (2007). Sustainable disaster recovery: Operationalizing an existing agenda. In H. Rodriguez, E. L. Quarantelli, & R. R. Dynes (Eds.), *Handbook of disaster research* (pp. 234–257). Springer.

Sou, G. (2019). Sustainable resilience? Disaster recovery and the marginalization of sociocultural needs and concerns. *Progress in Development Studies*, *19*(2), 144–159. https://doi.org/10.1177/1464993418824192

Sou, G., & Webber, R. (2019, August). Disruption and recovery of intangible resources during environmental crises: Longitudinal research on "home" in post-disaster Puerto Rico. *Geoforum. Elsevier*, *106*, 182–192. https://doi.org/10.1016/j.geoforum.2019.08.007

Stoddard, A. et al. (2007). *Cluster approach evaluation*. https://www.humanitarianresponse.info/sites/www.humanitarianresponse.info/files/documents/files/Cluster Approach Evaluation 1.pdf

Taiban, S., Lin, H. N., & Ko, C. C. (2020). Disaster, relocation, and resilience: Recovery and adaptation of Karamemedesane in Lily Tribal Community after Typhoon Morakot, Taiwan. *Environmental Hazards*, *19*(2), 209–222. https://doi.org/10.1080/17477891.2019.1708234

Takeuchi, K., Elmqvist, T., Hatakeyama, M., Kauffman, J., Turner, N., & Zhou, D. (2014). Using sustainability science to analyse social–ecological restoration in NE Japan after the great earthquake and tsunami of 2011. *Sustainability Science. The Hindu 'Floods Displace Over 4 Lakh in Assam'*, *9*(4), 513–526. https://www.thehindu.com/news/national/floods-displace-over-4-lakh-in-assam/article28391969.ece; https://doi.org/10.1007/s11625-014-0257-5

The Hindu. (2019). Floods displace over 4 lakh in Assam. Available at https://www.thehindu.com/news/national/floods-displace-over-4-lakh-in-assam/article28391969.ece

Thornley, L., Ball, J., Signal, L., Lawson-Te Aho, K., & Rawson, E. (2015). Building community resilience: Learning from the Canterbury earthquakes. *Kōtuitui. Taylor & Francis*, *10*(1), 23–35. https://doi.org/10.1080/1177083X.2014.934846

Twigg, J. et al. (2017, October). *Self-recovery from disasters: An interdisciplinary perspective*. https://www.researchgate.net/publication/352212481_Self-recovery_from_disasters_an_interdisciplinary_perspective

Ullberg, S. (2013). *Watermarks urban flooding and memoryscape in Argentina* [Internet]. Social Anthropology, Stokholm University. Available at http://www.diva-portal.org/smash/record.jsf?pid=diva2%3A618130&dswid=6065

United Nations Development Programme. (2001). *From relief to recovery: The Gujarat experience*. http://www.recoveryplatform.org/assets/publication/fromrelieftorecovery gujarat.pdf

United Nations Development Programme. (2008). *Capacity development—practice note*. UNDP.

UNDRR. (2016). *Report of the open-ended intergovernmental expert working group on indicators and terminology relating to disaster risk reduction*. https://www.preventionweb.net/files/50683_oiewgreportenglish.pdf

UNDRR. (2019). *Global assessment report on disaster risk reduction*. UNDRR.

Uson, M. A. M. (2017). Natural disasters and land grabs: The politics of their intersection in the Philippines following super typhoon Haiyan. *Canadian Journal of Development Studies/Revue Canadienne d'Études du Développement*, *38*(3), 414–430. https://doi.org/10.1080/02255189.2017.1308316

Vallance, S. (2015). Disaster recovery as participation: Lessons from the Shaky Isles. *Natural Hazards*, *75*(2), 1287–1301. https://doi.org/10.1007/s11069-014-1361-7

Von Meding, J., Chmutina, K., Forino, G., & Raju, E. (2020). Guest editorial. *Disaster Prevention and Management*, *29*(6), 829–830. https://doi.org/10.1108/DPM-11-2020-405

Walch, C. (2018). Typhoon Haiyan: Pushing the limits of resilience? The effect of land inequality on resilience and disaster risk reduction policies in the Philippines. *Critical Asian Studies*, *50*(1), 122–135. https://doi.org/10.1080/14672715.2017.1401936

Waugh, W. L., & Streib, G. (2006). Collaboration and leadership for effective emergency management. *Public Administration Review*, *66*(s1), 131–140. https://doi.org/10.1111/j.1540-6210.2006.00673.x

Wisner, B. et al. (2004). *At risk: Natural hazards, people's vulnerability and disasters*. Routledge.

World Bank. (2016). *Peru: A comprehensive strategy for financial protection against natural disasters*. World Bank.

Xu, J., & Shao, Y. (2020). The role of the state in China's post-disaster reconstruction planning: Implications for resilience. *Urban Studies*, *57*(3), 525–545. https://doi.org/10.1177/0042098019859232

Zavar, E. M., & Schumann, R. L. (2019). Patterns of disaster commemoration in long-term recovery. *Geographical Review*, *109*(2), 157–179. https://doi.org/10.1111/gere.12316

PART V

Disaster risk reduction and management

CHAPTER 14

Disaster risk reduction

Fig 14.1
Creating an understanding of DRR.
Source: photo courtesy of Dewald van Niekerk

This chapter will create an understanding of the concept of DRR (Figure 14.1). It traces the development of DRR since the 1990s and how it brought about a paradigm shift from disaster response and recovery to minimising the potential of hazards turning into disasters. This chapter addresses the various level at which DRR occurs and highlights the multidisciplinary and transdisciplinary nature of the concept. The major components of DRR are highlighted, and the linkage between DRR, sustainable development and CCA is made.

14.1 Introduction

Since the late 1980s, there has been a global shift in the way disasters are being managed. A gradual realisation set in that the management of the disastrous event does not yield enough benefits in saving lives, the environment and the systems on which we depend. Science in the field of disaster studies expanded from an initial focus on natural hazards turning into disasters, to be more socially oriented. In as early as 1976,

DOI: 10.4324/9781315469614-20

Box 14.1: Taking the naturalness out of natural disasters

To read one of the most influential and seminal works that heralded a change in how disasters are perceived, scan and read the article by O'Keefe, Westgate and Wisner in 1976 here.

Fig 14.2
QR code for Box 14.1. Available at https://www.nature.com/articles/260566a0

O'Keefe, Westgate and Wisner made the argument that disasters should not be viewed as natural (Box 14.1 and Figure 14.2). Forty-odd years later, the global community are only now coming to terms with the fact that all disasters are human-made (Chmutina & Von Mending, 2019). A paradigm shift in the management of disasters also occurred. The traditional acceptable thinking of managing disasters (and even later, disaster risk) through the application of the so-called disaster management cycle were questioned (Coetzee & Van Niekerk, 2012; Bosher et al., 2021). The gradual expansion of disaster studies within the sociology domain (notably by the scholars Quarantelli, Dynes and Haas) occurred since the 1960s. However, it was only in the mid-1990s with the mid-term review of the Yokohama Framework that the notion of disasters being social constructs gained traction. It is now widely accepted that for DRR to be effective, it must have a primary focus on issues of exposure, vulnerability and resilience.

This chapter will introduce the reader to the concept of DRR. It traces the development of the term since the 1990s, and argues for the multisectoral and multidisciplinary nature of DRR. Although one must acknowledge the immense impact that several global disasters had on the development of the field of DRR, this chapter adopts a more pragmatic approach linked to the statutory instruments that guided change within states. To this end, this DRR at various scales is described, and the focus is placed on the various international frameworks that shaped DRR as it is currently known. DRR strategies and policies define goals and objectives across different timescales and with concrete targets, indicators and time frames. In line with the SFDRR 2015–2030, these should be aimed at preventing the creation of disaster risk, the reduction of existing risk and the strengthening of economic, social, health and environmental resilience. The components of DRR are alluded to, showing the need for all-of-society risk reduction but not forsaking the need for DRR as an integral part of prevention and mitigation strategies. The chapter ends off by emphasising the need for DRR, CCA and sustainable development integration. Only through such integration can inroads towards safeguarding lives and livelihoods, and the protection of the environment, be made.

14.2 What is disaster risk reduction?

DRR, also sometimes referred to as disaster reduction, is all the actions and decisions to address all known risks that could lead to disasters but also to prevent any new disaster risks. DRR should be seen as policy objectives that lead to tangible actions though a DRM approach (UNDRR, 2015). DRR as a policy objective aims to address the underlying factors that contribute to disaster risks. Such factors include the underlying systems and processes that leads to the creation of, or sustaining, disaster risks (Box 14.2 and Figure 14.3). DRR is proactive and preventive (Van Niekerk, 2008). It considers three basic categories of disaster risk, which are (a) hazard and exposure, (b) capacity (and resilience), and (c) vulnerability. These categories not only relate to human beings but also to all the various systems on which we depend (e.g. environmental, political, economic and social systems). For example, inequality in many developing countries (driven by certain historical events) leads to inequitable access by people to the economic sector, which, in turn, contributes to poverty. It is widely known that poverty is one of the driving forces of vulnerability, and vulnerability, in turn, contributes to heightened disaster risk. It is important to understand that DRR is also not only about the disaster event. It is widely accepted by scientists and policymakers alike that certain global risk drivers are at the heart of understanding DRR. *Climate change* has, in the past 15 years, found significant attention within the domain of DRR. Similarly, there is a realisation that DRR is crucial towards climate adaptation and mitigation. *Environmental degradation* occurs on massive geographical scales and is mostly caused by human interventions. *Globalised economic development* linked to an inequitable *global economic system* forces developing countries to remain in states of neo-colonialism. In turn, *poverty* and *hunger* are rife. *Poor planning* leads to unsustainable urban centres, which are currently the boiling pots for urban disaster risk. Unfortunately, most of the earlier points are underscored by *weak governance* and institutions of governance in many developing states. These global risk drivers are intertwined, and their management cannot happen in isolation, and their separation is impossible if solutions are sought.

For DRR to be effective, a transdisciplinary and multisectoral approach is needed (Van Niekerk, 2012). Therefore, DRR must find its manifestation across sectors and development initiatives. DRR should be grounded in our development thinking. DRR is, thus, our insurance policy to safeguard our current and future development. If DRR is

Box 14.2: Disaster risk – the broader focus

This episode of *Disasters: Deconstructed* podcast is a conversation with DRR expert and #NoNaturalDisasters advocate Irasema Alcántara-Ayala, Professor of Natural Hazards and Risk at the National Autonomous University of Mexico. The podcast discusses how risk is created within society and how state responses so often focus on reacting to hazards rather than addressing root causes.

Fig 14.3
QR code for Box 14.2. Available at https://disastersdecon.podbean.com/e/s4e10-drr-in-latin-america-the-caribbean/

not present in how we manage our spaces (urban and rural planning), how we use our natural resources (sustainable development), how we safeguard our environment (conservation of biodiversity) or how we as humans develop, then we are in effect contributing to circumstances that could lead to disaster. To achieve the earlier points, certain capacities in our systems and institutions are needed. These capacities relate to the ability to identify hazards and vulnerability, to assess their possible contribution to risk, and to eliminate, reduce, mitigate or transfer these risks. For DRR to be effective certain institutional and operational abilities, therefore, need to be present. For instance, understanding how the economy impacts on peoples' lives and livelihoods, where people live and why they live there, which capabilities are present in various communities to address their level of risk, what are the traditional, cultural and gender-based practices, and which resources do they have access to, are all important in understanding disaster risk.

It is widely acknowledged that the responsibility to ensure DRR rests with governments. Various states should, therefore, ensure dedicated public sector institutions, which can lead and coordinate the inter-sectoral actions that will address disaster risks. Certain capacities towards DRR cannot be provided by individuals alone, and other scientific and technical skills are also needed. For instance, a multi-hazard warning system can only be effectively implemented through the cooperation of many private and public role-players, yet its optimal functioning in general cuts across all sectors and contributes to DRR.

DRR is, thus, a global concern that needs to be addressed at various levels of governance by diverse role-players. On a historical scale, however, 'disaster risk' reduction is still a new term and, as an academic discipline, is far from reaching any form of maturity. To better understand how DRR has come about, and how it functions within the international and national levels, the section that follow will allude to the development of the term linked to some significant international agreements and frameworks.

14.3 Development of disaster risk reduction

The works of UNDP (2010), Van Niekerk (2005) and Manyena et al. (2011) trace the development of the field of DRR since the 1990s. These studies point out that the field of DRR has undergone various stages of progression, and its evolutionary path is marked by several mentionable international events (see later). Collins (2009) argues that over decades, disciplines, such as geography, environmental studies, economics, sociology, public health, policy studies and planning, have contributed significantly to disaster studies. Most notably is the paradigm shift from a reactive disaster management perspective (disaster response and recovery) to that of reducing the risks of disaster occurring in the first place (Van Niekerk, 2005). Initially, the research focus within the disaster studies domain was on understanding the various processes and systems of Earth and the hazards that these create. Emphasis was placed on understanding the characteristics of natural hazards, and in doing so, the field of study developed a phenomena-centric approach. The argument was that once we understand the dynamics of natural hazards, we would be able to prevent and mitigate their impacts. Very little attention was given to the social perspective of disasters. Smith (2004, p. 1) specifically notes that, as the world population grows, more people are exposed to hazards, and as people become prosperous, more personal wealth is at risk. The growing gap between rich and poor where a handful of countries with power in politics, trade and culture dominate the world, has in itself contributed to feelings of alienation and hostility that occasionally find expression in hazards of mass violence. However, as national and global disasters continually occurred all over the world, killing thousands of people and destroying years of development investment (CRED, 2020), a realisation emerged that a one-sided focus on hazards is not adequate to understand the dynamics associated with disasters. From the mid-1990s, the human and social perspective of disaster risk, rooted in vulnerability and resilience, became more prominent.

14.4 Disaster risk reduction at different scales

The realisation of the goals outlined in the various international and national frameworks depends and requires implementation of DRR at various scales, where each of the components of DRR are also addressed. Diverse actors and institutions are involved in various roles and mandates at different scales contributing towards the success of DRR. Much of the international, regional and national actions in DRR were shaped over the past 30 years by several international agreements and frameworks. These frameworks, to a more and lesser extent, addressed the known body of knowledge on disasters and risk at the time. Mostly, they were guided by national and international interests, and the desire to find solutions to the complex problem that we call disaster.

At the global level, DRR is spearheaded through the United Nations Office for Disaster Risk Reduction (UNDRR), which is responsible for global agenda-setting, drawing from evidence-based scientific research, to contribute to global policy direction on DRR issues. The institution also disseminates risk data and information, such as lessons, best practices, experiences, emerging issues or changes in DRR, to inform risk-informed decision-making for sustainable development. The UNDRR fulfils its mandate through diverse avenues, such as the Global Platform for DRR (GPDRR), the World Conferences on Disaster Reduction (WCDR) and International Day of Disaster Reduction campaigns. The first WCDR was held in 1994, and its outcome was the Ten Principles of the Yokohama Strategy for a Safer World which were as follows:

a Risk assessment as a required step for disaster reduction policies and measures;
b Disaster prevention and preparedness are of primary importance in reducing the need for disaster relief;
c Disaster prevention and preparedness should be considered integral aspects of development policy and planning at national, regional, bilateral, multilateral and international levels;
d The development and strengthening of capacities to prevent, reduce and mitigate disasters is a top priority area to be addressed;
e Early warnings of impending disasters and their effective dissemination using telecommunications, including broadcast services, are key factors to successful disaster prevention and preparedness;
f Preventive measures are most effective when they involve participation at all levels, from the local community through the national government to the regional and international level;
g Vulnerability can be reduced by the application of proper design and patterns of development focused on target groups, by appropriate education and training of the whole community;
h The need to share the necessary technology to prevent, reduce and mitigate disaster;
i Environmental protection as a component of sustainable development consistent with poverty alleviation is imperative in the prevention and mitigation of natural disasters; and
j The international community should demonstrate strong political determination required to mobilise adequate and make efficient use of existing resources, including financial, scientific and technological means, in the field of natural disaster reduction, bearing in mind the needs of the developing countries, particularly the least developed countries.

The second WCDR was held in 2005, and its outcome was the HFA. The HFA was developed around five priorities that directly related the lessons learned from the Yokohama Strategy as well as the plethora of research that saw the light emanating from many disciplines. The priorities of the HFA were to do the following:

a. ensure that DRR is a national and local priority with an institutional basis for implementation;
b. identify, assess and monitor disaster risks and enhance early warnings;
c. use knowledge, innovation and education to build a culture of resilience and safety at all levels;
d. reduce the underlying risk factors; and
e. strengthen preparedness for effective response at all levels.

(UN, 2005)

In 2015, the third WCDR was held, and the SFDRR was its major outcome (Box 14.3 and Figure 14.4). Many authors (Kelman & Glantz, 2015; Coetzee et al., 2019; Wisner, 2020) expressed critique against yet another global framework at a time when many countries have not yet come to terms with the implementation of the HFA. Notwithstanding, the SFDRR has a refined version of the HFA, with four priority areas:

a. Understanding disaster risk;
b. Strengthening disaster risk governance to manage disaster risk;
c. Investing in DRR for resilience; and
d. Enhancing disaster preparedness for effective response, and to "Build Back Better" in recovery, rehabilitation and reconstruction.

(UN, 2015a)

A significant difference between the HFA and SFDRR is more tangible measuring instruments towards achieving the priorities. To this end, the SFDRR proposed seven global targets that must be achieved at regional and national levels:

a. Substantially reduce continental disaster mortality by 2030, aiming to lower the average per 100,000 continental mortality rate in the decade 2020–2030 compared to the period 2005–2015;
b. Substantially reduce the number of affected people continentally in Africa by 2030, aiming to lower the average continental figure per 100,000 in the decade 2020–2030 compared to the period 2005–2015;
c. Reduce direct disaster economic loss in relation to continental gross domestic product (GDP) by 2030;
d. Substantially reduce disaster damage to critical infrastructure and disruption of basic services, among them health and educational facilities, including through developing their resilience by 2030;

Box 14.3: What is the Sendai Framework for DRR?

Want to know more about the SFDRR? Watch this video by the United Nations Office for DRR.

Fig 14.4
QR code for Box 14.3. Available at https://www.youtube.com/watch?v=M9m6mb-blYM

e. Substantially increase the number of countries with national and sub-national/local disaster risk reduction strategies by 2020;
f. Substantially enhance international cooperation to developing countries through adequate and sustainable support to complement national actions for implementation of the Sendai Framework by 2030; and
g. Substantially increase the availability of and access to multi-hazard early warning systems and disaster risk information and assessments to people by 2030.

(UN, 2015a)

The UNDRR has oversight responsibility on implementation and monitoring of the international frameworks on DRR during their respective terms and communicates progress updates during respective terms. Through the implementation and review of the SFDRR, the UNDRR also coordinates and promotes integration with other global development frameworks. For instance, the SFDRR facilitates integration with the UN 2030 Agenda for Sustainable Development, which outlines SDGs, and the Paris Agreement on Climate Change (UNDRR, 2019). Other global instruments where DRR is relevant include the New Urban Agenda (UN, 2017), the Addis Ababa Action Agenda (UN, 2015b) and the Agenda for Humanity (UN, 2016). In promoting integration and mainstreaming of DRR in development the UNDRR works with other UN agencies (such as UNDP, UNFCCC, UN-Habitat and UNOCHA), international NGOs and finance institutions. Various platforms also exist for financing DRR at a global level, with one notable funding mechanism being the World Bank–managed Global Facility for Disaster Reduction and Recovery (GFDRR), through which DRR funding, research and technical assistance are provided. Generally, at the global level, the main focus is predominantly on minimisation of the creation of new risk and preparedness, which seeks to reduce existing and residual risk. There can be exceptional cases where global institutions get involved in response to national or regional disasters, mainly those focusing on humanitarian relief. Likewise, global institutions can be involved in post-disaster

Box 14.4: Views from the Frontline

In reaction to the implementation of the HFA, the GNDR launched the Views from the Frontline (VFL) initiative in 2009. Though a decentralised research approach, the VFL became a participatory multistakeholder engagement process designed to monitor, review and report on critical aspects of local governance considered essential to building disaster resilient communities (GNDR, 2009). In total, 48 countries in Africa, Asia and the Americas participated in this community-focused research project aimed at determining progress on the implementation of the HFA at local government level. The findings were that there existed significant gaps between national commitments and local actions. It is widely acknowledged that the VFL process shifted the DRR agenda globally towards local-level evidence of national policies. The VFL project was repeated in a number of iterations in 2011, 2013 and 2019.

Fig 14.5
QR code for Box 14.4. Available at https://www.gndr.org/programmes/views-from-the-frontline.html

recovery to guide how post-disaster reconstruction and recovery does not create new risks and ensures that affected Member States build back better. To fulfil its mandate on communication of research, progress updates and developments in the DRR field the UNDRR uses the Global Platform for DRR (GPDRR), which also publishes and communicates the Global Assessment Report (GAR) biennially since 2007. The GPDRR is a worldwide gathering of the DRR community representing more than 150 countries, made up of UN Agencies, international NGOs, civil society, private sector, practitioners, policymakers, researchers, academics, local communities, representatives from key population groups, such as indigenous people, women, youth, children and people living with disability. The GPDRR sets the stage for discussions and engagement on policy and strategy direction, and this is where the GAR is normally launched and disseminated.

The UNDRR also provides regional support to guide and direct continent-specific DRR activities; for example, in Africa through the African Union Commission. In such partnerships, the regional bodies are supported through project funding, capacity enhancement and scientific research to the contexts of each continent. There are also feedback loops that enable the input of risk data and information from national to regional to the UNDRR. The UNDRR also drives awareness on DRR through campaigns, the major one being the International Day of Disaster Reduction, held annually on 13 October since 1989. Each year, the institution comes up with a carefully selected theme to draw attention towards pertinent DRR issues. At the global level, there is also recognition of civil society organisations (CSOs) as key players in anticipatory DRR strategies due to their proximity to the experiences of, and relationships with, communities at the front line (Gibson, 2017). The strength of civil society in local community mobilisation enables CSOs to act as a bridge between global processes and the communities at risk. CSOs usually have access to community-level risk knowledge, which they often use to speak as a collective in lobbying and advocacy, and to also feed into global agenda-setting processes. They are also involved in post-disaster response and recovery, where their participatory community engagement approaches can help to identify and direct post-disaster needs and risk reduction priorities. The Global Network of Civil Society Organisations for Disaster Reduction (GNDR) is an example of civil society organisation at the global level (born out of the Kobe/HFA process) and is known to be the main coordinating network that brings together more than 1,000 local-level CSOs in more than 100 countries, who work in local communities directly impacted by disasters (Box 14.4 and Figure 14.5).

At the regional level, there are various institutions that provide platforms to discuss DRR issues, mainly about respective continents. The UNDRR works in partnership with these institutions to conduct preparatory events ahead of global meetings. Some of the meetings at the regional level bring together diverse stakeholders who contribute to policy, research, implementation and progress monitoring of DRR work in the continents, including transboundary issues and solutions. For example, in each continent, the UNDRR works through the respective regional platforms (RPs) viz. Africa Regional Platform for DRR; Regional Platform for DRR in the Americas; and Regional Platforms for the Arab States, Asia, Europe and the Pacifics. The RPs' work is furthered through regional organisations, such as the European Union, the Organization of American States or the African Union (Figure 14.6). Through regional ministerial and other similar meetings conducted ahead of significant global gatherings, different regions adopt contextual recommendations, outputs and positions from region-wide stakeholder consultations. The regional institutions also drive the domestication of global frameworks and strategies; coordinate DRR projects and activities, including capacity development and research; convene regional conferences; and produce region-specific DRR frameworks and policies. Regional approaches to DRR are underpinned by recognition of the critical role of sovereignty, subsidiarity, regional integration and cooperation in DRR (O'Donnell, 2017), and their involvement in DRR is mainly in response, preparedness, mitigation and recovery through policy direction, early warning systems, regional response teams, regional disaster management centres and funding support. Differences in the scope of engagement of regional actors in the different DRR components should be expected because of contextual differences in the different continents.

Fig 14.6
DRR deliberations at the African Union.
Source: photo courtesy of Dewald van Niekerk

At a sub-regional scale DRR work is further cascaded through institutions that are stand-alone entities, such as the Caribbean Disaster Emergency Management Agency (CDEMA) and the Coordination Centre for Natural Disaster Prevention in Central America (CEPREDENAC). In some cases, already existing platforms are used to drive the sub-regional DRR agenda, such as the Economic Community of West African States (ECOWAS), where DRR is under the Directorate on Humanitarian and Social Affairs, and the Southern African Development Community (SADC) has a DRR Unit. At this level, intergovernmental meetings are convened to formulate specific policies and strategies to address DRR and capacity-enhancement programmes for practitioners and policymakers from Member States. Evidence-based research and special studies are also conducted to inform DRR policy architecture in the respective sub-regions. At the sub-regional level, there is also the coordination of reporting on the global frameworks and formulation of collective positions informed by sub-regional disaster risk profiles and public goods to feed into the regional level. Conventions and conferences are held at the sub-regional level, bringing together diverse actors involved in DRR. Generally, at regional level, more focus is on mitigation and response in the case of regional disasters, but this greatly depends on the context of the region, and regional institutions are usually involved in all components of DRR to provide support to Member States.

At the national level, DRR activities are mainly centred around establishment of DRR institutions, formulation of relevant DRR-specific and related legislation and policies, capacity-development, political and financial support, mobilisation of resources for DRR, domestication and implementation of global and regional DRR frameworks, as well as monitoring and reporting according to international guidelines (Nemakonde et al., 2017). During the SFDRR period (2015–2030)

Member States are required to report to the Sendai Framework Monitor, which is coordinated by regional bodies. Disaster risk governance at the national level underpins the success of DRR and disaster risk management. Central governments are involved in all components of DRR, although state action still tends to be biased towards reactionary response. Given that governments are responsible for championing sustainable development and ensuring the safety of their citizens, there is growing emphasis on risk-informed development that requires governments to increasingly pay attention to preventing and reducing risks before disasters occur (Wilkinson & Kelman, 2017). At national level integration, mobilisation and coordination of DRR and non-DRR actors needs to be promoted through an all-of-society and whole-of-government approach of DRR mainstreaming (Opitz-Stapleton, 2019). CSOs can also be actively involved in lobbying and advocacy; for example, lobbying for improved disaster risk governance. Central government at the national level may enter into funding partnerships with donors for implementation of DRR work at national or identified local levels.

The sub-national level is the hub policy implementation where diverse actors come together for DRR (Figure 14.7). It is the front line of disasters, where complex interactions between vulnerabilities, coping capacities, risks and resilience are manifested. The sub-national level is the primary target where DRR needs to be operationalised, be it in stopping the creation of new risks or reducing existing and residual risk as well as underlying risk drivers. The sub-national level is where inequalities and injustices are expressed, and local voices and lived-out experiences emerge and can be shared to contribute to bottom-up engagement, lobbying and advocacy (Wilkinson & Kelman, 2017). The sub-national level is where there should be an actualisation of intent in DRR, and specific DRR interventions are identified at this level. Therefore, both central and local governments are required to operationalise risk-informed decisions and policies in development planning, provide DRR budgets and integrate expertise and resources to drive DRR. It is essential to include local communities at the sub-national level in the DRR processes, and local governance provisions may exist to cater for such.

The private sector is increasingly getting involved in DRR, as it realises that their infrastructure and investments are threatened by disasters and require protection. It is also their consumers and suppliers who are affected by disasters (providing a market for risk transfer mechanisms and other business products), and they are also creators of risk through their diverse industrial operations (Stevenson & Seville, 2017). Losses, such as labour, equipment, infrastructure and markets, often lead to business disruption and have motivated private sector interest in DRR, mainly in business continuity and supply chain resilience. The private sector is a key strategic partner to the public sector, especially given that, in some instances, the private sector

Fig 14.7
DRR planning and implementation requires multiple role players. Source: EAGER Project, North-West University

may have some financial and material resources that the latter may need. Likewise, the public sector may also have resources required by the private sector; hence, theirs is a mutual relationship in DRR. The involvement of private sector in DRR illustrates how shared responsibility that is outlined in the SFDRR can be achieved. The private sector can collectively advocate for a move away from reactionary response and recovery, towards increased investment in proactive DRR through prevention, mitigation and preparedness. The sector can also provide research funding to generate scientific evidence useful in steering policy direction, lobbying and advocacy, and risk-informed investment decisions in business development. The successful involvement of private sector in DRR requires senior-level political will. The UNDRR has private sector partners and advisors under the banner of Private Sector Partnership DRR and the Private Sector Advisory Group. For example, in Australia, the Australian Business Roundtable for Disaster Resilience and Safer Communities was formed in partnership with the Australian Red Cross. The Roundtable brings together senior executives from businesses and contributes resources to partner government efforts. More importantly, this private sector platform advocates for a move away from reactionary response and recovery, to increasingly invest in proactive DRR through mitigation and preparedness. They provide research funding to generate an evidence base that is useful in steering policy direction, lobbying and advocacy, and risk-informed investment decisions in business development (see: http://australianbusinessroundtable.com.au/). The involvement of private sector in DRR illustrates how shared responsibility that is outlined in the SFDRR can be achieved. This shared responsibility cut across sectors and disciplines. DRR can only be effective if it is aimed at solving complex real-world problems through a transdisciplinary lens.

14.5 Disaster risk reduction through multisectoral and transdisciplinary cooperation

An increase in disaster losses since the adoption of the concept and practice of DRR does not mean that DRR embodied in international frameworks is not fit for purpose. There might be several reasons for such increases, including lack of full implementation of international, regional and national frameworks for DRR. Indeed, international frameworks are a good basis on which DRR strategies and measures can be developed. However, there are tendencies of countries, particularly in the developing world, to focus their efforts on reacting to disasters rather than putting mitigation and preparedness measures in place to address existing risks and building community resilience to avoid creating new risks. The other challenge is that DRR knowledge is still fragmented within and amongst different stakeholders, and has not been adequately incorporated in different sectoral planning (Abedin & Shaw, 2015). The real ownership of DRR by national stakeholders remain blocked in many countries by conceptual, political, economic, governance and instrumental shortcomings and failures (Lavell & Maskrey, 2013). Lavell and Maskrey (2013) further state that under the paradigm of DRR, risk has become abstract and compartmentalised, and its dependent relationship with development processes has been blurred and obscured. Most importantly, development and DRR decision-making processes occur in silos, conducted by different agencies, institutions with differing priorities, perspectives and outlook time horizons (Thomalla et al., 2018).

Unfortunately, the silo approach to reducing the risk of disasters entrenches the bureaucratic approaches to DRR whereby institutions vested with mandates for DRR coordination are viewed as autonomous with no need for other sectors to get involved in DRR. In most countries, many sectors have not yet identified their roles in reducing the risk of disasters and, thus, do not participate in risk reduction measures within their sector. The notion of disasters and DRR as "everyone's business" does not seem to be fully comprehended and has not yet resulted in everyone fulfilling their role in reducing the risks of disasters in their sectors. It is, therefore, key to make DRR an underlying principle in all relevant development sectors, if we are to address existing risk and avoid creating new risk. The socio-cultural structures and processes, such as the links between disaster risk and poverty, gender, population growth, public awareness, and in certain cases, social exclusion (where certain minority groups are forced to live in risky locations due to the nature of their livelihoods or pure exclusion based on ethnicity, race or religion) must be taken into consideration when addressing disaster risk (World Bank, 2017). As Thomalla et al. (2018) argues, addressing the underlying drivers of risk inherent in the failures of development and DRR requires actions that challenge existing structures, power relations, vested interests and dominant narratives that persist within systems and maintain and perpetuate poverty, inequality and marginalisation.

Due to a combination of risk drivers, such as climate risk, resource scarcity and depletion, ecosystem degradation, livelihoods' impoverishment, demographic changes, urbanisations, and limited capacities to manage risks, amongst others, the extent of vulnerability and exposure to hazards are on the rise (UN, 2013). Therefore, the need to reduce the impending risk of disasters cannot be overemphasised. Moreover, this makes DRR an agenda for everyone. As alluded to in the SFDRR, "more dedicated action needs to be focused on tackling underlying disaster risk drivers, such as the consequences of poverty and inequality, climate change and variability, unplanned and rapid urbanization, poor land management and compounding factors such as demographic change, weak institutional arrangements, non-risk-informed policies, lack of regulation and incentives for private DRR investment, complex supply chains, limited availability of technology, unsustainable uses of natural resources, declining ecosystems, pandemics and epidemics" (UN, 2015a). In this regard, efforts and resources must be channelled towards transforming the underlying drivers that generate risk in the first place. In this way, reducing the risk of disasters becomes a cost-effective investment that will help prevent future losses (UN, 2015a). The UN (2015a) further argues that for DRR practices to be efficient and effective, they must be multisectoral, inclusive and accessible, and must focus on multi-hazards. This is so because DRR is an integrated domain that cuts across a wide range of disciplines and fields, embracing numerous disciplines as well as policy and practice (Takara, 2018).

The cross-cutting nature of DRR renders the reduction of disaster risks a multifarious endeavour requiring a wide range of sectors, institutions and disciplines to coordinate their efforts (Box 14.5). This is so because efforts and investments required to address all dimensions of disaster risk surpass the capacity of any one sector. Therefore, actors in different sectors, institutions and disciplines must pull together their respective strengths and capacities, including resources and expertise, as a strategy in pursuing DRR activities. As Bosher and Dainty (2011) argue, the emerging emphasis on DRR has broadened the range of experts and professions whose input must now be garnered, and who must pool their collective knowledge and expertise to resolve complex socio-technical challenges.

Disaster risk is conceptually approached and theorised from within six disciplinary schools of thought (Alexander, 1993), which interact or overlap with each other in one way or another (Benjamin, 2014). These include the geographical approach, anthropological approach, sociological approach, developmental studies approach, disaster medicine and epidemiological approach, and technical approach. Other disciplines, such as psychology, philosophy, management and economics, have also contributed to the disaster risk field (Benjamin, 2014). The diversity of these disciplines that shaped the disaster risk field reflects its transdisciplinary nature. Therefore, effective and efficient reduction of disasters risk will require an application of

Box 14.5: Institutional system for DRR

The World Bank (2017) has classified the institutional system for DRR into four categories which includes the following:

- Institution/s with primary responsibility for DRR coordination and policy guidance, such as National Disaster (Risk) Management Authority;
- Fully dedicated institutions with specific responsibilities on different aspects of DRR, such as Meteorological Services and research centres;
- Sectoral ministries and local governments that have a role in integrating DRR into development planning (land-use, safer construction, rangeland management, water conservation and management, awareness, and education), such as agriculture, environment, education, urban development, water, transport, social affairs, as this makes almost all governmental ministries in some countries have an existing or potential role in DRR; and
- Private sector and civil society organisations (CSOs), such as insurance companies, business associations, and international NGOs, community-based organisations and women's organisations.

multisectoral partnerships and transdisciplinary approaches to DRR. Multisectoral partnerships increase the efficiency and boost overall capacity for DRR efforts by encouraging sharing of information and experiences, reducing duplication through effective coordination of data, expertise, capabilities and resources (R3ADY Asia-Pacific, 2014). On the other hand, transdisciplinarity has the potential to effectively address increasingly complex social problems, such as disaster risk, that transcends boundaries between different orthodox disciplinary knowledge (Benjamin, 2014). Moreover, transdisciplinary approach is "an approach in which all players and stakeholders in various disciplines (natural, social and human sciences) and sectors (public, private, academia and civil society) work together to achieve a common goal" (Matsuura & Razak, 2019, p. 811). Matsuura and Razak (2019) further argue that the transdisciplinary approach requires different stakeholders (including non-academic) of various disciplines to collaboratively find solutions beyond the limit of single disciplinary knowledge and, in doing so, work outside of one's disciplinary boundaries (Table 14.1 explains the components of the transdisciplinary paradigm).

Transdisciplinarity requires that all stakeholders co-design, co-produce, co-deliver and co-implement (Matsuura & Razak, 2019) DRR initiatives to reduce the risks of disasters in a coherent and integrated manner. To increase the understanding of risks of natural, human-made, technological and complex hazards, and to apply appropriate measures to reduce such risks, Takara (2018) advocates studying lessons from past disasters and encourages interdisciplinary

Paradigm (World view which underlies our theories and methodologies)	Transdisciplinary Crosses disciplinary boundaries to create holistic approach – uses concepts and methods originally created in other disciplines. Relates scientific knowledge to real world problem solving.
Ontology (What do we know?)	Social/human systems are different to natural/ecological/environmental systems, but they have become inextricably coupled/interconnected.
Epistemology (Theory of knowledge – how do we know what we know?)	Co-producing of knowledge, collaborative, participatory, mutual learning, uncertainty.
Anthropology (Study of humans and society)	The humans (the social) are different from nonhumans (the natural) but are inextricably entangled.
Axiology (What elements can contribute to the intrinsic value of a state of affairs?)	Equal value of disciplinary and non-disciplinary knowledge systems (when facing complex real-world problems).
Methodology (How do we study?)	Integrating, bringing together, synthesising, merging.
Methods (How do we find/explore knowledge?)	Integrative methods: conceptual clarification, theoretical framing, research questions, hypothesis formulation, assessment measures and so forth.

Table 14.1
Transdisciplinary paradigm
Source: Van Breda (2012)

collaboration across a wide range of sciences, including engineering, natural sciences, social and health sciences. As Max-Neef (2005) posits, transdisciplinarity, more than a new discipline or super-discipline, is actually a different manner of seeing the world, more systemic and more holistic.

The involvement of stakeholders from different sectors and disciplines, and the need for coordination between these stakeholders necessitates putting governance mechanisms in place. The UNDP (2010) views governance as the umbrella under which DRR takes place, as it influences how national and sub-national actors (including governments, parliamentarians, public servants, the media, the private sector, and civil society organisations) are willing and able to coordinate their actions to manage and reduce disaster-related risk. Priority 2 of the SFDRR focuses on strengthening governance to manage disaster risk and outlines the activities to be carried out at global, regional, national and local levels for governance of disaster risk to work. Transdisciplinarity presents opportunities to break down the silo walls and foregone segmented or sector-specific approaches to DRR. Therefore, for DRR to have any effect, it requires a vast number of disciplines to find numerous solutions to real-world problems, through a transdisciplinary paradigm. After all, the SFDRR is underpinned by the principles of all-of-society and whole-of-government approach to DRR. As Takara (2018) argues, when optimised through inter- and transdisciplinary engagement, DRR can contribute to development progress beyond the SFDRR, and assist in attaining the aspirations of the SDGs, the Paris Agreement, the New Urban Agenda and the World Humanitarian Summit. Indeed, coordination within and between sectors, and with all relevant stakeholders, is needed for the successful implementation of DRR to reduce the risk and impacts of disasters. Transdisciplinarity provides a binding paradigm for DRR (Van Niekerk, 2012). If we view the reduction of disaster risk through the transdisciplinary lens, then addressing the underlying risk drivers in the social, economic, natural and the built environments should not be a difficult task for DRM.

14.6 Components of disaster risk management

When DRR policies, activities and processes are applied to manage, reduce or prevent different types of risk to strengthen resilience, it is referred to as *disaster risk management* (UNDRR, 2019). DRM can be classified into three categories based on the type of risk being managed in each respective category. *Prospective* DRM focuses on actions and decisions aimed at preventing the creation of *new* risk in development processes. *Corrective* DRM has its focus mainly on reducing known *pre-existing* risk, and *compensatory* DRM focuses on managing unavoidable *residual* risk to minimise the negative impacts on communities at risk. This shows that DRM is a complex process of interrelated actions, policies and processes (Box 14.6 and Figure 14.8).

The systemic nature of risk based on the realisation that risks are dynamic and interconnected, including the increasing climate risk and other already existing underlying risks, underpins DRM (UNDRR, 2019). There is also a growing need to reimagine risk management approaches considering the sustainable development agenda. To comprehensively tackle the developmental challenges faced by the world at different scales, there is a need to improve and modernise DRM. Innovative approaches to finding a solution to pressing disaster risk issues will be key in this regard. A paradigm shift in risk management that is informed by intent to holistically address the complex interrelated hazards and their equally complex interactions with exposure, inadequate coping capacities and increasing vulnerability is necessary.

One of the areas where transformation is required in DRM is the traditional dominant reliance on historical data as a preferred way to predict the future (UNDRR, 2019; Opitz-Stapleton, 2019). However, given the complex and dynamic interactions of risks, including climate

Box 14.6: What is transdisciplinary DRR?

Want to know more about the transdisciplinary nature of DRR? Watch this video.

Fig 14.8
QR code for Box 14.6. Available at https://www.youtube.com/watch?v=Mvswo1QFegw

risk, one must exercise caution because a myopic focus on historical trends may miss the emerging risk interactions with dynamic changes, such as population growth, climate change, urbanisation, infrastructural development and environmental degradation, amongst others. This also means that for DRM to bear any fruit, attention needs to be paid to the generation of people-centric risk data and information management. Such can contribute to risk-informed decision-making for sustainable development, including but not limited to budget provisions and infrastructural projects.

Another dimension to be taken in transforming DRM is to embrace an integrated multi-hazard approach to address the various forms of risks in development. This requires that DRM break down existing silos and adopt an all-encompassing approach where coherence permeates through actions, mandates, policies and institutions (UNDRR, 2019). Conventional hazard-by-hazard problem-solving approaches that previously informed DRM are insufficient in addressing the emerging complex developmental challenges that emanate from the systemic nature of risk. A holistic approach that brings together diverse actors, institutions, strategies and resources is desirable in each of the DRR components that are furthered through DRM. Key to successful DRM is engaging with the principles of disaster management, which still greatly influences how government and institutions perceive to engage with hazards and disasters. The known foci of prevention, mitigation, preparedness and response must also include DRM principles.

Prevention considers that steps and actions need to be taken to prevent the creation of risk (UNDRR, 2019), and ideally, this would be the most desirable to save lives, livelihoods and infrastructure. However, it is also possible that preventing the creation of new risk could also stifle development and derail ambitions to improve the social and economic well-being of populations, given that some development projects are associated with inherent risks. Nevertheless, it remains important to consider how the creation of new risks may be prevented as societies develop, societies become more urbanised, populations grow, and exploitation of environmental resources increases. Assessments, such as cost-benefit analyses and risk modelling, are useful in informing investment decisions on the suitability of development projects (Enarson & Dhar Chakrabarti, 2018).

Mitigation is described as an activity or a combination of activities undertaken before, or after, a disaster occurs to reduce its likelihood of occurrence, where complete prevention is not possible (UNISDR, 2009). Mitigation aims to protect lives and livelihoods, and limit property loss and damage. Pro-active decisions are made, and strategies are implemented that lessen disaster impacts, ultimately reducing fatalities and financial costs of disasters. Mitigation may be either structural or non-structural. Structural mitigation primarily focuses on the tangible hard, built environment. For example, raised levees, elevated buildings and land buyouts in floodplains, retrofits for seismic hazards, food grain reserves in drought mitigation, or consistent maintenance of public infrastructure. Non-structural mitigation focuses on the soft aspects, such as public education and awareness-raising, building codes and regulations, land-use planning, insurance, and relocations. Consultative and inclusive mitigation planning through vulnerability, risk and hazard assessments facilitates collaborations by diverse stakeholders, guides mitigation priority-setting and enables business continuity. A hazard mitigation plan that details recommended mitigation actions and assigns responsibilities and a timeframe is produced in the planning process. The mitigation plan also states goals, priorities and how the actions will be strategically implemented, and must be reviewed, evaluated and updated at certain intervals. By its nature, mitigation attracts high capital expenditure on infrastructural developments, which at times may give a false sense of comfort and refusal to evacuate by those at risk. When adequately conducted, mitigation measures enhance the integration of DRR with sustainable development and CCA. Where risks can neither be prevented nor mitigated in DRM, or when it is not cost-effective to do so, then risk transfer need to be considered. Market-based mechanisms, such as insurance and risk pooling, can be considered by both individuals and institutions, especially concerning "disasters that have a long return period" (Dhar Chakrabarti, 2017, p. 25).

Preparedness includes steps taken in anticipation of, and readiness for, an imminent event/disaster *before* it occurs (UNDRR, 2015). The success of preparedness largely depends on well-coordinated plans, early warning systems, time efficiency, good communication and information dissemination, to avoid unnecessary loss of lives and damage. Early warning information needs to be appropriately packaged for the diverse recipients using relevant platforms for dissemination. For example, consideration needs to be given to literacy, people living with disability, children, women, youth and people in marginal settlements who may not be within the conventional communication channels. Preparedness may be guided by vulnerability management plans, crisis modifiers, emergency preparedness plans, evacuation plans and business continuity plans, which are put in place before an event and are used to guide decision-making. It also includes the assembling of volunteer teams who can assist in evacuation. Simulation exercises and drills may be done from time to time to test the existing plans and identify areas for improvement. Maintenance of critical stock supplies for food, water and medication, as well as alternative off-grid electricity options, are all part of preparedness. Success of preparedness is underpinned by adequate budget, and institutional and legislative provisions, and in turn shapes the successful transition to response and recovery (UNDRR, 2015). When adequately performed, preparedness ensures that chaos, psychological torment and panic are minimised. However, when early warning messages lack accuracy, then they may not be taken seriously, and communities may refuse evacuation. Preparedness actions should be used as a feedback loop into the DRM process. This will assist in identifying areas that need converted DRM actions, which, once addressed, will enhance preparedness and mitigation.

Response involves actions undertaken during or immediately after a disaster (UNDRR, 2015). The main focus is on life-saving through search and rescue operations, as well as provision of critical emergency services, such as medical supplies and attention, food and non-food items, water and sanitation, shelter, and immediate restoration of basic services where possible. In emergency services provision, such provision must restore the dignity of affected communities, and response activities should avoid the creation of other risks, such as disease outbreaks and gender-based violence at holding camps or conflict in host communities when there has been displacement. Response is reactionary and requires humanitarian aid (relief) to provide financial, human and material resources, and its success is shaped by risk-informed preparedness planning, thereby facilitating speedy and effective recovery. There are standard guides for delivering relief, and various actors from local and international institutions, such as local authorities, UN Agencies and international NGOs, are involved in response as affected communities are often incapacitated to respond on their own. The demerits of response are that it is often associated with a high demand for financial and other resources, often leading to diversion of such resources from other development projects to relief. Response is also associated with trauma and chaos when actors are not well-coordinated. Similar to prevention and mitigation, after the initial chaotic period associated with a response to a disaster, the debriefing of the response actions is a valuable input to a new iteration of DRM. In such a way, the unknown or unforeseen risks can become part of the DRM process.

Recovery is a component of DRR that comprises activities conducted *after* a disaster has occurred to restore normalcy or initiate improvements, enabling those affected to build back better and contributing to sustainable development (UNDRR, 2015), or, as Manyena et al. (2011) put within a DRR context, "bounce forward". Recovery activities, such as reconstruction of damaged infrastructure, rehabilitation of facilities, psychosocial support and counselling, may be conducted and should also be informed by risk assessments to prevent instilling historical risk and creation of new risks, and mitigate unavoidable risk. Mitigation measures may be incorporated during recovery, such as retrofitting new buildings. Recovery is more development-focused in nature and includes both short- and long-term measures; hence, the transition from humanitarian to development programmes is not distinct. Recovery is often characterised by high costs of reconstruction and rebuilding livelihoods. Poor communities are rarely able to meet the costs on their own; hence, humanitarian flash appeals or emergency disaster funds are created. Nevertheless, recovery also offers the opportunity for development when improvements are made or new infrastructure is set up. When it is risk-informed, it allows communities to build back better and creates employment opportunities.

As can be seen from the previous section, the interlinkages between DRM and development are keys to DRR. On a macro/global scale, moreover, the notions of the integration of the two terms linked to CCA is presented as best practices.

14.7 Disaster risk reduction, sustainable development and climate change adaptation

Beyond 2015, the global development agenda is entrenched on the integration of DRR, CCA and sustainable development (SD) as a vehicle towards achievement of risk-informed development (Opitz-Stapleton, 2019). Each of the three global agendas is directed by a separate global framework, the SFDRR, Paris Agreement to combat climate change and the 2030 Agenda for Sustainable Development with its corresponding SDGs. The three global frameworks have important core policy alignments and common goals that make pursuing them in a unified manner to reduce risks and strengthen resilience desirable (Solecki et al., 2011). Wisner (2020) adds four more international frameworks that have a direct bearing on successful DRR and the integration with CCA and SD: the Addis Ababa Action Agenda (Financing for Development), the World Humanitarian Summit, the New Urban Agenda, and various regional peace and security initiatives. DasGupta and Shaw (2017) explain the link between DRR, CCA and SD by highlighting that CCA is a subset of DRR, which in turn is a subset of sustainable development. The basis of this relationship is that in combating climate change, disaster risk, including climate risk, can be reduced, while protecting and strengthening development, ultimately contributing to sustainable development (Banwell et al., 2018). Actually, without an integrated focus on risk reduction and CCA, sustainable development may remain nothing but a pipe dream, and likewise, unless DRR and CCA are framed within sustainable development, both are likely to fail.

Integration promotes coherence in implementation of the SFDRR, CCA and SDGs, and helps address the common underlying objective of resilience-building in all three agendas. All three seek to assist communities to build back better and leave no one behind in the face of crises by recognising the centrality of risk in development processes and decisions. Despite differences in how each agenda seeks to achieve resilience, pursuing DRR, CCA and sustainable development in an integrated manner is preferred as it promotes the application of DRM in development, shaping risk-informed sustainable development decisions (Pelling, 2010). Integration embraces multi-hazard approach and ensures reduction in loss of life and infrastructural damage when disasters occur, whilst also protecting economic gains. Although resilience-building is a key tenet of the drive towards integration, it is important to realise that many other diverse interconnections also contribute to the objectives of the three agendas.

Integration requires that various actors from the global level right up to the local level work together in the implementation of all three agendas at policy and institutional levels. At the policy level, integration is based on the identification of possible synergies to be harnessed, prioritisation of goals where necessary and elimination of possible policy contradictions. Finding policy cohesion becomes paramount. Diverse actors with common goals come together to pursue the achievement of these efficiently and effectively, whilst simultaneously addressing gaps and eliminating unnecessary redundancies. Integration of DRR, CCA and SD can take on various options depending on the contexts within which it is pursued. At various scales, integration of the three agendas may be done in a sequential step-wise manner or as a continuum from fragmentation to perfect merging. In selecting integration options, it is important to carefully consider that the desirable self-determined goals of each agenda are not watered down (Wallace, 2017). Therefore, in pursuing integration, it is worthwhile to maintain a degree of autonomy and separation that allows each agenda to still maintain its unique focus. Integration of DRR, CCA and SD recognises that different sectors focus on all three agendas to address the common theme of resilience. Throughout all the agendas, the significant role of ecosystems in the reduction of disaster risks, including climate risks and sustainable development, is acknowledged. For example, at least 7 out of the 17 SDGs are related to disaster risks, including climate risk. Also, in all three agendas, emphasis is made on the active involvement of diverse state and non-state actors across different scales internationally right down to local level. Multiscalar inter-institutional integration leverages on the stakeholders' capacities and expertise, ensuring resource-use optimisation where resources are scarce. Integration serves as a platform for information dissemination, sharing experiences and best practices and problem-solving strategies. Additionally, through integration, the programmes in each of the agendas may be better coordinated, engagement improved and new networks established, facilitating an iterative integration process. This is especially important between state and non-state actors, such as the private sector and civil society, who may complement state efforts and bring resources that could be lacking in the public sector. A multiscale, multisector integration also strengthens developmental planning and risk-informed decision-making at all levels of DRR, CCA and SD. Furthermore, integration is people-centred, especially when it is framed through an all-of-society and whole-of-government lens (UNDRR, 2019). The ultimate goal for all three agendas is to protect the lives and livelihoods of communities across the globe, and it is at the lowest household unit where benefit from DRR, CCA and SD integration is realised. It is at the community level that all policies for DRR, CCA and SD are operationalised. Actually, at the community level, there may be no distinction to all three, and as communities innovate, engage and mobilise, they may contribute to the agendas simultaneously. Table 14.2 illustrates some of the advantages and demerits of integrating DRR, CCA and SD.

Altogether, it is seen that the positives of integration of DRR, CCA and SD outweigh the demerits. Integration, when properly conducted and with adequate political support, offers a vehicle to achieve the post-2015 development agenda for risk-informed sustainable development.

14.8 Concluding remarks

DRR is all the strategies, policies, plans, actions, structures and capacities that enable addressing the underlying factors to disasters through a risk management approach. DRR provides a paradigm shift from the traditional management of a disaster to understanding the underlying and contributing factors leading to a hazard (natural or human-made) becoming a disaster. The emphasis is, therefore, on the reduction of the risk of a disaster occurring. This chapter aimed to provide an overview of DRR policies and practices. It initially identified the main principles and components of DRR before identifying their respective strengths and limitations. It also emphasise good practices at different geographical scales, from international policies and initiatives to national and local actions. It outlined the respective theoretical and actual roles of different stakeholders and tools in fostering DRR.

Previous chapters alluded to the elements that constitute disaster risk as being vulnerability, hazard, resilience, coping capacity, manageability and exposure. Taking all these variables into consideration, one can argue that a reduction on the one means an increase in the other. However, disaster risk is more complex than that. Therefore, for DRR to be effective, one needs a whole-of-society approach, which is underscored by several disciplines. Only through a transdisciplinary approach to disaster risk can one truly reduce the possibility of an event escalating into a disaster. This chapter emphasised the need for cross-scale and sectoral involvement and application. The need for the integration of DRR, CCA and sustainable development was emphasised because finding combined solutions to these problems has multiple gains. DRR should be seen as the insurance policy of our global development. It is not acceptable that countries lose years of development gains by the

Table 14.2 Advantages and disadvantages of integrating DRR, CCA and SD

Advantages	*Disadvantages*
Integration promotes efficiency, where there is scarcity of resources (human, finance and technology) to realise objectives of the agendas. Integration ensures sharing of data, information and resources; enhances progress monitoring; and facilitates learning across sectors and institutions (UNDRR, 2019).	Finance may be a limiting factor, especially in developing countries. However, this may be addressed by pursuing different funding opportunities available under each agenda.
Integration promotes effectiveness through efficient use of limited resources and increased coherence in policy and activity implementation (Mitchell & van Aalst, 2008). **Efforts to achieve goals of one agenda will simultaneously contribute to progress in the others. This is made possible through sharing of risk data, information and monitoring tools, and also encourages joint risk-informed decision-making.**	When the intent of integration has not been clearly articulated and understood, there may be confusion and chaos amongst the various actors (Street et al., 2018).
Integration ensures incorporation of traditional knowledge in all three agendas.	There may be dilution of issues in some agendas (Wallace, 2017). This may be avoided by ensuring that institutions involved in integration maintain a part of their autonomy, which would allow them to pursue their distinctive priorities and mandates.
Integration promotes inclusiveness in all agendas, as it brings together state and non-state actors, including local communities.	
Integration helps balance humanitarian and development programmes, ensuring sustainable resilience in all components of societies.	
Integration facilitates a holistic systems approach to DRR, CCA and SD.	

destructive nature of disasters. Although we live in nature, we need to learn to live with nature. This requires an understanding of how human actions change the various linked systems on which we depend (be it economic, social, environmental or political).

Take-away messages

1. Disasters are not natural;
2. There has been a gradual international shift in thinking about hazards and disasters from responding to disasters, to rather reducing the risks that these hazards pose;
3. DRR is all decisions and actions aimed at reducing the risks of natural and anthropogenic hazards turning into disasters;
4. Practical DRR requires a multisectoral and multidisciplinary approach at various levels;
5. The Sendai Framework for DRR 2015–2030 currently provides a global framework for DRR that all states should aim to implement and to achieve the targets by 2030;
6. DRR and CCA should not be treated as separate entities, but their implementation must be mainstreamed and integrated; and
7. Not all residual disaster risks can be fully reduced, and if this happens, we must be prepared for any eventuality through disaster preparedness and response actions.

To learn more about the topic discussed in this chapter, listen to the *Disasters: Deconstructed* interview with Dr Barbara Carby (Figure 14.9).

Fig 14.9
QR code for Chapter 14.

Further suggested reading

Bosher, L. S., Chmutina, K., & Van Niekerk, D. (2021). Stop going around in circles: Towards a reconceptualisation of disaster risk management phases. *Disaster Prevention and Management*, *30*(4–5), 525–537. https://doi.org/10.1108/DPM-03-2021-0071

Coetzee, C., & Van Niekerk, D. (2012). Tracking the evolution of the disaster management cycle: A general system theory approach. *Jàmbá: Journal of Disaster Risk Studies*, *4*(1). https://doi.org/10.4102/jamba.v4i1.54

Keast, R., Brown, K., & Mandell, M. (2007). Getting the right mix: Unpacking integration meanings and strategies. *International Public Management Journal*, *10*(1), 9–33. https://doi.org/10.1080/10967490601185716

Peters, K., Langston, L., Tanner, T., & Bahadur, A. (2016). *'Resilience' across the post-2015 frameworks: Toward coherence?* [Working paper]. Overseas Development Institute.
United Nations Climate Change Secretariat. (2017). *Opportunities and options for integrating climate change adaptation with the sustainable development goals and the Sendai framework for disaster risk reduction 2015–2030* [Technical paper]. UN Climate Change Secretariat.
Van Niekerk, D., Coetzee, C., & Nemakonde, L. D. (2020). Implementing the Sendai framework in Africa: Progress against the targets (2015–2018). *International Journal of Disaster Risk Science, 11*(2), 179–189. https://doi.org/10.1007/s13753-020-00266-x
World Bank Group. (2016). *Striving toward disaster resilient development in sub-Saharan Africa: Strategic framework 2016–2020.* World Bank Group.

References

Abedin, M. A., & Shaw, R. (2015). The role of university networks in disaster risk reduction: Perspective from coastal Bangladesh. *International Journal of Disaster Risk Reduction, 13*, 381–389. https://doi.org/10.1016/j.ijdrr.2015.08.001
Alexander, D. (1993). *Natural disasters.* UCL Press.
Banwell, N., Rutherford, S., Mackey, B., & Chu, C. (2018). Towards improved linkage of disaster risk reduction and climate change adaptation in health: A review. *International Journal of Environmental Research and Public Health, 15*(4). https://doi.org/10.3390/ijerph15040793
Benjamin, A. (2014). *Transdisciplinarity of the disaster risk discourse.* https://www.academia.edu/895602/Transdisciplinarity_of_the_Disaster_Risk_Discourse
Bosher, L. S., Chmutina, K., & van Niekerk, D. (2021). Stop going around in circles: Towards a reconceptualisation of disaster risk management phases. *Disaster Prevention and Management, 30*(4–5), 525–537. https://doi.org/10.1108/DPM-03-2021-0071
Bosher, L. S., & Dainty, A. R. J. (2011). Disaster risk reduction and "built-in" resilience: Towards overarching principles for construction practice. *The Journal of Disaster Studies, 35*(1), 1–18. https://doi.org/10.1111/j.1467-7717.2010.01189.x
Centre for Research on the Epidemiology of Disasters. (2020). *The international disaster database—EM-DAT.* Université Catholique de Louvain. https://www.emdat.be/emdat_db/
Chmutina, K., & Von Meding, J. (2019). A dilemma of language: "Natural disasters" in academic literature. *International Journal of Disaster Risk Science, 10*(3), 283–292. https://doi.org/10.1007/s13753-019-00232-2
Coetzee, C., & Van Niekerk, D. (2012). Tracking the evolution of the disaster management cycle: A general system theory approach. *Jàmbá:*

Journal of Disaster Risk Studies, *4*(1). https://doi.org/10.4102/jamba.v4i1.54

Coetzee, C., Van Niekerk, D., & Kruger, L. (2019). Building disaster resilience on the edge of chaos: A systems critique on mechanistic global disaster reduction policies, frameworks and models. In J. Kendra, S. Knowles, & T. Wachtendorf (Eds.), *Disaster research and the second environmental crisis, Environmental hazards*. Springer.

Collins, A. E. (2009). *Disaster and development*. Routledge Perspectives In Development Series. Routledge.

Dasgupta, R., & Shaw, R. (2017). Disaster risk reduction: A critical approach. In I. Kelman, J. Mercer, & J. C. Gaillard (Eds.), *The Routledge handbook of disaster risk reduction including climate change adaptation*. Routledge.

Dhar Chakrabarti, P. G. (2017). *Mainstreaming disaster risk reduction for sustainable development: A guidebook for Asia-Pacific*. UN Economic and Social Commission for Asia and the Pacific. https://www.unescap.org/sites/default/files/publication_WEBdrr02_Mainstreaming.pdf

Enarson, E., & Dhar Chakrabarti, P. G. (2018). *Women, gender and disaster: Global issues and initiatives*. Sage.

Gibson, T. (2017). NGOs doing disaster risk reduction including climate change adaptation. In I. Kelman, J. Mercer, & J. C. Gaillard (Eds.), *The Routledge handbook of disaster risk reduction including climate change adaptation*. Routledge.

Global Network for of Civil Society Organisations for Disaster Reduction. (2009). *"Clouds but little rain. . ." Views from the Frontline a local perspective of progress towards implementation of the hyogo framework for action*. Tearfund. https://www.gndr.org/wp-content/uploads/2021/11/View-from-the-Frontline-Report-2009-English-Version.pdf

Kelman, I., & Glantz, M. H. (2015). Analyzing the Sendai framework for disaster risk reduction. *International Journal of Disaster Risk Science*, *6*(2), 105–106. https://doi.org/10.1007/s13753-015-0056-3

Lavell, A., & Maskrey, A. (2013). *The future of disaster risk management: Draft synthesis document meeting notes, background papers and additional materials from a scoping meeting for gAR 2015*. United Nationals Office for Disaster Risk Reduction (UNISDR) & Facultad Latinoamericana de Ciencias Sociales (FLACSO). https://www.preventionweb.net/files/35715_thefutureofdisasterriskmanagement.pdf

Manyena, S., O'Brien, G., O'Keefe, P., & Rose, J. (2011). Disaster resilience: A bounce back or bounce forward ability? *Local Environment*, *16*(5), 417–424. https://doi.org/10.1080/13549839.2011.583049

Matsuura, S., & Razak, K. A. (2019). Exploring transdisciplinary approaches to facilitate disaster risk reduction. *Disaster Prevention and Management*, *28*(6), 817–830. https://doi.org/10.1108/DPM-09-2019-0289

Max-Neef, M. A. (2005). Foundations of transdisciplinarity. *Ecological Economics*, *53*(1), 5–16. https://doi.org/10.1016/j.ecolecon.2005.01.014

Mitchell, T., & van Aalst, M. (2008). *Convergence of disaster risk reduction and climate change adaptation. A review for DFID*. ODI.

Nemakonde, L. D., Van Niekerk, D., & Wentink, G. (2017). National and sub-national doing disaster risk reduction including climate change adaptation. In I. Kelman, J. Mercer, & J. C. Gaillard (Eds.), *The Routledge handbook of disaster risk reduction including climate change adaptation*. Routledge.

O'Donnell, I. (2017). Regional organisations doing disaster risk reduction including climate change adaptation. In I. Kelman, J. Mercer, & J. C. Gaillard (Eds.), *The Routledge handbook of disaster risk reduction including climate change adaptation*. Routledge.

O'Keefe, P., Westgate, K., & Wisner, B. (1976). Taking the naturalness out of natural disasters. *Nature*, *260*(5552), 566–567. https://doi.org/10.1038/260566a0

Opitz-Stapleton, S., Nadin, R., Kellett, J., Quevedo, A., Caldarone, M., & Peters, K. (2019). *Risk-informed development: From crisis to resilience*. Overseas Development Institute.

Pelling, M. (2010). *Adaptation to climate change: from resilience to transformation*. Routledge.

R3ADY Asia-Pacific. (2014). *Harnessing the full potential of multi-sectoral partnerships in disaster risk management*. Input in support of the process to develop the post-2015 framework for disaster risk reduction from multi-sectoral leaders in the U.S. and Japan. https://www.wcdrr.org/wcdrr-data/uploads/496/Final-Input-Post-2015-Framework-for-DRR.pdf

Smith, K. (2004). *Environmental hazards: Assessing risk and reducing disaster*. Routledge.

Solecki, W., Leichenko, R., & O'Brien, K. (2011). Climate change adaptation strategies and disaster risk reduction in cities: Connections, contentions, and synergies. *Current Opinion in Environmental Sustainability*, *3*(3), 135–141. https://doi.org/10.1016/j.cosust.2011.03.001

Stevenson, J. R., & Seville, E. (2017). Private sector doing disaster risk reduction including climate change adaptation. In I. Kelman, J. Mercer, & J. C. Gaillard (Eds.), *The Routledge handbook of disaster risk reduction including climate change adaptation*. Routledge.

Takara, K. (2018). Promotion of interdisciplinary and transdisciplinary collaboration in disaster risk reduction. *Journal of Disaster Research*, *13*(7), 1193–1198. https://doi.org/10.20965/jdr.2018.p1193

Thomalla, F., Boyland, M., Johnson, K., Ensor, J., Tuhkanen, H., Gerger Swartling, Å., Han, G., Forrester, J., & Wahl, D. (2018). Transforming development and disaster risk. *Sustainability*, *10*(5), 1458. https://doi.org/10.3390/su10051458

UNDRR. (2015). *Proposed updated terminology on disaster risk reduction: A technical review*. UNDRR.

UNDRR. (2019). *Global assessment report on disaster risk reduction*. UNDRR.

United Nations. (2005). *Hyogo framework for action 2005–2015: Building the resilience of nations and communities to disasters*. United Nations.

United Nations. (2013). *Implementation of the international strategy for disaster reduction: Report of the secretary-general*. United Nations.

United Nations. (2015a). *Sendai framework for disaster risk reduction 2015–2030*. United Nations.

United Nations. (2015b). *Addis Ababa action agenda of the third international conference on financing for development*. United Nations.

United Nations. (2016). *Outcome of the world humanitarian summit*. United Nations.

United Nations. (2017). *New urban agenda*. United Nations.

United Nations Development Programme (UNDP). (2010). *Disaster risk reduction, governance and mainstreaming*. United Nations Development Program.

United Nations International Strategy for Disaster Reduction (UNISDR). (2009). *Terminology*. United Nations International Strategy for Disaster Reduction.

Van Breda, J. (2012). *Transdisciplinarity*. Peri Peri U Conference, held at University of Stellenbosch, pp. 1–34.

Van Niekerk, D. (2005). *A comprehensive framework for multi-sphere disaster risk reduction in South Africa* [PhD thesis, North-West University].

Van Niekerk, D. (2008). From disaster relief to disaster risk reduction: A consideration of the evolving international relief mechanism. *Journal for Transdisciplinary Research in Southern Africa*, *4*(2), 355–376. https://doi.org/10.4102/td.v4i2.158

Van Niekerk, D. (2012). *Transdisciplinarity: The binding paradigm for disaster risk reduction*. Scientific contributions series H: Inaugural Address Nr. 254. North-West University.

Wallace, B. (2017). A framework for adapting to climate change risk in coastal cities. *Environmental Hazards*, *16*(2), 149–164. https://doi.org/10.1080/17477891.2017.1298511

Wilkinson, E., & Kelman, I. (2017). Private sector doing disaster risk reduction including climate change adaptation. In I. Kelman, J. Mercer, & J. C. Gaillard (Eds.), *The Routledge handbook of disaster risk reduction including climate change adaptation*. Routledge.

Wisner, B. (2020). Five years beyond Sendai—can we get beyond frameworks? *International Journal of Disaster Risk Science*, *11*(2), 239–249. https://doi.org/10.1007/s13753-020-00263-0

World Bank. (2017). *Cross-cutting sector – disaster risk reduction (English). PDNA guidelines volume B: Disaster risk reduction*. World Bank Group. http://documents.worldbank.org/curated/en/120541493102189066/Cross-cutting-sector-disaster-risk-reduction

CHAPTER 15

Disaster management

Fig 15.1
Monitoring room of the Brazilian Early Warning Center (Cemaden). Source: photo courtesy of Victor Marchezini (6 December 2015)

This chapter covers aspects of disaster management, especially disaster preparedness and response. One of the essential elements of preparedness is warning systems (Figure 15.1). Warning systems are composed of four axes (risk knowledge, monitoring, communication and response capability) and are implemented through two main approaches – known as last mile and first mile. Warning systems must be connected to contingency planning and disaster response measures. Disaster response interventions involve essential actions in water supply, sanitation and hygiene promotion (WASH), food security and nutrition, shelter and settlement and health. It is fundamental to have basic notions of disaster preparedness and response to reduce disaster impacts.

15.1 Introduction

Over the course of history, societies have dealt with disasters and calamities in varied ways (Tuan, 1979). In ancient Egypt, pharaohs were often concerned with natural events that could damage entire regions and threaten to disrupt the cosmic order. To avoid crop failure–related

DOI: 10.4324/9781315469614-21

famines caused by flooding, drought and pests, pharaohs kept food reserves, and when a crop failure occurred, they distributed food and clothing to the whole population and forgave taxes in the affected areas (Tuan, 1979). These types of events have become a hazard and an object of governmental concern, fostering the emergence of new types of knowledge to manage human populations using a set of techniques, security apparatuses and understandings of social life through statistical analysis, such as the rates of illness, death, marriage and birth (Foucault, 2007; Marchezini, 2015). The practical elements of governance and management include the administration of rights, allocation of resources, appointments to office and authorisation of certain practices (Lund, 2006). Disasters are examples of social events that demand various types of governance and forms of management. Throughout history, new knowledge, discourse and practices have been composing a new field of expertise called disaster management.

Disaster management can be defined as the organisation, planning and application of measures to prepare for, respond to and recover from disasters. Preparedness is based on an analysis of disaster risks and includes activities, such as warning systems, contingency planning, the stockpiling of equipment and supplies, the development of arrangements for coordination, evacuation and public information, and associated training and field exercises (UNISDR, 2017). Response is focused mainly on immediate and short-term measures to save lives, reduce health impacts, meet basic subsistence needs, ensure public safety and so forth. Response includes the provision of emergency services and public assistance by many actors from the private and public sectors, including volunteer participation. Temporary shelters, for instance, are one of the key examples of disaster response activities. Finally, disaster management is composed of recovery measures for restoring basic services and facilities to improve livelihoods and economic, social, cultural, physical and environmental assets (UNISDR, 2017). The next sections will discuss preparedness and response activities, whilst disaster recovery will be explained in Chapter 16.

15.2 From disaster preparedness to response

As discussed in the previous chapter, the concepts of disaster risk, vulnerability, capacity and mitigation are very important to the planning and implementation of actions towards DRM and also in disaster management (DM). This applies to preparedness, response and relief activities. More specifically, regarding preparedness, it is important to remember that disaster risk (DR) is a function of hazards (H), vulnerability (V), capacity (C) and larger-scale risk mitigation by preventive action and social protection (M) (Wisner et al., 2012). Preparedness can be represented by capacity (C) and mitigation (M) to reduce losses, damages and impacts. Previous chapters explained the importance of capacities and mitigation actions. But it is also vital to reflect on them in the context of disaster management and their different tools and actions.

Warning systems (WS) are an important example of disaster preparedness. They aim to be an integrated system that comprises disaster risk assessment, hazard forecast, prediction and monitoring, risk communication, and emergency preparedness activities (UNISDR, 2017). WS are defined as a set of capacities, data, information and knowledge that allow for the early action of individuals and communities exposed to hazards in order to prepare and evacuate in an appropriate manner and in adequate time to reduce the likelihood of loss of life, personal injury, property losses and infrastructure damage (UNISDR, 2009). In 2013, a group from the United Nations University (UNU) identified frameworks of national warning systems for different hydrometeorological hazards (severe storms, tornadoes, winter storms, cyclones, hurricanes, typhoons, floods, extreme heat, droughts and food insecurity, coastal high seas etc.), geological hazards (earthquakes, volcanic activity, tsunamis, landslides and snow avalanches), forest fires and biological hazards (influenza, locust swarms). These frameworks were classified according to the number of alert levels, ways to represent them (colours, terms) and criteria to define the alert levels – likelihood of event, magnitude/severity, time until event, period of return (Table 15.1) (Villagrán de León et al., 2013).

Table 15.1 Global survey of early warning systems
Source: adapted from **Villagrán de León et al. (2013**, p. 80)

Hazard	*Variation in number of levels*	*Most frequent number of levels*	*Most frequent type of labels*	*Criteria most often used to define levels*
General weather	1 to 5	4	Colours/terms	Potential severity of event
Tropical cyclone	2 to 6	4	Colours/terms	Interval of time remaining for arrival of event
Flood	1 to 5	3 and 4	Colours/terms	Potential severity of event
Drought	3 to 7	4	Colours/terms	Interval of time remaining for arrival of event
Heat wave	1 to 5	3	Terms	Threshold reached
Storm surge	1 to 4	1 and 3	Terms	Potential severity of event
Volcanic activity	3 to 6	4	Colours and numeric levels	Interval of time remaining for arrival of event; likelihood, thresholds and magnitude or severity

(*Continued*)

Table 15.1 (Continued)

Hazard	*Variation in number of levels*	*Most frequent number of levels*	*Most frequent type of labels*	*Criteria most often used to define levels*
Earthquake	3	3	Terms	Interval of time remaining for arrival of event
Tsunami	1 to 4	3 and 4	Terms	Mixed criteria for warning
Landslide	1 to 3	1	Terms	Likelihood of event, but only regional average
Avalanche	4 to 5	5	Colours/terms	Likelihood of size or type of event, again only regional average
Forest fire	3 to 6	5	Colours/terms	Magnitude/ severity
Multi-hazard	2 to 4	4	Colours	Potential severity of event

Technical parameters, strategic issues, institutional requirements and recommendations for strengthening early warning systems (EWS), including incorporating EWS into new policies and developmental frameworks, were also discussed in three international conferences in 1998, 2003 and 2006 (UNISDR, 2004, 2006a, 2006b). These international conferences and related scientific research promoted many discussions, leading to a growing consensus that the warning system frameworks should be composed of four complementary subsystems: risk knowledge, monitoring, communication and response capability (UNISDR, 2005, 2006a, 2006b, 2015; Basher, 2006; Kelman & Glantz, 2014). Risk knowledge can be defined as the data collection and analysis of hazards/threats and vulnerabilities – physical, social, economic, environmental and so forth – to generate knowledge about the risk scenarios across different spatial scales, in the short and long term. Monitoring refers to the resources and capacities for collecting and verifying dynamic data and information about hazards/threats and vulnerabilities in order to take decisions on the basis of prior risk knowledge – for instance, the (multi)hazard-prone areas and the people potentially exposed. Communication is the process of sharing data, information and knowledge about the risks (hazards and vulnerabilities), communicating them by using bulletins, alerts and so forth. Response capability is the preparedness capacity to know how to act and is often rooted in the resources, skills and networks that people have.

Capacity is not synonymous with capability. Capability is not just the personal abilities (capacities) but the freedom and access to resources (opportunities) created by the political, social and economic structures (Wisner, 2016). Local governments, for instance, can be more capable when they have sufficient personnel, clear structure, proper tasks, delegation and division of labour within the organisation (Kusumasari & Alam, 2012).

There are several challenges when implementing these four subsystems. Dávila (2016), for instance, analysed 21 flood-warning systems in Latin America and the Caribbean (LAC) and identified that organisations were more focused on monitoring. Risk knowledge, communication and response capability received less investment in terms of human resources and time. The lack of high-quality data, technical and technological capacity to generate forecasts, the deterioration of monitoring networks, inadequate communication of warnings by forecasters to end users, and poor accessibility of warning systems were other barriers reported in LAC (Dávila, 2016; Lumbroso et al., 2016), as well as in some countries in Africa (Lumbroso et al., 2016). Another challenge is how to build a people-centred approach in these four axes of warning systems (Marchezini et al., 2017, 2018) and to include a multi-hazard perspective – two key recommendations of the HFA (UNISDR, 2005) and the SFDRR (UNISDR, 2015).

The literature on warning systems has defined two main approaches in this area (Thomalla & Larsen, 2010; Garcia & Fearnley, 2012; Kelman & Glantz, 2014) (Figure 15.2). The last mile approach considers that all the relevant data, information and knowledge are outside the local communities. In this perspective, the people who need the EWS are not at the centre of the design but are the last to be involved in the system because it depends on external specifications and experts. In this top-down approach, people are involved as receivers of alert messages. The technical equipment and sensors (for example, radar and rainfall gauges) detect a hazard and issue alerts to people. The last mile approach frames the warning system as a linear chain with a focus on hazard prediction, monitoring and alerts communication. Also known as the end-to-end model (Basher, 2006), this approach does not directly engage the subjects in the four elements of the system. The emphasis is on risk knowledge and monitoring, whereas communication and response capability are the weaker elements (Garcia & Fearnley, 2012).

The first mile approach considers that local people should be involved as the central component of the design and operations of warning systems. To make this approach effective, actions must occur at different scales, involving multiple stakeholders in dialogue and collaboration at every stage of the process (Gaillard & Mercer, 2013). The warning system designers must consider features, such as demographics, age, gender, culture and the livelihood of the subjects involved. The system planners must be aware of the different forms and degrees of vulnerability and capability of different people (minorities,

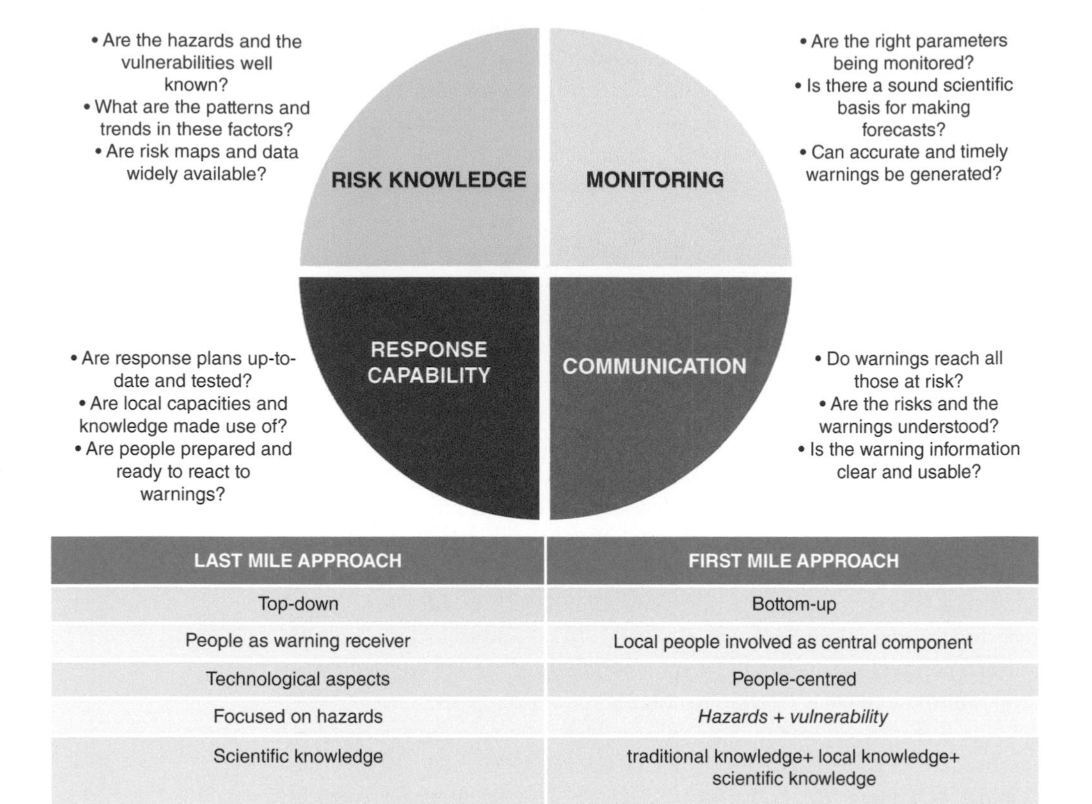

LAST MILE APPROACH	FIRST MILE APPROACH
Top-down	Bottom-up
People as warning receiver	Local people involved as central component
Technological aspects	People-centred
Focused on hazards	*Hazards + vulnerability*
Scientific knowledge	traditional knowledge+ local knowledge+ scientific knowledge

Fig 15.2
Four axes and types of approaches for warning systems.
Source: elaborated by authors, based on bibliographic and desk review

refugees, gender groups, age groups, people with disabilities and so on). One of these important groups is people with disabilities and the intersectionalities (as we have seen in Chapter 4) that they can accumulate depending on the context in which they live (Box 15.1).

As stated in the previous chapters, these degrees of vulnerability and capacities are dependent on root causes and dynamic pressures (Wisner et al., 2012) that shape differentiated access to rights and resources – healthcare services, education, access to information – and will contribute to the ability to prepare, cope with and recover in the aftermath of hazards. Warning systems are placed in different social contexts (Kelman & Glantz, 2014). Gender relations, for instance, refer to the politico-economically and socio-culturally constructed roles, and responsibilities ascribed to women and men, boys, girls and members of sexual and gender minorities, which are context- and history-specific, and are inseparable from power relations, and which change over time (Mustafa et al., 2015). Interestingly, the HFA recommended that a gender perspective should be integrated into all DRM policies, plans and decision-making processes, including those related to risk assessment,

Box 15.1: People with impairments

The United Nations Convention on the Rights of Persons with Disabilities defines people with disabilities as those who have long-term physical, mental, intellectual or sensory impairments, who, in interaction with various barriers, may have their full and effective participation in society on an equal basis hindered. More than 1 billion people – 15% of the world's population – live with some type of impairment, 80% of them in low- and middle-income countries (World Health Organization, 2011). A global survey conducted with 5,717 people with impairments across 137 countries (UNISDR, 2014) indicated this:

- 72% of respondents had no personal preparedness plan for disasters;
- 31% of respondents always had someone to help them evacuate;
- 21% answered that they could evacuate immediately without difficulty in the event of a sudden disaster;
- 17% of respondents were aware of a disaster management plan in their community; and
- 14% of respondents said they had been consulted on disaster management plans in their community.

People with disabilities is not a homogeneous group: 46% of people over the age of 60 years live with some type of disability; one in every five women is likely to experience disability in her life; one in every ten children lives with a disability (WHO, 2011). It is important to take these aspects into account when co-developing the disaster management strategies. There are a lot of important guidelines to help you.

Please access the "Guidance on strengthening disability inclusion in Humanitarian Response Plans": https://sites.unicef.org/disabilities/files/Guidance_on_strengthening_disability_inclusion_in_Humanitarian_Response_Plans_2019.pdf

Sources: UNISDR (2014); WHO (2011)

warning systems, information management and education and training (UNISDR, 2005) (Box 15.2). The SFDRR also states that the designing and implementation of warning systems "should occur through a participatory process, tailoring them to the needs of users, including social and cultural requirements, and in particular gender" (UNISDR, 2015, p. 21).

In the first mile approach, the aim of EWS is to be people-centred, horizontal and participatory (Basher, 2006; Villagrán de Léon, 2012). This perspective has been represented by a variety of terminologies and frameworks, such as people-centred EWS (UNISDR, 2005, 2006a), community early warning systems (CEWS) (IFRC, 2012), citizen-centred EWS (Mustafa et al., 2015), community-centric EWS (Baudoin et al., 2016), community-based EWS (Macherera & Chimbari, 2016a) and

Box 15.2: Challenges of tailoring warning systems considering gender issues

Afghan Abadi, a city in Pakistan, is home to around 5,000 families, including ethnic Pashtuns from different provinces and refugees from Afghanistan settled there for the last 30 years. Whilst men are employed in the local market or as daily wage labourers, women's mobility is excessively limited due to a stricter understanding of purdah. Discussing how to include gender aspects in flood EWS in Pakistan, Mustafa and colleagues (2015) stated that culturally appropriate and gender-specific warning systems need to go beyond blaring out a siren and tailoring risk messages. As put by one respondent interviewed by Mustafa et al. (2015, pp. 12–13),

> We are not allowed to go outside of our houses. We know about flood hazard from our men. There are announcements in the mosques but loud speakers do not work due to absence of electricity, we could not hear that warning. . . . Even during floods women are not allowed to go outside the homes without the permission of the males. We find safe places inside the home to save ourselves. We cannot move anywhere without the permission of our husbands and for the fear of punishment.

The case of Pakistan and other reported by scientific literature (Mulyasari & Shaw, 2013; Tyler & Fairbrother, 2018; Coles & Quintero-Angel, 2018) show the challenges of tailoring warning systems according to gender needs.

participatory EWS (Baudoin et al., 2016; Marchezini et al., 2017). Despite the differences between these terminologies, there are some shared principles, such as the importance of understanding local context, integrating local knowledge, considering local motivations when planning and implementing risk management activities (Baudoin et al., 2016), and finally, social participation.

Social participation is an essential value in warning systems and in other actions of disaster management. Social participation refers to the appropriation by individuals of their right to democratic construction of their own destiny. Participatory processes involve overcoming power imbalances and ensuring the exercise of citizenship, particularly concerning people in a situation of greater socio-environmental vulnerability (Loureiro & Layrargues, 2013). But it is also important to reflect on the types of participation. There are different modes of engagement and forms of participation that need to be evaluated before designing, planning and implementing disaster management actions (see Figure 15.3).

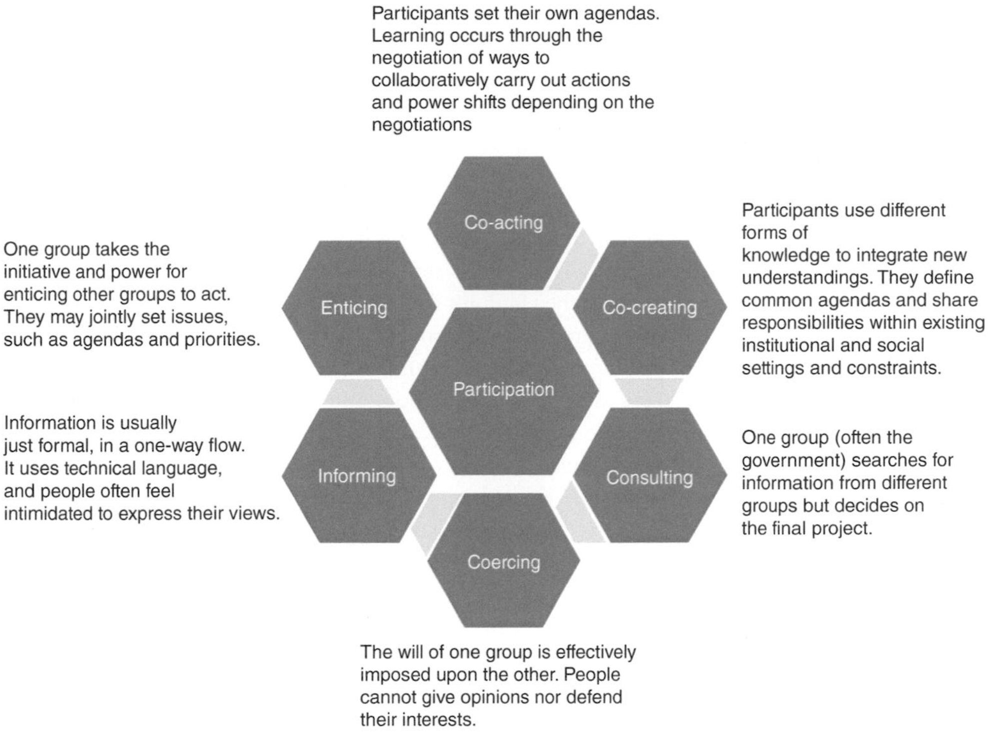

Participants set their own agendas. Learning occurs through the negotiation of ways to collaboratively carry out actions and power shifts depending on the negotiations

Co-acting

One group takes the initiative and power for enticing other groups to act. They may jointly set issues, such as agendas and priorities.

Enticing

Participation

Co-creating

Participants use different forms of knowledge to integrate new understandings. They define common agendas and share responsibilities within existing institutional and social settings and constraints.

Information is usually just formal, in a one-way flow. It uses technical language, and people often feel intimidated to express their views.

Informing

Consulting

One group (often the government) searches for information from different groups but decides on the final project.

Coercing

The will of one group is effectively imposed upon the other. People cannot give opinions nor defend their interests.

Fig 15.3
Types of participation in disaster management actions.
Source: elaborated by authors based on Dyball et al. (2009)

Social participation in warning systems is a challenge. Gender inequality (Mustafa et al., 2015), lack of transparency, confidence, financing and ownership (Šakić Trogrlić et al., 2017) can be barriers to citizen engagement in disaster management. Governance issues are also a key factor. There are countries, provinces and municipalities where formal warning systems do not exist or are largely malfunctioning. In 2009, during the Typhoon Morakot in Taiwan, only 13.8% of the residents of mountainous villages in Kaohsiung City received institutional (official) alerts to evacuate from areas prone to debris flow (Luo et al., 2014). Additionally, many village mayors re-assessed alerts from higher authorities, and did not forward warnings and evacuation orders to the people. According to Luo et al. (2014), the crucial point is that the current warning system in Taiwan does not permit the engagement of people, especially in community-based disaster risk management (CBDRM) initiatives. A similar diagnosis was found in 21 flood warning initiatives in LAC, where community-based frameworks existed only in 20% of them (Dávila, 2016). The magnitude and scale of the hazards can also influence the possibilities of participation and the governance structure that can promote it. For instance, the warning systems for tsunamis can involve several countries and agencies, creating complex environments of participation, especially in areas that receive foreign tourists (Kelman, 2006). This was the case of the 2004 tsunami in Asia and Africa. As exemplified in Chapter 3, the cascading disasters pose several challenges to warning systems and disaster management.

Despite the challenges of participation, there are remarkable initiatives of people-centred warning systems and community-based disaster management, combined with interdisciplinary and transdisciplinary approaches. During the 2006 eruption of Mount Merapi, Indonesia, Donovan et al. (2012) used social volcanology and ethnography to understand how different cultures influenced local community actions during eruptions. The authors stated that local people refused to evacuate because they had not received the traditional warnings, which include forest animals fleeing into the villages, an increase in rockfalls, a change in plume direction, dreams or premonitions. The authors mapped these and other information, and included them in GIS platform to subsidise EWS activities. In Ecuador, a citizen science project around volcano Tungurahua involved volunteers in data collection, analysis and interpretation. The volunteers "were given basic training from the scientists about what to observe, how to describe phenomena and how to communicate with OVT [Tungurahua Volcano Observatory]" (Stone et al., 2014, p. 7).

Some initiatives tried to cover the four subsystems of warning framework, tailoring it to different subjects and considering various hazards. In Zimbabwe, a community-based malaria warning system was designed to integrate indigenous knowledge and the conventional health system (Macherera & Chimbari, 2016b). Using participatory rural appraisals and workshops, communities increased their own knowledge about the malaria: they did a trend analysis of malaria from 1970 to 2011, and that of temperature and rainfall from 1960 to 2011. After it, they constructed malaria calendars with its causes and the season of occurrence. During focus group discussions (FGDs), the indicators used by the community to predict the occurrence of malaria were documented. FGDs indicated that the behaviour of lions and elephant could be used as indicators for malaria. Communities stated that if elephants or lions pass through the villages at night during the month of September, it means that the coming malaria season is going to be bad (Macherera & Chimbari, 2016b). After documenting the meaningful indicators for the residents (wind patterns, direction, variation etc.), participants agreed as to who would be the volunteers willing to carry out observations and monitoring. These FGDs also planned the risk communication strategy. They determined that communication of the indicators should be bidirectional – from the observers to the general population and also from the population to the observers. The warning should be communicated to the people through the village health workers.

In Sri Lanka, an initiative involving volunteer monitoring, communication and response capability was formulated to prepare a contingency plan and evacuation measures in the Matale district. Residents were trained to daily monitor rainfall data in portable fiberglass rain gauges and educated on how to communicate data to their neighbours. The rainfall thresholds were marked in the rain gauges and colours (green, yellow and red) were used to differentiate what types of actions should be taken. During severe rainfall in October and November 2010, "121 families used this method to evacuate to safer places during landslides" (Baudoin et al., 2016, p. 169).

In Brazil, a citizen science project developed interdisciplinary activities to engage high school students in basic scientific research to enable them to understand the notion of people-centred early warning systems. Guidelines were developed for adopting a bottom-up approach towards achieving the four elements of EWS with the help of school curricula (Marchezini et al., 2017) (Table 15.2). One of the activities in the pilot study in São Luiz do Paraitinga was related to participatory mapping.

Table 15.2 Four interrelated elements of EWS and interdisciplinary activities led by high schools
Source: **Marchezini et al. (2017)**

Risk knowledge	*Monitoring*	*Communication of warnings*	*Response capability*
– Oral history and disasters – Watershed mapping (using Google Earth) – Fieldwork about land use – Risk mapping using participatory social cartography – School vulnerability assessment	– Meteorological monitoring – Hydrometeorological monitoring – Water balance	– Risk communication using wall newspaper	– Protection Map Game to design a contingency plan (using social cartography methodology) – Two workshops to create a Committee for Disaster Prevention and the Protection of Life (*Com-VidAção*)
– Oral history and disasters – Watershed mapping (using Google Earth) – Fieldwork about land use	– Hydrometeorological monitoring	– Risk communication using wall newspaper – Warning issued by school – Awareness campaign	– Creation of a tree nursery

Students of the High School Monsenhor Ignacio Gioia engaged in the Protection Map Game (PMG), inspired by the Disaster Imagination Game, a Japanese method for disaster prevention. The participants took part in a brainstorming session and designed a local disaster plan related to a mission: each group had to formulate a plan to rescue a vulnerable group from an area at risk and lead them to a safe place. Within ten minutes, they had to choose a type of hazard (flood or landslide), identify a vulnerable group on their map (for example, children in a kindergarten or elderly people in a retirement home), find a shelter, draw two escape routes to safe areas and define the strategies needed to accomplish their mission. When asked about how the mapping was conducted during the evaluation, a student said "before the participatory mapping, we never stopped to think about risk zones and escape routes". Another commented, "when we looked at the map, we saw that the flood-prone area was much larger than we had imagined". Another student pointed out that "the map helps to understand the scale of the disaster, since you look at the entire city, not just your home" (Marchezini et al., 2017). This participatory, strategic decision-making game was an excellent learning tool for everybody, including the civil defence officers and teachers who took part in the workshop. Social cartography is a useful tool that can help to involve multiple stakeholders and encourage participation (Mitchell et al., 2008; López-Marrero & Tschakert, 2011).

Contingency planning is very important for disaster response (Box 15.3). It is defined as the process that analyses the disaster risks, establishes operational arrangements and actions, and defines clear responsibilities of organisations and institutions in advance to enable timely, appropriate and effective responses. Usually, these plans are based on scenarios of possible emergency conditions or hazardous events in order to think about the potential problems that can appear during disasters. These exercises to think about scenarios also demand identifying institutional roles and resources, as well as to consider specific actors at times of need. A successful example of contingency planning being put in practice was the 2014 Chilean response to the 8.2 magnitude earthquake that struck in the Pacific Ocean, 79 kilometres from the city of Iquique, Chile. The quake happened at night, and the ONEMI (National Emergency Office of the Ministry of Interior) coordinated the evacuation of the coastal regions of Tarapaca, Painacota and Arica. Ensuing waves reached a maximum height of 2.4 metres. About 900,000 people were evacuated, whilst only 6 people died (UNISDR, 2014).

Contingency planning needs to be regularly updated and exercised to test how it works in different scenarios. These plans can be formulated considering the magnitude, frequency, potential occurrence, speed of onset and spatial extent of different hazards, as well as the vulnerabilities and capacities. Sometimes these plans require the participation of various organisations from local, regional, state, national and international levels. It is essential to take into account that nations, and even their

Box 15.3: The miracle of Kamaishi

In the aftermath of the 11 March 2011, triple disaster in Japan, one story brought hope to many people: the survival of almost all the nearly 3,000 elementary and junior high school students of Kamaishi, Iwate Prefecture. More than 1,000 lives were lost in Iwate in the tsunami, but only 5 of them were school-age children who weren't at school when the tsunami happened. The story of the successful evacuation came to be known as "the miracle of Kamaishi". Immediately after the magnitude 9.0 earthquake struck that afternoon, the students of Kamaishi East Junior High School ran out of the school to higher ground. Their quick response prompted the children and teachers of the neighbouring Unosumai Elementary School to follow and, consequently, drew in many local residents. As they continued to run, older students supported the younger school children, and together, they reached a safe location, whilst behind them, the mega-tsunami swallowed their schools and the town (Katada & Kanai, 2016). According to Katada and Kanai (2016), it is fundamental to encourage children to take the initiative in any evacuation and not to put too much faith in hazard maps because they are based on past tsunamis. "In general, people don't evacuate even though they know they should. It's natural to be reluctant to escape when no one else is escaping. So, I told the students that they must be brave and be the first ones to evacuate. If you do, others will follow you and you can save their lives, too", he said. "And that is exactly what happened" (Government of Japan, 2013).

provinces, have similarities and differences in their governance structure. In Canada, for instance, the provinces of Ontario, Alberta and British Columbia have different numbers of alert levels, ways to represent them (colours, terms) and criteria to define alert levels for floods (Villagrán et al., 2013). These governance structures are also dependent on root causes of vulnerability (Wisner et al., 2004). Ideologies and values, such as militarism and post-war fragility, can be important drivers of top-down approaches in policymaking and decision-making before, during and after disasters. In many countries of the LAC, for instance, the civil protection units are usually commanded by military personnel, from the police and/or Armed Forces. In Brazil, the disaster management system is called the National System of Civil Defence (SINDEC) and is coordinated by the National Secretariat of Civil Defence (SEDEC) at the Ministry of National Integration, which has been traditionally commanded by military firefighters, retired army officers and/or military police agents. At the state level, there are state civil protection units that consist of military police or firefighter officers, who provide periodical training sessions for municipal civil defence. However, the Municipal Civil Defence system does not receive enough support from SINDEC, nor are the officers recognised as part of a formal career structure. Thus, the situation of institutional vulnerability is aggravated by job instability – after municipal elections, new mayors usually change their local civil protection team, which disrupts the continuity of the civil protection (Londe et al., 2015; Marchezini et al., 2017). Additionally, the lack of formal institutions (regulations, laws, codes, bureaucracy and procedures) and informal institutions (traditions, norms, culture) can pose barriers to contingency planning and disaster management. For example, do organisations prepare contingency planning for different hazards? Do they involve people in the formulation of these plans? Are the drills being practiced? How often are they practiced? Contingency planning is vital for emergency management, to prepare evacuation routes, to perform rescues and conduct operations, as well as for response and relief. The next section will discuss some elements of response and relief.

15.3 Disaster response and relief

The kinds of response actions vary according to the types of disasters and their short- and long-term impacts. Technological disasters related to oil spills, nuclear contaminants, tailing dams collapse may require complex interventions. Even more complex is the combination of technological and natural hazards – earthquakes, tsunamis and so forth – as verified in the 2011 triple disaster in Japan (see Chapter 3). This chapter will discuss some of the general disaster response and relief activities.

According to UNISDR (2017), the response includes the provision of emergency services and public assistance by many actors from private and public sectors. A lot of questions can be asked when planning disaster response and relief: how should disaster assistance be managed? How can outside aid be balanced with local self-help? What types of housing or shelter should be provided – permanent, transitional or emergency? How can the active participation of the affected community be mobilised during the post-hazard pressure? What are the main challenges? What are the policy recommendations?

SFDRR recommends the training of government personnel and voluntary workers in disaster preparedness and response, strengthening technical and logistical capacities, promoting regular evacuation drills, and training to ensure rapid and effective response to provide safe shelter, food and non-food relief supplies, and so forth (UNISDR, 2015). Several NGOs working in disaster response have discussed and designed practical guidelines and humanitarian standards on how to improve the quality of humanitarian networks. Since 1997, they have been developing the Sphere project, which aims to improve the quality of humanitarian work during disaster response in four main areas: (i) water supply, sanitation and hygiene promotion (WASH); (ii) food security and nutrition; (iii) shelter and settlement; and (iv) health. The Sphere

Box 15.4: Standards for humanitarian response

The Sphere initiative does not cover all aspects of humanitarian assistance but recommends complementary standards of its partners, such as the following:

- Livestock Emergency Guidelines and Standards (LEGS): LEGS Project;
- Minimum Standards for Child Protection in Humanitarian Action (CPMS): Alliance for Child Protection in Humanitarian Action;
- Minimum Standards for Education: Preparedness, Response, Recovery: Inter-Agency Network for Education in Emergencies (INEE);
- Minimum Economic Recovery Standards (MERS): Small Enterprise Education and Promotion (SEEP) Network;
- Minimum Standard for Market Analysis (MISMA): Cash Learning Partnership (CaLP); and
- Humanitarian Inclusion Standards for Older People and People with Disabilities: Age and Disability Consortium.

Handbook proposes several guidelines, recommendations, indicators and approaches to those topics (Sphere Association, 2018) (Box. 15.4).

Diseases related to water, sanitation and hygiene (WASH), such as diarrhoeal diseases, schistosomiasis, intestinal worms, typhoid and trachoma, trigger preventable sickness and death during disasters and crises – situations where people are usually more vulnerable to illness. WASH interventions aim to reduce public health risks working with the community in a two-way communication to understand the people's risk knowledge and disease prevention practices, to improve access to WASH facilities and materials. The provision of WASH facilities, services and materials requires the management of the entire water chain – water sourcing, treatment, distribution, collection, household storage and consumption – and sanitation in an integrated manner. The promotion of positive health behaviour and hygiene practices is based on ensuring access to safe drinking water, sanitation facilities (excreta management etc.), hygiene items (handwashing etc.), as well as the environmental and social conditions to live with health, dignity, comfort and safety (Sphere Association, 2018). The Sphere handbook recommends key WASH principles and priority interventions to plan and implement with the communities during outbreaks. To start, an inter-sectoral outbreak preparedness and response plan is needed. This plan may elaborate on and agree to indicators specific to the outbreak to permit monitoring and communication, as well as the agreed upon roles and responsibilities within and between sectors. The access to public health data is also necessary for planning, monitoring, communication and improvement of actions, which may be guided by principles, such as respect, quality, speed and appropriateness. (Figure 15.4). The Sphere handbook has

Fig 15.4
WASH principles during disaster management.
Source: elaborated by authors based on Sphere Association (2018)

detailed guidelines, indicators and a trove of information regarding each of the WASH components, including outbreak situations.

WASH components are interlinked with food security and nutrition. People need an adequate quantity of water to drink, to prepare food, and to care for themselves, their relatives and livelihoods. The inadequate water, repeated disease and food insecurity can lead to malnutrition, which, in turn, reduces (i) the immunity to diseases and chronic illness, (ii) the ability to engage within the community, (iii) the opportunity to maintain one's livelihood and (iv) the people's capacity to cope with short- and long-term consequences. Since international law has recognised the right to be free from hunger and to have adequate food, the assessments of food security and nutrition are required throughout a crisis, to show how the situations evolve. This type of assessment helps to understand current needs and how to meet them, to estimate how many survivors need assistance by identifying groups with higher vulnerability and to provide a baseline to monitor the impacts of response actions (Sphere Association, 2018). This type of assessment can be conducted in detailed analysis.

Food security assessments are carried out to evaluate the degree and extent of food security in disasters. There is food security when people have physical and economic access to sufficient, safe and nutritious food that meets their dietary needs and food preferences for an active and healthy life (Sphere Association, 2018). Disaster response actions towards food security should consider the adoption of coping strategies that reduce the negative impacts on natural resources, the environment and the local economy. It is important to conduct a market analysis to evaluate if local markets, both formal and informal, can protect livelihoods by supplying productive items – seeds and tools, for example – and the nutritional needs of a typical household. Some sources, tools and information systems can assist in this task. Satellite images, crop assessments, household assessments and other data provided by famine warning systems are some of the inputs (Sphere Association, 2018). This assessment can be complemented by the detailed nutrition assessment, which involves data collection and analysis to define the type, degree and extent of malnutrition, undernutrition and so forth. This detailed nutrition assessment is important to plan nutrition responses, which in turn are vital in reducing morbidity and mortality – the short-term consequences of the disaster. According to the Sphere Handbook, the selection of the nutrition response actions should be designed together with people and also requires an analysis of sex-disaggregated needs, household preferences, cost efficiency, specific type and quantity of food required, the methods of distribution, the seasonal changes, and so forth (Sphere Association, 2018).

Providing shelter is also costly in terms of time, energy and funding, and it is perceived to be too expensive for the voluntary sector because it usually lacks expertise in shelter management. Additionally, there is the need to work with national governments, and oftentimes, this involvement is protracted in different types of shelters, from emergency to permanent shelters (Davis et al., 2015). Emergency shelters, for instance, serve to protect people and their belongings against cold, heat, wind and rain, as well as allow for a minimum level of privacy.

One of the first global efforts regarding shelter administration was carried out in 1975 by the Office of the United Nations Disaster Relief (UNDRO), a predecessor of the United Nations Office for the Coordination of Humanitarian Affairs (UNOCHA). The UN initiative of *Shelter After Disaster* (Davis et al., 2015) was conducted as a research project to investigate and develop practical guidelines for assisting organisations in listening to and providing shelter for homeless disaster-affected people. The research project took seven years and investigated shelters in the aftermath of disasters around the world. The UN study aimed to remedy shelter problems and set out 16 basic principles regarding shelter administration.

To begin, a people-centred approach is necessary. The survivors', their friends' and their families' motivations are the primary resource in the provision of shelter. Usually, the locals are the first or even the zero responders when hazards strikes, performing rescues and other response and relief activities (Briones et al., 2019). Survivors usually want to remain as close as possible to their damaged and ruined homes, and their means of livelihood, improving temporary shelters or occupying buildings, emergency shelters, tents in campsites provided by external agencies (Davis et al., 2015). Family and friends, sometimes even the unknown community members, are important providers of humanitarian shelter to disaster survivors. The external responders, such as humanitarian agencies, must recognise, respect and support the hosting communities as a legitimate form of humanitarian shelter, avoiding the duplication of efforts that are being undertaken by survivors themselves, and the imposition of the external's cultural views and values to dismiss the local efforts as inappropriate (Davis et al., 2015). This respect can be accompanied by actions to increase or build capacity in assisting groups with shelter and housing reconstruction. This topic is usually neglected by NGOs and the donor community, especially because of their reluctance to provide human and financial resources throughout all the different time phases, including the "immediate relief period (impact to day 5), rehabilitation period (day 5 to 3 months), and reconstruction period (3 months onward)" (Davis et al., 2015, p. 42). The time and the scale of disaster are factors that limit the participation of external actors.

Another principle is how to expand the scale of operations over time and space. It is recognised that progress has been made in the area of user-built safe dwelling reconstruction (Davis et al., 2015). However, shelter management has failed to respond to vulnerability growth and urbanisation. In Mexico, the process of rebuilding shelters in the aftermath of floods in Angangueo City created new vulnerabilities, such as open sewage and unsafe drinking water (García, 2017). In Brazil, shelters were established in schools during the holiday season, but the emergency shelter extended beyond the period initially planned, which ultimately lead to conflicts with the hosting community who required the use of the school buildings for educational purposes (Marchezini, 2014). It is necessary to think about the impacts on urban areas, avoiding problems in response, relief and recovery because models of redevelopment can increase or create new vulnerabilities, and new urban dynamics. Urban safety, debris removal, land and site availability, tenure issues, housing finance and the need of renters are other issues that influence shelter and housing management.

Hazards, combined with other social conflicts, are happening more and more frequently, posing challenges to the humanitarian sector. There are areas where humanitarian shelter activities are becoming permanent, which will require devising new solutions to various problems, including rental shelter and support of social validation of occupancy rather than only the legal validation of ownership. Finding these new solutions can also contribute to expanding the potential choices of users. According to Davis and colleagues (2015), the value of cash grants and vouchers in lieu of delivering standardised shelters and housing units should be taken into account because they can be exchanged for building materials and other construction services that can assist survivors. Sometimes, the free distribution of donations can trigger effects on local trade, causing new indirect victims of the disaster, as well as creating problems of dependency. The authors consider that there are situations where cash grants may be an effective form of aid, such as for the poor craftsmen and labourers who need to replace the destroyed equipment essential to their livelihood, or even when they need access to land for housing and resettlement.

Cash grants and vouchers do not substitute the need for effective settlement planning and building design, as well as the engineering necessary to ensure safety and quality. Quality and safety of shelters and buildings, and land-use planning controls are extremely necessary, and they require the need to develop, apply and reinforce standards and regulations. Another important principle aligned with land-use planning controls, and quality and safety of shelter programmes, is their compatibility with the preservation of livelihoods and the environment. Compatibility means that temporary settlement or permanent reconstruction should not destroy valued environments, that they need to consider people's lifestyles, especially "in enabling the users and residents of such housing to reach their places of employment, carry on their remunerative activities and do so without further destruction of the environment that nurtures and supports such activities" (Davis et al., 2015, p. 39). Moreover, it is indispensable that those shelter and housing projects are driven towards a developmental scope rather than being focused on an emergency and/or relief context when, oftentimes, organisations use the opportunities to sell products or use the publicity of shelter images to increase their funding.

The people's lifestyles and their needs regarding remunerative activities, access to housing, public health and other vital dimensions of social life demonstrate that reality is a continuum that needs to be taken into account, analysed, understood and supported by agencies working in the sheltering process – which involves immediate, transitional and permanent shelters (Davis et al., 2015). Understanding the values and limitations of immediate and transitional shelters or housing can allow time for important steps of reconstruction and recovery, such as consultation, planning, the revision of building safety codes, site investigation and risk assessment of future risk. Technology can offer benefits to those important steps. It can also contribute to efforts related to the 16 principles recommended by *Shelter After Disaster*, including the need of accompanying strategies with communications plans that promote core messages and manage expectations in the short and long term. Communication plans also help in the challenge of thinking, planning and redefining the roles of the many actors in the light of new knowledge, shifts in human and financial resources, policy changes and so forth. Communication is also necessary to identify research gaps and to seek ways to fill them, improving the disaster response and relief activities. Figure 15.5 summarises the 16 principles shared in the *Shelter After Disaster* (Davis et al., 2015).

These 16 principles must be accompanied by policies that should be avoided and others that should be adopted in shelter and housing programmes (Table 15.3). Regarding the policies to avoid, there are basic

Fig 15.5
Sixteen principles for shelter management in disasters.
Source: elaborated by authors based on Davis et al. (2015)

statements that include the duplication of efforts pointed out previously, the importation of labourers, prefabricated solutions and materials that can be found locally, such as rubble and timber from damaged houses that are usually destroyed instead of recycled into new homes. Additionally, other actions that should be avoided concerns evacuation, forms of shelter and relocation. Compulsory evacuation, especially of women and children, can cause social misery and apathy (Davis et al., 2015). Moreover, survivors are usually relocated several times during their cycle of displacement: they are sent to large emergency campsites where there are risks of adverse social and environmental effects, as well as to areas remote from work, markets, schools and other social and economic needs (Marchezini, 2014; Davis et al., 2015; Portella & Oliveira, 2017; Vasquez & Marchezini, 2020).

Concerning the policies to be adopted, one of the first statements highlights the importance of encouraging local people to participate in the assessment of their own needs and resources in order to minimise dependency on outside support and to put them at the centre of the sheltering process. This recommendation applies not only to the affected people but also to local builders, engineers, architects and other professionals with knowledge of building traditions and resources. For instance, in the cultural heritage site of São Luiz do Paraitinga, Brazil, the external engineers proposed several flood mitigation projects that were interpreted as a type of violence against the displaced survivors. One of the projects proposed a wall along the riverbanks of Paraitinga River that would block the view of the river and forbid people from

Table 15.3
Policies to avoid and to adopt in shelter management
Source: elaborated by authors, based on Davis et al. (2015)

Policies to avoid	*Policies to adopt*
– Doing actions that duplicate the efforts of survivors; – Bulldozing rubble and burning timber from damaged houses; – Importing labour for reconstruction when there is ample labour to be found locally; – Importing building materials that can be obtained locally; – Building imported or prefabricated temporary shelters unnecessarily; – Compulsory evacuating, especially of women and children; – Relocating survivors on land that is remote from social and economic needs; and – Creating large emergency campsites.	– Encouraging people to participate in the assessment of their own needs and resources; – Providing materials and tools – Keeping stocks of robust winterised tents in cold climates or seasons; – Providing transport for voluntary evacuation; – Requisition public or community buildings; and – Providing cash grants and selling building materials.

fishing. The wall proposal was refused in the public hearings, and the locals criticised the way the rebuilding process was happening. Survivors participated in those public hearings because the prefecture provided temporary shelters in hotels close to the central plaza where those public audiences were organised (Marchezini, 2015).

Other recommendation made in *Shelter After Disaster* (Davis et al., 2015) relates to the requisition of public or community buildings – schools, churches and so forth – because they can provide emergency accommodation for homeless families. Interestingly, this important guide made reference to the importance of having qualified engineers to check the structural resistance of these buildings, but it does not mention the importance of their previous social uses and the potential conflicts that can emerge from using them for different purposes. For instance, in Jaboatao dos Guararapes, Brazil, displaced people were sheltered in schools during the seasonal holidays. However, the emergency shelter continued, and the schools had to return to their activities using only half of the classrooms. Students had to be accommodated in fewer rooms than usual, and also the families affected had to divide the remaining spaces with even more families. This situation accentuated the stigma on the displaced people because of the conflict between them and the hosting community (Valencio et al., 2009).

Other recommendations concern the importance of cash grants explained previously, and the provision of donations, materials, tools and transportation. It is essential to consider cultural styles before planning and providing these items. Usually, blankets, plastic sheeting, roofing sheets, and tools for building and clearing rubble are necessary, and it is important to check first if they are available locally and/or if there are traditional building materials that can substitute them. Regarding transportation, it is important to take into account that families wishing to leave the affected area to stay with friends or relatives should be accommodated (Davis et al., 2015). But evacuees living in emergency shelters in remote areas, campsites or transition housing also need transportation to move to areas where they are used to performing their remunerative activities. It is imperative to consider that saving lives, and the subsequent response and relief activities, is not limited to biological aspects of guaranteeing meals and water. The social aspects of life are an intrinsic part of human health, and the notion of shelter is not restricted to the physical structure. Suicidal behaviours and other psychosocial problems are part of the disaster cycle (Kõlves et al., 2013).

Despite these recommendations, there are several research gaps in the topic of shelter management. Regarding the resources of survivors, there are basic and detailed needs to be researched. For instance, it is necessary to identify and understand the preferences of people for the various modes of shelter, considering different types of hazards, climates, and urban and rural settings. In those various contexts, what are the functions of shelters? Filling these basic needs, research can move to detailed needs. Understanding the capacities of hosting families to accommodate displaced people is crucial, especially the special needs that those affected by disaster can have in terms of psychosocial conditions, injuries and other stressors that can make them more vulnerable. Usually, the host community also demands support to repair and retrofit buildings, shelters and houses – roof repair kits, energy-efficient stoves and so forth (Davis et al., 2015). It is always important to recognise and share the successful initiatives as well as the mistakes. For instance, two years after the 1976 earthquake in Guatemala, a committee of voluntary agencies wrote to the president recognising their mistakes and listed five main errors: (1) too much aid was given away, (2) too many of the houses constructed were merely of an emergency type, (3) some institutions/organisations used large numbers of foreign volunteers, (4) too much was done under pressure and without proper consultation so that the victims became mere spectators to the work being carried out rather than active participants and (5) a lot of reconstruction work was undertaken without first consulting the government's Reconstruction Committee (Norton, 1980).

15.4 Concluding remarks

This chapter discussed the management of disasters, including response and relief activities. It shared an overview of current practices amongst emergency and humanitarian actors, emphasising barriers and opportunities for addressing the needs of survivors and places. It included discussions on the contributions of different actors and tools in four main areas of disaster response and relief: (i) water supply, sanitation and hygiene promotion (WASH); (ii) food security and nutrition; (iii) shelter and settlement and (iv) health. The chapter provided a brief overview and suggested scientific and grey literature for further reading, including content related to implementation activities.

Take-away messages

1. Concepts are important to guide your research about disaster preparedness and response;
2. People must be involved in disaster preparedness and response;
3. There are various methodologies to promote engagement in disaster preparedness and response; and
4. Use quantitative and qualitative methods to improve your research of disaster preparedness and response.

To learn more about the topic discussed in this chapter, listen to the *Disasters: Deconstructed* interview with Kristin Lange.

Fig 15.6
QR code for Chapter 15.

Further suggested reading

Guadagno, L. (2016). Human mobility in the Sendai framework for disaster risk reduction. *International Journal of Disaster Risk Science*, 7(1), 30–40. https://doi.org/10.1007/s13753-016-0077-6

Harvey, P., Proudlock, K., Clay, E., Riley, B., & Jaspars, S. (2010). *Food aid and food assistance in emergencies and transitional contexts: A review of current thinking*. Humanitarian Policy Group.

Inter-agency network for education in emergencies (INEE). (2010). *INEE minimum standards for education: Preparedness, response, recovery*. www.ineesite.org

Peek, L. (2008). Children and disasters: Understanding vulnerability, developing Capacities, and promoting resilience—an introduction. *Children, Youth and Environments*, *18*(1), 1–29.

Ramesh, A., Blanchet, K., Ensink, J. H., & Roberts, B. (2015). Evidence on the effectiveness of water, sanitation, and hygiene (WASH) interventions on health outcomes in humanitarian crises: A systematic review. *PLoS One*, *10*(9), e0124688. https://doi.org/10.1371/journal.pone.0124688

References

Basher, R. (2006). Global early warning systems for natural hazards: Systematic and people-centred. *Philosophical Transactions. Series A, Mathematical, Physical, and Engineering Sciences*, *364*(1845), 2167–2182. http://doi.org/10.1098/rsta.2006.1819

Baudoin, M. A., Henly-Shepard, S., Fernando, N., Sitati, A., & Zommers, Z. (2016). From top-down to "community-centric" approaches to early warning systems: Exploring pathways to improve disaster risk reduction through community participation. *International Journal of Disaster Risk Science*, *7*(2), 163–174. https://doi.org/10.1007/s13753-016-0085-6

Briones, F., Vachon, R., & Glantz, M. (2019). Local responses to disasters: Recent lessons from zero-order responders. *Disaster Prevention and Management*, *28*(1), 119–125. https://doi.org/10.1108/DPM-05-2018-0151

Coles, A. R., & Quintero-Angel, M. (2018). From silence to resilience: Prospects and limitations for incorporating non-expert knowledge into hazard management. *Environmental Hazards*, *17*(2), 128–145. http://doi.org/10.1080/17477891.2017.1382319

Dávila, D. (2016). *Sistemas de alerta temprana ante inundaciones en América Latina*. Soluciones Prácticas.

Davis, I., Thompson, P., & Krimgold, F. (2015). *Shelter after disaster* (2nd ed.). IFRC/ OCHA. https://www.ifrc.org/Global/Documents/Secretariat/201506/Shelter_After_Disaster_2nd_Edition.pdf

Donovan, K., Suryanto, A., & Utami, P. (2012). Mapping cultural vulnerability in volcanic regions: The practical application of social volcanology at Mt Merapi, Indonesia. *Environmental Hazards*, *11*(4), 303–323. http://doi.org/10.1080/17477891.2012.689252

Dyball, R., Brown, V. A., & Keen, M. (2009). Towards sustainability: Five strands of social learning. In A. E. J. Wals (Ed.), *Social learning towards a sustainable world. Principles, perspectives, and praxis* (pp. 181–194). Wageningen Academic Publishers.

Foucault, M. (2007). *Security, territory, population: Lectures at the Collège de France, 1977–1978*. Palgrave Macmillan.

Gaillard, J. C., & Mercer, J. (2013). From knowledge to action: Bridging gaps in disaster risk reduction. *Progress in Human Geography*, *37*(1), 93–114. http://doi.org/10.1177/0309132512446717

Garcia, C., & Fearnley, C. J. (2012). Evaluating critical links in early warning systems for natural hazards. *Environmental Hazards*, *11*(2), 123–137. http://doi.org/10.1080/17477891.2011.609877

García, H. I. R. (2017). El proceso de desastre y reubicación en Angangueo, Michoacán, México o ¿dónde comienza la vulnerabilidad y el desastre? In V. Marchezini, B. Wisner, L. R. Londe, & S. M. Saito (Eds.), *Reduction of vulnerability to disasters: From knowledge to action* (pp. 235–252). Rima Editora.

Government of Japan. (2013). *The miracle of Kamaishi.* https://mnj.gov-online.go.jp/kamaishi.html

International Federation of Red Cross and Red Crescent Societies. IFRC. (2012). *Community early warning systems: Guiding principles*. IFRC.

Katada, T., & Kanai, M. (2016). The school education to improve the disaster response capacity: A case of "Kamaishi Miracle". *Journal of Disaster Research*, *11*(5), 845–856.

Kelman, I. (2006). Warning for the December 26, 2004 tsunamis. *Disaster Prevention and Management*, *15*(1), 178–189. https://doi.org/10.1108/09653560610654329

Kelman, I., & Glantz, M. H. (2014). Early warning systems defined. In Z. Zommers & A. Singh (Eds.), *Reducing disaster: Early warning systems for climate change* (pp. 89–108). Springer.

Kõlves, K., Kõlves, K. E., & De Leo, D. (2013). Natural disasters and suicidal behaviours: A systematic literature review. *Journal of Affective Disorders*, *146*(1), 1–14. https://doi.org/10.1016/j.jad.2012.07.037

Kusumasari, B., & Alam, Q. (2012). Bridging the gaps: The role of local government capability and the management of a natural disaster in Bantul, Indonesia. *Natural Hazards*, *60*(2), 761–779. http://doi.org/10.1007/s11069-011-0016-1

Londe, L. D. R., Soriano, E., & Coutinho, M. P. (2015). Capacidades das instituições municipais de proteção e defesa civil no Brasil: Desafios e perspectivas. *Geography Department University of Sao Paulo*, *30*, 77–95. http://www.revistas.usp.br/rdg/article/view/98715; https://doi.org/10.11606/rdg.v30i0.98715

López-Marrero, T., & Tschakert, P. (2011). From theory to practice: Building more resilient communities in flood-prone areas. *Environment and Urbanization*, *23*(1), 229–249. https://doi.org/10.1177/0956247810396055

Loureiro, C. F. B., & Layrargues, P. P. (2013). Ecologia política, justiça e educação ambiental crítica: Perspectivas de aliança contra-hegemônica. *Trabalho, Educação e Saúde*, *11*(1), 53–71. http://doi.org/10.1590/S1981-77462013000100004

Lumbroso, D., Brown, E., & Ranger, N. (2016). Stakeholders' perceptions of the overall effectiveness of early warning systems and risk assessments for weather-related hazards in Africa, the Caribbean and South Asia. *Natural Hazards*, *84*(3), 2121–2144. https://doi.org/10.1007/s11069-016-2537-0

Lund, C. (2006). Twilight institutions: Public authority and local politics in Africa. *Development and Change*, *37*(4), 685–705. https://doi.org/10.1111/j.1467-7660.2006.00497.x

Luo, Y., Shaw, R., Lin, H., & Joerin, J. (2014). Assessing response behaviour of debris-flows affected communities in Kaohsiung, Taiwan. *Natural Hazards*, *74*(3), 1429–1448. http://doi.org/10.1007/s11069-014-1258-5

Macherera, M., & Chimbari, M. J. (2016a). A review of studies on community based early warning systems. *Jamba*, *8*(1), 206. http://doi.org/10.4102/jamba.v8i1.206

Macherera, M., & Chimbari, M. J. (2016b). Developing a community-centred malaria early warning system based on indigenous knowledge: Gwanda District, Zimbabwe. *Jamba*, *8*(1), 289. http://doi.org/10.4102/jamba.v8i1.289

Marchezini, V. (2014). La producción silenciada de los “desastres naturales” en catástrofes sociales. *Revista Mexicana de Sociología*, *76*(2), 253–285.

Marchezini, V. (2015). Redução de vulnerabilidade a desastres: Dimensões políticas, científicas e socioeconômicas. *Waterlat-Gobacit Network*, *2*, 82–102.

Marchezini, V., Horita, F. E. A., Matsuo, P. M., Trajber, R., Trejo-Rangel, M. A., & Olivato, D. (2018). A review of studies on participatory early warning systems (P-EWS): Pathways to support citizen science initiatives. *Frontiers in Earth Science*, *6*, 184. https://doi.org/10.3389/feart.2018.00184

Marchezini, V., Trajber, R., Olivato, D., Muñoz, V. A., de Oliveira Pereira, F., & Oliveira Luz, A. E. (2017). Participatory early warning systems: Youth, citizen science, and intergenerational dialogues on disaster risk reduction in Brazil. *International Journal of Disaster Risk Science*, *8*(4), 390–401. https://doi.org/10.1007/s13753-017-0150-9

Mitchell, T., Haynes, K., Hall, N., Choong, W., & Ovenl, K. (2008). The roles of children and youth in communicating disaster risk. *Children, Youth and Environments*, *18*(1), 254–279.

Mulyasari, F., & Shaw, R. (2013). Role of women as risk communicators to enhance disaster resilience of Bandung, Indonesia. *Natural Hazards*, *69*(3), 2137–2160. http://doi.org/10.1007/s11069-013-0798-4

Mustafa, D., Gioli, G., Qazi, S., Waraich, R., Rehman, A., & Zahoor, R. (2015). Gendering flood early warning systems: The case of Pakistan. *Environmental Hazards*, *14*(4), 312–328. http://doi.org/10.1080/17477891.2015.1075859

Norton, R. (1980). Disasters and settlements. *Disasters*, *4*(3), 339–347. https://doi.org/10.1111/j.1467-7717.1980.tb00121.x

Portella, S., & Oliveira, S. S. (2017). Vulnerabilidades deslocadas e acirradas pelas políticas de habitação: A experiência do Terra nova. In V. Marchezini, B. Wisner, L. R. Londe, & S. M. Saito (Eds.), *Reduction of vulnerability to disasters: From knowledge to action* (pp. 499–516). RiMa.

Vasquez, J. R. S., & Marchezini, V. (2020). Procesos de recuperación posdesastre en contextos biopolíticos neoliberales: Los casos de Chile 2010 y Brasil 2011. *ÍCONOS Revista de Ciencias Sociales*, *66*, 131–148.

Šakić Trogrlić, R., Wright, G. B., Adeloye, A. J., Duncan, M. J., & Mwale, F. (2018). Taking stock of community-based flood risk management in Malawi: Different stakeholders, different perspectives.

Environmental Hazards, *17*(2), 107–127. https://doi.org/10.1080/17477891.2017.1381582
Sphere Association. (2018). *The sphere handbook: Humanitarian charter and minimum standards in humanitarian response* (4th ed). Sphere.
Stone, J., Barclay, J., Simmons, P., Cole, P. D., Loughlin, S. C., Ramón, P., & Mothes, P. (2014). Risk reduction through community-based monitoring: The vigías of Tungurahua, Ecuador. *Journal of Applied Volcanology*, *3*(1). http://doi.org/10.1186/s13617-014-0011-9
Thomalla, F., & Larsen, R. K. (2010). Resilience in the context of tsunami early warning systems and community disaster preparedness in the Indian Ocean Region. *Environmental Hazards*, *9*(3), 249–265. http://doi.org/10.3763/ehaz.2010.0051
Tuan, Y. F. (1979). *Landscapes of fear*. Pantheon Books.
Tyler, M., & Fairbrother, P. (2018). Gender, households, and decision-making for wildfire safety. *Disasters*, *42*(4), 697–718. http://doi.org/10.1111/disa.12285
UNISDR (United Nations International Strategy for Disaster Reduction). (2004). *Early warning as a matter of policy: The conclusions of the Second International Conference on Early Warning*. UNISDR; the German Disaster Reduction Committee (DKKV).
UNISDR (United Nations International Strategy for Disaster Reduction). (2005). *Hyogo Framework for Action 2005–2015: Building the resilience of nations and communities to disasters*. UNISDR.
UNISDR (United Nations International Strategy for Disaster Reduction). (2006a). *Developing early warning systems: A checklist*. UNISDR.
UNISDR (United Nations International Strategy for Disaster Reduction). (2006b). *Global survey of early warning systems: An assessment of capacities, gaps and opportunities towards building a comprehensive global early warning system for all natural hazards*. UNISDR.
UNISDR (United Nations International Strategy for Disaster Reduction). (2009). *Terminology: Basic terms of disaster risk reduction*. Retrieved March 1, 2018, from https://www.unisdr.org/we/inform/publications/7817
UNISDR (United Nations International Strategy for Disaster Reduction). (2014). Living with disability and disasters*: UNISDR 2013 survey on living with disabilities and disasters—key findings*. UNISDR.
UNISDR (United Nations International Strategy for Disaster Reduction). (2015). *Sendai framework for disaster risk reduction 2015–2030*. Retrieved August 2018, from http://www.wcdrr.org/uploads/Sendai_Framework_for_Disaster_Risk_Reduction_2015-2030.pdf
UNISDR (United Nations International Strategy for Disaster Reduction). (2017). *Terminology: Basic terms of disaster risk reduction*. Retrieved August 1, 2019, from https://www.unisdr.org/we/inform/terminology#letter-e
Valencio, N. F. L. S., Marchezini, V., & Siena, M. (2009). Desastre e indiferença social: O Estado perante os desabrigados. *Antropolítica (UFF)*, *23*, 223–254.

Villagrán de León, J. C. (2012). Early warning principles and practices'. In B. Wisner, J. C. Gaillard, & I. Kelman (Eds.), *Handbook of hazards and disaster risk reduction and management* (pp. 481–492). Routledge.
Villagrán de León, J. C., Pruessner, I., & Breedlove, H. (2013). *Alert and warning frameworks in the context of early warning systems: A comparative review. Intersections 12*. United Nations.
Wisner, B. (2016). Vulnerability as concept, model, metric, and tool. In *Oxford research encyclopedia of natural hazard scie nce. http://naturalhazardscience.oxfordre.com/view/10.1093/acrefore/9780199389407.001.0001/acrefore-9780199389407*-e-25
Wisner, B., Blaikie, P., Cannon, T., & Davis, I. (Eds.). (2004). *At risk: Natural hazards, people's vulnerability and disasters*. Routledge.
Wisner, B., Gaillard, J. C., & Kelman, I. (2012). Framing disaster: Theories and stories seeking to understand hazards, vulnerability and risk. In B. Wisner, J. C. Gaillard, & I. Kelman (Eds.), *The Routledge handbook of hazards and disaster risk reduction* (pp. 18–34). Routledge.
World Health Organization. (2011). *World report on disability*. WHO.

CHAPTER 16

Fostering disaster recovery

Fig 16.1
A tsunami memorial in Sri Lanka with the time when the tsunami struck on 26 December 2004.
Source: photo courtesy of Emmanuel Raju

DOI: 10.4324/9781315469614-22

> Recovery cannot be simply reduced to a universal or predefined strategy as if resilience will satisfy all groups across space and time. Successful post-disaster recovery ought to be more than simply rebuilding resilient housing and other assets to reduce disaster risk, it should also be measured by whether it facilitates the establishment of the lives that vulnerable people want to have and aspire to have. Taking such an approach, which listens to, and integrates local needs that may fall outside dominant conceptualizations of resilience, takes a step towards a normative recovery agenda that recovers and improves the lives of disaster-affected people, and in ways that are locally appropriate. It is also fundamentally important to ensure the sustainability of recovery and resilience policies more broadly.
>
> (Sou, 2019, p. 155)

16.1 Introduction

Disaster recovery is a crucial aspect of DRM. Although the boundaries cannot be clearly defined, the transition from disaster response to recovery raises questions about the nature and scope of coordination during recovery. Disasters bring together multiple actors who often differ in terms of their sector, expertise and nationality (Kettl, 2008), as well as their abilities, values, norms and goals (Alberts et al., 2010). Moreover, the sheer number of actors affects coordination (Comfort et al., 2001; Kruke & Olsen, 2005; Balcik et al., 2010). They come from both the public and private sector (Berke et al., 1993; Christoplos et al., 2010; Tierney and Oliver-Smith, 2012), and there is even competition to be the first organisation to respond to the disaster (Comfort et al., 2001). The Indian Ocean tsunami that devastated Asia in 2004 attracted massive international attention from the media and donors. In the wake of this disaster, and with the experience of the Gujarat earthquake of 2001, civil society organisations working in collaboration with the Government of Tamil Nadu, India, and United Nations (UN) agencies set up an initial coordination body in the worst-affected district of Nagapattinam. Following the immediate impact of the disaster, coordination structures were established at the district and state level to facilitate relief and recovery activities. In South Africa, repeated flooding in the Western Cape region has led many governmental departments and non-governmental agencies to become involved in recovery efforts, and coordination has been a crucial issue. Although there was immense media attention in the aftermath of the 2004 tsunami (Figure 16.1), there was a quantitative decline in the number of organisations involved in these coordination platforms in comparison to disaster response. Similarly, in South Africa the government machinery has dealt with repeated flooding through ad hoc planning bodies, and it is not clear how district municipalities can become involved and coordinate recovery efforts.

Box 16.1: Seven golden rules for recovery

1 Trust survivors and avoid paternalism;
2 Enable survivors to assess their own needs;
3 Provide survivors with cash rather than kind;
4 Think locally;
5 Give priority attention to vulnerable groups;
6 Think process, not product; sheltering, not shelters; housing, not houses; and
7 Whilst addressing short-term needs, adopt a long-term perspective.

Source: Davis and Alexander (2016, pp. 35–38)

One of the key issues in the aftermath of the Indian Ocean tsunami of 2004 was the enormous number of actors, making response coordination difficult, and one could even question its effectiveness (Telford & Cosgrave, 2007). These actors may be forced into new and unplanned roles and responsibilities by the unfamiliar, complex and dynamic character of the situation (Neal & Phillips, 1995; Scanlon, 1999; Drabek & McEntire, 2003). In other words, disasters give rise to new ways of functioning depending on the local context and the needs of the situation (Christoplos et al., 2010). In addition, these different actors have different organisational mandates and goals, and are thus engaged in various activities. Although the response to the 2004 tsunami has been the focus of numerous evaluations, which – at least in theory – may represent a shift towards greater accountability (Bennett et al., 2006), evaluations of the response to and recovery from subsequent disasters indicate that there has not been sufficient progress since then (Comfort, 2007; Nolte et al., 2012). An interesting question to be explored is what had happened to the coordination platforms that were established? With this question in mind, this chapter is aimed at building a better understanding of the factors affecting disaster recovery coordination. It aims to develop deeper knowledge in the field of multi-organisational coordination for disaster recovery (Box 16.1). Therefore, exploring the dimensions of disaster recovery coordination and collaboration is an important subject that has been neglected in disaster studies (Figure 16.2). There a number of stakeholders who are involved in disaster recovery. However, a collaborative approach is needed in order to make recovery sustainable (Raju, 2021).

16.2 Recovery coordination or collaboration?

Coordination poses a challenge during disasters that involve a multitude of different stakeholders (Katoch, 2006; Telford & Cosgrave, 2007). The nature of stakeholders during response and recovery changes based on their experience and the expertise of organisations and government

Fig 16.2
Lac bangles from India. During the COVID-19 pandemic, many Lac bangle makers have struggled with their livelihoods. Very often, the struggles of many livelihood forms go unrecognised.
Source: photo courtesy of Sameeksha Mehra, India (2020)

departments. However, it is known that even when all these organisations have a wealth of expertise and varied experience, coordination poses huge challenges (Granot, 1997). Along with the changing nature of stakeholders in response and recovery, furthermore, there is a difference in the type of information required by different stakeholders. During response, most information is related to damage and needs assessments, relief requirements, logistics and so forth, and most stakeholders require similar information. As recovery begins, information demands become more sector-specific with temporal and spatial complexity. Stakeholders need information that is relevant to their activities in specific sectors, and it is important that it is made available in a timely fashion. The role of the coordinating agencies in Tamil Nadu after the tsunami is a good example of information sharing in recovery. The difference between coordination in response and recovery lies in the focus and priority given to coordinating approaches during disaster recovery. Recovery coordination efforts are modified to suit the local context, which highlights the need for various approaches. In long-term recovery, coordinating approaches involves looking at different thematic areas (such as education, livelihoods and shelter) that need attention and deciding on the intervention that suits the context best. Like the problem of handling the many different perspectives and working patterns of government and multiple organisations, approach-based coordination offers a common way to deal with similar problems. This helps in building consensus on appropriate approaches that are in keeping with overall recovery goals. The involvement of a wide variety of stakeholders in disaster recovery from the tsunami case is not a novel finding, but this chapter contributes to extending coordination from more response oriented to coordinating approaches in disaster recovery. It is evident that stakeholders come with a certain expertise and experience. In mega disasters, like the tsunami that affected many regions in Asia or the Haiti disaster of 2010, recovery involves planning and envisaging rebuilding and addressing complex issues in a holistic manner. In order to be able to address the complex issues in disaster recovery, it is most likely that adopting tailor-made approaches that suit the local context is beneficial for disaster recovery.

16.3 Coordination: temporary or long term?

In a South African study, all the government officials who responded highlighted the lack of coordination during recovery and the importance of coordination as a requirement for effective recovery (Raju & Van Niekerk, 2013). At the same time, the different departments involved made little effort to coordinate their efforts. Further, the study indicates that whilst government officials perceived coordination as playing an important role (at least initially), it did not actively take ownership of the process at all levels.

It is not new that disaster response management takes precedence over recovery, and as already identified, disaster recovery has not received much attention from researchers (Smith & Wenger, 2007). The study also suggests that in South Africa, recovery coordination was not properly addressed, even after three consecutive years of flooding. Furthermore, Raju (2013a) highlights that coordination was affected by the lack of communication between the various departments involved in disaster recovery. One of the reasons for this lack of communication stems from the argument that disaster recovery is not seen as a priority by any of the departments. In normal planning and development conditions, DRM is not a priority. Therefore, disaster recovery reconstruction is considered to be a normal and simple procedure that does not recognise the role of other departments and the interdependencies involved.

Ideally, disaster recovery should be considered as a platform for stakeholders to develop a plan that addresses various forms of vulnerabilities (Wisner et al., 2004, 2012) and integrate it into development planning (van Riet, 2009). However, the lack of communication not only hampers coordination and collaboration between government departments that are responsible for recovery but is also an obstacle to mainstreaming disaster recovery into development. In a study of the effect of the tsunami on Sri Lanka and Indonesia, "the post-tsunami and recovery process implied major changes in organizational structures, the creation of new agencies and legislations for disaster management" (Birkmann et al., 2008). Similarly, in India, many coordination platforms were set up for the response and recovery efforts. However, the sustainability of these new institutions, their coordination and the involvement of stakeholders were crucial to the success of the recovery process. For effective post-disaster planning and coordination, the government must lay down the foundations for effective DRR (UNISDR, 2005).

Coordination was complicated by a general failure to institutionalise coordination structures and mechanisms and those of governmental bodies in particular. This meant that over time, constant changes in governmental structures and high staff turnover undermined any progress made in coordination. It is clear that a lot of money, time and effort were invested in creating a common knowledge base to be shared with the actors participating in one or more coordination structures. Further, it is equally important to sustain these results by creating an institutional repository for long-term disaster recovery learning. There are some good examples of this, such as the NGO Coordination and Resource Centre (NCRC) in Nagapattinam, and the Kanyakumari Rehabilitation Resource Centre (KRRC) in India, but these are unfortunately more the exception than the rule, and vast amounts of vital information have been lost or are no longer maintained, as coordination structures dwindled and died. Also, it is not only information that is lost but also vital knowledge and lessons learnt (or at least identified) – that is, knowledge and lessons that could have been used to facilitate better response and recovery in the future. It appears that after every big disaster, with excessive attention, there are coordination efforts that build new coordination platforms for a short time. Although this is truly needs-based, efforts to institutionalise this process as a truly collaborative one has not been done sufficiently.

16.4 What is coordination?

Like many other terms, coordination has been defined in many ways in different domains. It must also be noted that terms, such as 'coordination', 'collaboration' and 'cooperation' are used interchangeably by researchers (Drabek & McEntire, 2002; Kilby, 2008; Nolte et al., 2012). Coordination requires "a clearly articulated goal, a shared knowledge base, and a set of systematic information search, exchange and feedback processes" (Comfort et al., 2004, p. 64). According to Klein (2001, p. 70), "coordination is the attempt by multiple entities to act in concert in order to achieve a common goal by carrying out a script they all understand". It should, however, be noted that coordination at policy level is very different to coordination at field level (Bennett et al., 2006). Drabek and McEntire (2002, p. 199) define coordination as "a collaborative process through which multiple organisations interact to achieve common objectives". Looking at these definitions, the aspects that stand out are common tasks and goals. Therefore, 'coordination' is defined as "the act of managing interdependencies between activities performed to achieve a goal" (Malone & Crowston, 1990, p. 361).

The two elements that clearly define coordination are interdependencies and goals. It is worth mentioning that coordination means different things to different stakeholders. Although research identifies other key aspects of coordination, such as promoting neglected issues and prioritising resources, this study indicates a more or less complete focus on information sharing, and little attention is given to actual collaboration or joint planning and programmes (Raju, 2013). In other words, although coordination is deemed to be vital, it is limited to the most basic activities. The literature suggests that the best form of coordination leads to joint and collaborative efforts (IFRC, 2000), and it has been argued that "collaboration occurs over time as organisations interact formally and informally through repetitive sequences of negotiation, development of commitments, and execution of those commitments" (Thomson & Perry, 2006, p. 21). However, empirical studies show that there is a decline in coordination over time and that it does not lead to collaboration. Furthermore, coordination structures and the process of coordination itself do not emphasise a common goal (referred to by Thomson and Perry (2006) as 'commitments').

Empirical studies further indicate that there is a need to coordinate disaster recovery, although it is expressed in various ways. This is perhaps not surprising, given the immense focus on coordination and its challenges in current DRM discourse. The massive magnitude of the 2004 tsunami, as well as experience of past disasters, contributed to the consensus that coordination was key to the effectiveness of ensuing operations, especially as the number and variety of actors skyrocketed. It is also interesting to note that coordination started to dwindle as soon as immediate needs were met and the focus shifted towards recovery. For example, housing issues were almost the only focus of subsequent recovery coordination meetings. There may be different reasons for this (e.g. housing is a tangible and costly sector), but the consequences for coordination were significant. Actors stopped attending coordination meetings, funding for coordination declined and coordination structures were shut down. Nevertheless, it seems that actors that were embedded into the local context and had a long-term local presence were more likely to continue to take an active part in coordination.

In the wake of the tsunami of 2004, coordination was complicated by a plethora of coordination structures. A comparison with the definition of coordination and what happens on the ground makes it clear that parallel coordination structures at the same administrative level undermine its effectiveness. Therefore, it is clear from both the

South African (Raju and Van Niekerk, 2013) and the tsunami case studies (Raju & Becker, 2013) that managing interdependencies in order to achieve common goals was not the primary focus of coordination. They highlight the results of working in departmental silos: in the South African case, a lack of interaction with civil society; and in India, more dependencies than interdependencies, and a clear emphasis on simple information sharing between stakeholders. "Aligning one's actions with those of other relevant actors and organisations to achieve a shared goal" (Comfort, 2007, p. 194) entails having a shared goal and a common interface for communication. The harmonisation of the activities of diverse actors (Mcentire, 2007) requires a common platform for dialogue and action, which is limited when there are parallel platforms that lack sophisticated links between them. However, it is interesting to note that it was difficult to establish common goals not only between coordination structures but also within each structure. Given that not only interdependencies but also goals are crucial elements to the definition of coordination, it appears that managing interdependencies is influenced by the goals stakeholders adopt for disaster recovery. Stakeholders seem to establish goals that are very specific to their organisational mandates and only in relation to their expertise in different sectors. This may be a possible reason for not acknowledging interdependencies between different stakeholders, thereby reducing the need for collaborations and joint programming, which is suggested as a deeper form of coordination. Institutions are a very important part of disasters in general. Social institutions in recovery, per Bolin and Bolton (1986), highlight the structure of the church and the involvement of the church as the crucial factors affecting recovery. In my fieldwork in Tamil Nadu in India, the church and other religious institutions played a central role in recovery and had won the immense trust of the people (given the deeply religious society), also due to historical reasons. However, it was unclear whether the church or any faith-based organisations were taken seriously into the central state-level coordination or collaboration mechanisms.

With regard to coordination, Wildavsky (1973, p. 142) writes that "many of the world's ills are attributed to lack of coordination in government. Yet, so far as we know there has never been a serious effort to analyse the term". Coordination in the public sector is considered to be a longstanding problem (Bouckaert et al., 2010), as different governmental organisations and agencies have specific functions. At the same time, DRR and response, and recovery-related activities, are considered to be an additional function (Tempelhoff et al., 2009). Research into governmental coordination shows that "complex issues which do not fit neatly within a department portfolio, or span the interests of several departments, tend to be neglected" (Flinders, 2002, p. 57). A key issue that lies at the heart of this chapter is the point that there has been "little distinction between coordination at the operational level (who does what and where) and strategic coordination at the policy level (such as for joint advocacy)" (Telford & Cosgrave, 2007, p. 12). In this context, an Oxfam study of post-tsunami recovery efforts highlights that communities are keen to be equal partners in recovery. Furthermore, it goes on to say that "local knowledge, capacity and priorities were overlooked" (Oxfam International, 2009, p. 21). Participation and voice in disaster recovery are briefly discussed in Chapter 13.

Despite the general objective of coordination structures, which aim to bring actors together around a mutual platform, there are numerous examples of parallel structures being created in disaster situations (Bennett et al., 2006; Aubrey, 2010). Not only must actors be in contact with each other, they must also be willing to coordinate with each other. Telford and Cosgrave (Telford & Cosgrave, 2007 p. 12) attribute one of the major constraints on coordination to the "absence of any agreed representative mechanism". The most basic activity that facilitates coordination is to share information (IFRC, 2000). Although information sharing is vital for coordination, it has limited effects on the overall efficiency of the total operation if it is not combined with more collaborative efforts (ibid.). The deepest, most beneficial and also most difficult level of coordination is joint-planning and programming (ibid.). Here, actors join forces at many levels, not only sharing information and helping each other to solve particular problems, but planning and implementing joint activities to reach shared goals. Nolte et al. (2012) argue that there are differences in coordination and collaboration activities. According to them, "collaboration refers to activities that cross organizational boundaries" (Nolte et al., 2012, p. 709). In other words, the functioning of coordination structures can be summarised as facilitating clear and common goals, effective and ongoing information sharing, concrete inter-actor collaboration, and joint planning and programmes. Furthermore, project goals cannot be attained without interaction and collaboration (ibid.). The literature highlights the need for more research in this area, and Johnson and Olshansky (2013) pose the important question of why the same institutional problems repeatedly occur.

Donor-related issues, such as funding and project expectations, are also highlighted as factors that affect coordination (Kruke & Olsen, 2005; Balcik et al., 2010). In this regard, although external funding is required for disaster recovery, it is only effective when there is a certain amount of flexibility (Olshansky, 2006). It has been argued that shared incentives have a high impact on coordination (Nolte et al., 2012), which is another indication that donors have a substantial influence on the effectiveness of coordination. Moore et al. (2003, p. 314) exemplify this in their study of coordination following the Mozambique floods in 2000. They state that "international NGOs were sometimes under significant pressure to spend money in a short period of time, thus leading to 'short-term' thinking and fewer relevant projects with long-term benefits".

At the same time, the time pressure for recovery is very high (Olshansky et al., 2012). Olshansky refers to this phenomenon as time compression and argues that it may have implications for the power relationships between stakeholders, interaction patterns and the exchange of information as "the thirst for information is greater than the system can provide" (Olshansky et al., 2012, p. 176). Furthermore, time compression may have implications for "institutional design" (Olshansky et al., 2012, p. 177) during recovery, as new organisations are formed or existing organisations reorganise their functions (Wachtendorf, 2004; Johnson & Olshansky, 2013). Institutional reorganisation can happen at different levels: national government (when disasters cross geographical and political jurisdictions), regional governments (who handle response and recovery) and other organisations that may be created to coordinate between government departments (Johnson & Olshansky, 2013).

16.5 Values and social interfaces in recovery

Values are defined as "desirable trans-situational goals, varying in importance, that serve as guiding principles in the life of a person or other social entity" (Schwartz, 1994, p. 21). For example, communities attach great importance to cultural values and their continuity in post-disaster relocation and reconstruction (Oliver-Smith, 1991). This conceptualisation of values can be expressed as different social entities that have different priorities in disaster recovery.

A complex situation with different values and priorities may contribute to different forms of social interfaces. These are defined as "critical points of intersection between different social fields, domains or life worlds, where social discontinuities based upon differences in values, social interest and power are found", and furthermore, "interfaces typically occur at points where different, and often conflicting, life worlds or social fields intersect" (Long, 2001, p. 177). Further, Long contends that although interface interactions presuppose some degree of common interest, they also have a propensity to generate clashes due to conflicting interests and unequal power relations. Actors with different values, interests and power (e.g. government departments, local and international NGOs, fishing communities, community-based organisations, and other stakeholders) must interact during the recovery process. Therefore, "interface phenomena are often embedded in critical events that tie together a number of spatially distinct, institutionally complex and culturally distinct activities" (Long, 2001, p. 84).

When values are different, a "clash of cultural paradigms" occurs (Long, 2001, p. 70). As long ago as the 1980s, Rubin et al. (1985, p. 61) highlighted the role of values; they noted that "upholding community values in post-disaster setting was observed to be a difficult task". More recent research has noted that it is important to understand the different perspectives of stakeholders with respect to values as "having an explicit dialogue of what is valuable and important to protect also seems to mobilise stakeholders who may not usually consider themselves important for disaster risk management" (Becker & Tehler, 2013, p. 9). In disaster recovery, it is important to address not only stakeholders but also multiple other sectors. Dynes and Quarantelli (1989, p. 3) note that "the emergency phase then is a time period when things get done because values and priorities are clear and resource allocation is based on observable needs. The recovery period is characterized by conflicting priorities". Disaster recovery is complex, and Becker and Tehler's (2013, p. 9) analogy may be useful here, as "it is clear that each stakeholder only has one piece of this puzzle, and it is not until they come together and share their individual knowledge as the richer picture emerges".

Long (2001) characterises the key elements from an interface perspective, noting that social interfaces have a long-term impact on the community. Conflicting ideas and value systems arise from the multiplicity of actors in the process. Whenever these systems meet, there is potential for conflict or other social processes, such as negotiation, accommodation and cooperation. It is useful to adopt the interface perspective in disaster recovery in order to identify differences in the values of different stakeholders, as well as to highlight the complexity that arises from the participation of multiple stakeholders who are involved in housing relocation and disaster recovery. As social life is complex and heterogeneous, it is important to understand the long-term implications of short-term interventions (Pomeroy et al., 2006). Long (2001, p. 59) also uses the term 'social arenas' to highlight that these "are either spaces in which contestation associated with different practices and values of different domains takes place or they are spaces within a single domain where attempts are made to resolve discrepancies in value interpretations and incompatibilities between actor interests". Long's interface concept provides further material for the analysis of large-scale post-disaster interventions involving heterogeneous actors. The concept of the social interface has not been used in the disaster research literature, although it does highlight other aspects, such as stakeholder participation (UNDP, 2001), cultural conflicts (Oliver-Smith, 1991) and social aspects (Nigg, 1995). Research on disaster participation highlights that "little is known about how this participation occurs or can be facilitated under time-constrained circumstances such as post-disaster recovery" (Chandrasekhar, 2010, p. 6).

Disaster recovery is an example of a situation where there are multiple discourses (extending Long's 2001 framework). Given the multiplicity of actors involved in the process, conflicting ideas and value systems are brought together. When these systems meet, it creates a potential platform for conflict or other social processes, such as negotiation, accommodation and cooperation. The conflict interfaces that emerged between the state and the community in India (Raju, 2013b) were found to be important in the debate about reconstruction. According to a fishing community that was being subjected to relocation the natural habitat (i.e. living close to the sea) was of primary importance. The immense resistance from the community came about as the result of an absence of dialogue between them and the state over a long period. This communication gap was exacerbated when the community was not involved in the recovery process. In contrast, the state was guided by the principle that it must offer protection to communities and reduce future damage. Perceptions of what is important to protect differ between these stakeholders. The state argued that physical safety could only be ensured by building new houses away from the coast. The fishing community responded that they risked losing their livelihood if they moved away. These emerging interfaces may have long-term repercussions for both the state one the one hand, and the livelihood of the fishing community on the other. Similar issues were seen in Sri Lanka (Klein, 2008).

Chandrashekar's study identified four factors that affected stakeholder participation (Chandrasekhar, 2010). They may be summed up as power, legitimacy, trust and urgency of action. Furthermore, the study highlights the issue of power between the community and state government. This was also noted by Oliver-Smith (1996, p. 309) as "disasters can create contexts in which power relations and arrangements can be more clearly perceived and confronted, which transforms political consciousness, shapes individual actions, and strengthens or dissolves institutional power arrangements". There are power relations between different stakeholders also identified that affects stakeholder participation in disaster recovery (Chandrasekhar, 2010). This is made clear by the ongoing tension between the state and the community in debating and negotiating where reconstruction should take place. The bargaining power of the fishing communities in India has helped them to maintain a debate that has turned into negotiations with the government. These findings were echoed by Santha (Santha, 2007, p. 68); although in a different setting, the conflict between the fishing community and the state is the result of threatening or contradicting "those components of culture that serve the purpose of sustaining livelihood needs". Further, this research view that "interface situations often provide the means by which individuals or groups come to define their own cultural or ideological positions vis-a-vis those espousing or typifying opposing views" (Long, 2001, p. 70) is true in the context of recovery as well.

It is important to understand and analyse the social processes inherent in a community and their cultural values in the context of the external intervention (Figure 16.3). Given the plurality of actors, interfaces occur when conflicting interests and viewpoints come together. On the other hand, social processes may take the form of cooperation and accommodate external intervention. Researchers have written about the importance of cultural values to a community, and its continuity, in post-disaster reconstruction or relocation. Post-disaster reconstruction yields better results when it takes a holistic view and incorporates the will of the people in planning and implementation (Oliver-Smith, 1991). Recovery research highlights the absence of community consultation and participation, the lack of involvement of the affected community in

Fig 16.3
Livelihood is key to the recovery process. This picture represents the intangible loss of art and many other livelihoods during disasters.
Source: photo courtesy of Emmanuel Raju, Nepal (2019)

decision-making and the lack of information about the reconstruction process, which exacerbated the problem.

There are many forms of interfaces inherent between different stakeholders. The non-governmental stakeholders are highly dependent on the government for legitimacy. This is elaborated in detail in the section on interdependencies later in this chapter. This approval for legitimacy reflects the element of power that plays out as an interface is between the government and the other stakeholders. This has also been highlighted as crucial in disaster recovery by Chandrasekhar (2010), in her research on participation in disaster recovery. Although interface elements in the cases here do not present direct conflicts, disasters may present situations of conflict with new recovery programmes that the government may suggest. In a recent study in Monserrat, a Caribbean British Overseas Territory, Monteil et al. show that it is important to align objectives. Conflicting objectives can lead to a challenge with long-term development and recovery. Objectives of social stability and economic progress do not seem to go hand in hand in Monserrat. "The sustainability of the recovery process is still endangered by the lack of major government initiatives for promoting social cohesion. Yet, there is growing public acknowledgement among the island's politicians, similar to what has driven the grassroots initiatives, that social segregation adversely affects the functioning of society and may be harmful to the Montserrat's long-term development" (Monteil et al., 2020, p. 7).

There are interfaces between implementing stakeholders and donor agencies with a constant negotiation for funding, project approvals, implementation and evaluation of projects. Further, interfaces between different non-governmental organisations appear with their interactions in coordination meetings, information exchange, joint collaborations if undertaken and other forums of discussion. Further, it is also highlighted that there is immense competition on the field between different organisations for funding and even for project implementation areas in the aftermath of disasters (Bennett et al., 2006). This relationship between organisations depicting competition may most likely be an obstacle for collaboration between stakeholders.

Stakeholders' values may be determined by their mandates and the organisation they represent. Therefore, there may be discrepancies in what different stakeholders prioritise during recovery. Coordination may, therefore, be affected by the varying values of stakeholders, which are determined by many factors (e.g. organisational affiliations, community priorities, government policies). Discrepancies in values may highlight stakeholders not willing to engage in coordination at a larger level but to engage in discussions and coordinate only with like-minded stakeholders. This is likely to undermine inherent interdependencies in disaster recovery.

16.6 Interdependencies in disaster recovery

Malone and Crowston (1994, p. 91) highlight that "if there is no interdependence, there is nothing to coordinate". Earlier research on interdependencies has tended to focus on manufacturing settings and infrastructure studies (for example, Thompson, 1967; Rinaldi et al., 2001). Although the literature highlights the crucial role of coordination in disaster situations, many of the definitions given in the section on coordination directly or indirectly highlight interdependencies. Furthermore, it is known that stakeholders do not possess all the expertise required to handle a disaster.

The *Oxford English Dictionary* defines the word 'interdependent' as "(of two or more people or things) dependent on each other". Interdependencies are, in other words, relations of mutual dependence, in contrast to dependencies that are relations in which only one side is dependent on the other. However, the two sides may still be interdependent, even if each dependency is unidirectional, as long as there is at least one dependency in each direction between them (Rinaldi et al., 2001). Though the strength of such interdependence is determined by the strength of the weakest aggregated dependencies of one side in relation to the other.

According to Savage et al. (2010, p. 21), collaboration achieves results "that cannot be accomplished in any other way" and helps to deal with issues that cannot be handled by a single organisation. An organisation's commitment to collaboration is highly dependent on "the degree of interdependence the organisation perceives that it has with the other stakeholders in dealing with the problem" (Logsdon, 1991, p. 24).

Scholars argue that the degree of interdependence amongst organisations is dependent on "the specific task they perform and it varies across tasks" (Van Scotter et al., 2012, p. 284). The world is becoming increasingly interconnected. For example, Perrow stated that "everything is indeed connected, but most of the connections exhibit far more dependency than interdependency" (Perrow, 2007, p. 528). Webb (2007, p. 431) highlighted that even the definition of disaster according to Fritz (1961) sees "society as a system of interrelated and interdependent parts". Similarly, Perrow (2007) argued that we should not make the mistake of viewing dependencies as interdependencies.

Recovery may be optimised by taking into account interdependencies between infrastructures and stakeholders (Tierney, 2007). Stakeholders are "interdependent when each is dependent on the other" (Rinaldi et al., 2001, p. 14), and the study identified four types of infrastructure interdependencies: physical, cyber, geographic and logical. This classification may be extended to disaster settings. Extending their taxonomy, two stakeholders may be physically interdependent if "the state of each is dependent on the material outputs of the other" (ibid., p. 14). In disaster situations, physical dependencies may relate to the sharing and exchange of resources. These physical dependencies have also been highlighted by Perrow (2007, p. 529). The term 'cyber interdependency' used by Rinaldi et al. is being adapted as 'information interdependency' in Raju's thesis (2013). One actor may be dependent on another if they depend on information transmitted between them. Geographical interdependencies occur when there is physical proximity, and a local event that affects one party also has impacts on others. Logical interdependencies arise when the states of these actors are influenced by any of the three interdependencies mentioned earlier (i.e. physical, information or geographical). This type of interdependency may be hard to grasp as it includes issues, such as decisions taken about one or more of the stakeholders involved in recovery.

In order to deal with social problems collaboratively, Logsdon (1991) highlights that both interdependency and interest in solving the problem are important. However, "organisations may not be motivated to deal interdependently with social problems because they may not perceive their interdependence or, even if they do recognise that multiparty efforts are necessary, they may not be motivated to act" (ibid., p. 26). Although other authors have stated that "coordination occurs between actors with shared beliefs", interdependencies may be the result of cause-effect relationships or may be imposed by an authority (Zafonte & Sabatier, 1998, p. 475).

Raju (2013) reflects on organisational interdependencies between stakeholders in disaster recovery. Whilst it has been argued that recovery is a non-linear process in which activities cannot be undertaken by a single entity, there has not been much discussion of interdependencies. Research illustrates that stakeholders have differing dependencies and expectations and that many of them depend upon the government for legitimacy. The findings from Raju's thesis (2013) indicate different types of dependencies between stakeholders in disaster recovery (Box 16.2). First of all, it is interesting to note that dependencies in relation to issues of mandate and legitimacy were common. Others have highlighted dependence of international stakeholders on the government for legitimacy, especially in relation to governmental regulations (Martin, 2005; Bennett et al., 2006). Many of the non-governmental stakeholders were dependent on the government for a legitimate approval to be involved in the recovery process, and these logical dependencies are, thus, fundamental for the lawful participation of entire organisations. This has been indicated by Chandrasekhar (2010) and Mitchell et al. (1997) before but then in terms of legitimacy as such and not as logical dependencies. However, it could be helpful in framing such issues of legitimacy as logical dependencies to grasp and improve coordination between stakeholders in complex post-disaster settings, as it facilitates comparison of the different dependencies that make up the interdependencies that must be managed through coordination to achieve goals. However, governmental dependencies tend to be focused on internal bureaucracy and the departments that undertake different recovery tasks.

Dependencies may relate to logical dependencies (Rinaldi et al., 2001). Although studies highlight dependencies, there is not much evidence of collaborative efforts between stakeholders. Information-sharing was a key coordination issue in most of the tsunami-related reports. For example, the government of Tamil Nadu report acknowledged the role of different coordination platforms as they "helped by regularly disseminating government policies, programmes and orders achieving a two-way flow of information from district to state and from state to district" (Govt. of Tamil Nadu, 2008, p. 2). Raju (2013) and Raju, Becker and Tehler (2018) highlight a clear interdependency between the government and other stakeholders regarding information, which may be seen as an example of an information dependency (Rinaldi et al., 2001). This study also highlights physical dependencies in the housing sector, as there is a flow of material goods and services in the construction of permanent houses. This is consistent with the view of Perrow (2007), as according to stakeholders, there are more dependencies than interdependencies.

Van Scotter et al. (2012, p. 283) highlight that the literature on coordination consists of two main currents of thought. One tends towards the command and control approach (this is not the focus of this chapter), and the other is a "more networked view of coordination as the interaction of interdependent actors outside of traditional hierarchical structures" (O'Toole, 1997; Agranoff, 2006; Wise, 2006). Post-disaster coordination must take cognisance of the fact that disasters vary in their geographical extent, impact, scale and manageability (Figure 16.4). Disaster recovery is not a linear function and cannot be undertaken as an

Box 16.2: Lessons for local governments

- Municipalities must focus on capacity development initiatives for government officials and politicians to facilitate a better understanding of the role and function of coordinating disaster response and recovery, and its relation to DRR as a cross-cutting issue;
- The municipalities must work towards the creation of a culture of safety through awareness building and advocacy programmes during disaster recovery;
- Municipality must include sustainable disaster recovery in their contingency, DRM and integrated development plans;
- Communication between government departments must be improved through regular meetings and joint planning exercises to facilitate sustainable disaster recovery;
- Debriefings and post-disaster evaluations must be planned and held as part of the transition from disaster response to sustainable disaster recovery, involving all relevant role-players;
- More emphasis must be placed on integrated inter-departmental planning and cooperation for disaster recovery through the appropriate DRM structures in the municipality;
- The municipalities must work closer with civil society to ensure better planning and actions when needed; and
- National government must spell out the process of post-disaster funding application and allocation of funds.

Source: Raju and van Niekerk (2013)

Fig 16.4
Fishing communities facing disasters every year that go unrecognised.
Source: picture from Bangladesh; photo courtesy of Pablo van Holm-Neilsen (2011)

independent activity by one department or as independent activities by many departments. Research indicates that during the recovery process, connections are not made between issues of gender, culture, livelihoods and physical infrastructure. One must remember that without a solid understanding of the local and cultural context, recovery interventions are bound to fail.

The various definitions earlier highlight that when multiple stakeholders work in a common environment (in this case disaster recovery) with interdependent factors, it is important to have common goals. Furthermore, "interdependence requires acceptance of common goals, use of standardised procedures and language, and constant communication among specialized individuals/teams" (Van Scotter et al., 2012, p. 285). Coordination seemed to have a significant positive impact during the tsunami, and agencies were created with this as their mandate (Masyrafah & Mckeon, 2008). Although the establishment of coordination platforms aimed to create these linkages, they have dwindled over time. This can be attributed to the lack of a common goal: as Fawcett and Daugbjerg (2012, p. 199) note, a policy community is only established when there is a shared agenda and its members are "dependent on others to achieve their aims". It may be inferred indirectly from the post-tsunami studies that an opportunity was missed to create a regional- and national-level policy community for DRM. This does not discount the fact that a network and linkages between stakeholders were established: Long's (2001, p. 69) framework argues that "the interface itself becomes an organized entity of inter-locking relationships". However, a conversation with a government official following the tsunami in India was revealing. When asked what happens if another disaster occurred, the official answered, "we may have to re-invent the wheel".

16.7 Recovery coordination and collaboration: a governance issue?

It is known that studies "without any attempt to generalize can certainly be of value" in research (Flyvbjerg, 2001, p. 76). This chapter acknowledges that knowledge developed in one or two cases cannot be completely generalised; nevertheless, it may be possible to make analytical generalisations (ibid.). The term 'coordination' means different things to different stakeholders. Since there are fewer guidelines for disaster recovery compared to disaster response, it is more likely that different stakeholders perceive coordination differently compared to how they perceive it during immediate response. Also, the empirical data from the tsunami recovery studies also shows that there were many opinions amongst the stakeholders involved in recovery activities concerning what the term actually means (Raju, 2013). The tsunami of 2004 is not a unique event with respect to the multitude of stakeholders involved and the relatively unclear guidelines for disaster recovery. Therefore, it is likely that coordination is perceived differently in other disasters as well. It may probably be the case that coordination is primarily focused on information sharing and networking, whilst stakeholders continue to carry out different activities independently without acknowledging their role in collaborating for the larger recovery goal.

Stakeholders have a wide variety of values. Given that these stakeholders represent different organisations, they are most likely to carry the bag of values of what they perceive to be most important and prioritise activities based on their mandates. These set of values are likely to determine the focus of recovery activities in other disaster recovery settings. In such situations, there is a high probability that different interfaces occur between different stakeholders (e.g. between communities and government, between different organisations, between organisations and the government). This is due to differing power relationships, working mandates, priorities of different actors, funding available and so forth.

The examples from India and South Africa presented in this chapter and in Chapter 13 highlight a key role of the government in recovery coordination. When government agencies function independently without recognising expertise of the non-governmental stakeholders, there are more chances that resources and capacities of stakeholders are not effectively utilised for recovery coordination. It is very evident that the role of the government is crucial for effectiveness in disaster recovery. It is highly likely that most of the actors involved in disasters across may perceive the government as the main working partner responsible for many factors in disaster recovery. The affected communities in any disaster may also consider that recovery is the mandate of the government at all levels, as it is one of the most prominent players committed to long-term activities. Furthermore, the role of the government in taking a lead role in coordination is most likely to determine the interest of the other stakeholders to collaborate with others.

Governance must take into account different dependencies and inter-dependencies during recovery. There are different types of dependencies between stakeholders involved in disaster recovery. Also, the strength of these dependencies varies. It is most likely that different disaster settings will have different types of qualitative and quantitative dependencies between stakeholders. Although the type of dependencies may not exactly be the same in all settings, it may be possible to extend this argument that there may be different types of dependencies in disaster recovery settings across. Theoretically, we may argue that the non-governmental stakeholders and, more specifically, the international community may be highly dependent on the government in many countries to get permission to involve in recovery or any related activity. However, this may vary across settings, as a stronger state may impose more limitations and regulations on these stakeholders to obtain legitimacy. In countries where the government may be politically weak and economically highly aid dependent, the legitimacy argument may be more of a principle on paper, and the international community takes over more activities by playing a prominent role.

Goals and lack of articulation of common goals are often identified as key factors affecting recovery coordination. However, it is most likely that articulation of common goals may continue to be a challenge as it is closely linked with interdependencies. Further, the lack of common goals highlights an inclination towards achieving more independent sector-related or organisational goals. Furthermore, there may be a clash of goals as, traditionally, recovery is seen as a physical linear rebuilding activity versus the more holistic approach that is required (i.e. a clash of approaches). Differences in parlance between the international community and the government may also be a possible reason for lack of common goals. One such example is where government institutions continue to adopt a more phase-oriented approach to DRM and use the word 'rehabilitation' as a synonym for 'recovery'.

Whilst these processes of goals articulation happen in the formal spaces, it is also important to recognise and focus on the real governance and invisible governance forms that happen in daily life and shape the disaster process (Hilhorst et al., 2020) (Box 16.3). For example, gender is one such aspect that plays out in formal, informal and invisible ways:

> Addressing gender issues in the context of DRR and CCA does not merely refer to the inclusion of representatives from all groups or ensuring a head count from different groups. Addressing gender is to address deep-rooted socio-structural issues, such as patriarchal power dynamics and access to resources (such as economic credit). While synergies may be drawn between CCA and DRR strategies, both these processes are challenged with questions of gender inequalities on the ground. Policy dialogues on gender mainstreaming have been in vogue for a long time now. However, change with regard to gender aspects must be seen in the light of power shifts and institutionalisation of gender as a key issue in larger development processes, which in turn feed into DRR and CCA.
>
> (Raju, 2019, pp. 131–132)

16.8 Concluding remarks

So how do we foster disaster recovery?

Experts in DRM, who are active in the international arena, explain that there are fewer guidelines for disaster recovery when compared to response. Some examples of response guidelines are the cluster approach, and the Inter Agency Standing Committee Guidelines for natural disasters and the Humanitarian Charter and Minimum Standards in Humanitarian Response (the SPHERE standards). The cluster approach is a recent development in the global arena. It began in 2005 and addresses sectoral coordination "as a means to strengthen

Box 16.3: Disaster governance

1 Governance is a new concept in the study of disasters; research on governance issues related to hazards and disasters is in its infancy.
2 Disaster governance is conceptually but not always empirically related to environmental, risk, earth system and collaborative governance.
3 Disaster governance arrangements are shaped by social, economic and political forces, such as globalisation, political and economic trends within the world system, and sociodemographic changes (e.g. population migration into hazardous areas).
4 Efforts at disaster governance face many challenges, including global disparities in income, well-being and political empowerment, as well as the rapid expansion of disaster vulnerability, particularly in poor nations.
5 Disaster governance arrangements exist at different scales and often focus on specific phases of the hazards cycle (e.g. mitigation and response). Governance arrangements tend to be reactive and fragmented; most systems are neither risk based nor comprehensive.
6 Governance arrangements and capabilities vary considerably across societies as a function of such factors as overall state capacity, state-civil society relationships and economic organisation.

Source: Tierney (2012, p. 359)

predictability, response capacity, coordination and accountability by strengthening partnerships in key sectors of humanitarian response, and by formalising the lead role of particular agencies/organisations in each of these sectors" (Stoddard et al., 2007, p. 1). Although evaluations of the cluster approach (Stoddard et al., 2007; Steets et al., 2010) highlight the scope for improvement, it is a huge step forward in establishing a global humanitarian response. However, the problem with recovery is the lack of established guidelines (Figure 16.5). The few documents that are available are very vague guidelines set out by the United Nations Development Programme (UNDP). After the tsunami, the UN agencies launched the idea of a single programme to combine all UN forces in India between 2005 and 2008. Similar efforts were seen in other affected Asian countries. However, there was a clear problem in operationalising the idea.

There is a growing need to address issues of disaster recovery governance (Fung, 2006; Ikeda et al., 2008; Renn, 2008; Djalante, 2012). From my analysis and during my research, it appears that coordination may be approached as a governance problem that needs to be investigated in more detail. Governance has many definitions, which have been contested. It is not the same as 'government' (Lemos & Agarwal, 2006; Jordan, 2008), as it encompasses all stakeholders and should "cover the whole range of institutions and relationships"

Fig 16.5
Memorial from the Mont Blanc Tunnel disaster of 1999. It is important to learn from previous disasters. Learning is key to governance.
Source: photo courtesy of Emmanuel Raju

(Pierre & Peters, 2000, p. 1). Coordination is certainly an important issue, along with control, accountability and political power (Flinders, 2002). However, extending these arguments to disaster recovery, there have been very few, scattered efforts to develop recovery governance (Tierney & Oliver-Smith, 2012). Governance also takes into account the factors discussed earlier, such as power, interdependence, autonomous functioning and a complex set of actors (Stoker, 1998). Further, these issues of interdependencies and goals in disaster recovery must be directly linked to governance issues (Box 16.4).

A quick review of different documents and reports from organisations involved in tsunami recovery in various Asian countries highlight that recovery approaches differ in the affected countries. This chapter does not advocate universal recovery guidelines, as recovery is based on local conditions and socio-cultural, economic and political factors. Although recovery processes are highly context-dependent, the question arises of what may be learned from these different approaches. Duyne Barenstein shows that "policy-making processes, practices and outcomes depend on a number of contextual factors" (Barenstein, 2010, p. 173), including the relationships between stakeholders and their previous disaster experience. For example, in India, there were many coordination structures at different levels. For example, in Indonesia, the government required agencies with a four-year mandate to coordinate their activities. However, it was clear that during recovery, coordinating agencies were more involved in the implementation of activities than coordination as a task in itself (Masyrafah & Mckeon, 2008). It becomes extremely difficult to progress with development and recovery when a country

Box 16.4: Recovery is complex

And finally, theory and practice should be closely connected. When policy and practice are not based on a solid understanding of human behaviour in general, and social and cultural behaviour specifically, their chances of success are limited. Conversely, policy and its application can serve as an important proving ground for the relevance and predictive ability of theory. If policies and practices do not coordinate, and do not produce successful results, it is the programmes and their applications that are at fault, but it is always the people who suffer. At best, disaster management can be viewed as a type of complex adaptive system that can learn from experience, processing information and adapting according to local principles and actions. In other words, complex adaptive systems do not merely react; they also attempt to take advantage of circumstances.

Source: Tierney and Oliver-Smith (2012, p. 141)

continues to face multiple disasters at the same time or in a given short span of time (Finucane et al., 2020).

"Every post-disaster recovery manifest[s] tension between speed and deliberation" (Olshansky, 2006, p. 148). For example, post-tsunami reports suggested that affected communities measured speed by the time it took to construct permanent housing. Along with speed, it is essential to bring all stakeholders on board and address issues of holistic planning, interdependencies, community participation, goals and other related factors. This may create the tension that Olshansky (ibid.) refers to. It is, therefore, highly important to note that social processes and deliberations take enormous amounts of time.

The literature highlights that coordination is a problem; however, there has not been many efforts to address the question of why coordination and collaboration are a challenge in recovery. This raises the need to open the debate on disaster governance in order to explore coordination and collaboration as a challenge and address the practical realities. To sum up, coordination and collaboration in disaster recovery require the engagement of stakeholders across sectors, disciplines, and governmental and non-governmental organisations. This chapter has raised important issues for governmental and non-governmental stakeholders in relation to goals, mandates, autonomy, adopting common recovery approaches and the importance of learning from disasters. Furthermore, this chapter reiterates that coordination goes beyond information-sharing and the exchange of ideas. Collaborative efforts that acknowledge interdependencies are highlighted as a key area in recovery coordination. This must be reflected from very early stages of recovery planning and key to building an effective recovery framework. There is not much consensus on the definition of disaster recovery. However, considering different aspects and changing aspects of disasters, recovery must be seen a holistic process that encompasses "physical, social, economic and natural environment" (Smith & Wenger, 2007, p. 237), considering pre- and post-disaster events. DRR is everyone's business (Raju & da Costa, 2018) and must be factored into recovery. Further, to foster recovery, considering the variety of definitions, recovery needs to take into account the importance of interdependencies and goals. "Social interfaces" (Long, 2001) could be very useful to explain different relationships between stakeholders, which may be crucial for disaster recovery. This is dependent on the values that stakeholders possess, where values are referred to as "guiding principles" (Schwartz, 1994, p. 21) of what is important to different actors. These different factors of goals, interdependencies, values and participation must be taken seriously to foster sustainable recovery.

(This chapter is based on Raju's (2013a) PhD thesis.)

Take-away messages

1. Disaster recovery is a complex process and must not be seen as linear. It consists of many dependencies and inter-dependencies.
2. Disaster recovery must not build on existing vulnerabilities. Disaster recovery must challenge status quo.
3. Disaster recovery governance must focus not only on formal issues but also on invisible forms of governance that shape daily life.
4. It is important to recognise and focus on the values of the affected populations.
5. Disaster recovery must set the foundation for new risk-informed development visions and goals.

To learn more about the topic discussed in this chapter, listen to the *Disasters: Deconstructed* interview with Dr Daniel Aldrich and Dr Wes Cheek (Figure 16.6).

Fig 16.6
QR code for Chapter 16.

Further suggested reading

Aldrich, D. P. (2012). *Building resilience: Social capital in post-disaster recovery*. University of Chicago Press.

Berke, P. R., Kartez, J., & Wenger, D. (1993). Recovery after disaster: Achieving sustainable development, mitigation and equity. *Disasters*, *17*(2), 93–109. https://doi.org/10.1111/j.1467-7717.1993.tb01137.x, PubMed: 20958760

Boano, C. (2013). Post-disaster recovery planning. Introductory notes on its challenges and potentials. In A. López-Carresi, M. Fordham,

B. Wisner, I. Kelman, & J. Gaillard (Eds.), *Disaster management: International lessons in risk reduction, response and recovery*. Taylor & Francis.

Coetzee, C., Van Niekerk, D., & Raju, E. (2016). Disaster resilience and complex adaptive systems theory: Finding common grounds for risk reduction. *Disaster Prevention and Management*, *25*(2), 196–211. https://doi.org/10.1108/DPM-07-2015-0153

Cretney, R. M. (2017). Towards a critical geography of disaster recovery politics: Perspectives on crisis and hope. *Geography Compass*, *11*(1). https://doi.org/10.1111/gec3.12302

Kinnvall, C., & Rydstrom, H. (2019). Climate hazards, disasters, and gender ramifications. In *Climate hazards, disasters and gender ramifications*. Routledge.

Lauta, K. C. (2016). Legal scholarship and disasters. In R. Dahlberg, O. Rubin, & M. T. Vendelø (Eds.), *Disaster research: Multidisciplinary and international perspectives* Humanitarian Studies Series (pp. 97–109). Routledge. https://doi.org/10.4324/9781315724584

Johnson, L. A., & Hayashi, H. (2012). Synthesis efforts in disaster recovery Research. *International Journal of Mass Emergencies and Disasters*, *30*(2), 212–239.

Olshansky, R. B., Hopkins, L. D., & Johnson, L. A. (2012). Disaster and recovery: Processes compressed in time. *Natural Hazards Review*, *13*(3), 173–178. Retrieved April 11, 2013, from http://ascelibrary.org/doi/abs/10.1061/%28ASCE%29NH.1527-6996.0000077; https://doi.org/10.1061/(ASCE)NH.1527-6996.0000077

Raju, E. (2013). Exploring disaster recovery coordination. *Lund University. Exploring disaster recovery coordination*. https://lup.lub.lu.se/search/ws/files/4305738/4180232.pdf. Lunds Universitet.

Raju, E., & da Costa, K. (2018). Governance in the Sendai: A way ahead? *Disaster Prevention and Management*, *27*(3), 278–291. https://doi.org/10.1108/DPM-08-2017-0190

Raju, E., & Van Niekerk, D. (2013). Intra-governmental coordination for sustainable disaster recovery: A case-study of the Eden district municipality, South Africa. *International Journal of Disaster Risk Reduction*, *4*, 92–99. http://linkinghub.elsevier.com/retrieve/pii/S2212420913000162. https://doi.org/10.1016/j.ijdrr.2013.03.001

References

Agranoff, R. (2006). Inside collaborative networks: Ten lessons for public managers. *Public Administration Review*, *66*(s1), 56–65. https://doi.org/10.1111/j.1540-6210.2006.00666.x

Alberts, D. S., Huber, R. K., & Moffat, J. (2010). *NATO NEC C2 maturity model*. CCRP Publications.

Aubrey, D. (2010). Kenya: Can temporary shelter contribute to participatory reconstruction? In M. Lyons, T. Schilderman, &

C. Boano (Eds.), *Building back better: Delivering people-centred reconstruction to scale* (pp. 215–240). Practical Action Publishing.

Balcik, B., Beamon, B. M., Krejci, C. C., Muramatsu, K. M., & Ramirez, M. (2010). Coordination in humanitarian relief chains: Practices, challenges and opportunities. *International Journal of Production Economics*, *126*(1), 22–34. https://doi.org/10.1016/j.ijpe.2009.09.008

Barenstein, J. D. (2010). *Who governs reconstruction? Changes and continuity in policies, practices and outcomes*. https://www.researchgate.net/publication/265707613_Rebuilding_Housing_after_a_Disaster_Factors_for_Failure

Becker, P., & Tehler, H. (2013). Constructing a common holistic description of what is valuable and important to protect: A possible requisite for disaster risk management. *International Journal of Disaster Risk Reduction*, *6*, 18–27. https://doi.org/10.1016/j.ijdrr.2013.03.005

Bennett, J. et al. (2006). *Coordination of international humanitarian assistance in tsunami-affected countries*. Tsunami Evaluation Coalition.

Berke, P. R., Kartez, J., & Wenger, D. (1993). Recovery after disaster: Achieving sustainable development, mitigation and equity. *Disasters*, *17*(2), 93–109. https://doi.org/10.1111/j.1467-7717.1993.tb01137.x

Birkmann, J., Buckle, P., Jaeger, J., Pelling, M., Setiadi, N., Garschagen, M., Fernando, N., & Kropp, J. (2008). 'Extreme events and disasters: A window of opportunity for change? Analysis of organizational, institutional and political changes, formal and informal responses after mega-disasters. *Natural Hazards*, *55*(3), 637–655. https://doi.org/10.1007/s11069-008-9319-2

Bolin, R. C., & Bolton, P. (1986). *Race, religion, and ethnicity in disaster recovery* (p. 282). Program on Environment and Behavior Monograph.

Bouckaert, G., Peters, B., & Verhoest, K. (2010). *The coordination of public sector organizations—shifting patterns of public management*. Palgrave Macmillan.

Chandrasekhar, D. (2010). *Understanding stakeholder participation in post-disaster recovery*. University of Illinois. https://www.researchgate.net/publication/43939582_Understanding_stakeholder_participation_in_post-disaster_recovery_case_study_Nagapattinam_India

Christoplos, I., Rodríguez, T., Schipper, E. L., Narvaez, E. A., Bayres Mejia, K. M., Buitrago, R., Gómez, L., & Pérez, F. J. (2010). Learning from recovery after Hurricane Mitch. *Disasters*, *34*(s2), S202–S219. https://doi.org/10.1111/j.1467-7717.2010.01154.x

Comfort, L. K. (2007). Crisis management in hindsight: Cognition, communication, coordination, and control. *Public Administration Review* (Special Issue, *Administrative Failure in the Wake of Katrina*), *67*, 189–197. https://doi.org/10.1111/j.1540-6210.2007.00827.x

Comfort, L. K., Dunn, M., Johnson, D., Skertich, R., & Zagorecki, A. (2004). Coordination in complex systems: Increasing efficiency in disaster mitigation and response. *International Journal of Emergency Management*, *2*(1–2), 62. https://doi.org/10.1504/IJEM.2004.005314

Comfort, L. K., Sungu, Y., Johnson, D., & Dunn, M. (2001). Complex systems in crisis: Anticipation and resilience in dynamic environments. *Journal of Contingencies and Crisis Management*, *9*(3), 144–158. https://doi.org/10.1111/1468-5973.00164

Davis, I., & Alexander, D. (2016). *Recovery from disaster*. Routledge.

Djalante, R. (2012). Review Article: "Adaptive governance and resilience: The role of multi-stakeholder platforms in disaster risk reduction". *Natural Hazards and Earth System Sciences*, *12*(9), 2923–2942. https://doi.org/10.5194/nhess-12-2923-2012

Drabek, T. E., & McEntire, D. A. (2002). Emergent phenomena and multiorganisational coordination in disasters: Lessons from the research literature. *International Journal of Mass Emergencies and Disasters*, *20*(2), 197–224.

Drabek, T. E., & McEntire, D. A. (2003). Emergent phenomena and the sociology of disaster: Lessons, trends and opportunities from the research literature. *Disaster Prevention and Management*, *12*(2), 97–112. https://doi.org/10.1108/09653560310474214

Dynes, R. R., & Quarantelli, E. L. (1989). *Reconstruction in the Context of Recovery: Thoughts on the Alaskan earthquake*, *141*.

Fawcett, P., & Daugbjerg, C. (2012). Explaining governance outcomes: Epistemology, network governance and policy network analysis. *Political Studies Review*, *10*(2), 195–207. https://doi.org/10.1111/j.1478-9302.2012.00257.x

Finucane, M. L., Acosta, J., Wicker, A., & Whipkey, K. (2020). Short-term solutions to a long-term challenge: Rethinking disaster recovery planning to reduce vulnerabilities and inequities. *International Journal of Environmental Research and Public Health*, *17*(2). https://doi.org/10.3390/ijerph17020482

Flinders, M. (2002). Governance in Whitehall. *Public Administration*, *80*(1), 51–75. https://doi.org/10.1111/1467-9299.00294

Flyvbjerg, B. (2001). *Making social science matter: Why social inquiry fails and how it can succeed again*. Cambridge University Press.

Fritz, C. E. (1961). Disaster. In R. K. Merton, & R. A. Nisbet (Eds.), *Contemporary social problems* (pp. 651–694). Harcourt, Brace and World.

Fung, A. (2006). Varieties of participation in complex governance. *Public Administration Review*, *66*(s1), 66–75. https://doi.org/10.1111/j.1540-6210.2006.00667.x

Govt. of Tamil Nadu. (2008). *Tiding Over the Tsunami*. http://www.tn.gov.in/tsunami/digitallibrary/ebooks-web/TOTPART2/TidingoverPART2.pdf

Granot, H. (1997). Emergency inter-organizational relationships. *Disaster Prevention and Management*, *6*(5), 305–310. https://doi.org/10.1108/09653569710193736

Hilhorst, D., Boersma, K., & Raju, E. (2020). Research on politics of disaster risk governance: Where are we headed? *Politics and Governance*, *8*(4), 214–219. https://doi.org/10.17645/pag.v8i4.3843

IFRC. (2000). *Improving coordination (Disaster preparedness training programme)*. IFRC.
Ikeda, S., Sato, T., & Fukuzono, T. (2008). Towards an integrated management framework for emerging disaster risks in Japan. *Natural Hazards*, *44*(2), 267–280. https://doi.org/10.1007/s11069-007-9124-3
Johnson, L. A., & Olshansky, R. B. (2013). The road to recovery: Governing post-disaster reconstruction. *Land Lines*, *25*(3), 14–21.
Jordan, A. (2008). The governance of sustainable development: Taking stock and looking forwards. *Environment and Planning C*, *26*(1), 17–33. https://doi.org/10.1068/cav6
Katoch, A. (2006). The responders' cauldron: The uniqueness of international disaster response. *Journal of International Affairs*, *59*(2), 153–172.
Kettl, D. F. (2008). Contingent coordination: Practical and theoretical puzzles for homeland security. In A. Boin (Ed.), *Crisis management* (pp. 348–370). Sage.
Kilby, P. (2008). The strength of networks: The local NGO response to the tsunami in India. *Disasters*, *32*(1), 120–130. https://doi.org/10.1111/j.1467-7717.2007.01030.x
Klein, G. (2001). Features of team coordination. In M. McNeese, M. R. Endsley, & E. Salas (Eds.), *New trends in cooperative activities* (pp. 68–95). HFES.
Klein, N. (2008). *The shock doctrine*. Knopf.
Kruke, B. I., & Olsen, O. E. (2005). Reliability-seeking networks in complex emergencies. *International Journal of Emergency Management*, *2*(4), 275–291. https://doi.org/10.1504/IJEM.2005.008740
Lemos, M. C., & Agarwal, A. (2006). Environmental governance. *Annual Review of Environment and Resources*, *31*(1), 297–325. https://doi.org/10.1146/annurev.energy.31.042605.135621
Logsdon, J. M. (1991). Interests and interdependence in the formation of social problem-solving collaborations. *Journal of Applied Behavioral Science*, *27*(1), 23–37. https://doi.org/10.1177/0021886391271002
Long, N. (2001). *Development sociology: Actor perspectives*. Routledge.
Malone, T. W., & Crowston, K. (1990). *What is coordination theory and how can it help design cooperative work systems?* Proceedings of the 1990 ACM Conference on Computer-Supported Cooperative Work—CSCW'90 (pp. 357–370). ACM Press. https://doi.org/10.1145/99332.99367
Malone, T. W., & Crowston, K. (1994). The interdisciplinary study of coordination. *ACM Computing Surveys*, *26*(1), 87–119. https://doi.org/10.1145/174666.174668
Martin, M. (2005). A voice for the vulnerable groups in Tamil Nadu. *Forced Migration Review* (Spl Issue: Tsunami: Learning From the Humanitarian Response), 44–45. https://reliefweb.int/sites/reliefweb.int/files/resources/81F91CA445B11D03C125703600493905-rsc-tsu-06jul.pdf

Masyrafah, H., & Mckeon, J. M. J. A. (2008). *Post-tsunami aid effectiveness in Aceh. Proliferation and coordination in reconstruction.* USA: Wolfensohn Centre for Development. https://www.brookings.edu/research/post-tsunami-aid-effectiveness-in-aceh-proliferation-and-coordination-in-reconstruction/

Mcentire, D. A. (2007). *Disaster response and recovery*. Wiley.

Mitchell, R. K., Agle, B. R., & Wood, D. J. (1997). Toward a theory of stakeholder identification and salience: Defining the principle of who and what really counts. *Academy of Management Review*, *22*(4), 853–886. https://doi.org/10.5465/AMR.1997.9711022105

Monteil, C., Simmons, P., & Hicks, A. (2020). Post-disaster recovery and sociocultural change: Rethinking social capital development for the new social fabric. *International Journal of Disaster Risk Reduction*, *42*. https://doi.org/10.1016/j.ijdrr.2019.101356

Moore, S., Eng, E., & Daniel, M. (2003). International NGOs and the role of network centrality in humanitarian aid operations: A case study of coordination during the 2000 Mozambique floods. *Disasters*, *27*(4), 305–318. http://www.ncbi.nlm.nih.gov/pubmed/14725089; https://doi.org/10.1111/j.0361-3666.2003.00235.x

Neal, D. M., & Phillips, B. D. (1995). Effective emergency management: Reconsidering the bureaucratic approach. *Disasters*, *19*(4), 327–337. http://www.ncbi.nlm.nih.gov/pubmed/8564456; https://doi.org/10.1111/j.1467-7717.1995.tb00353.x

Nigg, J. M. (1995*). Disaster recovery as a social process*. Disaster Research Centre, University of Delaware. http://udspace.udel.edu/handle/19716/625

Nolte, I. M., Martin, E. C., & Boenigk, S. (2012). Cross-sectoral coordination of disaster relief. *Public Management Review*, *14*(6), 707–730. https://doi.org/10.1080/14719037.2011.642629

Oliver-Smith, A. (1991) Successes and failures in post-disaster resettlement. *Disasters*, *15*(1), 12–23. https://doi.org/10.1111/j.1467-7717.1991.tb00423.x

Oliver-Smith, A. (1996). Anthropological research on hazards and disasters. *Annual Review of Anthropology*, *25*(1), 303–328. https://doi.org/10.1146/annurev.anthro.25.1.303

Olshansky, R. B. (2006). Planning after Hurricane Katrina. *Journal of the American Planning Association*, 72(2), 147–153. https://doi.org/10.1080/01944360608976735

Olshansky, R. B., Hopkins, L. D., & Johnson, L. A. (2012). Disaster and recovery: Processes compressed in time. *Natural Hazards Review*, *13*(3), 173–178. https://doi.org/10.1061/(ASCE)NH.1527-6996.0000077

O'Toole, L. J. (1997). Treating networks seriously: Practical and research-based agendas in public administration. *Public Administration Review*, *57*(1), 45–52. https://doi.org/10.2307/976691

Oxfam International. (2009). *Collaboration in crises: Lessons in community participation from the Oxfam International tsunami research program.* https://www.oxfam.org/en/research/collaboration-crises

Perrow, C. (2007). Disasters ever more? Reducing US vulnerabilities. In H. Rodriguez, E. L. Quarantelli, & R. R. Dynes (Eds.), *Handbook of disaster research* (pp. 521–533). Springer.

Pierre, J., & Peters, B. G. (2000). *Governance, politics and the state*. Palgrave Macmillan.

Pomeroy, R. S., Ratner, B. D., Hall, S. J., Pimoljinda, J., & Vivekanandan, V. (2006). Coping with disaster: Rehabilitating coastal livelihoods and communities. *Marine Policy*, *30*(6), 786–793. https://doi.org/10.1016/j.marpol.2006.02.003

Raju, E. (2013). Exploring disaster recovery coordination. Lund University. Exploring disaster recovery coordination. Lunds Universitet. https://lup.lub.lu.se/search/ws/files/4305738/4180232.pdf

Raju, E. (2013a). *Exploring disaster recovery coordination*. Lund University. https://lup.lub.lu.se/search/ws/files/4305738/4180232.pdf

Raju, E. (2013b, April). Housing reconstruction in disaster recovery: A study of fishing communities post-tsunami in Chennai, India. *PLoS Currents*, *5*, 2004–2007. https://doi.org/10.1371/currents.dis.a4f34a96cb91aaffacd36f5ce7476a36

Raju, E. (2019). Gender as fundamental to climate change adaptation and disaster risk reduction: Experiences from South Asia. In C. Kinnvall & H. Rydstrom (Eds.), *Climate hazards, disasters and gender ramifications* (1st ed.). Routledge.

Raju, E. (2021). Moving from response to recovery- what happens to coordination? In J. Mendes, G. Kalonji, R. Jigyasu, & A. Chang-Richards (Eds.), *Strengthening disaster risk governance to manage disaster risk* (1st ed., pp. 69–76). Elsevier.

Raju, E., & Becker, P. (2013). Multi-organisational coordination for disaster recovery: The story of post-tsunami Tamil Nadu, India. *International Journal of Disaster Risk Reduction*, *4*, 82–91. https://doi.org/10.1016/j.ijdrr.2013.02.004

Raju, E., Becker, P., & Tehler, H. (2018). Exploring interdependencies and common goals in disaster recovery coordination. *Procedia Engineering*, *212*, 1002–1009. https://doi.org/10.1016/j.proeng.2018.01.129

Raju, E., & da Costa, K. (2018). Governance in the Sendai: A way ahead? *Disaster Prevention and Management*, *27*(3), 278–291. https://doi.org/10.1108/DPM-08-2017-0190

Raju, E., & Van Niekerk, D. (2013). Intra-governmental coordination for sustainable disaster recovery: A case-study of the Eden District Municipality, South Africa. *International Journal of Disaster Risk Reduction*, *4*, 92–99. https://doi.org/10.1016/j.ijdrr.2013.03.001

Renn, O. (2008). *Risk governance: Coping with uncertainty in a complex world*. Earthscan Publications.

Rinaldi, B. S. M., Peerenboom, J. P., & Kelly, T. K. (2001). Identifying, understanding, and analyzing critical infrastructure interdependencies. *IEEE Control Systems*, *21*(6), 11–25. https://doi.org/10.1109/37.969131

Rubin, C., Saperstein, M., & Barbee, D. (1985). *Community recovery from a major natural disaster* (pp. 61–63). FMHI Publication. http://scholarcommons.usf.edu/cgi/viewcontent.cgi?article=1086&context=fmhi_pub

Santha, D. S. (2007). State interventions and natural resource management: A study on social interfaces in a riverine fisheries setting in Kerala, India. *Natural Resources Forum*, *31*(1), 61–70. https://doi.org/10.1111/j.1477-8947.2007.00128.x

Savage, G. T., Bunn, M. D., Gray, B., Xiao, Q., Wang, S., Wilson, E. J., & Williams, E. S. (2010). Stakeholder collaboration: Implications for stakeholder theory and practice. *Journal of Business Ethics*, *96*(s1), 21–26. https://doi.org/10.1007/s10551-011-0939-1

Scanlon, J. (1999). Emergent groups in established frameworks: Ottawa Carleton's response to the 1998 ice disaster. *Journal of Contingencies and Crisis Management*, *7*(1), 30–37. https://doi.org/10.1111/1468-5973.00096

Schwartz, S. H. (1994). Are there universal aspects in the structure and contents of human values? *Journal of Social Issues*, *50*(4), 19–45. https://doi.org/10.1111/j.1540-4560.1994.tb01196.x

Smith, G., & Wenger, D. (2007). Sustainable disaster recovery: Operationalizing an existing agenda. In H. Rodriguez, E. L. Quarantelli, & R. R. Dynes (Eds.), *Handbook of disaster research* (pp. 234–257). Springer.

Sou, G. (2019). Sustainable resilience? Disaster recovery and the marginalization of sociocultural needs and concerns. *Progress in Development Studies*, *19*(2), 144–159. https://doi.org/10.1177/1464993418824192

Steets, J. et al. (2010, April). Cluster approach Evaluation 2 synthesis report. *Approach Evaluation*. https://www.humanitarianresponse.info/sites/www.humanitarianresponse.info/files/documents/files/Cluster2.pdf.

Stoddard, A. et al. (2007). *Cluster approach evaluation*. https://www.humanitarianresponse.info/sites/www.humanitarianresponse.info/files/documents/files/ClusterApproachEvaluation1.pdf

Stoker, G. (1998). Governance as theory: Five propositions. *International Social Science Journal*, *50*(155), 17–28. https://doi.org/10.1111/1468-2451.00106

Telford, J., & Cosgrave, J. (2007). The international humanitarian system and the 2004 Indian Ocean earthquake and tsunamis. *Disasters*, *31*(1), 1–28. https://doi.org/10.1111/j.1467-7717.2007.00337.x

Tempelhoff, J., Gouws, I., & Botha, K. (2009). The December 2004-January floods in the Garden Route region of the Southern Cape, South Africa. *Jàmbá: Journal of Disaster Risk Studies*, 2(2), 93–112.

Thompson, J. (1967). *Organisations in action: Social science bases of administrative theory*. McGraw-Hill.

Thomson, A. M., & Perry, J. L. (2006). Collaboration processes: Inside the black box. *Public Administration Review*, *66*(s1), 20–32. https://doi.org/10.1111/j.1540-6210.2006.00663.x

Tierney, K. J. (2007). From the margins to the mainstream? Disaster research at the crossroads. *Annual Review of Sociology*, *33*(1), 503–525. https://doi.org/10.1146/annurev.soc.33.040406.131743

Tierney, K. J. (2012). Disaster governance: Social, political, and economic dimensions. *Annual Review of Environment and Resources*, *37*(1), 341–363. https://doi.org/10.1146/annurev-environ-020911-095618

Tierney, K. J., & Oliver-Smith, A. (2012). Social dimensions of disaster recovery. *International Journal of Mass Emergencies and Disasters*, *30*(2), 123–146.

UNDP. (2001). *From relief to recovery: The Gujarat experience*. http://www.recoveryplatform.org/assets/publication/from relief to recovery gujarat.pdf

UNISDR. (2005). *Hyogo framework for action 2005–2015: Building the resilience of nations and communities to disasters—summary*. UNISDR.

van Riet, G. (2009). Disaster risk assessment in South Africa: Some current challenges. *South African Review of Sociology*, *40*(2), 194–208.

Van Scotter, J. R., Pawlowski, S. D., & Cu, T. (2012). An examination of interdependencies among major barriers to coordination in disaster response James R. Van Scotter, Suzanne D. *International Journal of Emergency Management*, *8*(4), 281–307.

Wachtendorf, T. (2004). *Improvising, 9/11. organizational improvisation in the world trade center disaster*. Disaster Research Centre, University of Delaware.

Webb, G. (2007). The popular culture of disaster: Exploring a new dimension of disaster research. In H. Rodríguez, E. L. Quarantelli, & R. R. Dynes (Eds.), *Handbook of disaster research* (pp. 430–440). Springer.

Wildavsky, A. (1973). If planning is everything, maybe it's nothing. *Policy Sciences*, *4*(2), 127–153. https://doi.org/10.1007/BF01405729

Wise, C. R. (2006). Organizing for homeland security after Katrina: Is adaptive management what's missing? *Public Administration Review*, *66*(3), 302–318. https://doi.org/10.1111/j.1540-6210.2006.00587.x

Wisner, B., Gaillard, J. C., & Kelman, I. (2012). Framing disaster: Theories and stories seeking to understand Hazards, vulnerability and risk. *Handbook of Hazards and Disaster Risk Reduction*, 18–34.

Wisner, B. et al. (2004). *At risk: Natural hazards, people's vulnerability and disasters*. Routledge.

Zafonte, M., & Sabatier, P. (1998). Shared beliefs and imposed interdependencies as determinants of ally networks in overlapping subsystems. *Journal of Theoretical Politics*, *10*(4), 473–505. https://doi.org/10.1177/0951692898010004005

What do we know and need to know?

The foregoing chapters provide students and newcomers to the field of disaster studies with a broad, although not exhaustive, overview of the different dimensions of disaster risks. These chapters aimed at piecing together the otherwise fragmented knowledge on hazards, vulnerabilities, capacities, DRR and recovery in a cohesive narrative that unpacks the causes of disasters and outlines possible actions for a safer world. This narrative builds on and summarises the latest research and also vintage seminal works in the field of disaster studies. As such, it emphasises ontological and epistemological tensions and convergences that have shaped the long tradition of scholarship in our field and have informed DRR policies and actions throughout history. This conclusion offers a summary of key points as well as take-away messages for the students and future scholars of disaster risk.

1. *Disasters are not natural:* disasters require a hazard to occur, whether it is natural, socio-natural or of anthropogenic origin; however, a hazard does not necessarily lead to a disaster to happen. In this perspective, the vulnerability of our societies is the defining factor of disasters and the critical determinant of their impact on people and their livelihoods. As such, we should take "the naturalness out of disasters" (O'Keefe et al., 1976) and consider the expression "natural disaster" a misnomer. A misnomer that reflects a hazard-focused, positivist and technocratic paradigm inherited from the Enlightenment, which has failed to provide long-term solutions to address the root causes of disaster risk but instead has been re-establishing the status quo.
2. *Not all disasters are large events and processes that stir media attention:* both our everyday and academic knowledge of disasters are skewed towards large events that make the media headlines and stir the attention of humanitarian agencies and research institutions. However, growing evidence shows that there is a much larger number of small, moderate and everyday disasters that matter more for those who are affected. These events or processes may not qualify to be officially labelled disasters by local organisations, international databases and the media, but their cumulative impact is very significant. Furthermore, the frequent recurrence of these

DOI: 10.4324/9781315469614-23

neglected disasters entails a ratchet process of marginalisation and *vulnerabilisation* for those who are affected and who struggle to recover, on their own, before the next disaster, whether small or large.

3 *Disaster risks are complex:* if understanding disasters requires to comprehend both nature and society, then they become very complex processes. Both nature and society need to be understood in their interactions. These interactions are multiple, at different times and geographical scales. Understanding why some people are more affected than others by a range of diverse hazards in a particular locality requires an understanding of the historical heritages that have contributed to their unique vulnerabilities. It also requires an understanding of the international processes and power relations that shape their access to resources and means of protection in facing hazards that may also be affected by global forces, such as changes in climate patterns.

4 *Hazards are diverse and complex:* this textbook primarily focuses on natural and socio-natural hazards but recognises that the range of threats that contribute to disaster risks is broader and include anthropogenic processes, too. Natural endogenous and exogenous processes, as well as climatic and hydrometeorological hazards, are complex in nature, and fully understanding them is a huge, continuing task. Yet all these hazards cannot be taken in isolation and have to be understood as a whole thread of threats that affect people's everyday life. For example, floods and landslides are catalysed by human actions, such as urbanisation and deforestation, and hence, are transformed into socio-natural hazards.

5 *Understanding vulnerability is a difficult but essential task to tackle disaster risk:* as hinted earlier, vulnerability is the critical determinant of disaster risk. It defines who is going to be most affected at different points in time in diverse places. Understanding vulnerability requires to comprehend the broader social fabric and its multiple dimensions. It thus suggests going beyond what many studies of disasters encompass that are evidence of suffering and the proximate causes of people's susceptibility to be harmed. Vulnerability is about understanding everyday life, people's livelihoods, power structures, attitudes and behaviours, cultural norms and values, social interactions, economic processes, settlement patterns, the nature and distribution of infrastructure, political interests, governance mechanisms, international relations, and the history of all these, amongst many other dimensions of how we live.

6 *People are not helpless victims:* those affected by hazards and disasters, even the most vulnerable, are always creative and resourceful. People use their intrinsic knowledge, skills and experience, as well as available resources to devise ingenious strategies to prevent hazards and cope with the immediate and long-term impacts of disasters. These actions and strategies mirror rational and coherent behaviours and responses to hazards and disasters. These

differ from common messages conveyed by popular culture and the media, who usually portray those affected by disasters as prone to panicking, anti-social and criminal behaviours or long-term paralysing trauma. Acknowledging that people affected by hazards are actually society's first resource and responders is essential to designing our actions and policies for reducing the risk of disaster.

7 *DRR is a complex process that should combine diverse strategies and commitments from a range of actors in the short and long term:* because disaster risks are complex, DRR is necessarily complex, too. DRR requires understanding and preventing hazards, reducing vulnerability, and enhancing capacities. Any of these three tasks taken separately will not suffice. They need to be integrated in a cohesive process that warrants a diverse array of stakeholders to collaborate. These include local people, especially those most at risk, whose lives and livelihoods are at stake; scientists, who provide understanding of rare and yet unknown hazards, vulnerabilities and capacities, as well as guidance to designing early warning systems and structural and non-structural actions to address hazards; the private sector, who should reduce the disaster risks created by their activities and work towards sharing resources with the most vulnerable; national and local government agencies and parliamentarians, whose roles in prioritising resources usage for development, legislating and implementing for widespread outcomes are crucial; and finally, NGOs and international organisations, which encourage and support local and international dialogues and good practices between stakeholders.
8 *Disaster recovery should aim to enhance people's well-being and reduce the risk of future disasters:* disasters offer a "window of opportunities" (Christoplos, 2006) to foster development and DRR. Disasters reveal structural flaws in societies and contribute to better understanding of natural processes. They, thus, provide an opportunity for change in the governance of disaster risk and development. In the case of large disasters, this momentum can further build upon the blank slate offered by the scope of damage and additional resources provided by outside stakeholders. However, for these promises to be fulfilled – which is, unfortunately, rarely observed in practice – attention needs to be paid to long-term strategies that strike a balance between continuity and change for the affected territories, and people's lives and livelihoods. Such fine balance is only to be achieved if governance is decentralised and builds upon a dialogue between the diverse range of stakeholders that contribute to DRR and development at large.

This brief summary of our knowledge of disaster risks that is covered in the different chapters of this textbook sets out essential stepping stones and promising pathways for future studies in our field of scholarship. Students and scholars of disasters are encouraged to build upon them, take them in new directions and be creative in their research endeavours.

Disasters are unique in the sense that they offer scholarly opportunities for anyone interested in virtually anything in the world. As such, the following take-away messages do not constitute a rigid roadmap but rather an overarching framework to guide future studies of disaster risk.

1. *Disaster studies is a field of scholarship with a long tradition:* as hinted to in the introduction of this textbook, scholars of disaster risk often overlook the breadth of existing studies. This is detrimental in the sense that our field is very well known for constantly reinventing the wheel. Of course, some contemporary issues that affect disaster risk mirror new physical and social processes that require original attention. However, many of the drivers of disaster risk, as uncovered in this textbook, are not new things, and many of these have been studied in the past. Building on past studies allows us to expand, rather than reproduce, existing knowledge of disaster risk, and thus, push the boundaries of knowledge to further inform DRR.
2. *Studying disasters requires a wide range of different expertise:* because disasters are complex processes that cross the blurry divide between natural, social, cultural, economic, political and built environments, it takes a diverse range of academic expertise to fully understand them. This span across natural and physical sciences, engineering, medicine, social sciences, law and the humanities in their broadest acceptance. The challenge is often to facilitate sharing across such a wide range of disciplines that reflect diverse and often conflicting ontologies and epistemologies. Such cross-disciplinary sharing requires humility and understanding amongst scholars to accept our differences and recognise the values of an array of worldviews and ways of knowing in understanding disaster risk in diverse geographical and cultural contexts.
3. *Understanding disasters entails recognising the value of, listening to and learning from different forms of knowledge:* beyond scientific knowledge outlined in the previous point, understanding disasters requires considering the knowledge of the people at risk in all their diversity, whether their knowledge is indigenous to the place, imported or hybrid. Recognising the value of different forms of knowledge, including scientific, does not mean that they all need to be forced into the same canvas to come up with some sort of overall comprehensive knowledge of disaster risk. Different forms of knowledge reveal diverse epistemologies and ontologies that, most often, cannot be reconciled. As such, understanding disaster risk to ultimately support DRR is about making different forms of knowledge to co-exist alongside each other so that anyone can make informed decisions based on all stakeholders' views.
4. *Disaster studies need further grounding in theory:* besides constantly reinventing the wheel, disasters studies are frequently known for lacking grounding in deeper theory, especially within the social sciences. Our ultimate and genuine goal to reduce disaster risk often leads scholarship to be driven by empirical problems and practical

needs that end up taking prominence over ontological foundations. The latter are, however, crucial to making sure that studies of disasters are epistemologically coherent to address complex and deep-seated issues in societies, as unpacked in this textbook. This is, therefore, a call, particularly addressed to social scientists and researchers from the humanities, to further build on critical social science theories and their philosophical foundations, and shed brighter light on many complex problems we are often trying to tackle from a narrow empirical perspective.

5 *Disaster studies require a fine understanding of local contexts:* to fully unpack the complexities of local realities, including local hazards as well as people's lives and livelihoods, it is essential to be grounded in the contexts we study. These realities further differ amongst households and from one individual to another. Understanding disaster risk, thus, requires a fine-grained and holistic approach to capture the micro-realities of everyday life and the complexity of the local hazardscape. In the eyes of people at risk, life is a whole that can hardly be fragmented through the lens of academic disciplines and their specific object(s) of studies (Jigyasu, 2005). As such, local people should also be considered valuable researchers to understanding disaster risk.

6 *We need to foster local ontologies and epistemologies:* to fully unpack local realities and foster the contribution of local people in understanding disaster risk, we need to move away from the dominant Eurocentric interpretations and framings of disasters, when outside of their context. This very textbook is framed from such Western perspectives of disasters because it is almost everything disaster studies, as a field of scholarship, has to provide to date (Gaillard, 2022). There is, therefore, a huge opportunity for students and future scholars to explore non-Eurocentric ontologies and epistemologies to understanding disasters in colonial and postcolonial contexts. In fact, this is probably where the future of disaster studies currently lies and where significant prospects will open up in order to further reduce the risk of disaster where Western approaches have proved insufficient or inappropriate.

7 *Study your own disasters and local contexts:* to understand local realities and build upon local ontologies and epistemologies, it seems sensible to encourage students and future scholars to study their own disasters – those that occur or may happen in the contexts they are familiar with. Contextual and theoretical understandings of places, natural landscapes and societies are essential here, but grounding also extends to existing relationships with local people and stakeholders that allow to better identify and address research needs. Such grounding further allows us to build long-term partnerships towards DRR, which importance we emphasised in the previous paragraphs. It also facilitates smooth relationships between researchers and research participants on a topic that is highly sensitive.

8 *Studying disasters is sensitive and requires careful ethical considerations:* disasters entail harm, loss, suffering and grief. As such, they require a tactful approach to research and delicate relationships between researchers and local people at risk or those affected by disasters. Again, the former need to recognise that the latter are not helpless and can take the lead in studying local realities and their own experiences. Scholars also need to acknowledge the priorities of local people's everyday life and the diversity of threats they face. Researchers must also consider the urgency of recovering following disasters, which means that contributing to outside-driven research may not be the survivors' primary goal in life. Finally, researchers need to collaborate with local organisations, in their diversity and other scholars to avoid duplication of efforts and research fatigue amongst participants (Gaillard & Peek, 2019).

Following the direction outlined in these eight overarching take-away messages should set out for relevant, original, theoretically grounded, epistemologically appropriate and ethically sound disaster studies for decades to come. Hopefully, these further studies will provide materials for a revised edition of this handbook that will cover more diverse materials, especially some framed from non-Eurocentric perspectives. These should capture a more detailed and diverse picture of local realities of disaster risks. Ultimately and optimistically, the next iteration of this handbook will cater for a smaller audience and a decreasing interest in less-frequent disasters. One that will reflect a safer world.

References

Christoplos, I. (2006, February 2–3). *The elusive "window of opportunity" for risk reduction*. ProVention Consortium Forum.

Gaillard, J. C. (2022). *The invention of disaster: Power and knowledge in discourses on hazard and vulnerability*. Routledge.

Gaillard, J. C., & Peek, L. (2019). Disaster-zone research needs a code of conduct. *Nature*, *575*(7783), 440–442. https://doi.org/10.1038/d41586-019-03534-z

Jigyasu, R. (2005). Disaster: A "reality" or. In R. W. Perry & E. L. Quarantelli (Eds.), *What is a disaster? New answers to old questions, constructivist? Perspective from the "East"* (pp. 49–59). Xlibris.

O'Keefe, P., Westgate, K., & Wisner, B. (1976). Taking the naturalness out of natural disasters. *Nature*, *260*(5552), 566–567. https://doi.org/10.1038/260566a0

List of key concepts

Concept	*Definition*	*Source*
Altruism	A belief or practice aimed at achieving common good.	This book
Anthropocene	The current epoch in which humans and our societies have become a global geophysical force.	Steffen et al., 2007
Anticipation	The future used in action – it is, thus, a forward-looking attitude and the use of the former's result for action.	This book
Capacity/ies	The set of diverse knowledge, skills and resources people can claim, access and resort to in dealing with hazards and disasters.	Gaillard et al., 2019
Cascading disasters	An approach that contradicts a simple linear model to comprehending disaster impacts, where vulnerabilities interact and overlap, and/or one hazard can act upon another, producing escalation points and secondary effects with greater impacts than the original trigger.	Alexander and Pescaroli, 2019
Characteristics of resilient systems	An *anticipative* system is a system that has (or can envisage) a future time-bound blueprint model of itself. Such a system can take action to achieve the envisaged future state.	Van Niekerk and Terblanche-Greeff, 2017

(*Continued*)

(Continued)

Concept	Definition	Source
	Adaptive capacity rests on people's agency (i.e. their ability to make informed choices and to develop and successfully execute their plans). Adaptive capacity is the property of a system in which structures are modified to prevent future disasters.	Levine et al., 201, Lorenz, 2013, Norris et al., 2007
	Transformability is the capacity to create a fundamentally new system when ecological, economic or social structures make the existing system untenable.	Walker et al., 2004
	The *absorptive* characteristic means the capacity of a system to absorb the impacts of negative events to preserve and restore its structure and basic functions. It involves intentional protective action against shocks.	Jeans et al., 2017
	A system that exhibits a *resistant* characteristic is a system able to withstand all stressors, shocks or impacts without suffering any loss.	Lake, 2012
	Reflective resilient systems create an opportunity for active learning and room for the review of the effectiveness and efficiency of existing processes in the face of new shocks. Resilient systems need to learn from their past to inform the future.	Kerner and Thomas, 2016; this book; and Béné et al., 2012
	In an environment where there are limited resources in terms of time, human, financial, technological and natural resources, resilient systems need to be *resourceful*. Such a situation requires that the little available resources are allocated and used effectively, and where possible, best alternative pathways for resource use are identified and pursued in resilience-building.	Kerner and Thomas, 2014

Concept	*Definition*	*Source*
	In being *inclusive*, resilient systems realise that whilst approaches and strategies may be developed and disseminated in a top-down route, there needs to be provision for bottom-up consultative processes that feed into the strategies, planning and decision-making for resilience-building. Inclusiveness also requires that at all rungs of society acknowledge the diversity of social groups and their specific needs.	Béné et al., 2012, and this book
	A resilient system that is *innovative* is one that can combine exploration and exploitation. Exploration is related to how well a system can identify developments to which it must adapt and transform, and in doing so, in an unconventional manner. Exploitation is about fine-tuning the functionality of the system under current conditions to ensure gains in effectiveness and efficiency.	Kroeze et al., 2017
	An *integrated* resilient system holistically brings together institutions, stakeholders and different actors across the diversity of their disciplines in polycentric management and implementation approach.	Béné et al., 2016
	If a system is *robust*, it means that its design is well-conceived, constructed and managed, and includes making provision to ensure failure is predictable, safe and not disproportionate to the cause. Thus, its assets and systems are designed to withstand shocks and hazards.	Walker and Salt, 2006

(Continued)

(Continued)

Concept	*Definition*	*Source*
	Flexibility refers to the willingness and ability to adopt alternative strategies in response to changing circumstances or crises. With flexibility, resilience systems can make appropriate adjustments relating to changes in the social, economic and environmental contexts. This allows resilient systems to accommodate newly developed knowledge and technological innovations.	Béné et al., 2012, and this book
	Redundancy in resilient systems is, thus, the ability to offer numerous options to achieve desired goals or functions. The presence of diverse options or actors with overlapping functions ensures that when one area or component fails, there is no detrimental system collapse, as other components may be able to compensate for the failure or loss.	Béné et al., 2012, and this book
	Self-organisation refers to a system's ability to make its structure more complex given its system's rules. Positive self-organisation allows for the creation of heterogeneity.	This book
	Connectedness describes the quantity and quality of relationships between system elements. It also relates to the paths of interaction between system elements, and other systems and their elements. Connectedness can be economic, social, psychological or physical.	O'Sullivan et al., 2013, and this book

Concept	*Definition*	*Source*
Climate change adaptation	An action taken to prepare for and adjust to both the current effects of climate change and the predicted impacts in the future. In simple terms, countries and communities need to develop adaptation solution and implement action to respond to the impacts of climate change that are already happening, as well as prepare for future impacts.	European Commission UNFCCC
Complexity	Something lacking simplicity but not being hard to understand.	This book
Complex adaptive systems	The systems with the ability to adapt to stimuli in their environment.	This book
Community-based disaster risk management	A process of DRM in which at-risk communities are actively engaged in the identification, analysis, treatment, monitoring and evaluation of disaster risks in order to reduce their vulnerabilities and enhance their capacities. This means that the people are at the heart of decision-making and implementation of DRM activities. The involvement of the most vulnerable is paramount, and the support of the least vulnerable is necessary. In CBDRM, local and national governments are involved and supportive.	Abarquez and Murshed, 2004
Complex adaptive systems	Systems with the ability to adapt to stimuli in their environment.	This book

(*Continued*)

(Continued)

Concept	*Definition*	*Source*
Contingency planning	A management process that analyses disaster risks and establishes arrangements in advance to enable timely, effective and appropriate responses. Contingency planning results in organised and coordinated courses of action with clearly identified institutional roles and resources, information processes, and operational arrangements for specific actors at times of need. Based on scenarios of possible emergency conditions or hazardous events, it allows key actors to envision, anticipate and solve problems that can arise during disasters. Contingency planning is an important part of overall preparedness. Contingency plans need to be regularly updated and exercised.	UNISDR, 2017
Crisis	A situation where there is a threat to human lives (and their properties) that needs to be urgently addressed using available information.	Boin et al., 2018
Cycle of displacement	The process where survivors are usually relocated several times: they are sent to large emergency campsites where there are risks of adverse social and environmental effects, as well as to areas remote from work, markets, schools, and other social and economic needs.	Marchezini, 2014; Davis, Thompson and Krimgold, 2015; Portella and Oliveira, 2017; Vasquez and Marchezini, 2020

Concept	*Definition*	*Source*
Disaster	The severe alterations in the normal functioning of a community or a society due to hazardous physical events interacting with vulnerable social conditions, leading to widespread adverse human, material, economic or environmental effects that require immediate emergency response to satisfy critical human needs and that may require external support for recovery.	Lavell et al., 2012
Disaster capitalism	A situation where social groups who are in advantageous positions use resources to maximise their capitalist interests in the wake of disasters.	Klein, 2008; Schuller, 2008
Disaster subculture	A memory of past disaster experiences, which influence risk perception and, if properly used in DRR, proved important and useful as basis of planning and actual response in times of disasters.	Granot, 1996
Disaster management	The organisation, planning and application of measures preparing for, responding to and recovering from disasters. Disaster management may not completely avert or eliminate the threats; it focuses on creating and implementing preparedness and other plans to decrease the impact of disasters and build back better. Failure to create and apply a plan could lead to damage to life, assets and lost revenue.	UNISDR, 2017

(*Continued*)

(Continued)

Concept	*Definition*	*Source*
Disaster response	The actions taken directly before, during or immediately after a disaster in order to save lives, reduce health impacts, ensure public safety and meet the basic subsistence needs of the people affected. Disaster response is predominantly focused on immediate and short-term needs, and is sometimes called disaster relief. Effective, efficient and timely response relies on disaster risk–informed preparedness measures, including the development of the response capacities of individuals, communities, organisations, countries and the international community.	UNISDR, 2017
Disaster risk	The risk derived from a combination of hazards and the vulnerabilities of exposed elements, and will signify the potential for severe interruption of the normal functioning of the affected society once it materialises as disaster. Disaster risk is the probability or a latent condition that expresses the potential impact of one or more hazards of diverse origin on a group of people (or a system) with different degrees of vulnerability, who occupy a territorial space exposed to the effects of such hazards.	Lavell et al., 2012 This book
Disaster risk communication	The process of sharing data, information and knowledge about the risks (hazards and vulnerabilities), communicating them through different ways and channels.	This book

Concept	*Definition*	*Source*
Disaster risk drivers	The ongoing dynamic and active conditions or processes that create or increase conditions of vulnerability and exposure, often linked to, or rooted in, models of development. Major drivers of disaster risk include urbanisation and overcrowding, deforestation, inequality, poverty, unsafe conditions, illiteracy, insalubrity, lack of planning, corruption, failed risk governance, and climate change.	This book
Disaster risk governance	The way in which the public authorities, civil servants, media, private sector and civil society coordinate at community, national and regional levels in order to manage and reduce disaster and climate-related risks.	This book
Disaster risk knowledge	The data collection and analysis of hazards/threats and vulnerabilities – physical, social, economic, environmental and so forth – to generate knowledge about the risk scenarios across different spatial scales, in the short and long term.	This book
Disaster risk management	This should encompass a series of processes for designing, implementing, and evaluating strategies, policies, and measures to improve the understanding of disaster risk, foster DRR and transfer, and promote continuous improvement in disaster preparedness, response and recovery practices, with the explicit purpose of increasing human security, well-being, quality of life and sustainable development.	Lavell et al., 2012

(*Continued*)

(Continued)

Concept	*Definition*	*Source*
Disaster risk monitoring	The resources and capacities for collecting and verifying dynamic data and information about hazards/threats and vulnerabilities in order to take decisions on the basis of prior risk knowledge.	This book
Early warning systems (EWS)	A social process aiming to address the need to avoid harm due to hazards. The social process occurs at a variety of spatial scales, from individuals in isolated villages without electricity through to the global UN processes working with governments.	Kelman and Glantz, 2014
Economic capital	The financial resources, including savings, income, investments and credit, that people use to achieve their livelihoods.	Mayunga, 2007
Exposure	The situation of people, infrastructure, housing, production capacities and other tangible human assets located in hazard-prone areas.	UNISDR, 2017
Extensive disasters	The high-frequency but low-severity losses, mainly recurrent, but not exclusively associated with highly localised hazards. The costs of these extensive disasters are not visible and tend to be underestimated, as they are usually absorbed by low-income households and communities and small businesses.	UNDRR, 2015
Extensive disaster risk	The risk of low-severity, high-frequency hazardous events and disasters, mainly but not exclusively associated with highly localised hazards. Extensive disaster risk is usually high where communities are exposed to, and vulnerable to, recurring localised floods, landslides, storms or drought. Extensive disaster risk is often exacerbated by poverty, urbanisation and environmental degradation.	UNISDR, 2017

Concept	*Definition*	*Source*
Forensic Investigations of Disasters, FORIN	A scientific approach that stresses the need to understand the social construction of risk through an understanding of underlying causes and disaster risk drivers. FORIN aims at promoting integrated and transdisciplinary research that engages all relevant DRR stakeholders to enable a more holistic comprehension of root causes and disaster risk.	Oliver-Smith et al., 2016
Hazard	A process, phenomenon or human activity that may cause loss of life, injury or other health impacts, property damage, social and economic disruption, or environmental degradation. Hazards may be natural, anthropogenic or socio-natural in origin.	UNISDR, 2017
Human capital	The aggregate of innate abilities; an individual's intrinsic potential to acquire skills.	This book
Intensive disaster	The high-severity, mid- to low-frequency disasters, mainly associated with major hazards. The economic losses from intensive disasters are usually evaluated by international organisations, governments and the insurance industry.	UNDRR, 2015
Intensive disaster risk	The risk of high-severity, mid- to low-frequency disasters, mainly associated with major hazards. Intensive disaster risk is mainly a characteristic of large cities or densely populated areas that are not only exposed to intense hazards, such as strong earthquakes, active volcanoes, heavy floods, tsunamis or major storms, but also have high levels of vulnerability to these hazards.	UNISDR, 2017

(*Continued*)

(Continued)

Concept	*Definition*	*Source*
Intersectionality	A framework that allows taking into account people's overlapping identities and experiences in order to understand the complexity of discrimination and privileges they face.	This book
Land degradation	The many human-caused processes that drive the decline or loss in biodiversity, ecosystem functions or ecosystem services in any terrestrial and associated aquatic ecosystems.	IPBES, 2018
Marginalisation	A social process that leads some groups to lack access to resources because of their marginal positions in the existing social structures (i.e. being economically poor, belonging to a cultural minority group, being politically neglected or segregated, etc.). People are "marginalized geographically because they live in hazardous places, socially because they are members of minority groups, economically because they are poor, and marginalized politically because their voice is disregarded by those with political power" (197).	Gaillard and Cadag, 2009
Mitigation	The lessening or minimising of the adverse impacts of a hazardous event. The adverse impacts of hazards, in particular natural hazards, often cannot be prevented fully, but their scale or severity can be substantially lessened by various strategies and actions. Mitigation measures include engineering techniques and hazard-resistant construction, as well as improved environmental and social policies and public awareness.	UNISDR, 2017
Natural capital	The capital essential in sustaining all forms of life.	Mayunga, 2007

Concept	*Definition*	*Source*
Netizens	The citizens of the internet or those people who engage in discussion or sharing of information through social media platforms (i.e. Facebook, YouTube and Twitter, amongst other platforms).	This book
Non-structural measures	The measures not involving physical construction that use knowledge, practice or agreement to reduce disaster risks and impacts, in particular through policies and laws, public awareness raising, training, and education. Common non-structural measures include building codes, land-use planning laws and their enforcement, research and assessment, information resources, and public awareness programmes.	UNISDR, 2017
People's behaviour	The behaviour in times of disasters, which may then refer to decisions, actions or conducts of individuals or groups in response to the threat of hazards or impacts of disasters in a specific place and time.	This book
People-centred warning systems	The people-centred warning systems consider that local people should be involved as the central component of the design and operations of warning systems at different scales, involving multiple stakeholders in dialogue and collaboration at every stage of the process.	This book
Physical capital	The basic infrastructure, which includes affordable transport, secure shelter, adequate water supplies and sanitation, and access to information and producer goods needed to support livelihoods, such as the tools and equipment that people use to function more productively.	Twigg, 2001

(*Continued*)

(Continued)

Concept	*Definition*	*Source*
Preparedness	The knowledge and capacities developed by governments, response and recovery organisations, communities, and individuals to effectively anticipate, respond to and recover from the impacts of likely, imminent or current disasters. Preparedness action is carried out within the context of DRM and aims to build the capacities needed to efficiently manage all types of emergencies and achieve orderly transitions, from response to sustained recovery.	UNISDR, 2017
Prevention	The activities and measures to avoid existing and new disaster risks. Prevention (i.e. disaster prevention) expresses the concept and intention to completely avoid potential adverse impacts of hazardous events. Whilst certain disaster risks cannot be eliminated, prevention aims at reducing vulnerability and exposure in such contexts where, as a result, the risk of disaster is removed. Prevention measures can also be taken during or after a hazardous event or disaster to prevent secondary hazards or their consequences, such as measures to prevent the contamination of water.	UNISDR, 2017
Protracted crisis	A situation where a significant proportion of the population is acutely vulnerable to death, disease and disruption of livelihoods over a prolonged period. Three criteria were used to classify a country or an area in protracted crisis: (i) duration or longevity of crisis, the threshold being eight years or more; (ii) aid flows, which is the proportion of	FAO, 2010

Concept	*Definition*	*Source*
	humanitarian assistance received by the country as a share of total assistance; and, (iii) economic and food security status, if the country appears on the list of Low-Income Food-Deficit Countries.	
Rapid- and slow-onset hazards	The rapid-onset hazards include earthquake and volcanic eruption hazards (i.e. tsunami, ground shaking, pyroclastic flow etc.) and hydrometeorological hazards (typhoon, floods, landslides, storm surges and tornadoes). Slow-onset hazards, such drought, sea level rise and other climate change–related stimuli (precipitation and temperature changes), take years or decades to occur, and impacts are not felt immediately.	This book
Recovery	A differential process of restoring, rebuilding and reshaping the physical, social, economic and natural environment through pre-event planning and post event actions.	Smith and Wenger, 2006
Residual risk	The disaster risk that remains in unmanaged form, even when effective DRR measures are in place, and for which emergency response and recovery capacities must be maintained. The presence of residual risk implies a continuing need to develop and support effective capacities for emergency services, preparedness, response and recovery, together with socio-economic policies, such as safety nets and risk transfer mechanisms, as part of a holistic approach.	UNISDR, 2017

(*Continued*)

(Continued)

Concept	*Definition*	*Source*
Risk perception	The subjective interpretation or understanding of an individual of a potential event that determines his or her adjustment and response to hazards (i.e. people's behaviour).	This book
Root causes or underlying causes of disasters	The processes or conditions related to historical aspects of development, associated with political, economic, and territorial decisions and practices that have occurred throughout history. Quite often, such processes precede by several decades the disaster itself.	This book
Social capital	The aggregate of the actual or potential resources that are linked to the ability to claim, access and possess durable networks of more or less institutionalised relationships of mutual acquaintance and recognition.	This book
Socio-natural hazards	The phenomena that appear to be typical of natural hazards but have an expression or incidence that is socially induced because they are produced or exacerbated by human intervention in nature.	Lavell, 1996
Structural measures	The physical construction to reduce or avoid possible impacts of hazards, or the application of engineering techniques or technology to achieve hazard resistance and resilience in structures or systems. Common structural measures for DRR include dams, flood levies, ocean wave barriers, earthquake-resistant construction and evacuation shelters.	UNISDR, 2017
Sustainable development	The development that meets the needs of the present without compromising the ability of future generations to meet their own needs.	Brundtland Commission, 1987

Concept	*Definition*	*Source*
Transdisciplinarity	An approach in which all players and stakeholders in various disciplines (natural, social and human sciences) and sectors (public, private, academia and civil society) work together to achieve a common goal.	This book
Vulnerability	The propensity or predisposition to be adversely affected by the impact of hazards. It includes the characteristics of a person or group and their situation that influences their capacity to anticipate, cope with, resist and recover from the adverse effects of hazards, and is a result of diverse historical, social, economic, political, cultural, institutional, natural resource and environmental conditions and processes.	Wisner et al., 2004, and Lavell et al., 2012

Index

Note: Page numbers in *italic* indicate a figure and page numbers in **bold** indicate a table or box on the corresponding page.